# Cancer Drug Discovery and Development



For further volumes:
http://www.springer.com/series/7625

Rakesh Kumar
Editor

# Nuclear Signaling Pathways and Targeting Transcription in Cancer

Humana Press

*Editor*
Rakesh Kumar
Department of Biochemistry and Molecular Medicine
School of Medicine and Health Sciences
The George Washington University
Washington, DC, USA

ISBN 978-1-4614-8038-9 ISBN 978-1-4614-8039-6 (eBook)
DOI 10.1007/978-1-4614-8039-6
Springer New York Heidelberg Dordrecht London

Library of Congress Control Number: 2013944683



Printed on acid-free paper

Humana Press is a brand of Springer
Springer is part of Springer Science+Business Media (www.springer.com)

# Preface

Regulatory nuclear pathways feeding into the transcription of cancer-relevant molecules have emerged as the next frontier in pathway-centered cancer therapeutics. The significance of nuclear signaling in cancer is also evident by the convergence of a large number of signal transduction pathways continuously sensing extracellular milieu. The cumulative outcome of deregulated cytoplasmic and nuclear signaling is to provide a favorable environment for a cancer cell to survive by overriding death signals, to sustain an excessive hyper-mitogenic activity, to feed into deregulated cell cycle progression, and to support a defective segregation of genetic material during mitosis leading to genomic instability, to name a few essential hallmarks of cancer progression. Among other processes, chromatin remodeling and epigenetic modifications are two important nuclear regulatory arms of transcription that have offered a battery of exciting therapeutic opportunities in terms of specificity by focusing on specific modification or modifications of histone or nonhistone proteins, domain-targeting, enzymatic activities such as histone deacetylases, histone acetyltransferases, histone demethylases, or splicing factors – as all of these activities are widely deregulated in multiple human cancers. A large body of work during the last decade has demonstrated that these are targetable areas of translational cancer medicine, and therefore, a large number of small molecules or agents targeting these biological processes are rapidly moving through the preclinical development pipeline to clinical studies.

Another compelling aspect of nuclear signaling in cancer therapeutics is its inherent role in controlling the transcription of cancer-promoting factors. As many of the target gene products of nuclear signaling have functional relevance in the cytoplasmic compartment, these molecules are also at the center of signaling cascades in the cytoplasm and influenced by extracellular environment. Interestingly, the barrier of cytoplasmic and nuclear signaling is broken by recent advances in the area of nuclear receptor tyrosine kinases, which were earlier thought to be limited to the cytoplasm. Another recent excitement in the field stems from the translational control of the elements of metabolism, such as lipogenesis. In addition to therapeutic value, many of the pathways and molecules described above are also becoming prognostic biomarkers to monitor the progression of cancer or to assess the therapeutic efficacy of a given targeted therapy. Since nuclear signaling represents the funneling junction of upstream signaling and because it imparts selectivity at the level of transcription, the overall goal of this book is to enhance our understanding of relevant processes in an integrated manner in the context of cancer medicine. The editor felt it timely to compile this book, written by the leading authorities in the field and contributing to an emerging therapeutic discipline of signaling control of transcription in cancer medicine.

Washington, DC, USA Rakesh Kumar

# Contents

# Contributors

**Angela Alexander** Department of Experimental Radiation Oncology, Radiation Oncology, University of Texas, MD Anderson Cancer Center, Houston, TX, USA

**Hamid Band** Departments of Genetics, Cell Biology and Anatomy; Biochemistry and Molecular Biology; Pathology and Microbiology; Pharmacology and Experimental Neuroscience, College of Medicine, Eppley Institute for Research in Cancer and Allied Diseases, University of Nebraska Medical Center, Omaha, NE, USA

**Vimla Band** Department of Genetics, Cell Biology and Anatomy; Eppley Institute for Research in Cancer and Allied Diseases, University of Nebraska Medical Center, Omaha, NE, USA

**Elspeth M. Beauchamp** Robert H. Lurie Comprehensive Cancer Center, Northwestern University, Chicago, IL, USA

**Mohita Bhagat** Department of Biochemistry, All India Institute of Medical Sciences, New Delhi, Delhi, India

**Jessica L. Brusgard** Department of Pathology, University of Maryland, Marlene & Stewart Greenebaum Cancer Center, Baltimore, MD, USA

**Maria Carmo-Fonseca** Instituto de Medicina Molecular, Faculdade de Medicina, Universidade de Lisboa, Lisbon, Portugal

**Pratik Chandrani** Advanced Centre for Treatment, Research and Education in Cancer, Tata Memorial Centre, Kharghar, Navi Mumbai, India

**Parthaprasad Chattopadhyay** Department of Biochemistry, All India Institute of Medical Sciences, New Delhi, Delhi, India

**Kunzang Chosdol** Department of Biochemistry, All India Institute of Medical Sciences, New Delhi, Delhi, India

**Bhawana Dikshit** Department of Biochemistry, All India Institute of Medical Sciences, New Delhi, Delhi, India

**Amit Dutt** Advanced Centre for Treatment, Research and Education in Cancer, Tata Memorial Centre, Kharghar, Navi Mumbai, India

**Ana Rita Grosso** Instituto de Medicina Molecular, Faculdade de Medicina, Universidade de Lisboa, Lisbon, Portugal

**Wenchang Guo** Cancer Center, University of California, Davis, Sacramento, CA, USA

**Jennifer L. Hsu** Department of Molecular and Cellular Oncology, The University of Texas MD Anderson Cancer Center, Houston, TX, USA

Center for Molecular Medicine and Graduate Institute of Cancer Biology, China Medical University, Taichung, Taiwan

Department of Biotechnology, Asia University, Taichung, Taiwan

**Chiu-Lien Hung** Cancer Center, University of California, Davis, Sacramento, CA, USA

**Mien-Chie Hung** Department of Molecular and Cellular Oncology, The University of Texas MD Anderson Cancer Center, Houston, TX, USA

Center for Molecular Medicine and Graduate Institute of Cancer Biology, China Medical University, Taichung, Taiwan

Department of Biotechnology, Asia University, Taichung, Taiwan

**Longfei Huo** Department of Molecular & Cellular Oncology, The University of Texas MD Anderson Cancer Center, Houston, TX, USA

**Yoshihiro Izumiya** Dermatology, University of California, Davis, School of Medicine, Sacramento, CA, USA

**Jun-Yuan Ji** Molecular and Cellular Medicine, College of Medicine, Texas A&M Health Science Center, College Station, TX, USA

**Khandan Keyomarsi** Department of Experimental Radiation Oncology, Radiation Oncology Division, University of Texas, MD Anderson Cancer Center, Houston, TX, USA

**Kevin Kim** Department of Biochemistry, Molecular, Cellular and Developmental Biology, University of California, Davis, Davis, CA, USA

**Stefan Knapp** Nuffield Department of Clinical Medicine, Structural Genomics Consortium, Medical Sciences Division, University of Oxford, Oxford, Oxon, UK

**Rakesh Kumar** Biochemistry and Molecular Medicine, School of Medicine and Health Sciences, The George Washington University, Washington, DC, USA

**Hsing-Jien Kung** Cancer Center, University of California, Davis, Sacramento, CA, USA

Biochemistry & Molecular Medicine, University of California, Davis, Sacramento, CA, USA

**Da-Qiang Li** Department of Biochemistry and Molecular Medicine, The George Washington University, Washington, DC, USA

**Jing Liang** Department of Biochemistry and Molecular Biology, School of Basic Medical Sciences, Peking University Health Science Center, Beijing, People's Republic of China

**Maritza Lienlaf-Moreno** Immunology, H. Lee Moffitt Cancer Center, Tampa, FL, USA

**Hannah Lingard** Nuffield Department of Medicine, Structural Genomics Consortium, Medical Sciences Division, University of Oxford, Oxford, Oxon, UK

**Yong Liu** Department of Radiation Oncology, Lineberger Comprehensive Cancer Center, School of Medicine, University of North Carolina at Chapel Hill, Chapel Hill, NC, USA

**Weiwen Long** Molecular and Cellular Biology, Baylor College of Medicine, Houston, TX, USA

**Evanka Madan** Biochemistry Department, All India Institute of Medical Sciences, New Delhi, Delhi, India

**Shakur Mohibi** Department of Genetics, Cell Biology and Anatomy, University of Nebraska Medical Center, Omaha, NE, USA

**Susanne Müller** Nuffield Department of Clinical Medicine, Structural Genomics Consortium, University of Oxford, Oxford, Oxon, UK

**Bert W. O'Malley** Molecular and Cellular Biology, Baylor College of Medicine, Houston, TX, USA

**Antonino Passaniti** Department of Pathology, University of Maryland, Marlene and Stewart Greenebaum Cancer Center, Baltimore, MD, USA

**Richard G. Pestell** Cancer Biology, Thomas Jefferson University, Philadelphia, PA, USA

**Leonidas C. Platanias** Robert H. Lurie Comprehensive Cancer Center, Northwestern University, Chicago, IL, USA

**Mamata Pochampalli** Cancer Center, University of California, Davis, Sacramento, CA, USA

**Ruth Serrano** Robert H. Lurie Comprehensive Cancer Center, Northwestern University, Chicago, IL, USA

**Edward Seto** Department of Molecular Oncology, Moffitt Cancer Center and Research Institute, Tampa, FL, USA

**Yongfeng Shang** Department of Biochemistry and Molecular Biology, School of Basic Medical Sciences, Peking University Health Science Center, Beijing, People's Republic of China

Department of Biochemistry and Molecular Biology, Tianjin Medical University, Beijing, People's Republic of China

**Subrata Sinha** National Brain Research Centre, Manesar, Haryana, India

**Eduardo Sotomayor** Department of Malignant Hematology, H. Lee Moffitt Cancer Center, Oncologic Sciences, University of South Florida, Tampa, FL, USA

**Shashank Srivastava** Department of Genetics, Cell Biology and Anatomy, University of Nebraska Medical Center, Omaha, NE, USA

**Lifeng Tian** Cancer Biology, Thomas Jefferson University, Philadelphia, PA, USA

**Alejandro Villagra** Immunology, H. Lee Moffitt Cancer Center and research Institute, Tampa, FL, USA

Oncologic Sciences, University of South Florida, Tampa, FL, USA

**Chenguang Wang** Cancer Biology, Thomas Jefferson University, Philadelphia, PA, USA

**Ling-Yu Wang** Cancer Center, University of California, Davis, Sacramento, CA, USA

**David M. Woods** Pathology, University of South Florida College of Medicine, Tampa, FL, USA

**Yanping Zhang** Department of Radiation Oncology, Lineberger Comprehensive Cancer Center, School of Medicine, University of North Carolina at Chapel Hill, Chapel Hill, NC, USA

Laboratory of Biological Cancer Therapy, Xuzhou Medical College, Xuzhou, Jiangsu, China

# Part I
# Gene Regulation and Cancer

# Chapter 1
# Steroid Receptor Coactivators (SRCs) as Integrators of Multiple Signaling Pathways in Cancer Progression

**Weiwen Long and Bert W. O'Malley**

**Abstract** Steroid receptor coactivators (SRCs), including SRC-1, SRC-2, and SRC-3, mediate transcriptional activities of nuclear receptors and other transcription factors. SRCs' activities and functions are regulated by multiple signaling pathways, including those of hormones, growth factors, and cytokines, and are determined by post-translational modifications, including phosphorylation, ubiquitination, sumoylation, acetylation, and methylation. SRCs integrate signals from a variety of pathways that regulate multiple cellular processes such as metabolism, reproduction, and growth. For the growth response, they regulate proliferation, survival, migration, and invasion, and promote tumor development and metastasis. SRCs are highly disregulated in many types of cancers at multiple levels including gene amplification, mutation, and mRNA/protein overexpression. Alterations of SRCs are frequently associated with advanced tumor progression and drug resistance. As such, SRCs are important prognostic cancer biomarkers and could serve as therapeutic targets for cancer therapy.

**Keywords** Steroid receptor coactivators (SRCs) • Nuclear receptor • Gene transcription • Posttranslational modifications • Signaling pathways • Tumorigenesis • Cancer metastasis • Drug targeting

---

W. Long, Ph.D. • B.W. O'Malley, M.D. (✉)
Molecular and Cellular Biology, Baylor College of Medicine, Houston, TX 77030, USA
e-mail: wlong@bcm.edu; berto@bmc.edu

R. Kumar (ed.), *Nuclear Signaling Pathways and Targeting Transcription in Cancer*,
Cancer Drug Discovery and Development, DOI 10.1007/978-1-4614-8039-6_1,

## 1.1 Introduction

The p160 steroid receptor coactivator (SRC) family, consisting of SRC-1, SRC-2, and SRC-3, were originally identified as transcriptional coactivators of nuclear hormone receptors (NRs) for estrogen, progesterone, and androgen. SRC-1, also known as nuclear receptor coactivator 1(NCOA1), was cloned and characterized as the first NR coactivator in 1995 [1]. SRC-2, also known as NCOA2, GRIP1 (glucocorticoid receptor interacting protein 1), and TIF2 (transcriptional intermediary factor 2), was identified soon after the cloning of SRC-1 [2, 3]. SRC-3 was then identified by several laboratories nearly in the same year of 1997, and was provided with different names [4–7]: AIB1 (amplified in breast cancer 1), p/CIP (p300/CBP interacting protein), RAC3 (RAR-associated coactivator 3), ACTR (activator of thyroid and retinoic acid receptor), and TRAM1 (thyroid receptor activator molecule 1). Since the discovery of the first coactivator (SRC-1) in 1995, a substantial number of studies have been conducted to elucidate the molecular actions of SRCs in normal physiology and pathology. In this chapter, we will focus on the cancer-related functions of SRCs and the underlying molecular mechanisms, thereby highlighting the molecular structures and functional interacting partners of SRCs, the regulation of SRCs' activities by posttranslational modifications (PTMs), and the integration of multiple oncogenic signaling pathways by SRCs that promote tumor development and progression.

## 1.2 Structures and Transcriptional Interacting Partners of SRCs

SRC proteins share a common structure that contains five functional domains/regions (Fig. 1.1): the N-terminal basic helix-loop-helix-Per/ARNT/Sim (bHLH/PAS) domain, the serine/threonine rich (S/T) domain, the nuclear receptor interacting domain (RID), the CBP(cAMP-response element binding protein-binding protein) interacting domain (CID) or activation domain 1 (AD1), and the activation domain 2 (AD2) or the histone acetyltransferase (HAT) domain at the C-terminus [10, 11]. Each domain has different interacting partners that confer various functions (Fig. 1.1). The bHLH/PAS domain harbors nuclear localization signals (NLS) and is the most conserved region. This domain is also termed activation domain 3 (AD3) as it is responsible for the interaction of SRCs with multiple co-coactivators and non-NR transcriptional factors. CoCoA was shown to interact with SRC-2 through the bHLH/PAS domain and work cooperatively with p300/CBP co-coactivators to regulate NR-mediated gene transcription [12]. hBrm-associated factor 57 (BAF57), a core component of SNI/SWF chromatin remodeling complex, binds to the bHLH/PAS domain of SRCs and bridges the SNI/SWF chromatin remodeling complex to ER/SRCs transcription complex to promote estrogen-responsive gene transcription [13]. Melanoma antigen gene protein-A11 (MAGE-11) interacts with both TIF2 and AR and potentiates AR transcriptional activity probably through stabilizing the AR-TIF2 transcription complex on the target gene

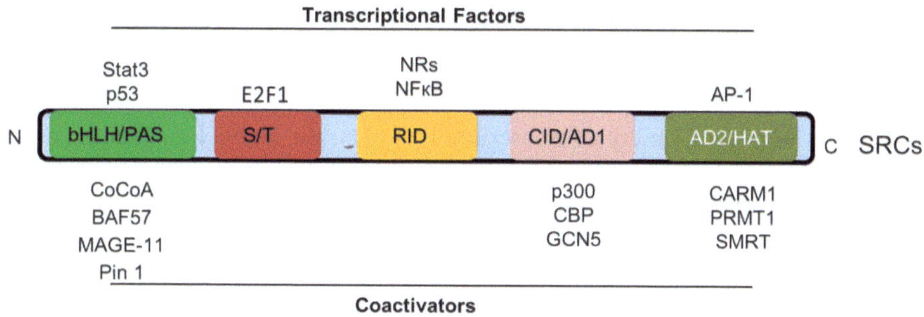


**Fig. 1.1** Structural domains and transcriptional interacting partners of SRCs. SRC proteins contain five functional domains: the N-terminal bHLH/PAS domain, the serine/threonine rich (*S/T*) domain, the nuclear receptor interacting domain (*RID*), the p300/CBP interacting domain (*CID*) or activation domain 1 (*AD1*), and the C-terminal activation domain 2 (*AD2*). SRC-1 and SRC-3 harbor a histone acetyltransferase domain (*HAT*) in the C-terminus. A representative list of transcriptional interacting partners within each domain of SRCs are indicated above (for interacting transcriptional factors) or below (for interacting coactivators) the structure. Interacting proteins are referenced from [8, 9] and as mentioned in the text

promoter [14]. The bHLH/PAS domain also mediates the interaction of SRCs with several non-NR transcription factors such as Stat3 [15] and p53 [16] that are important factors in cancer.

The S/T-rich domains of SRCs are frequently targeted by protein kinases and phosphatases, which regulate SRC protein stability and activity [17]. The S/T-rich domain of SRC-3 mediates its interaction with E2F1, an essential transcription factor in cell cycle control [18]. SRCs bind to NRs through the RID domain that harbors three "LXXLL" NR-binding motifs where "L" represents leucine residue and "X" denotes any amino acid [19, 20]. The interactions of SRCs with NRs are either hormone-dependent or hormone-independent based upon the NRs that SRCs are bound to and the growth conditions. Besides mediating the interaction with NRs, the RID domain of SRC-3 is important for its interaction with NFκB [21]. Following the binding to NRs, SRCs recruit p300/CBP histone acetyltransferases through the CID domain, which promotes chromatin remodeling and the recruitment of general transcription machinery [5, 22–24]. In the C-terminus of SRCs resides the activation domain 2 (AD2) that recruits CARM1 and PRMT1 methyltransferases [25, 26]. Interestingly, the C-terminus of SRC-1 and SRC-3 also contains a HAT domain [6, 27], but its functional substrates remain to be substantiated. In addition, the C-terminus was shown to mediate the interaction of SRCs with AP-1 transcription factors that play critical roles in cancer cell proliferation and invasion [28].

Of note, SRCs also have been shown to interact with other oncogenic transcription factors such as Rb [29] and HIF1α [30], and to potentiate activities of these transcription factors, although it is unclear which precise domains within SRCs are required for these interactions. Taken together, SRCs interact with a variety of transcriptional factors and coregulators through their five functional domains, suggesting that SRCs are important molecules that integrate diverse cellular processes.

## 1.3 Molecular Codes of SRCs: PTMs Targeted by Multiple Signaling Pathways

As coactivators of a variety of transcription factors, the activity and functions of SRCs are regulated by multiple signaling pathways. The molecular regulation of SRCs' activity and functions are determined by post-translational modifications (PTMs, Figs. 1.2 and 1.3), including phosphorylation, ubiquitination, sumoylation, acetylation, and methylation, all of which coordinately regulate SRCs' cellular localization, stability, and the interactions with their functional partners.

## 1.4 Phosphorylations

SRCs are phosphorylated by protein kinases in response to multiple signals including hormones, growth factors, and cytokines. These signals work independently or in concert to regulate SRCs' activities and functions.

## 1.5 Hormone-Induced Phosphorylations of SRCs

Hormones stimulate target gene transcription not only by activating hormone receptors via direct binding, but also by activating protein kinases that subsequently phosphorylate hormone receptors and coregulators including SRCs (Fig. 1.4). Hormones

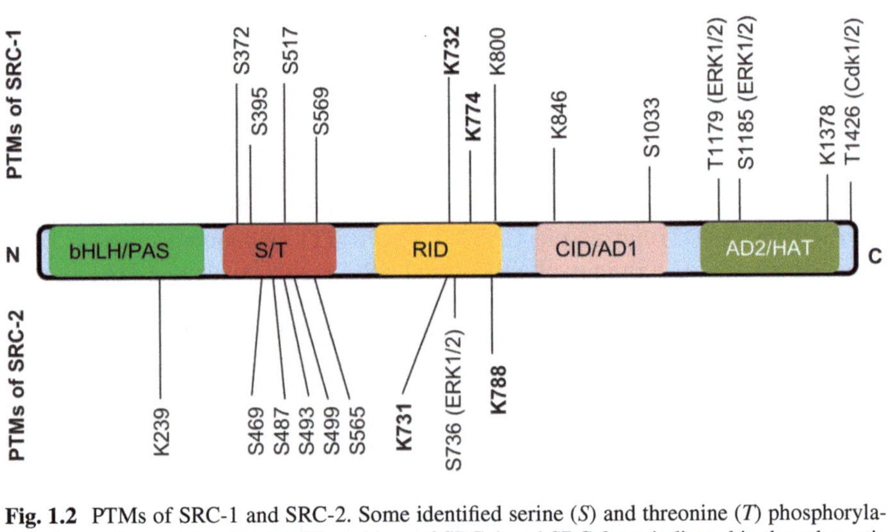


**Fig. 1.2** PTMs of SRC-1 and SRC-2. Some identified serine (*S*) and threonine (*T*) phosphorylation sites and sumoylated lysine (*K*) residues of SRC-1 and SRC-2 are indicated in the schematic structure. In the "( )" are shown certain kinases that target the specific phosphorylation residues. The two conserved sumoylation sites within RID domain of SRCs are shown in *bold*

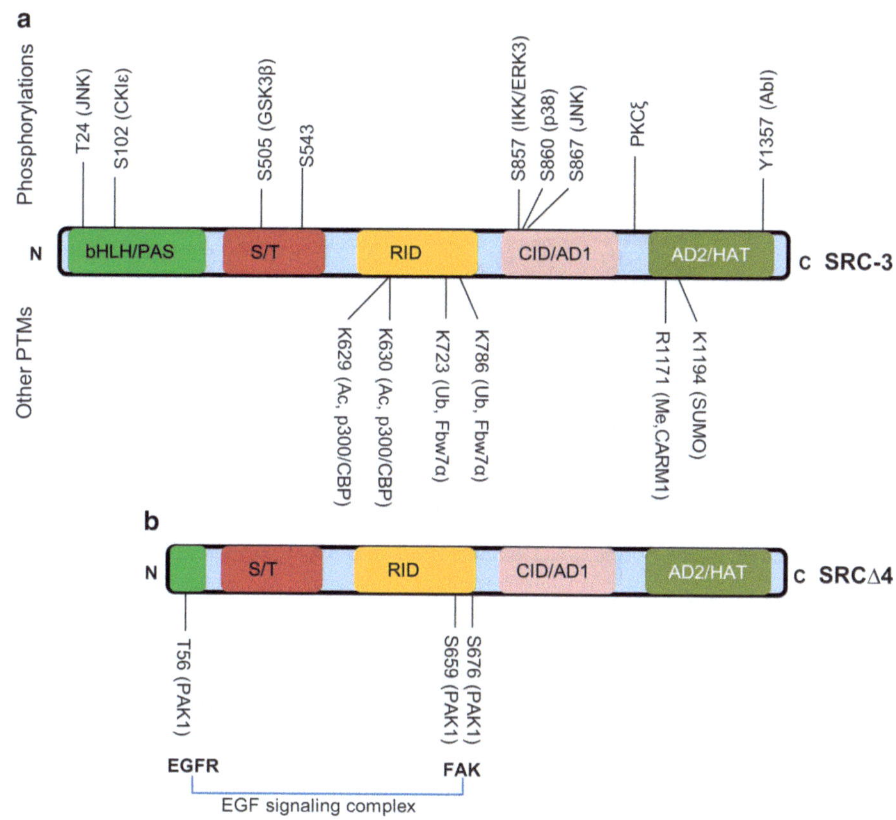


**Fig. 1.3** PTMs of SRC-3 and SRC-3Δ4. (**a**) Selectedphosphorylation sites of serine (*S*), threonine (*T*), or tyrosine (*Y*), lysine (*K*) residues with ubiquitination (*Ub*), sumoylation (*SUMO*), or acetylation (*Ac*), and arginine (*R*) residue with methylation (*Me*) of SRC-3 are indicated in the schematic structures. In the "( )" are also shown certain modifying enzymes for each specific PTM. Both K723 and K786 are sumoylation sites as well that are conserved in SRCs. (**b**) Phosphorylation codes of SRC-3Δ4 that are targeted by PAK1 for EGF signal transduction to FAK. PAK1 phosphorylates SRC-3Δ4 at T56, which mediates the interaction with EGFR, and at S569 and S676, which is important for the interaction with FAK. See the text for details (It should be noted that SRC-3 contains over 50 different PTMs that have been identified by a variety of techniques)

such as estrogen, progesterone, androgen, and glucocorticoid stimulate the activation of multiple kinases such as ERK1/2, Akt, p38, and JNK; rapid activations of these kinases by hormones are referred to as non-genomic signaling in contrast to direct actions of the receptors on the nuclear genome [31]. In response to E2 stimulation, SRC-3 is phosphorylated at multiple residues including T24 in the N-terminus, S505 and S543 in the S/T-rich region, and S857, S860, and S867 in the RID region [32]. E2-induced phosphorylation of SRC-3 occurs acutely (within minutes) and is dependent on ERα [33]. Phosphorylations at these residues promote the interaction of SRC-3 with ERα and CBP, and augment SRC-3's transcriptional activity.

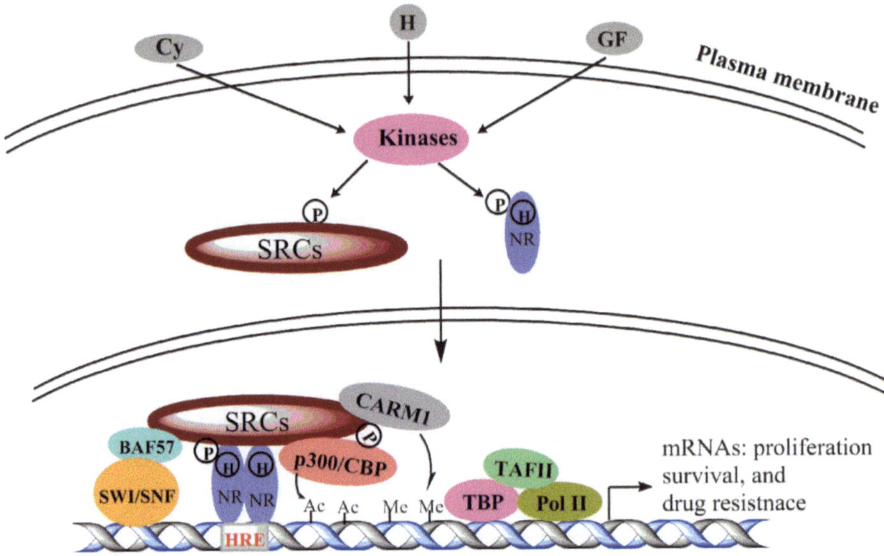


**Fig. 1.4** SRCs-mediated hormone signaling and the cross-talk with growth factor and cytokine signals in regulating NR target gene expression. Upon the binding of hormone (*H*), nuclear receptors (*NRs*) dimerize and bind to the hormone responsive element (*HRE*) of target genes. SRCs coactivate gene transcription by interacting with DNA-bound NRs and then recruiting secondary coactivators including p300/CBP histone acetyltransferases and CARM1/PRMT1 histone methyltranseferases. p300/CBP and CARM1/PRMT1 elicit histone acetylation (*Ac*) and methylation (*Me*), respectively, both of which facilitates the assembly of the general transcriptional machinery consisting of TBP, TAFIIs, and Pol II, and the subsequent gene transcription. SRCs also are capable of recruiting SWI/SNF chromatin remodeling complex via BAF57 and further potentiate gene transcription. The transcriptional activities of SRCs and NRs are regulated by multiple signaling pathways including those of cytokines (*Cy*), growth factors (*GF*), and non-genomic hormone actions, mainly through protein kinase-mediated phosphorylations (*P*) that are important for the interactions of SRCs with NRs and other coactivators

Interestingly, while androgen/AR induces SRC-3 phosphorylations similar to E2/ ERα, progesterone/PR is unable to do so, suggesting there is ligand/receptor/coactivator specificity in hormone-induced SRC phosphorylations. In agreement with this notion, SRC-2 is a primary coactivator for glucocorticoid receptor (GR) and is phosphorylated upon the stimulation of dexamethason eat five residues (S469, S487, S493, S499, and S565) in the S/T-rich region and one residue (S736) in the RID region [34]. Phosphorylations of SRC-2 facilitate GR transcriptional activity. Similarly, progesterone/PR stimulates SRC-1 phosphorylations that are important for SRC-1 transcription activity (Weiwen Long and Bert O'Malley, unpublished data). Collectively, these findings demonstrate that phosphorylations of SRCs serve as an integrating link between non-genomic and genomic actions of hormones.

## 1.6 Phosphorylations of SRCs Induced by Cytokines and Growth Factors

SRCs also are phosphorylated upon stimulation by growth factors and/or cytokines, which facilitates their functions in coactivating NR-mediated gene expression in both ligand-dependent and ligand-independent mechanisms; these stimulations serve as important molecular mechanisms for anti-hormone resistance during cancer therapy. SRC-1 is phosphorylated at T1179 and S1185 by MAPK upon stimulation by interleukin 6 (IL-6)and co-activates AR in a ligand-independent manner in prostate cancer cells [35]. Interestingly IL-4promotes PP2A-directed dephosphorylation of SRC-1 which is important for SRC-1/Stat6-regulated IL-4 target gene expression in Ramos B lymphoma cells [36]. Cytokines such as TNFa stimulate IKK-directed phosphorylation of SRC-3 which potentiates NFκB-mediated gene expression in breast cancer cells [32, 37].

SRC-1, SRC-2, and SRC-3 all are known to be regulated by growth factor signals. Epidermal growth factor (EGF) stimulates ERK2-directed phosphorylations of SRC-1 that are important for the interaction of SRC-1 with CBP in PR-dependent transactivation [38]. T1426 in AD2 of SRC-1 is targeted by Cdk1 and Cdk2, and this phosphorylation is important for PR/SRC-1-mediated cell cycle control [39]. SRC-2 is phosphorylated at S736 by EGF stimulation, and S736-phosphorylated SRC-2 promotes AR-dependent but ligand-independent transactivation, suggesting a potential role of SRC-2-mediated cross-talk between growth factor signaling and AR signaling in recurrent prostate cancer progression [40]. SRC-3 activity is tightly regulated by growth factor signaling. Both EGF and IGF-1 stimulate SRC-3 phosphorylation at tyrosine 1357 (Y1357) that is directed by AbI kinase in breast cancer and lung cancer cells [41]. Phosphorylation of Y1357 is critical for SRC-3 oncogenic activity in co-activating ERα and NFκB and promoting anchorage-independent cancer cell growth. Heregulin 1β, an EGF-like growth factor, stimulates SRC-3 phosphorylation through ERBB2 oncogenic kinase [42] which is implicated in the breast cancer tamoxifen resistance. cAMP/PKA signaling also induces SRC-1 and SRC-2 phosphorylation and potentiates NR-mediated transactivation in a ligand-independent manner [43, 44].

## 1.7 Phosphorylation Codes for SRC-3Δ4 Acting as an EGF Signaling Adaptor

SRC-3Δ4 is a splicing variant of SRC-3 with the deletion of exon 4 (Fig. 1.3b) [45, 46]. In comparison with full-length SRC-3 protein, SRC-3Δ4 lacks an N-terminal bHLH/PAS region that harbors the NLS. Consequently, SRC-3Δ4 primarily localizes in the cell cytosol due to its absence of an NLS. Upon EGF stimulation, PAK1 phosphorylates SRC-3Δ4 on T56 at the N-terminus and S659 and S676 within the RID region. Phosphorylations of SRC-3Δ4 promote its localization to the plasma membrane region where it interacts with EGFR through the N-terminus, which is mediated by T56 phosphorylation, and with FAK through the RID region, which is mediated by phosphorylations of S659 and S676(Fig. 1.3b). SRC-3Δ4 mediates the interaction between EGFR and FAK, thereby promoting EGF-induced c-Src activation and FAK phosphorylation on Y925, which in turn drives cancer cell migration and metastasis.

## 1.8 Ubiquitination and Its Regulation by Phosphorylation

Ubiquitination plays an essential role in regulating the stability and functions of SRC proteins. Ubiquitination of SRCs is frequently dependent on phosphorylations of specific residues that mediate the interaction of SRCs with the ubiquitin E3 ligases. The stability and activity of SRC-3 have been shown to be regulated by several protein kinase signals that are coupled to different E3 ligase complexes. GSK3β phosphorylates SRC-3 at S505 in the S/T-rich region, which is required for the binding of Fbw7α, a component of the SCF (Skp, Cullin, F-box protein containing) E3 ligase complex. SRC-3 is then ubiquitinated by $SCF^{Fbw7\alpha}$ at lysine 723 (K723) and K786 that are within the receptor-interacting domain. Interestingly, mono-ubiquitinations at K723 and K786 enhance the interaction of SRC-3 with ERα, ERα's phosphorylation at S118, and expression of the target genes in response to E2 stimulation [47]. Subsequent poly-ubiquitinations then lead to the degradation of SRC-3 and termination of SRC-3/ERα-regulated gene transcription. Ubiquitination-coupled activation of SRC-3 is also manifested by retinoic acid (RA) signaling, which involves the phosphorylation of SRC-3 at S860 by RA-activated p38, the subsequent phospho-S860-mediated recruitment of Cullin 3-based E3 ligase, ubiquitination of SRC-3, and finally activation of SRC-3/RAR-regulated gene transcription [48, 49]. In addition, SRC-3 protein stability and activity are regulated by a phospho-dependent degron that is located in the N-terminal bHLH/PAS region [50, 51]. Ser102 in the degron is phosphorylated by CKI (casein kinase I), which is required for the recruitment of a speckle-type POZ protein (SPOP)-based E3 ligase complex and the subsequent ubiquitination and turnover of SRC-3. Interestingly, phosphatases PP2A and PP1 target phosphorylated Ser102 and inhibit SRC-3 protein ubiquitination [50].

In addition to ubiquitin-dependent proteasomal degradation pathways, SRC-3 protein turnover is regulated by an ubiquitin-independent mechanism [52]. Importantly, PKCζ phosphorylates multiple residues in an acidic fragment that is important for the interaction of SRC-3 with the C8 subunit of the 20S proteasome, and enhances SRC-3 protein stability by inhibiting both ubiquitin-dependent and ubiquitin-independent proteolytic pathways [53].

In contrast with the extensive study on SRC-3, much less is known about the regulation of ubiquitination and stability of SRC-1 and SRC-2. However, a single nucleotide polymorphism (SNP) P1272S was shown to increase SRC-1 protein stability by disrupting a potential GSK3β-directed phospho-dependent degradation code in the AD2 region of SRC-1 [54]. Although no specific phosphorylation and ubiquitination sites were revealed, an ubiquitination-coupled activation mechanism for SRC-2 was shown to be regulated by cAMP/PKA signaling [44].

## 1.9 Sumoylation

Sumoylation is a type of PTM that involves an addition of small ubiquitin-like modifier (SUMO) to the lysine residues of proteins [55]. In spite of its structural and enzymatic processing similarities to ubiquitination, sumoylation often alters a protein's binding affinity with the associating partners or its subcellular localization rather than degradation. All of the SRCs were shown to be sumoylated on two conserved lysine residues of the nuclear receptor-interacting domains (Figs. 1.2 and 1.3); sumoylations alter the interaction of SRCs with NRs and their transcriptional activities [56]. However, while sumoylations enhance the interaction of SRC-1 and SRC-2 with PR and AR respectively [57, 58], and their retention and transactivity in the nucleus, SRC-3 transactivity is attenuated by sumoylations at K723 and K786 [47]. As aforementioned, K723 and K786 of SRC-3 are ubiquitination sites as well, and the ubiquitinations on these sites are important for the interaction of SRC-3 with ERα and their transactivity. As such, the attenuation of SRC-3 transactivity by sumoylations at K723 and K786 is likely due to the competitive blocking of ubiquitinations on these two sites. Although the regulation of SRCs' sumoylations by phosphorylations have not been reported to date, both SRC-2 and SRC-3 harbor a phosphorylation-dependent sumoylation motif (PDSF) ψKxExxSP (where ψ is a large hydrophobic residue)that has been characterized in heat shock factor 1 (HSF-1) and MEF2A transcription factors [59, 60]. S736 within the PDSF of SRC-2 is a phosphorylation site targeted by EGF/ERK1/2 signaling [40], and K731 is a sumoylation site that enhances SRC-2 transactivity [58]. It would be of interest to determine whether this is a *bona fide* PDSF that plays a role in SRC-2 activity.

## 1.10 Acetylation and Methylation

SRCs act as bridging factors to recruit histone acetyltransferases such as p300/CBP and histone methyltransferases such as CARM1/PRMT1 to DNA-bound NRs to remodel chromatin and regulate gene transcription (Fig. 1.4). Interestingly, these histone modifying enzymes target not only histones but also SRCs and NRs [61–64]. SRCs were shown to be acetylated by p300 [61] and methylated by CARM1 upon E2 stimulation [63, 64]. While histone acetylation and methylation facilitate the assembly of the transcription machinery and subsequent gene transcription, acetylation or methylation of SRC-3 leads to the dissociation of NRs/cofactors complex and the termination of gene transcription.

## 1.11 Molecular Actions of SRCs in Cancer Cells In Vitro and in Mouse Tumor Models In Vivo: SRCs as Integrators of Multiple Signaling Pathways

SRCs, in particular SRC-3, function as important mediators and integrators of a variety of oncogenic signaling (e.g., hormones, growth factors, and cytokines) pathways to regulate virtually every aspect of cellular processes: proliferation, survival, migration, and invasion (Figs. 1.4 and 1.5). Hormone signaling acts independently

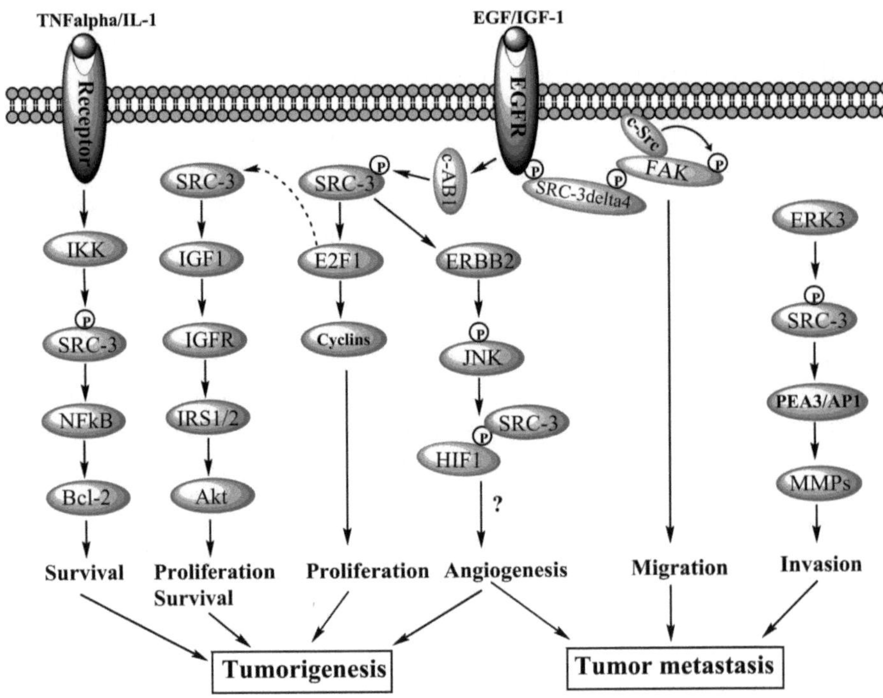


**Fig. 1.5** SRC-3 integrates multiple signaling pathways to promote tumorigenesis and metastasis. Besides its regulation of hormone signaling as illustrated in Fig. 1.4, SRC-3 integrates multiple growth factor and cytokine signaling pathways to regulate a variety of cancer cell processes. Only a subset of these pathways are illustrated in Fig 1.5. In response to TNFα/IL-1 signals, SRC-3 is phosphorylated by IKKs and coactivates NFκB-mediated Bcl-2 expression for cancer cell survival. SRC-3 upregulates the expression of multiple components of IGF1-IGFR-PI3K/Akt pathway that is important for both cancer cell proliferation and survival. Upon EGF or IGF-1 stimulation, SRC-3 is targeted by protein kinases including c-AB1, and then coactivates E2F1-mediated expression of cell cycle genes including cyclins E and A. SRC-3 itself is a target gene of E2F-1 (indicated by a *dashed arrow*), and upregulation of SRC-3 by E2F1 might boost other signaling pathways regulated by SRC-3, for example, the IGF1/Akt pathway. SRC-3 is also important for the activation (phosphorylations) of ERBB2 and EGFR and the downstream kinases such as JNK, and plays a role in tumor angiogenesis, likely by coactivating HIF1. In addition, SRC-3Δ4 acts as an EGF signaling adaptor by bridging EGFR to FAK, and promotes EGF-induced FAK phosphorylations and cancer cell migration. Furthermore, SRC-3 is targeted by ERK3 kinase and coactivates PEA-3/AP-1-mediated MMP gene expression for promoting cell invasion

or synergistically with growth factor or cytokine signaling to regulate these cellular processes in which SRCs play important, integrating roles (Fig. 1.4). The actions of SRCs are exquisitely regulated by PTMs (mainly phosphorylations) that are stimulated by the oncogenic signals. Increased growth factor and/or cytokine signaling hijack SRCs to drive the progression of cancer cells from hormone-dependent to hormone-independent growth and elicit anti-hormone resistance, for example, anti-estrogen resistance in breast cancer. Based on the expression status of hormone receptors (mainly ER and PR) and growth factor tyrosine kinase receptors (mainly ERBB2 and EGFR) and the origin of cancer cells, breast cancers can be classified as 4 subtypes [65, 66]: luminal A ($ER^+ PR^+ERBB2^-$), luminal B ($ER^+ PR^+ERBB2^+$), ERBB2-enriched ($ER^-PR^-ERBB2^{++}$), and basal-like (also known as triple-negative, $ER^-PR^-ERBB2^-$). Accompanied by enhanced growth factor signaling, luminal A subtype can progress to luminal B and further to ERBB2-enriched subtype, which leads to advanced cancer phenotype and increased anti-estrogen resistance. Basal-like breast cancer cells often have upregulated EGFR expression and high aggressiveness. As discussed below, SRC-1 and SRC-3 play critical functions in all four subtypes of breast cancer, whereas the roles of SRC-2 are minor.

## 1.12 SRCs with Hormones/NRs-Mediated Signaling

### *1.12.1 Estrogen/ER Signaling in Breast Cancer*

Both SRC-1 and SRC-3 act as coactivators of ERα to mediate estradiol signaling in promoting breast cancer cell proliferation and survival. SRC-1 is important for estradiol-induced cell proliferation ofMCF-7, a breast cancer cell line of luminal A subtype [67–69]. Depletion of SRC-1 differentially affected E2-inducible genes: with a significant decrease in the expression of pS2 and stromal cell-derived factor 1 (SDF-1) but little effect on c-Myc [69]. Both MCF-7 and T47D (another breast cancer cell line of luminal A subtype) overexpress SRC-3. Depletion of SRC-3in these cells diminishes the expression of estrogen/ER target genes including cyclin D1, c-Myc, and Bcl-2, and inhibits cell proliferation but increases apoptosis [70, 71]. Consequently, depletion of SRC-3 in MCF-7 cells inhibits estrogen-induced colony formation in soft agar and xenograft tumor growth in nude mice [72]. E2 exerts non-genomic signaling by activating multiple kinases including ERK1/2 and IKKα. E2-induced phosphorylations of SRC-3 via ERK1/2 and IKKα are critical for SRC-3's activity in promoting the expression of E2 target genes such as cyclin D1 and c-Myc [32, 73]. The functional relationship between SRC-3 and ERα in tumorigenesis was revealed in a mouse mammary tumor virus (MMTV)-SRC-3 transgenic mouse model [74]. Transgenic overexpression of SRC-3 induced tumors primarily in mammary glands but also in other organs including uterus and lung, suggesting SRC-3 is a *bona fide* oncogene. Ovariectomy in MMTV-SRC-3 mice greatly decreased mammary tumor formation, and genetic deletion of ERα in MMTV-SRC-3 mice by crossing with ERα-null mice completely abolished mammary tumor formation [75]. Taken together, these findings suggest that SRC-3 and possibly SRC-1 are critical for E2/ER signaling in promoting breast cancer progression.

### *1.12.2 Androgen/AR Signaling in Prostate Cancer*

While estrogen/ER signaling is critical for the progression of breast cancer, androgen/AR signaling plays an essential role in hormone responsive prostate cancer. All of three SRC members have been shown to coactivate AR to meditate androgen signaling in prostate cancer cells. Depletion of either SRC-1 [76], SRC-2 [77, 78], or SRC-3 [79, 80] significantly decreases androgen-dependent AR transcriptional activity and LNCaP prostate cancer cell growth in culture and in xenograft tumor mouse models. On the contrary, overexpression of SRC-3 or SRC-2 in LNCap greatly enhances the responsiveness of AR to androgen or to ligand-independent stimulation and promotes cell growth.

## 1.13 Interplay of SRCs with ERBB2 Signaling: Anti-estrogen Resistance in Breast Cancer

ERBB2, a member of the EGFR family, is an oncogenic protein that is frequently overexpressed in advanced breast cancer. ERBB2 overexpression promotes tumor progression, and is highly associated with the progression of ER-positive breast cancer cells from estrogen-dependent to estrogen-independent growth; the overexpression is concomitant with the gain of resistance to anti-estrogen drugs [81]. Both SRC-1 and SRC-3, in concert with ERBB2, play a critical role during these processes. Genetic loss of SRC-3 completely suppressed MMTV-ERBB2 induced mouse mammary tumor formation in association with a remarkable decrease in phosphorylations of ERBB2 and the downstream kinases JNK and Akt [82]. Interestingly, tumor angiogenesis also was curtailed significantly due to the loss of SRC-3. SRC-3 was confirmed as a pro-angiogenic factor recently [83] and acts as a coactivator of HIF1$\alpha$ [30]. While loss of SRC-1 did not significantly decrease the average tumor number formed in MMTV-ERBB2 transgenic mice, it increased tumor latency and greatly inhibited tumor metastasis to the lung [84]. Under cell culture conditions, stable exogenous expression of ERBB2 in MCF-7 elicits an estrogen-independent cell growth and resistance to tamoxifen, both of which were significantly inhibited by depletion of SRC-3 [42]. A similar synergistic role of ERBB2 with SRC-3 or SRC-1 was shown in BT474, a luminal-B subtype of breast cancer cell line with overexpression of ERBB2 [85, 86]. ERBB2 overexpression activates the downstream kinases ERK1/2, JNK, and Akt that phosphorylate ER$\alpha$ and SRCs, which leads to activation of ER$\alpha$ under low concentration or even in the absence of estrogen [42, 87]. In addition to ER$\alpha$, SRC-3 and SRC-1 interact with and coactivate other transcription factors such as Ets and PEA3 to promote cancer cell growth and invasion in response to enhanced ERBB2 signaling [88, 89]. Conversely, SRC-3 positively regulates ERBB2 expression by competing with PAX-2 (a repressor of ERBB2 gene expression) for the binding to the ER$\alpha$-bound site of the ERBB2 gene [90]. These findings suggest a positive feedback between ERBB2 and SRC-3 that regulates breast cancer cell proliferation and anti-estrogen resistance.

## 1.14 Interplay Between SRCs and EGFR Signaling: From the Membrane to the Nucleus

EGF/EGFR signaling is implicated in multiple cancers including those of breast, prostate, and lung. EGF signaling stimulates SRCs' phosphorylation and activation; activated SRCs then work with transcriptional factors including E2F-1, ETS, and AP-1 to regulate cell proliferation, migration, and invasion (Fig. 1.5).

In breast cancer, increased EGFR signaling is frequently associated with the invasive and triple-negative phenotype. In a triple negative breast cancer cell line MDA-MB231, SRC-3 and its N-terminus-deleted isoform SRC-3Δ4 elegantly regulate distinct aspects of the EGF signaling at different cellular locations (Fig. 1.5). SRC-3Δ4 acts as a signaling adaptor bridging EGFR to FAK at the plasma membrane to mediate activation of FAK and promote cell migration upon the EGF signal [46]. SRC-3 is phosphorylated by Abl kinase atY1357 in response to EGF stimulation, and this phosphorylation is important for SRC-3's function in coactivating AP-1 and E2F1 in the nucleus and promoting cell growth [41]. Intriguingly, EGFR tyrosine phosphorylation and activity is regulated by SRC-3. Depletion of SRC-3 greatly decreased EGF-induced EGFR tyrosine phosphorylation, which led to decreased JNK kinase activity and growth inhibition of MDAMB231 cells [91]. While the detailed molecular mechanism is unclear, downregulation of tyrosine phosphatases by SRC-3 partly contributes to an increase of EGFR phosphorylation.

In prostate cancer, enhanced EGFR signaling is highly associated with castration-resistant (androgen-independent) cancer progression. On one hand, EGF signaling stimulates AR tyrosine phosphorylation (Y534) via c-Src kinase [92], and AR tyrosine phosphorylation promotes its nuclear localization and transactivity. On the other hand, EGF signaling targets SRCs via MAPK-directed phosphorylations. For example, SRC-2 is phosphorylated at S736 by EGF stimulation, and S736-phosphorylated SRC-2 promotes AR-dependent but androgen-independent transactivation [40]. Together, EGF signaling stimulates AR/SRCs' transactivity and promotes prostate cancer cell growth in culture and tumor growth in castrated mice [40, 92].

## 1.15 Regulation of IGF-1/Akt Signaling by SRC-3 and SRC-1

IGF-1/Akt signaling is another molecular pathway that is regulated by SRCs and is critical for SRCs' oncogenic functions (Fig. 1.5). The IGF-1/Akt signaling pathway initiates with the binding of IGF-1 to its receptor (IGF-1R) on the cell membrane and the subsequent activation of IGF-1R by auto-tyrosine phosphorylation, followed by the recruitment of insulin receptor substrate (IRS) proteins and IRS-mediated activation of the phosphatidylinositol 3-kinase (PI3K)/Akt pathway [93]. SRC-3 regulates the expression levels of multiple components of the IGF-1/Akt pathway, includingIGF-1/IGF-2, IGF-1R, IRS-1/IRS-2, and Akt in breast cancer cells [94] and prostate cancer cells [95], thereby regulating the activities/

phosphorylations of IGF-1R and Akt and promoting cancer cell proliferation and anchorage-independent cell growth. The regulation of IGF-1/Akt signaling by SRC-3 was corroborated in vivo. Increased IGF-1 expression and activation of IGF-1R, Akt, and mTOR was seen in MMTV-AIB1 (SRC-3) transgenic mice, which led to spontaneous mammary tumor formation [74]. In contrast, genetic depletion of SRC-3 in v-Ha-Ras transgenic mice caused a remarkable decrease in tumor initiation and metastasis to the lung, partly due to decreased expression of IRS-1 and IRS-2 and the attenuated Akt activity [96]. Loss of SRC-1 did not affect primary mammary tumor growth, but greatly decreased tumor metastasis to the lung in MMTV-polyoma middle T antigen (PyMT) transgenic mice by downregulating ERBB2 expression and Akt phosphorylation [97].

## 1.16 Interplay of SRCs with Cytokine Signaling in Promoting Cancer Cell Aggressiveness

IL-6 signaling promotes castration-resistant prostate cancer progression by activating AR in an androgen-independent mechanism [98–100]. Besides the direct effect on AR, IL-6 signaling modulates SRCs' activity as well. SRC-1 phosphorylation by MAPK in response to IL-6 is important for ligand-independent activation of AR [35]. In addition, upregulation of SRC-2 in LNCaP cells upon long-term treatment with IL-6 is associated with acquired resistance to bicalutamide, an anti-androgen drug [101]. Conversely, cytokine production can be regulated by SRCs. For example, SRC-1 upregulates colony-stimulating factor-1 (CSF-1) expression and promotes the recruitment of macrophages to the tumor site, which contributes to tumor metastasis [97].

SRC-3 is phosphorylated upon the stimulation of cytokines such as TNFα and IL-1β [32, 37]. Phosphorylations of SRC-3 stimulate the interactions of SRC-3 with ERα and NFκB, which are important for TNFα-induced cyclin D1 expression and cell proliferation [102]. Interestingly, SRC-3 was shown to interact with translational repressors TIA1 (T cell restricted antigen 1) and TIAR (TIA1 related homologue) and regulate the translation of TNFα and interlukin 1 mRNAs [103], for which the implication in cancer is unclear at present.

A recent study revealed an intriguing interplay between a cleaved isoform of SRC-1 and TNFα signaling during endometriosis [104] which shed new light on the roles of SRCs in inflammation-associated diseases such as cancers. Endometriosis is an inflammation-driven disease that is initiated by the migration of endometrial cells to distal sites. While TNFα is well-known as a critical driving factor for endometriosis, it is less clear how the intrinsic pro-apoptotic activity of TNFα is silenced during this process. It is shown in this interesting study that TNFα-induced MMP9 cleaves SRC-1 at Pro-790, and the c-terminus of cleaved SRC-1 promotes endometriosis by blocking caspase8-mediated apoptosis and by stimulating epithelial-to-mesenchymal transition (EMT) for increased invasiveness. Given the positive associations of both SRC-1 and inflammation with advanced and metastatic tumor stages, cytokine signaling and SRC-1 conceivably could synergistically promote cancer progression and metastasis following similar mechanisms as shown in endometriosis. Indeed, SRC-1 was shown to positively regulate TWIST, a master regulator of EMT, thereby promoting tumor cell migration/invasion and metastasis [89]. It would be interesting to determine whether SRC-1 undergoes proteolytic cleavage to produce the cleaved isoform during cancer progression and whether this cleaved form of SRC-1 is responsible for the upregulation of TWIST and EMT-associated tumor cell migration and invasion.

## 1.17 Phospho-dependent Regulation of SRC-3 by an Atypic MAPK for Cancer Cell Invasion

SRC-3 promotes cancer cell invasion by coactivating PEA-3- and AP-1-regulated matrix metalloproteinase (MMP) expression [105–107], but the invasive signals to SRC-3 and the molecular regulation of SRC-3 proinvasive activity were not elucidated until a recent study that revealed a phospho-dependent regulation of SRC-3 proinvasive activity by an atypical MAPK ERK3 (Fig. 1.5) [108]. ERK3 was identified as an interacting partner of SRC-3 by immunoprecipitation-mass spectrometry (IP-MS) analyses. ERK3 phosphorylates SRC-3 at serine 857 (S857), and this ERK3-mediated phosphorylation at S857 is essential for SRC-3's interaction with the ETS transcription factor PEA3, promoting upregulation of matrix metalloproteinase (MMP) gene expression and proinvasive activity in lung cancer cells. ERK3/SRC-3 signaling drives cancer cells to invade and form tumors in the lung. As such, this study not only revealed a molecular mechanism for regulating SRC-3 proinvasive activity, but also identified a novel oncogenic function for ERK3 in promoting lung cancer invasiveness.

## 1.18 Alterations of SRCs and the Clinical Implications in Cancers

As transcriptional coactivators integrating multiple signaling pathways to regulate cancer cell proliferation, survival, migration, and invasion, SRCs are highly dysregulated in a variety of cancers including breast cancer, prostate cancer, endometrial cancer, ovarian cancer, lung cancer, colorectal cancer, liver cancer, and pancreatic cancer. Alterations of SRCs, including primarily gene amplification, mRNA or protein overexpression, are implicated in cancer progression and metastasis and tumor resistance to therapeutic interventions.

## 1.19 Gene Amplification

Gene amplification is one of the fundamental features for defining an oncogene. SRC genes, in particular SRC-3 (also known as AIB1-Amplified in breast cancer gene 1), are amplified in multiple human cancers. Gain of copy numbers of SRC-1 and SRC-2 genes were reported in a recent study with a cohort of 218 prostate tumors consisting of 181 primaries and 37 metastases [109]. Of particular note, SRC-2 gene amplification was detected in 8 % of primary tumors and 37 % of metastases. Gain of SRC-2 expression is associated with increased rates of prostate tumor recurrence. SRC-3 gene amplification has been shown in multiple cancers including breast cancer (with 5–10 % frequency [4, 110]), ovarian cancer (with 7–25 % frequency [110, 111]), colorectal cancer (with 10–32 % frequency [112, 113], lung cancer (with 8.2–27 % frequency [114, 115]), and hepatocellular cancer (40 % frequency, [116]). SRC-3 gene amplification contributes to upregulation of SRC-3 mRNA and protein in cancers, and is positively correlated with advanced tumor stages.

## 1.20 Mutations

In contrast to high frequencies of gene amplification and mRNA/protein overexpression, gene mutation of SRCs is rarely detected in cancers. Even though a few point mutations of SRC-1 [117] and SRC-2 [109] were identified in tumors, the frequency is very low (~1 %) and the pathological association is undetermined. Interestingly, a fusion between MOZ (monocytic leukemia zinc finger) gene and TIF2 (SRC-2) gene has been repeatedly detected in acute myeloid leukemia [118–120]. MOZ-TIF2 fusion protein retains the PHD zinc finger domain and the MYST domain of MOZ and the CBP-interacting domain (CID) and activation domain 2 (AD2) of TIF2. The recruitment of p300/CBP via the CID of MOZ-TIF2 is essential for leukemogenesis of the fusion gene [120].

## 1.21 Upregulation of mRNA and Proteins

Expression of SRCs, in particular SRC-3 and SRC-1, are frequently upregulated in a variety of cancers. SRC-3 is the second most overexpressed oncogene among all human cancers, second only to c-myc. Upregulation of SRCs is often co-current with elevated protein kinase signaling, which indicates high tumor grade, increased tumor invasiveness, and tumor resistance to therapeutic treatments. A substantial number of studies have investigated the expression and clinical implication of SRCs in a variety of cancers, with focus mostly on breast cancer and prostate cancer.

## 1.22 Breast Cancer

Breast cancer progression relies on two major signaling pathways: hormone (e.g. E2/ERα) signaling and epidermal growth factor signaling pathways [81]. Antihormonal drugs (e.g. tamoxifen, an estradiol antagonist, and letrozole, an anti-aromatase inhibitor) have been commonly used for treating ER/PR-positive breast cancers. However, antihormone resistance (either naïve or acquired resistance after therapy) and associated cancer recurrence have been major obstacles for curing breast cancer patients. The acquisition of resistance to antihormonal drugs is often associated with a transition of hormone signaling-dependent to growth factor signaling-dependent tumor growth [121, 122]. SRC-1 and SRC-3 are overexpressed at high frequency in breast cancer, and their overexpression is implicated both in hormone-dependent and in hormone-independent breast cancer progression and metastasis.

SRC-1 expression has been shown to be significantly increased in around 19–30 % breast tumors [123–125]. In line with its primary role in promoting cancer cell migration and invasion in vitro and tumor metastasis in animal models, upregulation of SRC-1 is highly correlated with lymph node metastasis and poor disease-free survival (DFS) of breast cancer patients [88, 126]. Another important role for SRC-1 is the regulation of cancer cell sensitivity to anti-hormone drugs. Indeed, SRC-1 expression has been demonstrated as a predictor of anti-estrogen resistance and/or tumor recurrence following therapy [125, 126].

The implication of SRC-3 expression in breast cancer progression has been extensively studied. From various separate studies [4, 72, 124, 127, 128], overexpression of SRC-3 mRNA and protein has been shown in the range of from 13 % to 74 % of breast tumors, with an average of around 50 % overexpression rate. In agreement with the broad roles of SRC-3 in regulating cancer cell proliferation, survival, migration, and invasion in cultured cells and in animal models, overexpression of SRC-3 is positively associated with advanced tumor grade, increased tumor invasiveness and metastasis, and worse DFS in both ERα-positive and ERα-negative breast tumors [4, 87, 124, 127]. SRC-3 overexpression is frequently associated with enhanced protein kinase signaling. Tyrosine kinase receptor ERBB2 is frequently amplified (gene copy) and overexpressed (mRNA and protein) in cancers [129], with the highest frequency in breast cancer (~25 %). Simultaneous overexpression of SRC-3 and ERBB2 are reported in several studies [87, 128, 130, 131], and their co-overexpression indicates increased tumor resistance to tamoxifen treatment and increased tumor recurrence. Although the regulation of the IGF1R/Akt signaling pathway by SRC-3 has been well demonstrated by cell culture studies and mouse mammary gland tumor models, the correlation between these two has not been shown in tumor studies of breast cancer patients. Given the frequent alterations of both of these two factors in breast cancer, it is of high clinical significance to investigate the association of SRC-3 and IGF1R/Akt signaling pathways and their implications in cancer progression in more detail.

In contrast with ample evidence for the critical roles of SRC-1 and SRC-3 in breast cancer progression and metastasis, few studies have provided conclusive data to support a definite role for SRC-2 in breast cancer.

## 1.23 Prostate Cancer

Androgen/AR signaling plays a critical role in the initiation and progression of prostate cancer, and has been a major therapeutic target for treating this disease [132, 133]. Androgen ablation therapy (mainly by chemical castration) effectively inhibits tumor growth during the initial treatment. Unfortunately, most tumors relapse and become resistant to androgen ablation therapy. Castration-resistant cancer progression or recurrence is frequently associated with an advanced and metastatic tumor phenotype and has been a major obstacle for curing prostate cancer. AR and its target genes are commonly expressed in and are believed to drive recurrent prostate cancers [134]. In addition to AR gene mutations and AR overexpression, alterations in AR coactivators including SRCs and upregulation of growth factor signaling are two other major molecular mechanisms for castration-resistant tumor progression [135]. While genetic alterations of the SRC-1 gene are rare in prostate tumors, upregulation of SRC-1 has been shown by a few studies [76, 117, 136]. Upregulation of SRC-1 protein is associated with lymph node metastasis [76] and tumor recurrence after androgen deprivation therapy [117, 136]. Like SRC-1, SRC-2 expression is highly increased in recurrent prostate tumors following androgen deprivation therapy [117, 136]. More importantly, a recent study showed that elevated SRC-2 expression, probably due to SRC-2 gene amplification, was detected in both primary and metastatic prostate tumors [109]. Similar findings were reported for SRC-3 in prostate cancer [79, 106, 137, 138]. Upregulation of SRC-3 is positively correlated with increased Akt activity in prostate tumors [79, 138], which affirms a positive regulatory role of SRC-3 in Akt signaling in prostate cancer.

## 1.24 Lung Cancer

SRC-3 functions as an oncogene in lung cancer. Transgenic overexpression of SRC-3 in mice causes spontaneous lung tumor formation [74]. SRC-3 gene amplification and protein overproduction were shown in as high as 27 % of non-small cell lung cancers (NSCLCs) in one study [115]; overexpression correlates with poor disease free and overall survival. Interestingly, ERK3, a kinase that phosphorylates SRC-3 and confers SRC-3 pro-invasive activity in lung cancer cells, was shown to be highly upregulated in lung cancer [108]. Overexpression of SRC-3 protein in lung cancer also was reported in two other studies [139, 140]. The implications of SRC-1 and SRC-2 in lung cancer are not known.

In addition to the cancer types discussed above, SRC-3 has been shown to be overexpressed in many other cancers: a 64 % overexpression rate in high grade ovarian tumors [141], a 67 % overexpression rate in hepatocellular carcinomas [116], a 35 % overexpression rate in colorectal carcinomas [112], and a nearly 70 % overexpression frequency in pancreatic tumors [142].

## 1.25 Tumor Suppressor Functions of SRC-3 and SRC-2 in Specific Tissue Context

In contrast with the ample evidence for SRC-3′s function as an oncogene in a multitude of cancers, deletion of the SRC-3 gene promotes proliferation of lymphocytes and induces spontaneous malignant B-cell lymphoma upon aging in mice [143]. Similarly, while SRC-2 has been identified as an oncogene in prostate cancer, a recent study revealed a tumor suppressor role for SRC-2 in liver cancer [144]. Although the molecular mechanisms underlying the unexpected tumor-suppressing roles of SRC-2 and SRC-3 are not understood, these findings suggest that SRCs, as transcriptional coactivators, can regulate cell proliferation and survival either positively or negatively, depending upon the cell- and tissue context.

## 1.26 SRCs as Prognostic Biomarkers and Therapeutic Drug Targets

Increased expression and activities of SRCs not only are implicated in cancer progression and metastasis, but also are positively associated with drug resistance, including anti-estrogen resistance in breast cancer [87, 123, 125, 145, 146], resistance to EGFR inhibitors in lung cancer [115]), and resistance to chemotherapeutic drugs such as cisplatin and doxorubicin [30, 147, 148]. Hence, novel therapeutic drugs targeting SRCs might be utilized individually or in combination with other therapeutic drugs for treating cancers in different subtypes and at different stages.

## 1.27 Targeting SRCs by Intervening NR-SRC Interactions

The binding of SRCs with NRs via SRCs' 'LXXLL' NR interacting motifs is critical for their transcriptional activity. In addition, the flanking amino acids of the 'LXXLL' motif are shown to confer an order of specificity on differential NR/SRCs interactions. On the basis of these molecular mechanisms, there have been efforts to develop peptides containing LXXLL motifs or identifying small molecules that can disrupt the interactions between NRs and SRCs. A screening of the phage display library identified peptides that specifically and effectively inhibit ERα or ERβ transactivity [149–151], but it was unclear what specific ER/SRC interaction(s)was affected by these peptides. Soon after this, a peptide specifically blocking the interaction of SRC-2 with TRβ [152] and other peptides preventing the binding of SRC-1 with ERα or ERβ [153, 154] were identified utilizing a similar strategy. In addition, small molecule inhibitors (SMIs) were identified for targeting SRCs' interaction with ERα [155] and TRβ [156]. Although these identified peptides or SMIs were shown to efficiently inhibit NRs' transactivity, their efficacy on NR/SRCs-mediated cell functions were not evaluated in animals.

## 1.28 Small Molecule Inhibitors Targeting SRCs for Degradation

Since overexpression of SRC proteins, in particular SRC-3, is highly associated with advanced tumor progression and drug resistance, a potentially more effective strategy would be to identify small molecules that directly target SRCs and downregulate SRCs' protein stability and activities. Based on this idea, a recent study identified a SMI for SRC-1 and SRC-3 by both activity-based and stability-based screening assays [157]. This proof-of-principle drug, gossypol, downregulates the stability and activities of both SRC-1 and SRC-3 via direct binding, but is less selective forSRC-2 and other cofactors. Importantly, gossypol greatly increases the response of cancer cells to inhibitors of growth factor signaling, including MEK inhibitor AZD6244, EGFR inhibitor AG1478, and IGF-1R inhibitor AG1024. This study demonstrates that SRCs are accessible therapeutic targets to SMIs and encourages additional high throughout screenings for identifying drugs targeting SRCs.

SRCs also can serve as diagnostic and prognostic biomarkers, as they are altered in cancers at multiple levels including gene amplification, mRNA/protein overexpression, and protein posttranslational modifications. Of particular interest is the phosphorylation of SRC-3 at S857. S857 appears to be a hotspot that is targeted by multiple kinases and confers to SRC-3 and SRC-3Δ4 a variety of oncogenic functions: augmentation of cancer cell motility when targeted by PAK-1 [46], increase of cancer cell invasion when targeted by ERK3 [108], and gain of chemotherapeutic drug resistance when targeted by IKK [30]. It is of significant clinical interest to test whether phosphorylation on this site is positively associated with human cancer progression and metastasis, thereby serving as a diagnostic and/or prognostic tumor biomarker.

## 1.29 Conclusion and Perspective

Cancer cells must acquire a variety of capabilities for tumor initiation, uncontrolled outgrowth, invasion of surroundings, and metastasis to the distant organs. Hanahan and Weinberg have summarized these required capabilities as eight hallmarks of cancer: sustaining proliferative signaling, evading growth suppressors, resisting cell death, enabling replicative immortality, inducing angiogenesis, activating invasion and metastasis, and two more recently recognized hallmarks-reprogramming of energy metabolism and evading immune destruction [158]. Despite little information on the roles of SRCs in those two emerging hallmarks, substantial evidence exists that SRCs act as integrators for the other hallmarks, thereby making great contributions to cancer cells for acquisition of these hallmarks. It has been recognized that cancer cells reprogram energy metabolism to provide fuels and biosynthetic intermediates (nucleosides and amino acids) for uncontrolled cell growth and/or invasion by enhancing energy consumption from glycolysis [159]. Although the roles of SRCs in reprogramming energy metabolism of cancer cells have not been revealed, a number of studies, mostly by using SRC knock-out animals, have shown that SRCs are critical regulators of energy metabolism of glucose and lipids [160–166]. SRC-1 has distinct functions in cell metabolism of white and brown adipose tissues: loss of SRC-1 renders animals more susceptible to high fat diet-induced obesity, whereas loss of SRC-2 or SRC-3 confers protection against it. As such, future work on SRCs in cancer cell metabolism is warranted and should provide new insights on the molecular mechanisms by which SRCs alter the cues for uncontrolled growth and aggressiveness of cancers.

Evading immune destruction and harnessing tumor-associated inflammation is another hallmark that is important for cancer progression. The roles of SRCs in this process have been scarce but surely are worthy of investigations. Although their exact roles in tumor-associated inflammation remain to be determined, SRCs are engaged in cytokine signaling and inflammation. As mentioned above, SRCs are activated (phosphorylated) by cytokine signaling and function as coactivators of NFκB and/or Stats to positively regulate cytokine signaling in cancer cells [32, 35, 37, 101, 102] and in animal models of inflammation-associated disease [104, 167–169]. Hence, it would be of clinical significance to determine in more detail the roles of SRCs in the underlying molecular mechanisms of tumor-associated inflammation. Finally, as our recent proof-of-principle study demonstrates that SRCs are accessible therapeutic targets for SMIs, more effort should be put on future high throughput screenings for therapeutic drugs targeting SRCs for cancers.

## References

1. Onate SA, Tsai SY, Tsai MJ, O'Malley BW (1995) Sequence and characterization of a coactivator for the steroid hormone receptor superfamily. Science 270:1354–1357
2. Voegel JJ, Heine MJ, Zechel C, Chambon P, Gronemeyer H (1996) TIF2, a 160 kDa transcriptional mediator for the ligand-dependent activation function AF-2 of nuclear receptors. EMBO J 15:3667–3675
3. Hong H, Kohli K, Garabedian MJ, Stallcup MR (1997) GRIP1, a transcriptional coactivator for the AF-2 transactivation domain of steroid, thyroid, retinoid, and vitamin D receptors. Mol Cell Biol 17:2735–2744
4. Anzick SL, Kononen J, Walker RL, Azorsa DO, Tanner MM, Guan XY, Sauter G, Kallioniemi OP, Trent JM, Meltzer PS (1997) AIB1, a steroid receptor coactivator amplified in breast and ovarian cancer. Science 277:965–968

5. Torchia J, Rose DW, Inostroza J, Kamei Y, Westin S, Glass CK, Rosenfeld MG (1997) The transcriptional co-activator p/CIP binds CBP and mediates nuclear-receptor function. Nature 387:677–684
6. Chen H, Lin RJ, Schiltz RL, Chakravarti D, Nash A, Nagy L, Privalsky ML, Nakatani Y, Evans RM (1997) Nuclear receptor coactivator ACTR is a novel histone acetyltransferase and forms a multimeric activation complex with P/CAF and CBP/p300. Cell 90:569–580
7. Takeshita A, Cardona GR, Koibuchi N, Suen CS, Chin WW (1997) TRAM-1, A novel 160-kDa thyroid hormone receptor activator molecule, exhibits distinct properties from steroid receptor coactivator-1. J Biol Chem 272:27629–27634
8. Johnson AB, O'Malley BW (2012) Steroid receptor coactivators 1, 2, and 3: critical regulators of nuclear receptor activity and steroid receptor modulator (SRM)-based cancer therapy. Mol Cell Endocrinol 348:430–439
9. York B, O'Malley BW (2010) Steroid receptor coactivator (SRC) family: masters of systems biology. J Biol Chem 285:38743–38750
10. Xu J, Wu RC, O'Malley BW (2009) Normal and cancer-related functions of the p160 steroid receptor co-activator (SRC) family. Nat Rev Cancer 9:615–630
11. Leo C, Chen JD (2000) The SRC family of nuclear receptor coactivators. Gene 245:1–11
12. Kim JH, Li H, Stallcup MR (2003) CoCoA, a nuclear receptor coactivator which acts through an N-terminal activation domain of p160 coactivators. Mol Cell 12:1537–1549
13. Belandia B, Orford RL, Hurst HC, Parker MG (2002) Targeting of SWI/SNF chromatin remodelling complexes to estrogen-responsive genes. EMBO J 21:4094–4103
14. Askew EB, Bai S, Hnat AT, Minges JT, Wilson EM (2009) Melanoma antigen gene protein-A11 (MAGE-11) F-box links the androgen receptor NH2-terminal transactivation domain to p160 coactivators. J Biol Chem 284:34793–34808
15. Giraud S, Bienvenu F, Avril S, Gascan H, Heery DM, Coqueret O (2002) Functional interaction of STAT3 transcription factor with the coactivator NcoA/SRC1a. J Biol Chem 277:8004–8011
16. Lee SK, Kim HJ, Kim JW, Lee JW (1999) Steroid receptor coactivator-1 and its family members differentially regulate transactivation by the tumor suppressor protein p53. Mol Endocrinol 13:1924–1933
17. Wu RC, Smith CL, O'Malley BW (2005) Transcriptional regulation by steroid receptor coactivator phosphorylation. Endocr Rev 26:393–399
18. Louie MC, Revenko AS, Zou JX, Yao J, Chen HW (2006) Direct control of cell cycle gene expression by proto-oncogene product ACTR, and its autoregulation underlies its transforming activity. Mol Cell Biol 26:3810–3823
19. Heery DM, Kalkhoven E, Hoare S, Parker MG (1997) A signature motif in transcriptional co-activators mediates binding to nuclear receptors. Nature 387:733–736
20. Darimont BD, Wagner RL, Apriletti JW, Stallcup MR, Kushner PJ, Baxter JD, Fletterick RJ, Yamamoto KR (1998) Structure and specificity of nuclear receptor-coactivator interactions. Genes Dev 12:3343–3356
21. Gao Z, Chiao P, Zhang X, Zhang X, Lazar MA, Seto E, Young HA, Ye J (2005) Coactivators and corepressors of NF-kappaB in IkappaB alpha gene promoter. J Biol Chem 280:21091–21098
22. Dilworth FJ, Chambon P (2001) Nuclear receptors coordinate the activities of chromatin remodeling complexes and coactivators to facilitate initiation of transcription. Oncogene 20:3047–3054
23. Kraus WL, Manning ET, Kadonaga JT (1999) Biochemical analysis of distinct activation functions in p300 that enhance transcription initiation with chromatin templates. Mol Cell Biol 19:8123–8135
24. Demarest SJ, Martinez-Yamout M, Chung J, Chen H, Xu W, Dyson HJ, Evans RM, Wright PE (2002) Mutual synergistic folding in recruitment of CBP/p300 by p160 nuclear receptor coactivators. Nature 415:549–553
25. Koh SS, Chen D, Lee YH, Stallcup MR (2001) Synergistic enhancement of nuclear receptor function by p160 coactivators and two coactivators with protein methyltransferase activities. J Biol Chem 276:1089–1098

26. Chen D, Ma H, Hong H, Koh SS, Huang SM, Schurter BT, Aswad DW, Stallcup MR (1999) Regulation of transcription by a protein methyltransferase. Science 284:2174–2177
27. Spencer TE, Jenster G, Burcin MM, Allis CD, Zhou J, Mizzen CA, McKenna NJ, Onate SA, Tsai SY, Tsai MJ, O'Malley BW (1997) Steroid receptor coactivator-1 is a histone acetyltransferase. Nature 389:194–198
28. Lee SK, Kim HJ, Na SY, Kim TS, Choi HS, Im SY, Lee JW (1998) Steroid receptor coactivator-1 coactivates activating protein-1-mediated transactivations through interaction with the c-Jun and c-Fos subunits. J Biol Chem 273:16651–16654
29. Batsche E, Desroches J, Bilodeau S, Gauthier Y, Drouin J (2005) Rb enhances p160/SRC coactivator-dependent activity of nuclear receptors and hormone responsiveness. J Biol Chem 280:19746–19756
30. Wu MY, Fu J, Xu J, O'Malley BW, Wu RC (2012) Steroid receptor coactivator 3 regulates autophagy in breast cancer cells through macrophage migration inhibitory factor. Cell Res 22:1003–1021
31. Ordonez-Moran P, Munoz A (2009) Nuclear receptors: genomic and non-genomic effects converge. Cell Cycle 8:1675–1680
32. Wu RC, Qin J, Yi P, Wong J, Tsai SY, Tsai MJ, O'Malley BW (2004) Selective phosphorylations of the SRC-3/AIB1 coactivator integrate genomic responses to multiple cellular signaling pathways. Mol Cell 15:937–949
33. Zheng FF, Wu RC, Smith CL, O'Malley BW (2005) Rapid estrogen-induced phosphorylation of the SRC-3 coactivator occurs in an extranuclear complex containing estrogen receptor. Mol Cell Biol 25:8273–8284
34. Dobrovolna J, Chinenov Y, Kennedy MA, Liu B, Rogatsky I (2012) Glucocorticoid-dependent phosphorylation of the transcriptional coregulator GRIP1. Mol Cell Biol 32:730–739
35. Ueda T, Mawji NR, Bruchovsky N, Sadar MD (2002) Ligand-independent activation of the androgen receptor by interleukin-6 and the role of steroid receptor coactivator-1 in prostate cancer cells. J Biol Chem 277:38087–38094
36. Munz T, Litterst CM, Pfitzner E (2011) Interaction of STAT6 with its co-activator SRC-1/NCoA-1 is regulated by dephosphorylation of the latter via PP2A. Nucleic Acids Res 39:3255–3266
37. Wu RC, Qin J, Hashimoto Y, Wong J, Xu J, Tsai SY, Tsai MJ, O'Malley BW (2002) Regulation of SRC-3 (pCIP/ACTR/AIB-1/RAC-3/TRAM-1) Coactivator activity by I kappa B kinase. Mol Cell Biol 22:3549–3561
38. Rowan BG, Weigel NL, O'Malley BW (2000) Phosphorylation of steroid receptor coactivator-1. Identification of the phosphorylation sites and phosphorylation through the mitogen-activated protein kinase pathway. J Biol Chem 275:4475–4483
39. Moore NL, Weigel NL (2011) Regulation of progesterone receptor activity by cyclin dependent kinases 1 and 2 occurs in part by phosphorylation of the SRC-1 carboxyl-terminus. Int J Biochem Cell Biol 43:1157–1167
40. Gregory CW, Fei X, Ponguta LA, He B, Bill HM, French FS, Wilson EM (2004) Epidermal growth factor increases coactivation of the androgen receptor in recurrent prostate cancer. J Biol Chem 279:7119–7130
41. Oh AS, Lahusen JT, Chien CD, Fereshteh MP, Zhang X, Dakshanamurthy S, Xu J, Kagan BL, Wellstein A, Riegel AT (2008) Tyrosine phosphorylation of the nuclear receptor coactivator AIB1/SRC-3 is enhanced by Abl kinase and is required for its activity in cancer cells. Mol Cell Biol 28:6580–6593
42. Shou J, Massarweh S, Osborne CK, Wakeling AE, Ali S, Weiss H, Schiff R (2004) Mechanisms of tamoxifen resistance: increased estrogen receptor-HER2/neu cross-talk in ER/HER2-positive breast cancer. J Natl Cancer Inst 96:926–935
43. Rowan BG, Garrison N, Weigel NL, O'Malley BW (2000) 8-Bromo-cyclic AMP induces phosphorylation of two sites in SRC-1 that facilitate ligand-independent activation of the chicken progesterone receptor and are critical for functional cooperation between SRC-1 and CREB binding protein. Mol Cell Biol 20:8720–8730

44. Hoang T, Fenne IS, Cook C, Borud B, Bakke M, Lien EA, Mellgren G (2004) cAMP-dependent protein kinase regulates ubiquitin-proteasome-mediated degradation and subcellular localization of the nuclear receptor coactivator GRIP1. J Biol Chem 279:49120–49130
45. Reiter R, Wellstein A, Riegel AT (2001) An isoform of the coactivator AIB1 that increases hormone and growth factor sensitivity is overexpressed in breast cancer. J Biol Chem 276:39736–39741
46. Long W, Yi P, Amazit L, LaMarca HL, Ashcroft F, Kumar R, Mancini MA, Tsai SY, Tsai MJ, O'Malley BW (2010) SRC-3Delta4 mediates the interaction of EGFR with FAK to promote cell migration. Mol Cell 37:321–332
47. Wu RC, Feng Q, Lonard DM, O'Malley BW (2007) SRC-3 coactivator functional lifetime is regulated by a phospho-dependent ubiquitin time clock. Cell 129:1125–1140
48. Gianni M, Parrella E, Raska I Jr, Gaillard E, Nigro EA, Gaudon C, Garattini E, Rochette-Egly C (2006) P38MAPK-dependent phosphorylation and degradation of SRC-3/AIB1 and RARalpha-mediated transcription. EMBO J 25:739–751
49. Ferry C, Gaouar S, Fischer B, Boeglin M, Paul N, Samarut E, Piskunov A, Pankotai-Bodo G, Brino L, Rochette-Egly C (2011) Cullin 3 mediates SRC-3 ubiquitination and degradation to control the retinoic acid response. Proc Natl Acad Sci USA 108:20603–20608
50. Li C, Liang YY, Feng XH, Tsai SY, Tsai MJ, O'Malley BW (2008) Essential phosphatases and a phospho-degron are critical for regulation of SRC-3/AIB1 coactivator function and turnover. Mol Cell 31:835–849
51. Li C, Ao J, Fu J, Lee DF, Xu J, Lonard D, O'Malley BW (2011) Tumor-suppressor role for the SPOP ubiquitin ligase in signal-dependent proteolysis of the oncogenic co-activator SRC-3/AIB1. Oncogene 30:4350–4364
52. Li X, Lonard DM, Jung SY, Malovannaya A, Feng Q, Qin J, Tsai SY, Tsai MJ, O'Malley BW (2006) The SRC-3/AIB1 coactivator is degraded in a ubiquitin- and ATP-independent manner by the REGgamma proteasome. Cell 124:381–922
53. Yi P, Feng Q, Amazit L, Lonard DM, Tsai SY, Tsai MJ, O'Malley BW (2008) Atypical protein kinase C regulates dual pathways for degradation of the oncogenic coactivator SRC-3/AIB1. Mol Cell 29:465–476
54. Hartmaier RJ, Richter AS, Gillihan RM, Sallit JZ, McGuire SE, Wang J, Lee AV, Osborne CK, O'Malley BW, Brown PH, Xu J, Skaar TC, Philips S, Rae JM, Azzouz F, Li L, Hayden J, Henry NL, Nguyen AT, Stearns V, Hayes DF, Flockhart DA, Oesterreich S (2012) A SNP in steroid receptor coactivator-1 disrupts a GSK3beta phosphorylation site and is associated with altered tamoxifen response in bone. Mol Endocrinol 26:220–227
55. Gareau JR, Lima CD (2010) The SUMO pathway: emerging mechanisms that shape specificity, conjugation and recognition. Nat Rev Mol Cell Biol 11:861–871
56. Li S, Shang Y (2007) Regulation of SRC family coactivators by post-translational modifications. Cell Signal 19:1101–1112
57. Chauchereau A, Amazit L, Quesne M, Guiochon-Mantel A, Milgrom E (2003) Sumoylation of the progesterone receptor and of the steroid receptor coactivator SRC-1. J Biol Chem 278:12335–12343
58. Kotaja N, Karvonen U, Janne OA, Palvimo JJ (2002) The nuclear receptor interaction domain of GRIP1 is modulated by covalent attachment of SUMO-1. J Biol Chem 277:30283–30288
59. Hietakangas V, Anckar J, Blomster HA, Fujimoto M, Palvimo JJ, Nakai A, Sistonen L (2006) PDSM, a motif for phosphorylation-dependent SUMO modification. Proc Natl Acad Sci USA 103:45–50
60. Shalizi A, Gaudilliere B, Yuan Z, Stegmuller J, Shirogane T, Ge Q, Tan Y, Schulman B, Harper JW, Bonni A (2006) A calcium-regulated MEF2 sumoylation switch controls postsynaptic differentiation. Science 311:1012–1017
61. Chen H, Lin RJ, Xie W, Wilpitz D, Evans RM (1999) Regulation of hormone-induced histone hyperacetylation and gene activation via acetylation of an acetylase. Cell 98:675–686
62. Fu M, Wang C, Zhang X, Pestell RG (2004) Acetylation of nuclear receptors in cellular growth and apoptosis. Biochem Pharmacol 68:1199–1208

63. Feng Q, Yi P, Wong J, O'Malley BW (2006) Signaling within a coactivator complex: methylation of SRC-3/AIB1 is a molecular switch for complex disassembly. Mol Cell Biol 26:7846–7857
64. Naeem H, Cheng D, Zhao Q, Underhill C, Tini M, Bedford MT, Torchia J (2007) The activity and stability of the transcriptional coactivator p/CIP/SRC-3 are regulated by CARM1-dependent methylation. Mol Cell Biol 27:120–134
65. Eroles P, Bosch A, Perez-Fidalgo JA, Lluch A (2012) Molecular biology in breast cancer: intrinsic subtypes and signaling pathways. Cancer Treat Rev 38:698–707
66. Neve RM, Chin K, Fridlyand J, Yeh J, Baehner FL, Fevr T, Clark L, Bayani N, Coppe JP, Tong F, Speed T, Spellman PT, DeVries S, Lapuk A, Wang NJ, Kuo WL, Stilwell JL, Pinkel D, Albertson DG, Waldman FM, McCormick F, Dickson RB, Johnson MD, Lippman M, Ethier S, Gazdar A, Gray JW (2006) A collection of breast cancer cell lines for the study of functionally distinct cancer subtypes. Cancer Cell 10:515–527
67. Tai H, Kubota N, Kato S (2000) Involvement of nuclear receptor coactivator SRC-1 in estrogen-dependent cell growth of MCF-7 cells. Biochem Biophys Res Commun 267:311–316
68. Cavarretta IT, Mukopadhyay R, Lonard DM, Cowsert LM, Bennett CF, O'Malley BW, Smith CL (2002) Reduction of coactivator expression by antisense oligodeoxynucleotides inhibits ERalpha transcriptional activity and MCF-7 proliferation. Mol Endocrinol 16:253–270
69. Kishimoto H, Wang Z, Bhat-Nakshatri P, Chang D, Clarke R, Nakshatri H (2005) The p160 family coactivators regulate breast cancer cell proliferation and invasion through autocrine/paracrine activity of SDF-1alpha/CXCL12. Carcinogenesis 26:1706–1715
70. Planas-Silva MD, Shang Y, Donaher JL, Brown M, Weinberg RA (2001) AIB1 enhances estrogen-dependent induction of cyclin D1 expression. Cancer Res 61:3858–3862
71. Karmakar S, Foster EA, Smith CL (2009) Unique roles of p160 coactivators for regulation of breast cancer cell proliferation and estrogen receptor-alpha transcriptional activity. Endocrinology 150:1588–1596
72. List HJ, Lauritsen KJ, Reiter R, Powers C, Wellstein A, Riegel AT (2001) Ribozyme targeting demonstrates that the nuclear receptor coactivator AIB1 is a rate-limiting factor for estrogen-dependent growth of human MCF-7 breast cancer cells. J Biol Chem 276:23763–23768
73. Park KJ, Krishnan V, O'Malley BW, Yamamoto Y, Gaynor RB (2005) Formation of an IKKalpha-dependent transcription complex is required for estrogen receptor-mediated gene activation. Mol Cell 18:71–82
74. Torres-Arzayus MI, Font de Mora J, Yuan J, Vazquez F, Bronson R, Rue M, Sellers WR, Brown M (2004) High tumor incidence and activation of the PI3K/AKT pathway in transgenic mice define AIB1 as an oncogene. Cancer Cell 6:263–274
75. Torres-Arzayus MI, Zhao J, Bronson R, Brown M (2010) Estrogen-dependent and estrogen-independent mechanisms contribute to AIB1-mediated tumor formation. Cancer Res 70:4102–4111
76. Agoulnik IU, Vaid A, Bingman WE III, Erdeme H, Frolov A, Smith CL, Ayala G, Ittmann MM, Weigel NL (2005) Role of SRC-1 in the promotion of prostate cancer cell growth and tumor progression. Cancer Res 65:7959–7967
77. Agoulnik IU, Vaid A, Nakka M, Alvarado M, Bingman WE III, Erdem H, Frolov A, Smith CL, Ayala GE, Ittmann MM, Weigel NL (2006) Androgens modulate expression of transcription intermediary factor 2, an androgen receptor coactivator whose expression level correlates with early biochemical recurrence in prostate cancer. Cancer Res 66:10594–10602
78. Shi XB, Xue L, Zou JX, Gandour-Edwards R, Chen H, deVere White RW (2008) Prolonged androgen receptor loading onto chromatin and the efficient recruitment of p160 coactivators contribute to androgen-independent growth of prostate cancer cells. Prostate 68:1816–1826
79. Zhou HJ, Yan J, Luo W, Ayala G, Lin SH, Erdem H, Ittmann M, Tsai SY, Tsai MJ (2005) SRC-3 is required for prostate cancer cell proliferation and survival. Cancer Res 65:7976–7983

80. Zou JX, Zhong Z, Shi XB, Tepper CG, deVere White RW, Kung HJ, Chen H (2006) ACTR/AIB1/SRC-3 and androgen receptor control prostate cancer cell proliferation and tumor growth through direct control of cell cycle genes. Prostate 66:1474–1486
81. Higgins MJ, Baselga J (2011) Targeted therapies for breast cancer. J Clin Invest 121:3797–3803
82. Fereshteh MP, Tilli MT, Kim SE, Xu J, O'Malley BW, Wellstein A, Furth PA, Riegel AT (2008) The nuclear receptor coactivator amplified in breast cancer-1 is required for Neu (ErbB2/HER2) activation, signaling, and mammary tumorigenesis in mice. Cancer Res 68:3697–3706
83. Al-Otaiby M, Tassi E, Schmidt MO, Chien CD, Baker T, Salas AG, Xu J, Furlong M, Schlegel R, Riegel AT, Wellstein A (2012) Role of the nuclear receptor coactivator AIB1/SRC-3 in angiogenesis and wound healing. Am J Pathol 180:1474–1484
84. Han JS, Crowe DL (2010) Steroid receptor coactivator 1 deficiency increases MMTV-neu mediated tumor latency and differentiation specific gene expression, decreases metastasis, and inhibits response to PPAR ligands. BMC Cancer 10:629
85. Chen B, Wang Y, Kane SE, Chen S (2008) Improvement of sensitivity to tamoxifen in estrogen receptor-positive and herceptin-resistant breast cancer cells. J Mol Endocrinol 41:367–377
86. Su Q, Hu S, Gao H, Ma R, Yang Q, Pan Z, Wang T, Li F (2008) Role of AIB1 for tamoxifen resistance in estrogen receptor-positive breast cancer cells. Oncology 75:159–168
87. Osborne CK, Bardou V, Hopp TA, Chamness GC, Hilsenbeck SG, Fuqua SA, Wong J, Allred DC, Clark GM, Schiff R (2003) Role of the estrogen receptor coactivator AIB1 (SRC-3) and HER-2/neu in tamoxifen resistance in breast cancer. J Natl Cancer Inst 95: 353–361
88. Fleming FJ, Myers E, Kelly G, Crotty TB, McDermott EW, O'Higgins NJ, Hill AD, Young LS (2004) Expression of SRC-1, AIB1, and PEA3 in HER2 mediated endocrine resistant breast cancer; a predictive role for SRC-1. J Clin Pathol 57:1069–1074
89. Qin L, Liu Z, Chen H, Xu J (2009) The steroid receptor coactivator-1 regulates twist expression and promotes breast cancer metastasis. Cancer Res 69:3819–3827
90. Hurtado A, Holmes KA, Geistlinger TR, Hutcheson IR, Nicholson RI, Brown M, Jiang J, Howat WJ, Ali S, Carroll JS (2008) Regulation of ERBB2 by oestrogen receptor-PAX2 determines response to tamoxifen. Nature 456:663–666
91. Lahusen T, Fereshteh M, Oh A, Wellstein A, Riegel AT (2007) Epidermal growth factor receptor tyrosine phosphorylation and signaling controlled by a nuclear receptor coactivator, amplified in breast cancer 1. Cancer Res 67:7256–7265
92. Guo Z, Dai B, Jiang T, Xu K, Xie Y, Kim O, Nesheiwat I, Kong X, Melamed J, Handratta VD, Njar VC, Brodie AM, Yu LR, Veenstra TD, Chen H, Qiu Y (2006) Regulation of androgen receptor activity by tyrosine phosphorylation. Cancer Cell 10:309–319
93. Pollak M (2012) The insulin and insulin-like growth factor receptor family in neoplasia: an update. Nat Rev Cancer 12:159–169
94. Oh A, List HJ, Reiter R, Mani A, Zhang Y, Gehan E, Wellstein A, Riegel AT (2004) The nuclear receptor coactivator AIB1 mediates insulin-like growth factor I-induced phenotypic changes in human breast cancer cells. Cancer Res 64:8299–8308
95. Yan J, Yu CT, Ozen M, Ittmann M, Tsai SY, Tsai MJ (2006) Steroid receptor coactivator-3 and activator protein-1 coordinately regulate the transcription of components of the insulin-like growth factor/AKT signaling pathway. Cancer Res 66:11039–11046
96. Kuang SQ, Liao L, Zhang H, Lee AV, O'Malley BW, Xu J (2004) AIB1/SRC-3 deficiency affects insulin-like growth factor I signaling pathway and suppresses v-Ha-ras-induced breast cancer initiation and progression in mice. Cancer Res 64:1875–1885
97. Wang S, Yuan Y, Liao L, Kuang SQ, Tien JC, O'Malley BW, Xu J (2009) Disruption of the SRC-1 gene in mice suppresses breast cancer metastasis without affecting primary tumor formation. Proc Natl Acad Sci USA 106:151–156
98. Drachenberg DE, Elgamal AA, Rowbotham R, Peterson M, Murphy GP (1999) Circulating levels of interleukin-6 in patients with hormone refractory prostate cancer. Prostate 41: 127–133

99. Adler HL, McCurdy MA, Kattan MW, Timme TL, Scardino PT, Thompson TC (1999) Elevated levels of circulating interleukin-6 and transforming growth factor-beta1 in patients with metastatic prostatic carcinoma. J Urol 161:182–187
100. Lee SO, Lou W, Hou M, de Miguel F, Gerber L, Gao AC (2003) Interleukin-6 promotes androgen-independent growth in LNCaP human prostate cancer cells. Clin Cancer Res 9:370–376
101. Feng S, Tang Q, Sun M, Chun JY, Evans CP, Gao AC (2009) Interleukin-6 increases prostate cancer cells resistance to bicalutamide via TIF2. Mol Cancer Ther 8:665–671
102. Rubio MF, Werbajh S, Cafferata EG, Quaglino A, Colo GP, Nojek IM, Kordon EC, Nahmod VE, Costas MA (2006) TNF-alpha enhances estrogen-induced cell proliferation of estrogen-dependent breast tumor cells through a complex containing nuclear factor-kappa B. Oncogene 25:1367–1377
103. Yu C, York B, Wang S, Feng Q, Xu J, O'Malley BW (2007) An essential function of the SRC-3 coactivator in suppression of cytokine mRNA translation and inflammatory response. Mol Cell 25:765–778
104. Han SJ, Hawkins SM, Begum K, Jung SY, Kovanci E, Qin J, Lydon JP, DeMayo FJ, O'Malley BW (2012) A new isoform of steroid receptor coactivator-1 is crucial for pathogenic progression of endometriosis. Nat Med 18:1102–1111
105. Qin L, Liao L, Redmond A, Young L, Yuan Y, Chen H, O'Malley BW, Xu J (2008) The AIB1 oncogene promotes breast cancer metastasis by activation of PEA3-mediated matrix metalloproteinase 2 (MMP2) and MMP9 expression. Mol Cell Biol 28:5937–5950
106. Yan J, Erdem H, Li R, Cai Y, Ayala G, Ittmann M, Yu-Lee LY, Tsai SY, Tsai MJ (2008) Steroid receptor coactivator-3/AIB1 promotes cell migration and invasiveness through focal adhesion turnover and matrix metalloproteinase expression. Cancer Res 68:5460–5468
107. Li LB, Louie MC, Chen HW, Zou JX (2008) Proto-oncogene ACTR/AIB1 promotes cancer cell invasion by up-regulating specific matrix metalloproteinase expression. Cancer Lett 261:64–73
108. Long W, Foulds CE, Long W, Foulds CE, Qin J, Liu J, Ding C, Lonard DM, Solis LM, Wistuba II, Qin J, Tsai SY, Tsai MJ, O'Malley BW (2012) ERK3 signals through SRC-3 coactivator to promote human lung cancer cell invasion. J Clin Invest 122:1869–1880
109. Taylor BS, Schultz N, Hieronymus H, Gopalan A, Xiao Y, Carver BS, Arora VK, Kaushik P, Cerami E, Reva B, Antipin Y, Mitsiades N, Landers T, Dolgalev I, Major JE, Wilson M, Socci ND, Lash AE, Heguy A, Eastham JA, Scher HI, Reuter VE, Scardino PT, Sander C, Sawyers CL, Gerald WL (2010) Integrative genomic profiling of human prostate cancer. Cancer Cell 18:11–22
110. Bautista S, Valles H, Walker RL, Anzick S, Zeillinger R, Meltzer P, Theillet C (1998) In breast cancer, amplification of the steroid receptor coactivator gene AIB1 is correlated with estrogen and progesterone receptor positivity. Clin Cancer Res 4:2925–2929
111. Tanner MM, Grenman S, Koul A, Johannsson O, Meltzer P, Pejovic T, Borg A, Isola JJ (2000) Frequent amplification of chromosomal region 20q12-q13 in ovarian cancer. Clin Cancer Res 6:1833–1839
112. Xie D, Sham JS, Zeng WF, Lin HL, Bi J, Che LH, Hu L, Zeng YX, Guan XY (2005) Correlation of AIB1 overexpression with advanced clinical stage of human colorectal carcinoma. Hum Pathol 36:777–783
113. Lassmann S, Weis R, Makowiec F, Roth J, Danciu M, Hopt U, Werner M (2007) Array CGH identifies distinct DNA copy number profiles of oncogenes and tumor suppressor genes in chromosomal- and microsatellite-unstable sporadic colorectal carcinomas. J Mol Med (Berl) 85:293–304
114. He LR, Zhao HY, Li BK, Zhang LJ, Liu MZ, Kung HF, Guan XY, Bian XW, Zeng YX, Xie D (2010) Overexpression of AIB1 negatively affects survival of surgically resected non-small-cell lung cancer patients. Ann Oncol 21:1675–1681
115. Cai D, Shames DS, Raso MG, Xie Y, Kim YH, Pollack JR, Girard L, Sullivan JP, Gao B, Peyton M, Nanjundan M, Byers L, Heymach J, Mills G, Gazdar AF, Wistuba I, Kodadek T, Minna JD (2010) Steroid receptor coactivator-3 expression in lung cancer and its role in the regulation of cancer cell survival and proliferation. Cancer Res 70:6477–6485

116. Wang Y, Wu MC, Sham JS, Zhang W, Wu WQ, Guan XY (2002) Prognostic significance of c-myc and AIB1 amplification in hepatocellular carcinoma. A broad survey using high-throughput tissue microarray. Cancer 95:2346–2352
117. Maki HE, Waltering KK, Wallen MJ, Martikainen PM, Tammela TL, van Weerden WM, Vessella RL, Visakorpi T (2006) Screening of genetic and expression alterations of SRC1 gene in prostate cancer. Prostate 66:1391–1398
118. Carapeti M, Aguiar RC, Goldman JM, Cross NC (1998) A novel fusion between MOZ and the nuclear receptor coactivator TIF2 in acute myeloid leukemia. Blood 91:3127–3133
119. Carapeti M, Aguiar RC, Watmore AE, Goldman JM, Cross NC (1999) Consistent fusion of MOZ and TIF2 in AML with inv(8)(p11q13). Cancer Genet Cytogenet 113:70–72
120. Deguchi K, Ayton PM, Carapeti M, Kutok JL, Snyder CS, Williams IR, Cross NC, Glass CK, Cleary ML, Gilliland DG (2003) MOZ-TIF2-induced acute myeloid leukemia requires the MOZ nucleosome binding motif and TIF2-mediated recruitment of CBP. Cancer Cell 3:259–271
121. Osborne CK, Schiff R (2011) Mechanisms of endocrine resistance in breast cancer. Annu Rev Med 62:233–247
122. Ariazi EA, Ariazi JL, Cordera F, Jordan VC (2006) Estrogen receptors as therapeutic targets in breast cancer. Curr Top Med Chem 6:181–202
123. Fleming FJ, Hill AD, McDermott EW, O'Higgins NJ, Young LS (2004) Differential recruitment of coregulator proteins steroid receptor coactivator-1 and silencing mediator for retinoid and thyroid receptors to the estrogen receptor-estrogen response element by beta-estradiol and 4-hydroxytamoxifen in human breast cancer. J Clin Endocrinol Metab 89:375–383
124. Hudelist G, Czerwenka K, Kubista E, Marton E, Pischinger K, Singer CF (2003) Expression of sex steroid receptors and their co-factors in normal and malignant breast tissue: AIB1 is a carcinoma-specific co-activator. Breast Cancer Res Treat 78:193–204
125. McBryan J, Theissen SM, Byrne C, Hughes E, Cocchiglia S, Sande S, O'Hara J, Tibbitts P, Hill AD, Young LS (2011) Metastatic progression with resistance to aromatase inhibitors is driven by the steroid receptor coactivator SRC-1. Cancer Res 72:548–559
126. Myers E, Fleming FJ, Crotty TB, Kelly G, McDermott EW, O'Higgins JN, Hill AD, Young LS (2004) Inverse relationship between ER-beta and SRC-1 predicts outcome in endocrine-resistant breast cancer. Br J Cancer 91:1687–1693
127. Zhao C, Yasui K, Lee CJ, Kurioka H, Hosokawa Y, Oka T, Inazawa J (2003) Elevated expression levels of NCOA3, TOP1, and TFAP2C in breast tumors as predictors of poor prognosis. Cancer 98:18–23
128. Bouras T, Southey MC, Venter DJ (2001) Overexpression of the steroid receptor coactivator AIB1 in breast cancer correlates with the absence of estrogen and progesterone receptors and positivity for p53 and HER2/neu. Cancer Res 61:903–907
129. Arteaga CL, Sliwkowski MX, Osborne CK, Perez EA, Puglisi F, Gianni L (2011) Treatment of HER2-positive breast cancer: current status and future perspectives. Nat Rev Clin Oncol 9:16–32
130. Kirkegaard T, McGlynn LM, Campbell FM, Muller S, Tovey SM, Dunne B, Nielsen KV, Cooke TG, Bartlett JM (2007) Amplified in breast cancer 1 in human epidermal growth factor receptor – positive tumors of tamoxifen-treated breast cancer patients. Clin Cancer Res 13:1405–1411
131. Lahusen T, Henke RT, Kagan BL, Wellstein A, Riegel AT (2009) The role and regulation of the nuclear receptor co-activator AIB1 in breast cancer. Breast Cancer Res Treat 116:225–237
132. Agoulnik IU, Weigel NL (2006) Androgen receptor action in hormone-dependent and recurrent prostate cancer. J Cell Biochem 99:362–372
133. Tsao CK, Small AC, Galsky MD, Oh WK (2012) Overcoming castration resistance in prostate cancer. Curr Opin Urol 22:167–174
134. Pienta KJ, Bradley D (2006) Mechanisms underlying the development of androgen-independent prostate cancer. Clin Cancer Res 12:1665–1671

135. Heemers HV, Tindall DJ (2007) Androgen receptor (AR) coregulators: a diversity of functions converging on and regulating the AR transcriptional complex. Endocr Rev 28:778–808
136. Gregory CW, He B, Johnson RT, Ford OH, Mohler JL, French FS, Wilson EM (2001) A mechanism for androgen receptor-mediated prostate cancer recurrence after androgen deprivation therapy. Cancer Res 61:4315–4319
137. Gnanapragasam VJ, Leung HY, Pulimood AS, Neal DE, Robson CN (2001) Expression of RAC 3, a steroid hormone receptor co-activator in prostate cancer. Br J Cancer 85:1928–1936
138. Ayala G, Yan J, Li R, Ding Y, Thompson TC, Mims MP, Hayes TG, MacDonnell V, Lynch RG, Frolov A, Miles BJ, Wheeler TM, Harper JW, Tsai MJ, Ittmann MM, Kadmon D (2008) Bortezomib-mediated inhibition of steroid receptor coactivator-3 degradation leads to activated Akt. Clin Cancer Res 14:7511–7518
139. Marquez-Garban DC, Chen HW, Fishbein MC, Goodglick L, Pietras RJ (2007) Estrogen receptor signaling pathways in human non-small cell lung cancer. Steroids 72:135–143
140. Wang H, Zhang D, Wu W, Zhang J, Guo D, Wang Q, Jing T, Xu C, Bian X, Yang K (2010) Overexpression and gender-specific differences of SRC-3 (SRC-3/AIB1) immunoreactivity in human non-small cell lung cancer: an in vivo study. J Histochem Cytochem 58:1121–1127
141. Yoshida H, Liu J, Samuel S, Cheng W, Rosen D, Naora H (2005) Steroid receptor coactivator-3, a homolog of Taiman that controls cell migration in the Drosophila ovary, regulates migration of human ovarian cancer cells. Mol Cell Endocrinol 245:77–85
142. Henke RT, Haddad BR, Kim SE, Rone JD, Mani A, Jessup JM, Wellstein A, Maitra A, Riegel AT (2004) Overexpression of the nuclear receptor coactivator AIB1 (SRC-3) during progression of pancreatic adenocarcinoma. Clin Cancer Res 10:6134–6142
143. Coste A, Antal MC, Chan S, Kastner P, Mark M, O'Malley BW, Auwerx J (2006) Absence of the steroid receptor coactivator-3 induces B-cell lymphoma. EMBO J 25:2453–2464
144. O'Donnell KA, Keng VW, York B, Reineke EL, Seo D, Fan D, Silverstein KA, Schrum CT, Xie WR, Mularoni L, Wheelan SJ, Torbenson MS, O'Malley BW, Largaespada DA, Boeke JD (2012) A sleeping beauty mutagenesis screen reveals a tumor suppressor role for Ncoa2/Src-2 in liver cancer. Proc Natl Acad Sci USA 109:E1377–E1386
145. Smith CL, Nawaz Z, O'Malley BW (1997) Coactivator and corepressor regulation of the agonist/antagonist activity of the mixed antiestrogen, 4-hydroxytamoxifen. Mol Endocrinol 11:657–666
146. Redmond AM, Bane FT, Stafford AT, McIlroy M, Dillon MF, Crotty TB, Hill AD, Young LS (2009) Coassociation of estrogen receptor and p160 proteins predicts resistance to endocrine treatment; SRC-1 is an independent predictor of breast cancer recurrence. Clin Cancer Res 15:2098–2106
147. He LR, Liu MZ, Li BK, Rao HL, Deng HX, Guan XY, Zeng YX, Xie D (2009) Overexpression of AIB1 predicts resistance to chemoradiotherapy and poor prognosis in patients with primary esophageal squamous cell carcinoma. Cancer Sci 100:1591–1596
148. Chen Q, Li W, Wan Y, Xia X, Wu Q, Chen Y, Lai Z, Yu C, Li W (2012) Amplified in breast cancer 1 enhances human cholangiocarcinoma growth and chemoresistance by simultaneous activation of Akt and Nrf2 pathways. Hepatology 55:1820–1829
149. Norris JD, Paige LA, Christensen DJ, Chang CY, Huacani MR, Fan D, Hamilton PT, Fowlkes DM, McDonnell DP (1999) Peptide antagonists of the human estrogen receptor. Science 285:744–746
150. Chang C, Norris JD, Gron H, Paige LA, Hamilton PT, Kenan DJ, Fowlkes D, McDonnell DP (1999) Dissection of the LXXLL nuclear receptor-coactivator interaction motif using combinatorial peptide libraries: discovery of peptide antagonists of estrogen receptors alpha and beta. Mol Cell Biol 19:8226–8239
151. Hall JM, Chang CY, McDonnell DP (2000) Development of peptide antagonists that target estrogen receptor beta-coactivator interactions. Mol Endocrinol 14:2010–2023

152. Geistlinger TR, Guy RK (2001) An inhibitor of the interaction of thyroid hormone receptor beta and glucocorticoid interacting protein 1. J Am Chem Soc 123:1525–1526
153. Galande AK, Bramlett KS, Burris TP, Wittliff JL, Spatola AF (2004) Thioether side chain cyclization for helical peptide formation: inhibitors of estrogen receptor-coactivator interactions. J Pept Res 63:297–302
154. Galande AK, Bramlett KS, Trent JO, Burris TP, Wittliff JL, Spatola AF (2005) Potent inhibitors of LXXLL-based protein-protein interactions. Chembiochem 6:1991–1998
155. Rodriguez AL, Tamrazi A, Collins ML, Katzenellenbogen JA (2004) Design, synthesis, and in vitro biological evaluation of small molecule inhibitors of estrogen receptor alpha coactivator binding. J Med Chem 47:600–611
156. Arnold LA, Estebanez-Perpina E, Togashi M, Jouravel N, Shelat A, McReynolds AC, Mar E, Nguyen P, Baxter JD, Fletterick RJ, Webb P, Guy RK (2005) Discovery of small molecule inhibitors of the interaction of the thyroid hormone receptor with transcriptional coregulators. J Biol Chem 280:43048–43055
157. Wang Y, Lonard DM, Yu Y, Chow DC, Palzkill TG, O'Malley BW (2012) Small molecule inhibition of the steroid receptor coactivators, SRC-3 and SRC-1. Mol Endocrinol 25:2041–2053
158. Hanahan D, Weinberg RA (2011) Hallmarks of cancer: the next generation. Cell 144:646–674
159. Dang CV (2012) Links between metabolism and cancer. Genes Dev 26:877–890
160. Picard F, Gehin M, Annicotte J, Rocchi S, Champy MF, O'Malley BW, Chambon P, Auwerx J (2002) SRC-1 and TIF2 control energy balance between white and brown adipose tissues. Cell 111:931–941
161. Wang Z, Qi C, Krones A, Woodring P, Zhu X, Reddy JK, Evans RM, Rosenfeld MG, Hunter T (2006) Critical roles of the p160 transcriptional coactivators p/CIP and SRC-1 in energy balance. Cell Metab 3:111–122
162. Louet JF, Coste A, Amazit L, Tannour-Louet M, Wu RC, Tsai SY, Tsai MJ, Auwerx J, O'Malley BW (2006) Oncogenic steroid receptor coactivator-3 is a key regulator of the white adipogenic program. Proc Natl Acad Sci USA 103:17868–17873
163. Coste A, Louet JF, Lagouge M, Lerin C, Antal MC, Meziane H, Schoonjans K, Puigserver P, O'Malley BW, Auwerx J (2008) The genetic ablation of SRC-3 protects against obesity and improves insulin sensitivity by reducing the acetylation of PGC-1{alpha}. Proc Natl Acad Sci USA 105:17187–17192
164. Chopra AR, Louet JF, Saha P, An J, Demayo F, Xu J, York B, Karpen S, Finegold M, Moore D, Chan L, Newgard CB, O'Malley BW (2008) Absence of the SRC-2 coactivator results in a glycogenopathy resembling Von Gierke's disease. Science 322:1395–1399
165. Louet JF, Chopra AR, Sagen JV, An J, York B, Tannour-Louet M, Saha PK, Stevens RD, Wenner BR, Ilkayeva OR, Bain JR, Zhou S, DeMayo F, Xu J, Newgard CB, O'Malley BW (2010) The coactivator SRC-1 is an essential coordinator of hepatic glucose production. Cell Metab 12:606–618
166. York B, Reineke EL, Sagen JV, Nikolai BC, Zhou S, Louet JF, Chopra AR, Chen X, Reed G, Noebels J, Adesina AM, Yu H, Wong LJ, Tsimelzon A, Hilsenbeck S, Stevens RD, Wenner BR, Ilkayeva O, Xu J, Newgard CB, O'Malley BW (2012) Ablation of steroid receptor coactivator-3 resembles the human CACT metabolic myopathy. Cell Metab 15:752–763
167. Ma X, Xu L, Wang S, Cui B, Li X, Xu J, Ning G (2011) Deletion of steroid receptor coactivator-3 gene ameliorates hepatic steatosis. J Hepatol 55:445–452
168. Li J, Liu YH, Ou S, Dai XM, Wang JP, Su YP (2012) Steroid receptor coactivator-3 differentially regulates the inflammatory response in peritoneal macrophages. Mol Med Report 5:1099–1105
169. Li J, Niu J, Ou S, Ye ZY, Liu DQ, Wang FC, Su YP, Wang JP (2012) Effects of SCR-3 on the immunosuppression accompanied with the systemic inflammatory response syndrome. Mol Cell Biochem 364:29–37

# Chapter 2
# Role of Alteration/Deficiency in Activation (ADA) Complex in Cell Cycle, Genomic Instability and Cancer

**Shakur Mohibi, Shashank Srivastava, Hamid Band, and Vimla Band**

**Abstract** In eukaryotes, DNA wraps around histone proteins to form highly condensed chromatin structures that usually remain inert and inaccessible to proteins involved in DNA-related processes. Thus, multitudes of important DNA-related biological processes, including transcription, replication, DNA repair, apoptosis, chromosome condensation, and segregation, are dependent upon alteration of this chromatin structure so that proteins involved in these processes can access the DNA. This required change in chromatin structure is brought about by binding of various chromatin modifying proteins that loosen the chromatin by distinct mechanisms, one of which is covalent histone modification. Various histone post-translational modifications, specifically acetylation, play a major role in opening up of this highly condensed chromatin allowing access to proteins involved in the several important processes. Histone acetyl transferases (HATs) and histone deacetylases (HDACs) are important for maintaining a steady-state level of this particular post-translational modification in cells and are present in multi-subunit complexes. One such multi-subunit HAT complex is the alteration/deficiency in

S. Mohibi, M.Sc. • S. Srivastava, B.Tech.
Department of Genetics, Cell Biology and Anatomy, University of Nebraska Medical Center, Omaha, NE, USA

H. Band, M.D., Ph.D.
Department of Genetics, Cell Biology and Anatomy; Biochemistry and Molecular Biology; Pathology and Microbiology; Pharmacology and Experimental Neuroscience, College of Medicine, and the Eppley Institute for Research in Cancer and Allied Diseases, University of Nebraska Medical Center, Omaha, NE, USA

V. Band, Ph.D. (✉)
Department of Genetics, Cell Biology and Anatomy; Eppley Institute for Research in Cancer and Allied Diseases, University of Nebraska Medical Center, Omaha, NE, USA
e-mail: vband@unmc.edu

R. Kumar (ed.), *Nuclear Signaling Pathways and Targeting Transcription in Cancer*, Cancer Drug Discovery and Development, DOI 10.1007/978-1-4614-8039-6_2,

activation (ADA) complex, which was originally discovered in yeast and is now known to be also present in mammalian cells as part of much larger HAT complexes. In this chapter, we discuss various components of the ADA complex with a special focus on the adaptor proteins Ada3 and Ada2 (Ada2a and Ada2b) for their role in important physiological processes, such as the cell cycle, genomic integrity, DNA repair response, and in pathology such as cancer. Further, we discuss recent developments using various inhibitors to target the HAT enzymes and disrupt HAT complex function as an anti-cancer strategy.

**Keywords** Ada3 • Ada2 • HATs • Cell cycle • DNA repair • p53 • Nuclear receptors • Viral oncogenes • Cancer • HAT inhibitors

## 2.1 Introduction

Precisely regulated cell proliferation is essential for embryonic development as well as adult tissue homeostasis, and uncontrolled cell proliferation is a hallmark of cancer [1, 2]. Coordination of cell-cycle progression with chromosomal duplication maintains genomic stability, a critical cancer-associated trait [3]. Deregulated cell-cycle components have now also emerged as key biomarkers and therapeutic targets in cancer [4]. Thus, a better understanding of the cell-cycle machinery and its aberrations in cancer are of fundamental importance in cell and cancer biology. In eukaryotes, DNA is wrapped tightly around histone proteins to form chromatin that facilitates higher order folding of DNA [5]. This greatly limits the accessibility of DNA by various proteins involved in transcription, replication, cell division, and DNA repair [6, 7]. Post-translational modifications of histones play an important role in modifying the folding of chromatin and affect the functions involving chromatin [8]. Acetylation of histones is one of the most important and widely studied post translational modification and it has emerged as a conserved mechanism that is invariably altered in cancer [9, 10] as it plays key roles in chromatin assembly, accessibility to transcription and replication machineries, and genome stability (Fig. 2.1) [11]. Acetylation of histones loosens the chromatin structure allowing proteins involved in various processes to bind to DNA (Fig. 2.1) [8]. Steady-state levels of histone acetylation and its dynamic changes represent a balance between histone acetyl transferases (HATs) and histone deacetylases (HDACs) [12].

HATs usually function in multi-subunit complexes and are evolutionary conserved [13]. One such complex is the yeast alteration/deficiency in activation (ADA) complex that consists of the HAT general control non derepressible 5 (Gcn5, originally called Ada4), ADA HAT complex component 1 (Ahc1) and adaptor proteins Ada2 and Ada3 [14]. The ADA genes were initially discovered in yeast based on mutations in them conferring resistance to GAL4-VP16 toxicity [15–18]. These genes included *Ada1/Hfi1*, *Ada2*, *Ada3*, *Ada4* and *Ada5* (*Ada4* and *Ada5* are commonly referred to as *Gcn5* and *Spt20*, respectively). The ADA complex has been shown to act as a co-activator complex in yeast and is involved in transcription [19, 20].

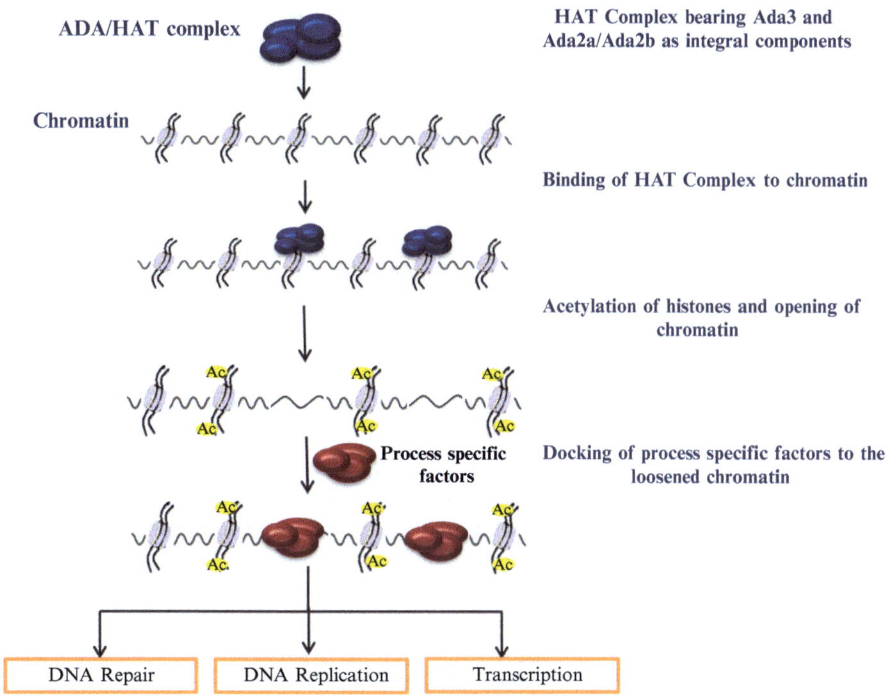


**Fig. 2.1** A schematic model showing different cellular processes regulated by ADA complex or by HAT complexes, that contain Ada2a/Ada2b and Ada3 as core components, through histone acetylation and chromatin remodeling. ADA/HAT complex binds to chromatin at promoters/enhancers/DNA repair sites/origins of replication and aid in loosening up of chromatin by acetylating histones at the sites. This allows various factors involved in distinct processes such as transcription, DNA replication and DNA repair to have access to DNA and perform their functions

In addition to the ADA complex, Gcn5, Ada2 and Ada3 proteins are also a part of Spt-Ada-Gcn5 acetyltransferase (SAGA) and SAGA-like (SLIK) complexes in yeast [19–21]. These proteins and the ADA complex are highly conserved from yeast to mammals where they usually form a HAT module of large multi-functional complexes such as the Spt3/Taf9/Gcn5 acetyltransferase complex (STAGA) (human homolog of yeast SAGA complex), the Ada2a-containing complex (ATAC), and the TBP-free TAF complex (TFTC) [13, 22].

The mammalian cells are more complex and contain multiple HATs (e.g., p300, CREB-binding protein (CBP), p300/CBP-associated factor (PCAF), Tip60 and MOF) in addition to Gcn5. The mammals also contain adaptor proteins Ada3, and two different homolog proteins for yeast Ada2 – Ada2a and Ada2b – that, along with various HATs, form essential and functional module of several multi-subunit HAT complexes, as mentioned above [13, 23, 24]. Notably, *Ada2a* and *Ada2b* are two different homologs of the same yeast *Ada2* gene, and are present in separate complexes in higher eukaryotes; however, these two homologs are unable to complement each other functionally indicating that both have a distinct set of functions [25–27].

Although the main function of the ADA complex is in histone acetylation, the components of this complex have been shown to regulate functions of various non-histone proteins such as nuclear hormone receptors (e.g., estrogen receptor, retinoic acid receptor, retinoic X receptor, and androgen receptor), p53, c-myc, retinoblastoma protein (Rb), and various E2Fs [28–39]. As the role of various mammalian HAT proteins in various physiological and pathological processes is the focus of multiple reviews [11–13, 23], in this chapter we have kept our focus on Ada proteins, particularly focusing on Ada3 and Ada2 (Ada2a and Ada2b), as these adaptor proteins have been shown to be indispensible for the functions of various HATs. We will discuss their potential roles in cell cycle, genomic stability, and their disruption in pathology focusing on cancer. Given the current knowledge that adaptor proteins have no independent enzymatic function and only function as part of the HAT complex, we will also briefly discuss functions of various HATs as we go through reviewing functions of Ada2a, Ada2b, and Ada3. Lastly, as the therapeutic efficacy of relatively general HDAC inhibitors in cancers has recently gained importance [40], we will discuss current knowledge of various HAT inhibitors considering that it is likely that more-targeted agents to alter acetylation in cancer cells will provide improved anticancer strategies.

## 2.2 The ADA Complex, Histone Acetylation and Chromatin Remodeling

As mentioned earlier, histone acetylation plays a fundamental role in modeling of chromatin structures in order for proteins involved in important DNA-related processes to access the DNA. The primary function of the ADA complex has been shown to be transcriptional regulation of genes by modulating histone acetylation at gene promoters [19, 20]. Although yeast Gcn5 alone can acetylate free core histones in vitro, various studies in yeast have conclusively proven that Gcn5 requires both Ada2 and Ada3 for efficient acetylation of nucleosomes, both, in vitro and in vivo [19, 41, 42]. Thus, even though Ada2 and Ada3 do not possess intrinsic HAT activity, they are essential for acetylation of nucleosome histones by Gcn5. Accordingly, it has been shown that yeast Ada2, Ada3, and Gcn5 form a catalytic core of the ADA and SAGA HAT complexes, which is necessary and sufficient in vitro for nucleosomal HAT activity and lysine specificity of the intact HAT complexes [19, 42]. The yeast ADA complex has been shown to preferentially acetylate lysine residues 9, 14, and 18 (and to a lesser extent lysine 23) of histone H3; however, the yeast ADA complex is unable to acetylate histone H4 in vitro [42]. The authors further demonstrated that Ada2 enhances catalytic activity of Gcn5. Moreover, they demonstrated that Ada3 is necessary for Gcn5-dependent nucleosomal HAT activity in yeast extracts and is important for expanding the lysine specificity of the ADA complex [42]. Similar to yeast, studies with mammalian Gcn5, Ada2b (present in STAGA complex), and Ada3 proteins have shown that these proteins can form a heterotrimer in vitro and can efficiently acetylate nucleosomal arrays [43]. Unlike yeast proteins, neither Ada2b nor Ada3 was demonstrated to enhance the acetylation of free-core histones by mammalian Gcn5 in vitro. However; efficient acetylation of chromatin by Gcn5 was shown to require both Ada2b and Ada3 [43]. Interestingly, unlike Ada2b, the Ada2a homolog of the yeast Ada2 was unable to facilitate acetylation of nucleosomal histone H3 in HAT assays in vitro, even though it could form a complex with Gcn5 and Ada3 both in vitro and in vivo [43]. Similar to earlier reports in *Drosophila*, the authors convincingly demonstrated that Ada2a and Ada2b have non-redundant functional roles in mammalian cells. Contrary to this report, which indicated that the mammalian Ada3 protein is unable to enhance HAT activity of Gcn5 on free core histones in vitro, a recent report from our laboratory demonstrated that mammalian Ada3 is able to enhance HAT activity of p300 even on free core histones [44].

Similar to important roles of these proteins in vitro for histone acetylation, several studies have shown that these proteins are important for histone acetylation in vivo. In yeast, it has been shown that depletion of Ada3 or Ada2 drastically affects the histone acetylation in vivo in cells and this has been linked to defects in replication and DNA damage repair in yeast cells (see later sections). Furthermore, *Drosophila* null for *Ada2b*, had reduced H3K9 and H3K14 acetylation during development; however, *Ada2a* deletion did not have any effect on acetylation of these residues [27]. Although, deletion of either *Ada2a* or *Ada2b* was lethal for *Drosophila* development, both proteins were shown to have non-redundant functions in *Drosophila*. Similarly, a later study demonstrated requirement of Ada3 for viability of *Drosophila* embryos [45]. These *dAda3* mutants were shown to be defective in histone acetylation at H3 K9, H3 K14, and H4 K12, whereas there were no defects observed in acetylation of H3 K18 or H4 K5, K8, or K16. These defects in histone acetylation were shown to affect the position effect variegation at certain loci and in the transcription of specific genes. Additionally, we recently demonstrated that depletion of Ada3 from mammalian cells results in drastic down-regulation of histone acetylation at various lysine residues such as H2A K5, H2B K5, H3 K9, H3 K56, and H4 K8 [44]. This dramatic down-regulation in various histone acetylations underscores the important role of the ADA complex in histone acetylation.

In order for appropriate proteins to bind DNA and carry out their function, the process of histone acetylation routinely needs to be coupled with chromatin remodeling, which occurs by nucleosome sliding leading to removal of nucleosomes at promoter regions or at DNA damage sites [46, 47]. The chromatin remodeling complex SWItch/Sucrose Non Fermentable (SWI/SNF) is an important complex that has been shown to be involved in nucleosome sliding [46, 47]. Interestingly, this complex has been shown to work in concert with HAT complexes at promoters of various genes and has been shown to be involved in gene activation [48]. More importantly, the recruitment of SWI/SNF complex onto various promoters is believed to be dependent on acetylation of nucleosomal histones by the SAGA complex [49]. Acetylated histones form a prerequisite for the recruitment of SWI/SNF complex through bromo domains present in Swi2/Snf2 [50]. Consistent with this observation, it has been shown that the SWI/SNF complex is capable of efficiently displacing nucleosomes from chromatin that are specifically acetylated by SAGA complex compared to nucleosomes that are not acetylated [51]. Additionally, it has been shown that yeast Gcn5 directly regulates the binding of SWI/SNF complex to chromatin, through acetylation of Snf2 component of SWI/SNF complex [52]. Snf2 acetylation by Gcn5 results in the dissociation of SWI/SNF complex from acetylated histones, thus inhibiting SWI/SNF complex function [52]. Furthermore, the ATAC HAT complex in *Drosophila* has been shown to stimulate nucleosome sliding by stimulating the ISWI, SWI/SNF, and RSC chromatin remodeling complexes [53]. These findings highlight a fundamental and functional link between histone acetylating complexes and complexes involved in chromatin remodeling and also underline the importance of HAT complexes in chromatin remodeling as histone acetylation acts as a pre-requisite for chromatin modeling through nucleosome sliding.

## 2.3 The ADA Complex Functions as a Co-activator for Nuclear Hormone Receptor-Mediated Transcription

The ability of nuclear hormone receptors (NR) to up-regulate or down-regulate the target gene expression is determined by their association with cofactors that may fall under the category of co-activator or co-repressor [54]. When bound to a co-activator, nuclear receptors up-regulate the gene expression whereas binding of a co-repressor leads to the down-regulation of target gene expression. Over the past two decades a number of co-activators have been studied extensively by different laboratories and an important class of co-activators was identified as steroid receptor co-activators (SRC-1, -2 and -3) by Bert O'Malley's group [55]. X-ray crystallography studies have demonstrated that a typical co-activator contains α helical LXXLL binding motif (where L is leucine and X is any amino acid) referred to as NR box through which it binds to a groove on the surface of ligand binding domain of nuclear receptor [56, 57]. In addition to SRCs, several novel nuclear receptor co-regulators, such as BCAS3, PELP1, and DLC1, have also been identified and characterized [58].

In regard to the role of the ADA complex in NR activation, initial observations that the ADA complex plays a role in NR-mediated transcription came from yeast Ada3 (yAda3) [59]. Though yeasts do not have NRs, the yAda3 protein was found to be associated with exogenously expressed NRs. In this context, it was shown that the ADA complex is required for the transactivation function of the glucocorticoid receptor (GR) [59]. The investigators further demonstrated that deletion of any of the components of the ADA complex reduces the activity of the GR responsive-lacZ reporter compared to the wild type. Notably, deletion of *Ada3* was found to cause a greater reduction in this activity than deletion of either *Ada2* or *Gcn5* alone [59]. Furthermore, Ada2 was shown to enhance the activity of the GR responsive reporter in mammalian cells [59].

The yAda3 also interacts with other NRs such as ERα, RXRα, and TRα, but not with RARα [60]. Reporter assays in yeast have demonstrated that yAda3 potentiates the AF-2 activity of ERα and RXRα when overexpressed in yeast and mammalian cells [60]. The authors further showed that, other subunits of the ADA complex, Ada2 and Gcn5, are also required for ERα- and RXRα-mediated transactivation [60]. Subsequently, our laboratory demonstrated that mammalian Ada3 interacts with ERα and RXRα and augments their transactivation and increases the levels of target gene expression [29–31]. More importantly, shRNA-mediated knockdown of *Ada3* significantly down-regulated estrogen-responsive genes and as a result suppressed ER-mediated cell proliferation, thus supporting an important role of co-activators in the NR-mediated functions [29, 30]. Subsequently, other investigators performed mutational analyses of Ada3, and observed that similar to classical NR co-activator, Ada3 binds to RAR through its LXXLL motifs [32]. In summary, these studies provide significant evidence that Ada3 functions as a co-activator in NR signaling.

## 2.4 Interaction of the ADA Complex with Non-Nuclear Hormone Receptor Proteins

The p53 protein is a tumor suppressor protein that transactivates stress responsive genes and regulates the cell cycle in response to DNA damage [61]. The activation domain of p53 possesses notable similarity with the activation domain of other transcriptional activators such as the activator of herpes simplex virus, VP16 [61]. Like other transcription factors, VP16 also requires co-activators for its activity, and, in yeast, the requirement of the ADA complex was first shown for VP16 transactivation [15, 16, 62]. The similarity of the p53 activation domain sequence with the activation domain of VP16 and the requirement of the ADA complex for its activator function generated the rationale to study the interaction of ADA components with p53 [63]. Investigators identified two activation subdomains (ASD-1, -2) in the p53 amino-terminus that require yeast adaptor complex Ada2/Ada3/Gcn5 for transcriptional activation [63]. ASD-1 was less dependent on the ADA complex than ASD2, and Ada3 was the most critical component in the complex for the function of p53 [63]. Subsequently, work from our laboratory, and that of other laboratories, demonstrated a direct interaction of Ada3 with p53 and its function as a co-activator for p53-mediated transactivation [34, 35, 64].

Full transcriptional activation of p53 requires its C-terminal acetylation by p300/CBP and PCAF [65, 66], and we subsequently demonstrated that Ada3 recruits p300 to acetylate p53 and regulates its transcriptional activity [35]. In this context, shRNA-mediated knockdown of Ada3 dramatically down-regulated p53 target genes. Most importantly, loss of Ada3 led to inhibition of DNA damage-induced p53 acetylation and cell-cycle arrest [35]. Subsequently, another group delineated the role of Ada2 in the function of p53 [67]. Using chromatin immunoprecipitation assay, the authors demonstrated that Ada2b, but not Ada2a, gets recruited to the p53 response element on promoters of target genes [67]. Indeed, the study revealed that Ada2a and Ada2b function in a non-redundant manner and only Ada2b is found to be the component of STAGA complex in humans [67].

In addition to NRs and p53, Ada3 also interacts with IL-1α and β-catenin [36, 68]. Typically IL-1α mediates its action in a secreted form. However, a proteolytic maturation product of IL-1α, known as IL-1α N-terminal peptide (IL-1NTP) found in the nucleus, acts as a transcription factor and is involved in variety of cellular processes such as control of cell proliferation and apoptosis [69]. A study performed in yeast and mammalian systems delineated the interaction of IL-1NTP with p300, PCAF, Gcn5, and Ada3 [68]. In yeast, the fusion protein Gal4BD/IL-1NTP was found to have a growth inhibitory effect that requires an intact SAGA complex [68]. More importantly, deletion of any of the components of SAGA complex was found to completely attenuate the suppressive effect, confirming the necessity of an intact SAGA complex for the action of IL-1NTP [68]. In the mammalian system, IL-1NTP was found to interact with p300, PCAF, Gcn5, and Ada3, and eventually integrate into the p300-PCAF complex, thus enhancing the transcriptional activation of this complex [68].

The role of the ADA complex in the activation of β-catenin, which is involved in developmental processes through the Wnt pathway, has also been demonstrated [36]. The Wnt pathway is crucial for development and proliferation, and abrogation of this pathway is linked to cancer development. Both Ada2a and Ada3 have been shown to interact directly with β-catenin and mediate its acetylation [36]. As a consequence, Ada2a and Ada3 regulate the target gene expression of β-catenin. Also, reduced levels of these proteins lead to repression in β-catenin-dependent cell proliferation [36]. Further studies in this context are warranted to conclusively address the role of the ADA complex in the Wnt-β-catenin pathway.

## 2.5 The ADA Complex and Cell Cycle

The eukaryotic cell cycle consists of the following four phases: G1, S, G2, and M [70]. During the G1 phase, cells accumulate nutrients, grow, and duplicate various cell organelles, except chromosomes, which occurs later [70]. Before entering the S phase, cells examine their size, determine the availability of appropriate nutrients and growth factors, and ensure that there is no DNA damage [70]. The process of DNA replication occurs during the S phase, and it provides a means for duplication of genetic material that can then be equally segregated into daughter cells during the process of mitosis [70]. The eukaryotic cell-cycle progression thus depends on proper coordination of DNA replication and segregation of duplicated chromosomes to daughter cells, a process precisely regulated by modification of chromatin that allows accessibility to factors involved in these processes. Thus, the HAT complexes involved in modulating the structure of chromatin, as mentioned earlier, play an important role in the cell-cycle progression. Consistent with this, various ADA complex components have been shown to play indispensible roles in various phases of cell cycle.

Recently, our laboratory demonstrated an important role of Ada3 in the G1-S phase transition as well as in mitotic progression of cell cycle [44]. To elucidate the physiological function of Ada3, we generated a conditional knockout mouse for the *Ada3* gene. We observed homozygous $Ada3^{FL/FL}$ mice were viable, fertile, and exhibited no gross abnormalities compared to *Ada3*$^{FL/+}$ or *Ada3*$^{+/+}$ controls, whereas Ada3$^{-/-}$ mice were lethal at E3.5 stage [44]. The failure of *Ada3*$^{-/-}$ embryos to remain viable beyond E3.5 suggested a potential role of Ada3 in cell proliferation because extensive cellular proliferation occurs during this early stage of embryogenesis.

Subsequently, by using *Ada3* deletion in *Ada3*$^{FL/FL}$ mouse embryonic fibroblasts (MEFs) we showed that Ada3 is required for efficient cell-cycle progression through the G1 to S phase transition as well as for proper mitosis [44]. Detailed analyses in this system revealed that an Ada3-c-myc-Skp2-p27 axis controls the progression of the G1 phase to the S phase and partly contributes to cell-cycle delay upon deletion of *Ada3* [44]. Microarray analysis showed that loss of *Ada3* resulted in several changes in gene expression that were involved in mitosis [44]. Consistent with this, *Ada3* deletion led to severe mitotic defects and formation of multi-nucleated cells. Also, the transition from the G2/M phase to the G1 phase was delayed upon deletion of *Ada3* [44]. Thus, Ada3, a core component of the ADA complex, is important in G1 phase as well as in mitosis during the cell-cycle progression.

Another group also showed a role of the ATAC complex in mitosis [71], where knockdown of ATAC complex components, such as *Ada2a* and *Ada3*, led to severe mitotic defects. These defects included centrosome multiplication, defective spindle and midbody formation, generation of binucleated cells, and a slow transition from G2/M to G1 phase [71]. The mitotic defects were attributed to the inefficient acetylation of the Cyclin A/Cdk2 complex by Gcn5 due to knockdown of *Ada3* or *Ada2a* [71]. Similar to mammalian *Ada3*, deletion of *Ada3* as well as *Gcn5* in flies leads to defective H3S10 phosphorylation, an event that marks the initiation of mitosis. This suggests a role of the ADA complex in the process of mitosis in flies as well as mammals [45].

Several reports have shown the role of the ADA complex component Gcn5 in replication, which is consistent with the important role of histone acetylation in DNA replication. In yeast, it was shown that Gcn5 is required for replication-coupled nucleosome assembly [72]. *Gcn5* deletion mutants in yeast showed a reduced level of H3K56 acetylation, a mark linked to replication-coupled nucleosome assembly in yeast [72]. Similar to Gcn5, deletion mutants of *Ada3* and *Ada2* showed defects in replication suggesting an important role of these components in the replication process [72]. In mammals, Gcn5 has also been shown to play an important role in the process of replication by controlling the acetylation of Cdc6, an important replication licensing factor [73]. Although, the role of other ADA complex components in replication needs to be explored extensively, these initial reports show promising results for a role of the ADA complex in replication. Taken together, these studies unequivocally support a critical role of the ADA complex in cell-cycle progression.

## 2.6 Role of the ADA Complex in DNA Damage Response

In addition to metabolic and transcriptional processes, the chromatin structure plays an important role in the DNA damage response (DDR) process. The DDR is manifested by assembly of DNA damage repair proteins at the site of damage [74].

Histone modifying enzymes such as HATs along with ATP-dependent chromatin remodeling complexes allow these DNA damage proteins to access DNA at the damage sites [74]. In the context of DDR, the role of Gcn5 and Ada2 has been elucidated [75]. This study focused the role of Gcn5 and Ada2 in nucleotide excision repair of yeast *MET16*, a gene regulated by these two components of the SAGA/ADA complex [75]. The role of Gcn5 and Ada2 in nucleotide excision repair was revealed by the finding that deletion of either *Ada2* or *Gcn5* delays the cyclobutane pyrimidine dimer removal on the *MET16* locus [75]. In another study, investigators showed that the STAGA complex interacts with UV-damaged-DNA binding factors DDB1 and DDB2 and this interaction facilitates the recruitment of nucleotide excision repair machinery through HAT activity of Gcn5 [76]. Furthermore, another role of STAGA complex in p53-dependent gene activation through Gcn5 and its recruitment on *p21* and *GADD45* promoters upon UV damage was shown [67]. Besides STAGA, the TFTC HAT complex is also reported to have an important role in DDR [77]. Researchers identified SPT130 as an integral subunit of the TFTC complex. Interestingly, SPT130 possesses homology with the UV-damaged DNA binding factor [77]. Given the presence of SPT130 in TFTC, the investigators found that TFTC is recruited on UV-damaged DNA and brings about the acetylation of histone H3 on the UV-damaged site, clearly suggesting a role of TFTC in DDR [77]. The role of p300 in DDR is also documented where it has been shown to stabilize and transactivate p53 in response to DNA damage [78]. Other HATs, such as MOF, acetylate H4 K16 and mediate the recruitment of repair proteins, such as Mdc1, 53BP1, and Brca1, upon ionizing radiation-induced DNA damage [79]. Similarly, Tip60 also acetylates core histones, and inactive Tip60 has been found to be associated with late double strand breaks [80–82].

Although a fairly good number of studies have delineated the role of various HATs in DDR, the role of the ADA complex per se had not been studied until recently. We demonstrated that loss of *Ada3* leads to dramatic genomic instability as seen by various chromosomal aberrations, which were further enhanced upon DNA damage [83]. Loss of *Ada3* led to an increase in the levels of DDR proteins, such as pATM, p53BP1, pRAD51, and γH2AX [83]. Significantly, *Ada3*-null cells exhibited a delay in the disappearance of the DNA damage foci for γH2AX, 53BP1, and CtIP after ionizing radiation, suggesting the important role of Ada3 in DDR [83]. Together these findings reveal a new role of Ada3 in the DNA repair process and maintenance of genomic stability and warrant further research to determine if other components of the ADA complex also regulate genomic stability and repair foci disappearance.

## 2.7 The ADA Complex and Cancer

As described above, components of the ADA complex are fundamental in the cell-cycle progression, regulation of various transcriptional factors, and in maintaining genomic stability. Not surprisingly, several of the components of the ADA complex are hijacked by viruses and are known to interact with viral onco-proteins, such as human papilloma virus 16 (HPV16) E6, simian virus 40 (SV40) large T, or adenoviral protein E1A, thus linking the ADA complex to oncogenesis [64, 84–91].

Our laboratory identified human Ada3 as a HPV16-E6-binding protein [64]; importantly, HPV16 is the most common HPV associated with human cancers. Significantly, Ada3 bound to immortalization-competent E6 mutants, and also to mutants that were incapable of binding to p53 [64]. We further demonstrated that E6 targets Ada3 for degradation, thus abrogating the function of p53 through an alternate pathway [64]. Subsequently, other investigators showed that degradation of Ada3 by E6 abrogated p14ARF-p53-mediated senescence pathway and led to E6-induced immortalization [92]. Further, p300/CBP have been shown to be associated with HPV16 E6/E7, adenoviral E1A, as well as SV40 large T antigen [84–88]. Also, the yeast SAGA complex has been shown to be important in adenoviral E1A induced growth inhibition [89, 90]. Recently, the HAT Gcn5 was shown to functionally interact with the adenoviral E1A protein [91]. Together, these studies underscore the important role of the ADA complex in viral oncogenesis.

Moreover, functions of several important cellular oncogenes (e.g., c-myc, E2Fs) and/or tumor suppressors (e.g., p53 [see above], Rb, BRCA1) have been shown to be regulated by various HATs and HAT complexes [37–39, 93–98]. c-myc oncogene was shown to be associated with TRRAP, a subunit of the Ada2b and Ada3 containing mammalian STAGA complex [93]. Upon mitogenic stimulation, c-myc, a sequence specific transcription factor, was shown to induce histone acetylation at its target gene promoters through recruitment of TRRAP [93]. More importantly, the C-terminal ATM-related domain of TRRAP has been shown to be required for c-myc driven transformation [94]. Furthermore, another study demonstrated that c-myc recruits TRRAP as well as Gcn5, by interacting with the STAGA HAT complex [95]. Both TRRAP and Gcn5 were shown to enhance the transcriptional activation of c-myc through its N-terminal activation/transformation domain [95]. Accordingly, an N-terminal deletion mutant of c-myc was unable to bind to STAGA complex and showed reduced transcription activation potential [95]. Interestingly, a similar naturally occurring truncated form of c-myc has been shown to be deficient in transforming primary cells [96]. These studies reveal an essential role of the STAGA HAT complex in c-myc induced oncogenic transformation.

E2F family of transcription factors regulate several cell cycle associated genes and have been shown to be regulated by various HATs [38, 39]. Two independent studies demonstrated that PCAF, p300 and CBP HATs bind to and acetylate E2F-1, -2 and -3 [38, 39]. This reversible acetylation of E2Fs by various HATs was shown to enhance their stability and increase their DNA binding ability as well as transactivation potential [38, 39]. Furthermore, transactivation domains of E2F1 and E2F4 were shown to directly bind and recruit Gcn5 and TRRAP, most likely as subunits of HAT complexes [97]. This study suggests that E2F transcription factors stimulate their transcriptional activation by recruiting the HAT complex components TRRAP and Gcn5, thus providing a mechanism to relieve the transcriptional repression at E2F target gene promoters [97]. Similar to E2Fs, p300 and PCAF have been shown to acetylate and regulate the function of the important cell-cycle regulator and tumor suppressor retinoblastoma (Rb) protein [37, 98]. This acetylation event has also been shown to be essential in nuclear localization of Rb and plays an important role in cellular differentiation. Thus, de-regulation of these HATs could potentially disrupt the function of Rb and contribute to oncogenesis. Additionally, the tumor suppressor BRCA1 has been shown to bind Gcn5 and TRRAP-containing HAT complexes [99]. This interaction has been shown to be indispensible for BRCA-1 mediated transcriptional regulation as well DNA repair. These studies emphasize an essential role of HAT complexes in regulating various functions of important cellular oncogenes as well as tumor suppressors.

p300 and CBP HATs have also been shown to be involved in leukemogenesis [100]. Somatic mutations of p300 and CBP have been reported in hematological malignancies [100]. These mutations include translocations of p300 and CBP genes that result in their fusion with the monocytic leukemia zinc finger (MOZ) gene or with the mixed lineage leukemia (MLL) gene [100]. These translocations lead to the formation of MOZ-p300/CBP or MLL-p300/CBP fusion proteins that can have aberrant loss-of-function or gain-of-function properties and can play an important role in cellular transformation [100]. Also, germline mutation of CBP causes Rubinstein-Taybi syndrome, a condition that predisposes its patients to cancer [101, 102]. Interestingly, recent reports from two different laboratories show that high expression of p300 in hepatocellular carcinoma correlates with poor survival and aggressive features in HCC, such as epithelial to mesenchymal transitions [103, 104]. Although further investigation is required in this area of research, the above observations highlight the importance of HAT complex components in the regulation of oncogenesis.

Our previous findings demonstrated Ada3 is a critical component of HAT complexes that regulate ER function [30]. These findings and the observations that overexpression of other ER co-activators, such as SRC-3 predicts clinical outcomes in breast cancer patients [105, 106], prompted us to examine potential significance of Ada3 expression/localization in human breast cancer patients [107]. Using immunohistochemical analysis of Ada3 expression in breast cancer tissue specimens from a large cohort of patients with known clinico pathological parameters and survival data, we reported that predominant nuclear Ada3 expression correlated with ER expression and predicted a favorable clinical outcome while predominant Ada3 expression in the cytoplasm correlated with ErbB2/EGFR expression and poor patient survival. These studies suggest an important role of Ada3 in breast cancer progression. Further studies are needed to examine the molecular mechanism of differential localization of Ada3 (and other components of the ADA complex) in the promotion of breast oncogenesis.

## 2.8 New Emerging Functions of the ADA Complex

In the previous sections of the chapter, we discussed well documented functions of the ADA complex. However, several laboratories have recently demonstrated an important role of the ADA complex and its components in endoplasmic reticulum stress. Endoplasmic reticulum is a cellular organelle that is involved in proper folding and post-translational modifications of secretory and transmembrane proteins and thus houses many chaperone proteins [108, 109]. Interestingly, endoplasmic reticulum has also been shown to act as a sensor of cellular stress [108]. Various forms of cellular stress cause an increase in the demand for protein folding, challenging the capacity of chaperones present in the endoplasmic reticulum [110–112]. This leads to accumulation of unfolded and misfolded proteins in the endoplasmic reticulum lumen that causes cells to initialize a cascade of signaling events which are collectively called the unfolded-protein response (UPR) [110–112]. Three distinct endoplasmic reticulum localized transmembrane protein sensors, inositol requiring 1α (IRE1 α), double-stranded RNA-dependent protein kinase (PKR)-like ER kinase (PERK) and activating transcription factor 6 (ATF6) initiate three different UPR signaling cascades in the cells [110–112]. The ADA complex and its components have been shown to be involved in at least two of the three UPR signaling cascades [113–117]. In this context, it was first shown in yeast that Gcn5 interacts with Ire1 protein kinase and that ADA complex is specifically required for UPR but not for heat shock protein response [113]. The authors proposed that Ire1 activation recruits the ADA complex specifically to the promoters of the genes involved in UPR [113]. In a follow-up study, the authors demonstrated that ADA complex components Gcn5, Ada2, Ada3 and Ada5 interact with Ire1 and HacI [114]. HacI is a transcription factor that is required for UPR and its translation is regulated by Ire1 [111]. HacI is constitutively expressed; however, its mRNA is not efficiently translated [111]. Activation of Ire1 leads to formation of a new processed form of HacI mRNA that gets efficiently translated and induces UPR [111]. The authors showed that the transcriptional activation of endoplasmic reticulum stress related genes was reduced upon deletion of *Gcn5*, *Ada2* or *Ada3*, whereas deletion of *Ada5* (Spt20) completely abrogated the transcriptional activation [114]. The same group further demonstrated that Ada5 is essential for Ire1 dependent processing of HacI mRNA in vivo, thus proving an important role of ADA complex components in the UPR in yeast [115]. Similar to yeast, it was shown that mammalian SAGA complex plays an important role in endoplasmic reticulum stress related genes [116]. The authors demonstrated that mammalian Spt20, like its yeast counterpart, was indeed a subunit of the SAGA complex [116]. By chromatin immunoprecipitation studies, the authors showed that Spt20 and other SAGA complex components are recruited onto the endoplasmic reticulum stress response genes and knockdown of Spt20 abrogates the endoplasmic reticulum stress response [116]. Again, similar to yeast, the recruitment of Spt20 was shown to be specific to endoplasmic reticulum response genes but not to other stresses [116]. Furthermore, in a recent study it was shown by multidimensional protein identification technology (MudPIT), that mammalian ATF6 transcription factor recruits the SAGA and ATAC complexes onto the endoplasmic reticulum stress response enhancer elements present on endoplasmic reticulum stress response genes and thus are involved in controlling the transcription of these genes [117]. Taken together, these studies point towards a potential role of the ADA complex and its components in UPR.

Further, the STAGA HAT complex has been shown to interact with spliceosome-associated protein 130 (SAP130), a component of the SF3b splicing factor [76].

SF3b gets recruited to pre-spliceosomal complexes in association with U2 snRNP. This points to an important role of STAGA complex in mRNA splicing [76]; however, further studies are required to provide more insights into this function. Another phenomenal study in yeast revealed the role of the SAGA complex and its components Ada2 and Sus1 in tethering of transcriptionally active genes to the nuclear envelope [118]. It has been known that certain genes come closer to the nuclear periphery upon their transcriptional activation. In this study, the authors demonstrated that yeast SAGA complex components Ada2 and Sus1 are involved in confinement of active *GAL* reporter genes to the nuclear periphery [118]. Ada2 and Sus1 achieve this by physically linking the active *GAL* genes to the nuclear pore complex component, Nup1 [118]. Accordingly, deletion of *Ada2* or *Nup1* was shown to abrogate this confinement of *GAL* genes to nuclear periphery [118] suggesting an important role of the SAGA complex in regulating transcription of genes by a novel mechanism of nuclear periphery tethering.

Recent evidence demonstrates mammalian Gcn5 and SAGA complex to be involved in telomere maintenance by controlling the ubiquitination of a component of the shelterin protein complex [119]. Shelterin is a multi-subunit protein complex involved in structural maintenance of telomeres [120]. The authors show that *Gcn5* deletion leads to telomere dysfunction in mammalian cells [119]. This phenomenon was shown to be dependent upon the deubiquitination module of the SAGA complex [119]. The authors demonstrated that the SAGA complex component, ubiquitin-specific protease 22 (Usp22), is involved in deubiquitinating TRF1 (a shelterin complex component) and thus plays an important role in controlling the stability of TRF1 [119]. Interestingly, Gcn5 was shown to be required for association of the Usp22 deubiquitinating module to the SAGA complex and is thus, involved in the maintenance of proper telomere structure through TRF1 [119]. These studies demonstrate the role of the ADA complex components in various important cellular processes and suggest further studies must be carried out to gain more insights into role of the ADA complex components in maintaining genomic integrity.

## 2.9 The ADA Complex and HAT Inhibitors

As discussed above, acetylation of histones and other proteins plays an important role in a variety of physiological processes in cells, and deregulation of the proteins that regulate acetylation leads to oncogenesis. Consistent with this idea, various laboratories have focused on discovering new synthetic or natural drugs that inhibit enzymes such as HDACs and HATs, which are involved in maintaining homeostasis in acetylation. Inhibitors targeting HDACs have been studied extensively [40, 121, 122]. Many of those inhibitors have shown promising anti-cancer activities without affecting non-cancerous cells, and, accordingly, these inhibitors are currently involved in ongoing clinical trials [40, 121, 122]. Recently, two HDAC inhibitors (HDACi), Vorinostat and Depsipeptide, were approved by the FDA for use as nti-cancer agents after their validation in cancer patients [122]. Vorinostat was the first HDAC inhibitor to be approved by FDA for the treatment of cutaneous T-cell lymphoma. Many other HDAC inhibitors are in clinical trials for use as anti-cancer drugs, either alone or in combination with other drugs [122]. On the other hand, inhibitors of HATs have seen a slow development, and only in recent years have considerable efforts been made to identify various HAT inhibitors (HATi). The HATi identified till now can be classified into the following three categories: (1) bi-substrate inhibitors, (2) small molecule synthetic inhibitors, and (3) natural compounds [123]. Bi-substrate-based inhibitors include the spermidinyl-CoA-based HAT inhibitors. These inhibitors induce a transient block in DNA replication and impair DNA repair in cancer cells but not in normal cells [124]. However, these compounds alone have been found to be incapable of affecting cancer cell proliferation [124]. Interestingly, these inhibitors were shown to provide cancer-specific chemo- and radio-sensitization due to their ability to affect the DNA repair process [124].

Various natural compounds have been shown to have HAT inhibitory properties. These include anacardic acid (potent inhibitor of p300 and PCAF HAT activity), garcinol (also inhibits HAT activity of both p300 and PCAF), Epigallo Cathenin (present in green tea) and curcumin (a specific inhibitor of p300/CBP HAT activity). Out of these inhibitors, curcumin has been extensively studied for its anti-cancer activities, and various clinical trials involving curcumin are in progress [123]. Furthermore, several small molecule synthetic inhibitors have been designed to inhibit HAT activity such as γ-butyrolactone MB-3 (a GCN5-specific inhibitor), isothiazolones (p300 and PCAF-specific inhibitor) as well as various quinoline derivatives.

Recently, an isothiazole inhibitor NU9056 (Tip60 inhibitor) was shown to have anti-cancer effects on prostate cancer cells [125]. Notably, several of these small molecule inhibitors were designed as analogs of naturally occurring HATi including anacardic acid and garcinol [123]. Although much progress has been made in discovering novel HATi, our knowledge of anti-cancer activities of HATi is limited in comparison with HDACi. This could be attributed to the fact that HATi are less efficient than HDACi, and also because the current HATi doses are not physiologically feasible. Clearly, other than curcumin, which is in clinical trials, HATi need to be improved, and further studies are required to accept these as anti-cancer agents.

## 2.10 Conclusions and Future Perspectives

Studies carried out at the end of the twentieth century and in the beginning of the twenty-first century have shown the importance of the ADA complex and its components in several important cellular processes in organisms ranging from yeast to humans (Fig. 2.2), thus indicating an indispensible role of these components during evolution. In this chapter we focused on the ADA complex and its components

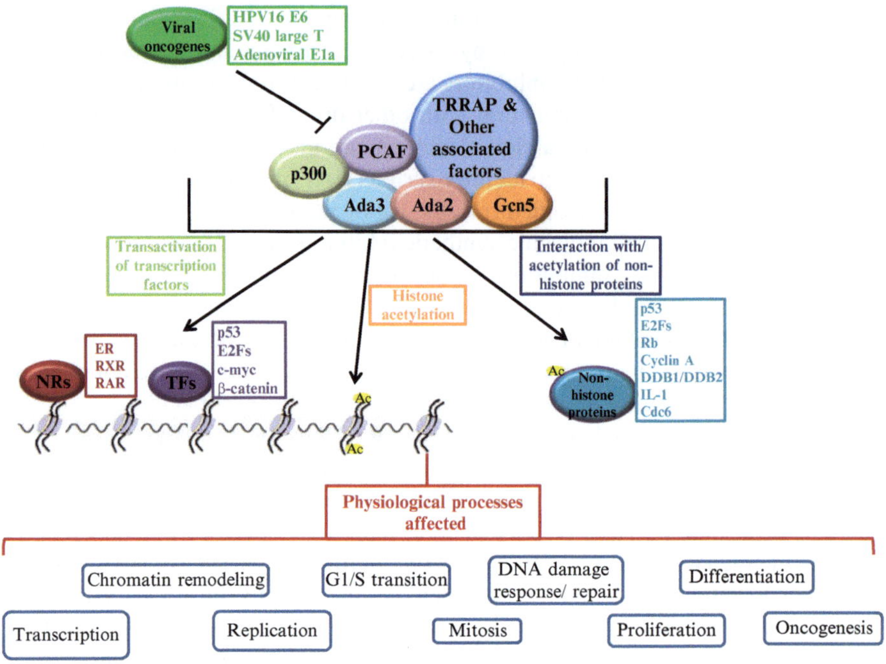


**Fig. 2.2** The ADA complex and its components regulate several cellular processes by associating with and/or acetylating various transcription factors (*TFs*), nuclear receptors (*NRs*), histones and non-histone proteins. Viral oncogenes, by associating with different components of ADA complex, disrupt its function

Ada2 and Ada3, which do not seem to have intrinsic HAT activity. We also discussed various important HATs that are present in ADA complex. Although, a distinct functional ADA complex has been shown to be present in yeast, there is no in vivo evidence of such a complex in mammals, clearly indicating that more work is required in this context. Importantly, most known functions of the proteins Ada2 and Ada3 are dependent on their association with HATs. However, it remains possible that these proteins could have HAT-independent functions. Based on the role of these complex components in various important processes, including cancer, several laboratories are now focusing on making and testing new HAT inhibitors (HATi) that could prove useful in treating cancer as well as other diseases. However, novel HATi need to be designed to be specific, to have lower IC50, and to be potent against tumor cells sparing normal cells. Importantly, development of chemical inhibitors that can prevent protein-protein interaction of Ada proteins with HATs is expected to be an alternative strategy to treat cancer. Taken together, although we have made tremendous progress in understanding role of mammalian ADA complex in various physiological processes, more studies particularly in animal models need to be carried out to understand the role of each component in vivo.

## References


1. Sherr CJ (1996) Cancer cell cycles. Science 274(5293):1672–1677
2. Kastan MB, Bartek J (2004) Cell-cycle checkpoints and cancer. Nature 432(7015):316–323. doi:10.1038/nature03097, nature03097 [pii]
3. Blow JJ, Tanaka TU (2005) The chromosome cycle: coordinating replication and segregation. Second in the cycles review series. EMBO Rep 6(11):1028–1034. doi:10.1038/sj.embor.7400557, 7400557 [pii]
4. Lapenna S, Giordano A (2009) Cell cycle kinases as therapeutic targets for cancer. Nat Rev Drug Discov 8(7):547–566. doi:10.1038/nrd2907, nrd2907 [pii]
5. Luger K, Mader AW, Richmond RK, Sargent DF, Richmond TJ (1997) Crystal structure of the nucleosome core particle at 2.8 A resolution. Nature 389(6648):251–260. doi:10.1038/38444
6. Li B, Carey M, Workman JL (2007) The role of chromatin during transcription. Cell 128(4):707–719. doi:10.1016/j.cell.2007.01.015, S0092-8674(07)00109-2 [pii]
7. Kouzarides T (2007) Chromatin modifications and their function. Cell 128(4):693–705. doi:10.1016/j.cell.2007.02.005, S0092-8674(07)00184-5 [pii]
8. Strahl BD, Allis CD (2000) The language of covalent histone modifications. Nature 403(6765):41–45. doi:10.1038/47412
9. Grunstein M (1997) Histone acetylation in chromatin structure and transcription. Nature 389(6649):349–352. doi:10.1038/38664
10. Archer SY, Hodin RA (1999) Histone acetylation and cancer. Curr Opin Genet Dev 9(2):171–174, doi:S0959437X99800264 [pii]
11. Carrozza MJ, Utley RT, Workman JL, Cote J (2003) The diverse functions of histone acetyltransferase complexes. Trends Genet 19(6):321–329, doi:S016895250300115X [pii]
12. Kouzarides T (1999) Histone acetylases and deacetylases in cell proliferation. Curr Opin Genet Dev 9(1):40–48, doi:S0959-437X(99)80006-9 [pii]
13. Lee KK, Workman JL (2007) Histone acetyltransferase complexes: one size doesn't fit all. Nat Rev Mol Cell Biol 8(4):284–295. doi:nrm2145 [pii] 10.1038/nrm2145
14. Eberharter A, Sterner DE, Schieltz D, Hassan A, Yates JR 3rd, Berger SL, Workman JL (1999) The ADA complex is a distinct histone acetyltransferase complex in Saccharomyces cerevisiae. Mol Cell Biol 19(10):6621–6631
15. Berger SL, Pina B, Silverman N, Marcus GA, Agapite J, Regier JL, Triezenberg SJ, Guarente L (1992) Genetic isolation of ADA2: a potential transcriptional adaptor required for function of certain acidic activation domains. Cell 70(2):251–265, doi:0092-8674(92)90100-Q [pii]
16. Pina B, Berger S, Marcus GA, Silverman N, Agapite J, Guarente L (1993) ADA3: a gene, identified by resistance to GAL4-VP16, with properties similar to and different from those of ADA2. Mol Cell Biol 13(10):5981–5989
17. Marcus GA, Silverman N, Berger SL, Horiuchi J, Guarente L (1994) Functional similarity and physical association between GCN5 and ADA2: putative transcriptional adaptors. EMBO J 13(20):4807–4815
18. Marcus GA, Horiuchi J, Silverman N, Guarente L (1996) ADA5/SPT20 links the ADA and SPT genes, which are involved in yeast transcription. Mol Cell Biol 16(6):3197–3205
19. Grant PA, Duggan L, Cote J, Roberts SM, Brownell JE, Candau R, Ohba R, Owen-Hughes T, Allis CD, Winston F, Berger SL, Workman JL (1997) Yeast Gcn5 functions in two multi subunit complexes to acetylate nucleosomal histones: characterization of an Ada complex and the SAGA (Spt/Ada) complex. Genes Dev 11(13):1640–1650
20. Grant PA, Sterner DE, Duggan LJ, Workman JL, Berger SL (1998) The SAGA unfolds: convergence of transcription regulators in chromatin-modifying complexes. Trends Cell Biol 8(5):193–197, doi:S0962-8924(98)01263-X [pii]
21. Pray-Grant MG, Schieltz D, McMahon SJ, Wood JM, Kennedy EL, Cook RG, Workman JL, Yates JR 3rd, Grant PA (2002) The novel SLIK histone acetyltransferase complex functions in the yeast retrograde response pathway. Mol Cell Biol 22(24):8774–8786

22. Spedale G, Timmers HT, Pijnappel WW (2012) ATAC-king the complexity of SAGA during evolution. Genes Dev 26(6):527–541. doi:10.1101/gad.184705.111, 26/6/527 [pii]
23. Roth SY, Denu JM, Allis CD (2001) Histone acetyl transferases. Annu Rev Biochem 70:81–120. doi:10.1146/annurev.biochem.70.1.81, 70/1/81 [pii]
24. Nagy Z, Tora L (2007) Distinct GCN5/PCAF-containing complexes function as co-activators and are involved in transcription factor and global histone acetylation. Oncogene 26(37):5341–5357. doi:10.1038/sj.onc.1210604, 1210604 [pii]
25. Muratoglu S, Georgieva S, Papai G, Scheer E, Enunlu I, Komonyi O, Cserpan I, Lebedeva L, Nabirochkina E, Udvardy A, Tora L, Boros I (2003) Two different Drosophila ADA2 homologues are present in distinct GCN5 histone acetyltransferase-containing complexes. Mol Cell Biol 23(1):306–321
26. Kusch T, Guelman S, Abmayr SM, Workman JL (2003) Two Drosophila Ada2 homologues function in different multi protein complexes. Mol Cell Biol 23(9):3305–3319
27. Pankotai T, Komonyi O, Bodai L, Ujfaludi Z, Muratoglu S, Ciurciu A, Tora L, Szabad J, Boros I (2005) The homologous Drosophila transcriptional adaptors ADA2a and ADA2b are both required for normal development but have different functions. Mol Cell Biol 25(18):8215–8227. doi:10.1128/MCB.25.18.8215-8227.2005, 25/18/8215 [pii]
28. Benecke A, Gaudon C, Garnier JM, Vom Baur E, Chambon P, Losson R (2002) ADA3-containing complexes associate with estrogen receptor alpha. Nucleic Acids Res 30(11): 2508–2514
29. Meng G, Zhao Y, Nag A, Zeng M, Dimri G, Gao Q, Wazer DE, Kumar R, Band H, Band V (2004) Human ADA3 binds to estrogen receptor (ER) and functions as a coactivator for ER-mediated transactivation. J Biol Chem 279(52):54230–54240. doi:10.1074/jbc.M404482200, M404482200 [pii]
30. Germaniuk-Kurowska A, Nag A, Zhao X, Dimri M, Band H, Band V (2007) Ada3 requirement for HAT recruitment to estrogen receptors and estrogen-dependent breast cancer cell proliferation. Cancer Res 67(24):11789–11797. doi:10.1158/0008-5472.CAN-07-2721, 67/24/11789 [pii]
31. Zeng M, Kumar A, Meng G, Gao Q, Dimri G, Wazer D, Band H, Band V (2002) Human papilloma virus 16 E6 onco-protein inhibits retinoic X receptor-mediated transactivation by targeting human ADA3 coactivator. J Biol Chem 277(47):45611–45618. doi:10.1074/jbc.M208447200M208447200 [pii]
32. Li CW, Ai N, Dinh GK, Welsh WJ, Chen JD (2010) Human ADA3 regulates RAR alpha transcriptional activity through direct contact between LxxLL motifs and the receptor coactivator pocket. Nucleic Acids Res 38(16):5291–5303. doi:10.1093/nar/gkq269, gkq269 [pii]
33. Zhao Y, Lang G, Ito S, Bonnet J, Metzger E, Sawatsubashi S, Suzuki E, Le Guezennec X, Stunnenberg HG, Krasnov A, Georgieva SG, Schule R, Takeyama K, Kato S, Tora L, Devys D (2008) A TFTC/STAGA module mediates histone H2A and H2B deubiquitination, coactivates nuclear receptors, and counteracts heterochromatin silencing. Mol Cell 29(1):92–101. doi:10.1016/j.molcel.2007.12.011, S1097-2765(07)00885-4 [pii]
34. Wang T, Kobayashi T, Takimoto R, Denes AE, Snyder EL, El-Deiry WS, Brachmann RK (2001) hADA3 is required for p53 activity. EMBO J 20(22):6404. doi:10.1093/emboj/20.22.6404
35. Nag A, Germaniuk-Kurowska A, Dimri M, Sassack MA, Gurumurthy CB, Gao Q, Dimri G, Band H, Band V (2007) An essential role of human Ada3 in p53 acetylation. J Biol Chem 282(12):8812–8820. doi:10.1074/jbc.M610443200, M610443200 [pii]
36. Yang M, Waterman ML, Brachmann RK (2008) hADA2a and hADA3 are required for acetylation, transcriptional activity and proliferative effects of beta-catenin. Cancer Biol Ther 7(1):120–128, doi:5197 [pii]
37. Nguyen DX, Baglia LA, Huang SM, Baker CM, McCance DJ (2004) Acetylation regulates the differentiation-specific functions of the retinoblastoma protein. EMBO J 23(7):1609–1618. doi:10.1038/sj.emboj, 76001767600176 [pii]
38. Martinez-Balbas MA, Bauer UM, Nielsen SJ, Brehm A, Kouzarides T (2000) Regulation of E2F1 activity by acetylation. EMBO J 19(4):662–671. doi:10.1093/emboj/19.4.662

39. Marzio G, Wagener C, Gutierrez MI, Cartwright P, Helin K, Giacca M (2000) E2F family members are differentially regulated by reversible acetylation. J Biol Chem 275(15): 10887–10892
40. Bolden JE, Peart MJ, Johnstone RW (2006) Anticancer activities of histone deacetylase inhibitors. Nat Rev Drug Discov 5(9):769–784. doi:10.1038/nrd2133, nrd2133 [pii]
41. Syntichaki P, Thireos G (1998) The Gcn5.Ada complex potentiates the histone acetyltransferase activity of Gcn5. J Biol Chem 273(38):24414–24419
42. Balasubramanian R, Pray-Grant MG, Selleck W, Grant PA, Tan S (2002) Role of the Ada2 and Ada3 transcriptional coactivators in histone acetylation. J Biol Chem 277(10):7989–7995. doi:10.1074/jbc, M110849200 M110849200 [pii]
43. Gamper AM, Kim J, Roeder RG (2009) The STAGA subunit ADA2b is an important regulator of human GCN5 catalysis. Mol Cell Biol 29(1):266–280. doi:10.1128/MCB.00315-08, MCB.00315-08 [pii]
44. Mohibi S, Gurumurthy CB, Nag A, Wang J, Mirza S, Mian Y, Quinn M, Katafiasz B, Eudy J, Pandey S, Guda C, Naramura M, Band H, Band V (2012) Mammalian alteration/deficiency in activation 3 (Ada3) is essential for embryonic development and cell cycle progression. J Biol Chem 287(35):29442–29456. doi:10.1074/jbc.M112.378901, M112.378901 [pii]
45. Grau B, Popescu C, Torroja L, Ortuno-Sahagun D, Boros I, Ferrus A (2008) Transcriptional adaptor ADA3 of *Drosophila melanogaster* is required for histone modification, position effect variegation, and transcription. Mol Cell Biol 28(1):376–385. doi:10.1128/MCB.01307-07, MCB.01307-07 [pii]
46. Martens JA, Winston F (2003) Recent advances in understanding chromatin remodeling by Swi/Snf complexes. Curr Opin Genet Dev 13(2):136–142, doi:S0959437X03000224 [pii]
47. Saha A, Wittmeyer J, Cairns BR (2006) Chromatin remodelling: the industrial revolution of DNA around histones. Nat Rev Mol Cell Biol 7(6):437–447. doi:10.1038/nrm1945, nrm1945 [pii]
48. Cosma MP, Tanaka T, Nasmyth K (1999) Ordered recruitment of transcription and chromatin remodeling factors to a cell cycle- and developmentally regulated promoter. Cell 97(3):299–311, doi:S0092-8674(00)80740-0 [pii]
49. Mitra D, Parnell EJ, Landon JW, Yu Y, Stillman DJ (2006) SWI/SNF binding to the HO promoter requires histone acetylation and stimulates TATA-binding protein recruitment. Mol Cell Biol 26(11):4095–4110. doi:10.1128/MCB.01849-05, 26/11/4095 [pii]
50. Hassan AH, Prochasson P, Neely KE, Galasinski SC, Chandy M, Carrozza MJ, Workman JL (2002) Function and selectivity of bromodomains in anchoring chromatin-modifying complexes to promoter nucleosomes. Cell 111(3):369–379, doi:S009286740201005X [pii]
51. Chandy M, Gutierrez JL, Prochasson P, Workman JL (2006) SWI/SNF displaces SAGA-acetylated nucleosomes. Eukaryot Cell 5(10):1738–1747. doi:10.1128/EC.00165-06, 5/10/1738 [pii]
52. Kim JH, Saraf A, Florens L, Washburn M, Workman JL (2010) Gcn5 regulates the dissociation of SWI/SNF from chromatin by acetylation of Swi2/Snf2. Genes Dev 24(24):2766–2771. doi:10.1101/gad.1979710, 24/24/2766 [pii]
53. Suganuma T, Gutierrez JL, Li B, Florens L, Swanson SK, Washburn MP, Abmayr SM, Workman JL (2008) ATAC is a double histone acetyltransferase complex that stimulates nucleosome sliding. Nat Struct Mol Biol 15(4):364–372. doi:10.1038/nsmb.1397, nsmb.1397 [pii]
54. McKenna NJ, O'Malley BW (2002) Combinatorial control of gene expression by nuclear receptors and coregulators. Cell 108(4):465–474, doi:S0092867402006414 [pii]
55. York B, O'Malley BW (2010) Steroid receptor coactivator (SRC) family: masters of systems biology. J Biol Chem 285(50):38743–38750. doi:10.1074/jbc.R110.193367, R110.193367 [pii]
56. McInerney EM, Rose DW, Flynn SE, Westin S, Mullen TM, Krones A, Inostroza J, Torchia J, Nolte RT, Assa-Munt N, Milburn MV, Glass CK, Rosenfeld MG (1998) Determinants of coactivator LXXLL motif specificity in nuclear receptor transcriptional activation. Genes Dev 12(21):3357–3368

57. He B, Wilson EM (2003) Electrostatic modulation in steroid receptor recruitment of LXXLL and FXXLF motifs. Mol Cell Biol 23(6):2135–2150
58. O'Malley BW, Kumar R (2009) Nuclear receptor coregulators in cancer biology. Cancer Res 69(21):8217–8222. doi:10.1158/0008-5472.CAN-09-2223, 0008-5472.CAN-09-2223 [pii]
59. Henriksson A, Almlof T, Ford J, McEwan IJ, Gustafsson JA, Wright AP (1997) Role of the Ada adaptor complex in gene activation by the glucocorticoid receptor. Mol Cell Biol 17(6):3065–3073
60. Vom Baur E, Harbers M, Um SJ, Benecke A, Chambon P, Losson R (1998) The yeast Ada complex mediates the ligand-dependent activation function AF-2 of retinoid X and estrogen receptors. Genes Dev 12(9):1278–1289
61. Oren M (2003) Decision making by p53: life, death and cancer. Cell Death Differ 10(4):431–442. doi:10.1038/sj.cdd, 44011834401183 [pii]
62. Berger SL, Cress WD, Cress A, Triezenberg SJ, Guarente L (1990) Selective inhibition of activated but not basal transcription by the acidic activation domain of VP16: evidence for transcriptional adaptors. Cell 61(7):1199–1208, doi:0092-8674(90)90684-7 [pii]
63. Candau R, Scolnick DM, Darpino P, Ying CY, Halazonetis TD, Berger SL (1997) Two tandem and independent sub-activation domains in the amino terminus of p53 require the adaptor complex for activity. Oncogene 15(7):807–816. doi:10.1038/sj.onc.1201244
64. Kumar A, Zhao Y, Meng G, Zeng M, Srinivasan S, Delmolino LM, Gao Q, Dimri G, Weber GF, Wazer DE, Band H, Band V (2002) Human papillomavirus oncoprotein E6 inactivates the transcriptional coactivator human ADA3. Mol Cell Biol 22(16):5801–5812
65. Gu W, Roeder RG (1997) Activation of p53 sequence-specific DNA binding by acetylation of the p53 C-terminal domain. Cell 90(4):595–606, doi:S0092-8674(00)80521-8 [pii]
66. Sakaguchi K, Herrera JE, Saito S, Miki T, Bustin M, Vassilev A, Anderson CW, Appella E (1998) DNA damage activates p53 through a phosphorylation-acetylation cascade. Genes Dev 12(18):2831–2841
67. Gamper AM, Roeder RG (2008) Multivalent binding of p53 to the STAGA complex mediates coactivator recruitment after UV damage. Mol Cell Biol 28(8):2517–2527. doi:10.1128/MCB.01461-07, MCB.01461-07 [pii]
68. Buryskova M, Pospisek M, Grothey A, Simmet T, Burysek L (2004) Intracellular interleukin-1alpha functionally interacts with histone acetyltransferase complexes. J Biol Chem 279(6):4017–4026. doi:10.1074/jbc, M306342200 M306342200 [pii]
69. Yin H, Morioka H, Towle CA, Vidal M, Watanabe T, Weissbach L (2001) Evidence that HAX-1 is an interleukin-1 alpha N-terminal binding protein. Cytokine 15(3):122–137. doi:10.1006/cyto.2001.0891, S1043-4666(01)90891-9 [pii]
70. Nurse P (2000) A long twentieth century of the cell cycle and beyond. Cell 100(1):71–78, doi:S0092-8674(00)81684-0 [pii]
71. Orpinell M, Fournier M, Riss A, Nagy Z, Krebs AR, Frontini M, Tora L (2010) The ATAC acetyl transferase complex controls mitotic progression by targeting non-histone substrates. EMBO J 29(14):2381–2394, doi:emboj2010125 [pii]
72. Burgess RJ, Zhou H, Han J, Zhang Z (2010) A role for Gcn5 in replication-coupled nucleosome assembly. Mol Cell 37(4):469–480. doi:10.1016/j.molcel.2010.01.020, S1097-2765(10)00071-7 [pii]
73. Paolinelli R, Mendoza-Maldonado R, Cereseto A, Giacca M (2009) Acetylation by GCN5 regulates CDC6 phosphorylation in the S phase of the cell cycle. Nat Struct Mol Biol 16(4):412–420. doi:10.1038/nsmb.1583, nsmb.1583 [pii]
74. Van Attikum H, Gasser SM (2009) Crosstalk between histone modifications during the DNA damage response. Trends Cell Biol 19(5):207–217. doi:10.1016/j.tcb.2009.03.001, S0962-8924(09)00060-9 [pii]
75. Ferreiro JA, Powell NG, Karabetsou N, Mellor J, Waters R (2006) Roles for Gcn5p and Ada2p in transcription and nucleotide excision repair at the Saccharomyces cerevisiae MET16 gene. Nucleic Acids Res 34(3):976–985. doi:10.1093/nar/gkj501, 34/3/976 [pii]

76. Martinez E, Palhan VB, Tjernberg A, Lymar ES, Gamper AM, Kundu TK, Chait BT, Roeder RG (2001) Human STAGA complex is a chromatin-acetylating transcription coactivator that interacts with pre-mRNA splicing and DNA damage-binding factors in vivo. Mol Cell Biol 21(20):6782–6795. doi:10.1128/MCB.21.20.6782-6795.2001
77. Brand M, Moggs JG, Oulad-Abdelghani M, Lejeune F, Dilworth FJ, Stevenin J, Almouzni G, Tora L (2001) UV-damaged DNA-binding protein in the TFTC complex links DNA damage recognition to nucleosome acetylation. EMBO J 20(12):3187–3196. doi:10.1093/emboj/20.12.3187
78. Kim MK, Shin JM, Eun HC, Chung JH (2009) The role of p300 histone acetyltransferase in UV-induced histone modifications and MMP-1 gene transcription. PLoS One 4(3):e4864. doi:10.1371/journal.pone.0004864
79. Li X, Corsa CA, Pan PW, Wu L, Ferguson D, Yu X, Min J, Dou Y (2010) MOF and H4 K16 acetylation play important roles in DNA damage repair by modulating recruitment of DNA damage repair protein Mdc1. Mol Cell Biol 30(22):5335–5347. doi:10.1128/MCB.00350-10, MCB.00350-10 [pii]
80. Yamamoto T, Horikoshi M (1997) Novel substrate specificity of the histone acetyltransferase activity of HIV-1-Tat interactive protein Tip60. J Biol Chem 272(49):30595–30598
81. Kimura A, Horikoshi M (1998) Tip60 acetylates six lysines of a specific class in core histones in vitro. Genes Cells 3(12):789–800, doi:229 [pii]
82. Ikura T, Ogryzko VV, Grigoriev M, Groisman R, Wang J, Horikoshi M, Scully R, Qin J, Nakatani Y (2000) Involvement of the TIP60 histone acetylase complex in DNA repair and apoptosis. Cell 102(4):463–473, doi:S0092-8674(00)00051-9 [pii]
83. Mirza S, Katafiasz BJ, Kumar R, Wang J, Mohibi S, Jain S, Gurumurthy CB, Pandita TK, Dave BJ, Band H, Band V (2012) Alteration/deficiency in activation-3 (Ada3) plays a critical role in maintaining genomic stability. Cell Cycle 11(22):4266–4274. doi:22613[pii]10.4161/cc.22613
84. Patel D, Huang SM, Baglia LA, McCance DJ (1999) The E6 protein of human papillomavirus type 16 binds to and inhibits co-activation by CBP and p300. EMBO J 18(18):5061–5072. doi:10.1093/emboj/18.18.5061
85. Bernat A, Avvakumov N, Mymryk JS, Banks L (2003) Interaction between the HPV E7 oncoprotein and the transcriptional coactivator p300. Oncogene 22(39):7871–7881. doi:10.1038/sj.onc.1206896, 1206896 [pii]
86. Arany Z, Sellers WR, Livingston DM, Eckner R (1994) E1A-associated p300 and CREB-associated CBP belong to a conserved family of coactivators. Cell 77(6):799–800, doi:0092-8674(94)90127-9 [pii]
87. Eckner R, Ewen ME, Newsome D, Gerdes M, DeCaprio JA, Lawrence JB, Livingston DM (1994) Molecular cloning and functional analysis of the adenovirus E1A-associated 300-kD protein (p300) reveals a protein with properties of a transcriptional adaptor. Genes Dev 8(8):869–884
88. Eckner R, Ludlow JW, Lill NL, Oldread E, Arany Z, Modjtahedi N, DeCaprio JA, Livingston DM, Morgan JA (1996) Association of p300 and CBP with simian virus 40 large T antigen. Mol Cell Biol 16(7):3454–3464
89. Kulesza CA, Van Buskirk HA, Cole MD, Reese JC, Smith MM, Engel DA (2002) Adenovirus E1A requires the yeast SAGA histone acetyltransferase complex and associates with SAGA components Gcn5 and Tra1. Oncogene 21(9):1411–1422. doi:10.1038/sj.onc.1205201
90. Shuen M, Avvakumov N, Walfish PG, Brandl CJ, Mymryk JS (2002) The adenovirus E1A protein targets the SAGA but not the ADA transcriptional regulatory complex through multiple independent domains. J Biol Chem 277(34):30844–30851. doi:10.1074/jbc, M201877200 M201877200 [pii]
91. Ablack JN, Cohen M, Thillainadesan G, Fonseca GJ, Pelka P, Torchia J, Mymryk JS (2012) Cellular GCN5 is a novel regulator of human adenovirus E1A-conserved region 3 transactivation. J Virol 86(15):8198–8209. doi:10.1128/JVI.00289-12, JVI.00289-12 [pii]
92. Shamanin VA, Sekaric P, Androphy EJ (2008) hAda3 degradation by papillomavirus type 16 E6 correlates with abrogation of the p14ARF-p53 pathway and efficient immortalization of human mammary epithelial cells. J Virol 82(8):3912–3920. doi:10.1128/JVI.02466-07, JVI.02466-07 [pii]

93. Frank SR, Schroeder M, Fernandez P, Taubert S, Amati B (2001) Binding of c-Myc to chromatin mediates mitogen-induced acetylation of histone H4 and gene activation. Genes Dev 15(16):2069–2082. doi:10.1101/gad.906601
94. Park J, Kunjibettu S, McMahon SB, Cole MD (2001) The ATM-related domain of TRRAP is required for histone acetyltransferase recruitment and Myc-dependent oncogenesis. Genes Dev 15(13):1619–1624. doi:10.1101/gad.900101
95. Liu X, Tesfai J, Evrard YA, Dent SY, Martinez E (2003) c-Myc transformation domain recruits the human STAGA complex and requires TRRAP and GCN5 acetylase activity for transcription activation. J Biol Chem 278(22):20405–20412. doi:10.1074/jbc, M211795200 M211795200 [pii]
96. Hirst SK, Grandori C (2000) Differential activity of conditional MYC and its variant MYC-S in human mortal fibroblasts. Oncogene 19(45):5189–5197. doi:10.1038/sj.onc.1203904
97. Lang SE, McMahon SB, Cole MD, Hearing P (2001) E2F transcriptional activation requires TRRAP and GCN5 cofactors. J Biol Chem 276(35):32627. doi:10.1074/jbc, M102067200 M102067200 [pii]
98. Pickard A, Wong PP, McCance DJ (2010) Acetylation of Rb by PCAF is required for nuclear localization and keratinocyte differentiation. J Cell Sci 123(Pt 21):3718–3726. doi:10.1242/jcs.068924, jcs.068924 [pii]
99. Oishi H, Kitagawa H, Wada O, Takezawa S, Tora L, Kouzu-Fujita M, Takada I, Yano T, Yanagisawa J, Kato S (2006) An hGCN5/TRRAP histone acetyltransferase complex co-activates BRCA1 transactivation function through histone modification. J Biol Chem 281(1):20–26. doi:10.1074/jbc.M510157200, M510157200 [pii]
100. Iyer NG, Ozdag H, Caldas C (2004) p300/CBP and cancer. Oncogene 23(24):4225–4231. doi:10.1038/sj, onc.1207118 1207118 [pii]
101. Miller RW, Rubinstein JH (1995) Tumors in Rubinstein-Taybi syndrome. Am J Med Genet 56(1):112–115. doi:10.1002/ajmg.1320560125
102. Petrij F, Giles RH, Dauwerse HG, Saris JJ, Hennekam RC, Masuno M, Tommerup N, van Ommen GJ, Goodman RH, Peters DJ et al (1995) Rubinstein-Taybi syndrome caused by mutations in the transcriptional co-activator CBP. Nature 376(6538):348–351. doi:10.1038/376348a0
103. Li M, Luo RZ, Chen JW, Cao Y, Lu JB, He JH, Wu QL, Cai MY (2011) High expression of transcriptional coactivator p300 correlates with aggressive features and poor prognosis of hepatocellular carcinoma. J Transl Med 9:5. doi:10.1186/1479-5876-9-5, 1479-5876-9-5 [pii]
104. Yokomizo C, Yamaguchi K, Itoh Y, Nishimura T, Umemura A, Minami M, Yasui K, Mitsuyoshi H, Fujii H, Tochiki N, Nakajima T, Okanoue T, Yoshikawa T (2011) High expression of p300 in HCC predicts shortened overall survival in association with enhanced epithelial mesenchymal transition of HCC cells. Cancer Lett 310(2):140–147. doi:10.1016/j.canlet.2011.06.030, S0304-3835(11)00387-9 [pii]
105. Osborne CK, Bardou V, Hopp TA, Chamness GC, Hilsenbeck SG, Fuqua SA, Wong J, Allred DC, Clark GM, Schiff R (2003) Role of the estrogen receptor coactivator AIB1 (SRC-3) and HER-2/neu in tamoxifen resistance in breast cancer. J Natl Cancer Inst 95(5):353–361
106. Lahusen T, Henke RT, Kagan BL, Wellstein A, Riegel AT (2009) The role and regulation of the nuclear receptor co-activator AIB1 in breast cancer. Breast Cancer Res Treat 116(2):225–237. doi:10.1007/s10549-009-0405-2
107. Mirza S, Rakha EA, Alshareeda A, Mohibi S, Zhao X, Katafiasz BJ, Wang J, Gurumurthy CB, Bele A, Ellis IO, Green AR, Band H, Band V (2013) Cytoplasmic localization of alteration/deficiency in activation 3 (ADA3) predicts poor clinical outcome in breast cancer patients. Breast Cancer Res Treat. doi:10.1007/s10549-012-2363-3
108. Kaufman RJ (1999) Stress signaling from the lumen of the endoplasmic reticulum: coordination of gene transcriptional and translational controls. Genes Dev 13(10):1211–1233
109. Braakman I, Bulleid NJ (2011) Protein folding and modification in the mammalian endoplasmic reticulum. Annu Rev Biochem 80:71–99. doi:10.1146/annurev-biochem-062209-093836
110. Schroder M, Kaufman RJ (2005) The mammalian unfolded protein response. Annu Rev Biochem 74:739–789. doi:10.1146/annurev.biochem.73.011303.074134

111. Ron D, Walter P (2007) Signal integration in the endoplasmic reticulum unfolded protein response. Nat Rev Mol Cell Biol 8(7):519–529. doi:10.1038/nrm2199, nrm2199 [pii]
112. Zhang K, Kaufman RJ (2008) From endoplasmic-reticulum stress to the inflammatory response. Nature 454(7203):455–462. doi:10.1038/nature07203, nature07203 [pii]
113. Welihinda AA, Tirasophon W, Green SR, Kaufman RJ (1997) Gene induction in response to unfolded protein in the endoplasmic reticulum is mediated through Ire1p kinase interaction with a transcriptional coactivator complex containing Ada5p. Proc Natl Acad Sci USA 94(9):4289–4294
114. Welihinda AA, Tirasophon W, Kaufman RJ (1999) The cellular response to protein misfolding in the endoplasmic reticulum. Gene Expr 7(4–6):293–300
115. Welihinda AA, Tirasophon W, Kaufman RJ (2000) The transcriptional co-activator ADA5 is required for HAC1 mRNA processing in vivo. J Biol Chem 275(5):3377–3381
116. Nagy Z, Riss A, Romier C, le Guezennec X, Dongre AR, Orpinell M, Han J, Stunnenberg H, Tora L (2009) The human SPT20-containing SAGA complex plays a direct role in the regulation of endoplasmic reticulum stress-induced genes. Mol Cell Biol 29(6):1649–1660. doi:10.1128/MCB.01076-08, MCB.01076-08 [pii]
117. Sela D, Chen L, Martin-Brown S, Washburn MP, Florens L, Conaway JW, Conaway RC (2012) Endoplasmic reticulum stress-responsive transcription factor ATF6alpha directs recruitment of the Mediator of RNA polymerase II transcription and multiple histone acetyltransferase complexes. J Biol Chem 287(27):23035–23045. doi:10.1074/jbc.M112.369504, M112.369504 [pii]
118. Cabal GG, Genovesio A, Rodriguez-Navarro S, Zimmer C, Gadal O, Lesne A, Buc H, Feuerbach-Fournier F, Olivo-Marin JC, Hurt EC, Nehrbass U (2006) SAGA interacting factors confine sub-diffusion of transcribed genes to the nuclear envelope. Nature 441(7094):770–773. doi:10.1038/nature04752, nature04752 [pii]
119. Atanassov BS, Evrard YA, Multani AS, Zhang Z, Tora L, Devys D, Chang S, Dent SY (2009) Gcn5 and SAGA regulate shelterin protein turnover and telomere maintenance. Mol Cell 35(3):352–364. doi:10.1016/j.molcel.2009.06.015, S1097-2765(09)00426-2 [pii]
120. de Lange T (2005) Shelterin: the protein complex that shapes and safeguards human telomeres. Genes Dev 19(18):2100–2110. doi:10.1101/gad.1346005, 19/18/2100 [pii]
121. Dokmanovic M, Clarke C, Marks PA (2007) Histone deacetylase inhibitors: overview and perspectives. Mol Cancer Res 5(10):981–989. doi:10.1158/1541-7786.MCR-07-0324, 5/10/981 [pii]
122. Kim HJ, Bae SC (2011) Histone deacetylase inhibitors: molecular mechanisms of action and clinical trials as anti-cancer drugs. Am J Transl Res 3(2):166–179
123. Manzo F, Tambaro FP, Mai A, Altucci L (2009) Histone acetyltransferase inhibitors and preclinical studies. Expert Opin Ther Pat 19(6):761–774. doi:10.1517/13543770902895727
124. Bandyopadhyay K, Baneres JL, Martin A, Blonski C, Parello J, Gjerset RA (2009) Spermidinyl-CoA-based HAT inhibitors block DNA repair and provide cancer-specific chemo- and radiosensitization. Cell Cycle 8(17):2779–2788, doi:9416 [pii]
125. Coffey K, Blackburn TJ, Cook S, Golding BT, Griffin RJ, Hardcastle IR, Hewitt L, Huberman K, McNeill HV, Newell DR, Roche C, Ryan-Munden CA, Watson A, Robson CN (2012) Characterisation of a Tip60 specific inhibitor, NU9056, in prostate cancer. PLoS One 7(10):e45539. doi:10.1371/journal, pone.0045539 PONE-D-12-11619 [pii]

# Chapter 3
# RUNX2 Transcriptional Regulation in Development and Disease

**Jessica L. Brusgard and Antonino Passaniti**

**Abstract** RUNX2, a member of the Runt family of transcription factors, plays important roles in embryonic development to promote osteogenesis and angiogenesis. RUNX2 has been implicated in the promotion of disease, including cleidocranial dysplasia, in cancer progression, and in metastasis of breast and prostate tumors. Its aberrant expression in disease states may be the result of several mechanisms such as haploinsufficiency, mutation, or amplification. In osteogenesis and cancer progression, interactions with core-binding factor-β (Cbf-β) and other cofactors are responsible for the regulation of target gene expression including, but not limited to, VEGF, osteopontin, osteocalcin, MMPs, and BMPs. RUNX2 transcriptional function within cells is regulated by signal transduction events leading to activation of ERK, Smads, cdks, and Akt, which result in phosphorylation, DNA binding, and transcriptional activation or repression of target genes. Constitutive activation of signaling pathways in tumor cells results in aberrant expression and activation of RUNX2. Specific RUNX2 targeting agents, therefore, may bypass the effects of redundant signal transduction pathways within cancer cells and be an effective therapeutic strategy for treatment of RUNX2-positive cancer patients.

**Keywords** Runx2 • Osteoblast • Cancer • Metastasis • Transcription • Cell cycle • TGF-β • Vitamin D3

---

J.L. Brusgard, Ph.D. Student • A. Passaniti, Ph.D. (✉)
Department of Pathology, University of Maryland, Marlene and Stewart Greenebaum Cancer Center, Baltimore, MD, USA
e-mail: jbenn003@umaryland.edu; apass001@umaryland.edu

R. Kumar (ed.), *Nuclear Signaling Pathways and Targeting Transcription in Cancer*, Cancer Drug Discovery and Development, DOI 10.1007/978-1-4614-8039-6_3,

## Abbreviations

| | |
|---|---|
| 1,25-OH D3 | 1,25-OH Vitamin D3 |
| AA | Amino acid |
| Akt | v-akt murine thymomaviral oncogene homolog 1 (aka: protein kinase B or PKB) |
| AML | Acute myeloid leukemia |
| AR | Aldose reductase (Context dependent. Not to be confused with androgen receptor) |
| AR | Androgen receptor (Context dependent. Not to be confused with aldose reductase) |
| Atf6 | Activating transcription factor 6 |
| BMP | Bone morphogenetic protein |
| BMP-2 | Bone morphogenetic protein-2 |
| BR-DIM | 3,3′-diindolylmethane |
| BSP | Bone sialoprotein (aka: OPN) |
| C/EBPβ | CCAAT/enhancer binding protein beta |
| CBFα | Core-binding factor alpha |
| Cbf-β | Core-binding factor subunit beta |
| Cbfβ-SMMHC | Core binding factor β – smooth muscle myosin heavy chain |
| CBP | CREB-binding protein |
| CCD | Cleidocranial dysplasia |
| Cdk | Cyclin-dependent kinase |
| CML | Chronic myeloid leukemia |
| c-myc | v-myc avian myelocytomatosis viral oncogene homolog |
| CoAA | Co-activator activator |
| CSC | Cancer stem cell |
| CSF2 | Colony stimulating factor 2 |
| CST7 | Cystatin-7 (aka: cystatin-F or leukocystatin) |
| CTGF | Connective tissue growth factor |
| Cx43 | Connexin 43 |
| D-ELISA | Deoxyribonucleic acid binding-enzyme-linked immunosorbent assay |
| DNA | Deoxyribonucleic acid |
| EC | Endothelial cell |
| EMT | Epithelial mesenchymal transition |
| ERK | Extracellular signal-regulated kinase |
| ERα | Estrogen receptor alpha |
| EWS-FLI | Ewing's sarcoma breakpoint region 1 t(1122)(q24q12) |
| FGF2 | Fibroblast growth factor 2 (basic) |
| FGF2R | Fibroblast growth factor 2 receptor |
| FoxO1 | Forkhead box O1 |
| G9a | Histone-lysine N-methyltransferase H3 lysine 9 specific 3 (aka: euchromatic histone-lysine N-methyltransferase) |

| | |
|---|---|
| GPCR | G-protein coupled receptor |
| HBME | Human bone marrow endothelial cell |
| HDAC | Histone deacetylase |
| HG | Hyperglycemia |
| HIF1-α | Hypoxia inducible factor 1 alpha subunit |
| HUVEC | Human umbilical vein endothelial cell |
| IGF-1 | Insulin-like growth factor-1 |
| IGF-1R | Insulin-like growth factor 1 receptor |
| Ihh | Indian hedgehog |
| IL8 | Interleukin 8 |
| JNK | c-jun-N-terminal kinase |
| MAPK | Mitogen-activated protein kinase |
| MET | Mesenchymal epithelial transition |
| miRNA | Micro-ribonucleic acid |
| MLK3 | MAP3K mixed-lineage kinase 3 |
| MM | Multiple myeloma |
| MMP | Matrix metalloproteinase |
| MORF | Monocyte leukemia zinc finger related factor |
| MOZ | Monocyte leukemia zinc finger protein |
| mRNA | Messenger ribonucleic acid |
| mSin3a | Mammalian transcriptional corepressors Sin3a |
| NBIF | Neobavaisoflavone |
| NLS | Nuclear localization signal |
| NMTS | Nuclear matrix targeting signal |
| OPN | Osteopontin (aka: BSP) |
| $p21^{CIP1}$ | Cyclin-dependent kinase inhibitor 1A |
| p27 | Cyclin-dependent kinase inhibitor 1B |
| p300 | E1A binding protein p300 |
| p38 | p38 mitogen activated protein kinase |
| PEBP2α | Polyomavirus enhancer-binding protein 2 alpha |
| PI3K | Phosphoinositide-3-kinase |
| PIP | Prolactin-induced protein |
| PKC | Protein kinase C |
| PlxnA2 | Plexin A2 |
| PSA | Prostate specific antigen |
| PTEN | Phosphatase and tensin homolog |
| PTHrP | Parathyroid hormone-related protein |
| RANKL | Receptor activator of nuclear factor kappa-B ligand |
| RNA | Ribonucleic acid |
| Rorβ | Retinoid-related orphan receptor beta |
| rRNA | Ribosomal ribonucleic acid |
| RUNX1/2/3 | Runt-related transcription factor 1/2/3 |
| SDC2 | Syndecan 2 |
| SDF1 | Stromal cell-derived factor 1 |
| SH3PXD2A | SH3 and PX domain-containing protein 2A |

| | |
|---|---|
| shRNA | Small hairpin ribonucleic acid |
| siRNA | Small interfering ribonucleic acid |
| Smad | Sma and mad related proteins |
| Smurf | Smad specific E3 ubiquitin protein ligase |
| SNAI2 | Snail homolog 2/slug |
| SNP | Single nucleotide polymorphism |
| Sox9 | Sex determining region Y-box 9 transcription factor |
| Suv39h1 | Suppressor of variegation 3–9 homolog 1 |
| TAZ | Transcriptional coactivator with a PDZ-binding motif |
| TGF-β | Transforming growth factor beta |
| TLE-1 | Transducin-like enhancer protein 1 |
| TLE2/3 | Transducin-like enhancer protein 2/3 |
| UV | Ultraviolet |
| VDR | Vitamin D receptor |
| VEGF | Vascular endothelial growth factor |
| Wip1 | Protein phosphatase magnesium-dependent 1 delta |
| WWOX | WW domain containing oxidoreductase |
| WWP1 | WW domaincontaining E3 ubiquitin protein ligase 1 |
| YAP | Yes-associated protein |

## 3.1 Introduction

The *RUNX* genes are a family of transcription factors originally identified in *Drosophila* [1, 2]. There are three mammalian *RUNX* genes encoding the proteins RUNX1, RUNX2, and RUNX3. Historically, the literature contains many different nomenclatures for the mammalian RUNX family proteins. RUNX2(Runt related transcription factor 2) is also AML-3 (acute myeloid leukemia-3), PEBP2α (polyoma enhancer binding protein 2α), and CBFα (core binding factor α) [2–5]. The *RUNX* genes control many normal cellular processes including hematopoiesis (*RUNX1*), osteogenesis (*RUNX2*), and epithelial and neuronal development (*RUNX3*) [1]. *RUNX* genes encode several evolutionarily conserved proteins. There are four *RUNX* genes in zebra fish, four in *drosophila*, one in sea urchins, and one in *C. elegans* to name a few [2]. High conservation of proteins through evolutionary history generally suggests an extremely essential biological function.

The RUNX proteins are members of a heterodimeric complex composed of an α and β subunit. The *runt* domain genes contain the DNA binding α subunit [1]. The β subunit, consisting of core-binding factor-β (Cbf-β), binds to the runt domain within RUNX proteins (Fig. 3.1) to help stabilize RUNX-DNA interactions [1]. In addition, Cbf-β protects RUNX proteins from phosphorylation and degradation via the proteasome [1]. The RUNX proteins share many common protein domains (Fig. 3.1). The runt domain, which is responsible for binding DNA, is located in the N-terminus of the protein and is composed of 128 amino acids [3]. The runt domain is the most highly conserved domain among members of the RUNX family including the *Drosophila* ortholog *runt*. Located C-terminal to the runt domain is the nuclear

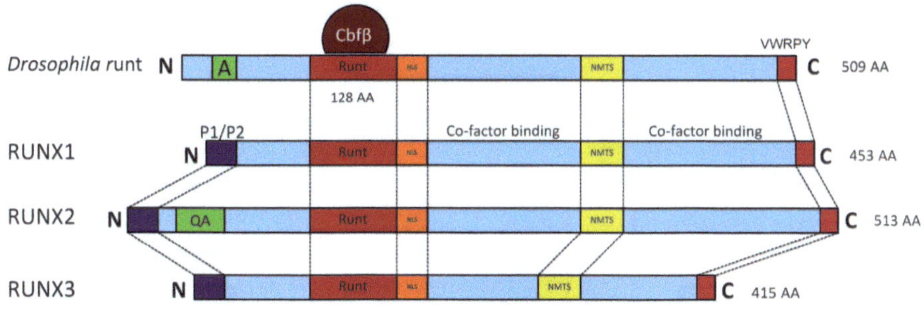


**Fig. 3.1 Mammalian RUNX isoforms protein structure**. The three mammalian RUNX proteins share domains with *drosophila runt*. N terminal (*purple*) are the P1 and P2 promoters. The Runt domain (*red*) is the DNA binding and Cbf-β interaction domain. C-terminal of the Runt domain (*orange*) is the nuclear localization sequence (NLS) controlling nuclear translocation of RUNX proteins. C-terminus (*light blue*) are a variety of sequences mediating co-factor binding to either activate or repress RUNX transcription of target genes. The nuclear matrix target sequence (NMTS; *yellow*) controls sub-nuclear localization of RUNX proteins

matrix targeting signal (NMTS) which is responsible for sub nuclear localization of RUNX proteins [1, 6]. This domain is comprised of 38 amino acids [5] folding into a loop-turn-loop tertiary structure [6]. Mutations within this domain reveal the essential nature of its function: without a functional NMTS, RUNX proteins cannot be transactivated or localized in foci within the nucleus [5]. Mutations within the second loop of the NMTS have been shown to inhibit RUNX2 interaction with the nuclear matrix [6] leading to compromised gene regulation. The nuclear localization signal/sequence (NLS) is comprised of 9 amino acids [5] and maintains RUNX localization to the nucleus. Furthermore, the C-terminus of RUNX proteins contains binding sites for corepressors and coactivators [1], which modulate RUNX activity. Depending on the cell stimulus or the cell type, different modulators of transcription are able to bind the RUNX proteins and either enhance or repress transcription.

All of the RUNX proteins have been implicated in disease (see Fig. 3.2 for diseases in which RUNX2 has been shown to play a role). Their normal functions are altered via mutations, epigenetic silencing, chromosomal translocation, cellular mislocalization, or by gene amplification. There is evidence that the RUNX proteins function as both tumor suppressors and as oncogenes depending on the disease context.

Although RUNX1 is not the largest protein isoform in mammals, it does encompass the largest genomic coding and regulatory region of 260 kb of DNA, comprised of 11 exons [4] encoding for 453 amino acids [1]. Within cells, RUNX1 functions to maintain normal hematopoiesis [1]. It is the target of numerous mutations and chromosomal translocations in hematological malignancies such as leukemia. In most reported cases of leukemia (specifically acute myeloid leukemia) where RUNX1 translocations are discovered, there appears to be a dominant negative function resulting from the new RUNX1-fusion protein. RUNX1 translocations and mutations are seen in acute myeloid leukemia (AML), blast crisis of chronic myeloid leukemia (CML), and acute lymphoblastic leukemia (ALL) [4]. There have been a few reports of gain of function mutations in RUNX1 as a result of an extra copy of

**Fig. 3.2 RUNX2 in development and disease.** Normal functions of RUNX2 (*yellow*) and diseases in which RUNX2 has been shown to play a role (*blue*)

the RUNX1 gene. This has been reported in Down Syndrome-related acute megakaryoblastic leukemia [4]. Knockout mice display a wide range of defects including, but not limited to, megakaryocyte defects, T-cell defects, myeloproliferative diseases, as well as T-cell lymphomas [2]. New studies have suggested a role for RUNX1 in endochondral ossification to mediate fracture healing in bone [7].

The smallest of the three mammalian RUNX proteins, RUNX3, is also thought to be the most primitive in evolutionary history, spanning 67 kb of DNA and composed of six exons [4] translating to 415 amino acids [1]. Expressed ubiquitously throughout the body, it can be found within the epithelia, mesenchyma, blood cells, dorsal root ganglion neurons, and predominantly in the gut epithelia [4]. RUNX3 has been shown to be essential for proper gut epithelial and neuronal development [8]. RUNX3 is essential for proprioceptive neuron axon path finding in the spinal cord [2] and there have also been reports of RUNX3 regulating CD4 silencing in T-cells [2]. Within the gut epithelia, research demonstrates that inactivation of RUNX3 leads to hyperplasia with a loss of response to transforming growth factor-β (TGF-β) inhibition [4] resulting in gastric cancers [8]. Inactivation has been described to occur through mutation, epigenetic silencing, hemizygous deletion, or cytoplasmic mislocalization.

Human RUNX2, located on the short arm of chromosome 6 at position 21 (6p21) [3], is the largest family member containing 513 amino acids [1] and has unique domains not present in the other mammalian isoforms (RUNX1 and RUNX3); one in the N-terminus and one in the C-terminus (see Fig. 3.1). RUNX2 is expressed early in embryonic development in mesenchymal stem cells [5]. During mouse embryogenesis, RUNX2 mRNA has been detected as early as E11.5 in the limb buds and the condensation of the humerus [5]. There is very weak expression observed as early as E9.5 in the notochord [5]. RUNX2 is the master regulator of osteoblast differentiation and chondrocyte maturation in a process called osteogenesis [1, 5]. RUNX2 controls the commitment of mesenchymal stem cells to the osteoblast lineage and has been shown to be abnormally expressed in adult tissues, leading to disease. Haploinsufficiency of RUNX2 promotes cleidocranial dysplasia (CCD) [9, 10]. In addition, there have been a few reports of RUNX2 mutations occurring within the runt domain which also result in CCD [5]. The oncogenic potential of RUNX2 was first identified from its ability to synergize with c-myc in T-cell lymphoma development [8, 127]. Aberrantly expressed RUNX2, normally at non-detectable to low levels in epithelial tissue, is thought to promote bone metastasis through activation of genes in malignancies such as breast and prostate cancer [8]. These target genes include, but are not limited to, vascular endothelial growth factor (VEGF), osteopontin (OPN), osteocalcin (OC), and matrix metalloproteinases (MMP's) [8]. The rest of this review will focus on RUNX2.

## 3.2 RUNX2: A Master Transcription Factor

### *3.2.1 Function in Osteogenesis and Angiogenesis*

Osteogenesis consists of intramembranous ossification (bone) and endochondral ossification (cartilage) [5]. Bone homeostasis is an important process that requires a balance between bone formation (osteoblasts) and resorption (osteoclasts). Osteoblasts are responsible for laying down new bone matrix in addition to the mineralization of the new bone matrix [11]. Osteoblasts also stimulate the differentiation of osteoclasts while osteoclasts produce factors which digest the mineralized bone matrix [11]. RUNX2 is the master regulator of osteoblast differentiation and osteogenesis [132]. RUNX2 expression is controlled by two promoters: P1 and P2 early in osteoblast differentiation [12]. As differentiation progresses RUNX2 protein levels do not increase, but rather, the transcriptional activity level increases [11]. Experiments in knockout mice show the essential role of RUNX2 in bone formation: knockout mice die soon after birth because of asphyxiation as a consequence of a lack of skeletal formation [5, 11]. In addition, analyses reveal that these mice lack mature osteoblasts thereby inhibiting the formation of any bone matrix or osteoclast differentiation.

Since osteoblasts lay down bone matrix to form mineralized bone and osteoclasts break down the matrix to resorb bone; RUNX2 indirectly controls osteoclast differentiation. Receptor activator of nuclear factor kappa-b ligand (RANKL) promotes osteoclast maturation and is also a RUNX2 target gene [5]. Cells lacking RUNX2 express less RANKL and, therefore, there is less osteoclast maturation [5] and less bone resorption. Research has further shown that while endochondral ossification is delayed in RUNX2 knockout mice, it does eventually occur [5, 130, 134]. Therefore, there is redundancy and other factors are able to compensate for the lack of RUNX2 during endochondral ossification.

The RUNX2 transcription factor is also a regulator of angiogenesis in bone development [13, 14], is expressed in vascularizing adult tissues [15], and promotes tumor metastasis [16, 17]. It interacts with its heterodimeric partner, Cbf-β, and with hypoxia-inducible factor 1-α (Hif1α) to activate the major angiogenic factor, VEGF [18]. RUNX2 is a transcriptional activator of specific target genes that promote angiogenesis, such as MMPs [19]. Conversely, it represses the cell cycle inhibitor $p21^{Cip1}$ and increases endothelial cell (EC) or cancer cell proliferation [20, 21]. Our laboratory has found that glucose metabolism, autocrine IGF-1 signaling, and phosphorylation by cyclin-dependent kinases, regulate RUNX2 DNA-binding activity, angiogenic target genes, EC proliferation, tube formation, and wound healing [20, 22–25, 138]. However, exposure of EC to hyperglycemia (HG) activated the aldose reductase (AR) polyol pathway, which increased oxidative stress and inhibited RUNX2 DNA binding [22].

### 3.2.2 RUNX2 in Disease

Although RUNX2 regulates osteogenesis and angiogenesis [1, 12], aberrant expression of RUNX2 can lead to disease (Fig. 3.2). Cleidocranial dysplasia (CCD) is a disease in which there are abnormalities in bone and dental development [5, 9, 26, 27]. Haploinsufficiency [1, 9] of RUNX2 has been shown to be a leading cause of CCD. In addition, there are mutations in RUNX2 that lead to CCD [26]. A heterozygous single-base deletion resulting in a premature stop codon in the runt domain produces a truncated form of RUNX2 [26] that is unable to bind DNA and control transcription of essential target genes. This mutation was not found in normal individuals or non-CCD subjects.

RUNX2 has been proposed as a biomarker in numerous cancers [1–4, 16, 17, 28, 29, 36–38, 41, 43, 45–49, 53, 54, 61, 94, 127–129, 131, 135] to evaluate the promotion of cancer cell metastasis to the bone. Sase et al. showed that RUNX2 expression was significantly associated with human colon carcinoma progression [28, 29]. In colorectal cancer RUNX2 is not only amplified, but the *RUNX2* gene also contains genetic variations termed single nucleotide polymorphisms or SNPs. In cases of colorectal carcinoma, Slattery et al. found a total of 19 SNPs in *RUNX2* [29].

One tumor in which both mRNA and protein levels of RUNX2 have been shown to be elevated is osteosarcoma [24, 30–35]. Osteosarcoma is a very aggressive pediatric cancer of the bone with a highly heterogeneous phenotype [32, 33, 35]. In osteosarcoma and chondrosarcomas, RUNX2 expression was found to positively correlate with bone morphogenetic protein-2 (BMP-2) mRNA levels [27]. Cell cycle deregulation of RUNX2 led to osteosarcoma pathogenesis [24]. High levels of RUNX2 in osteosarcoma resulted in high rates of metastasis and a poor survival rate, supporting the notion of RUNX2 as a good prognostic marker [35]. RUNX2 was also the only upregulated marker in osteosarcoma that exhibited a positive correlation with chemotherapeutic resistance [33].

RUNX2 expression correlates with unfavorable prognoses in prostate cancer [36] and has been found to be upregulated in both breast and prostate cancers [8, 16, 25, 37–46, 128, 129]. In breast and prostate cancer cell lines RUNX2 was shown to enhance cell motility [41]. Unfortunately, many of the RUNX2 functions are cell type specific making it difficult to discern a universal function in disease progression. In MDA-MB-231 breast cancer cells, knockdown of RUNX2 had no effect on cell growth and proliferation whereas in MCF7 breast cancer cells RUNX2 enhanced cell proliferation upon growth factor deprivation [41]. In MCF-10A cells RUNX2 disrupts normal mammary acini formation in suspension culture [47]. Using electron microscopy, an absence of lumen formation was noted, possibly due to an increase in cell proliferation, decreased apoptosis, and a loss of basement membrane formation which is dependent on RUNX2 expression [47]. These phenotypes can be reversed in MDA-MB-231 cells using siRNA to deplete RUNX2 [47]. The consensus of data supports an oncogenic function for RUNX2 in breast cancer. However, there is a report suggesting RUNX2 functions as a tumor suppressor in breast cancer [48]. There have been a few reports implicating RUNX2 in promoting the formation of hematological malignancies [131] such as myeloid leukemia [49]. For example, RUNX2 cooperates with the fusion protein, Cbf-β-SMMHC, to promote leukemia development [49] and haploinsufficiency of RUNX2 delays the onset of acute myeloid leukemia [49].

## 3.3 Transcriptional Regulation: Target Genes and Cofactors

### *3.3.1 RUNX2 Target Genes*

As a DNA-binding factor RUNX2 controls the transcription (through activation and repression) of numerous genes important in normal tissue homeostasis (Table 3.1). Many of these genes are abnormally activated in disease states, including cancers, and enable cancer cells to survive and metastasize to distant niches.

In osteoblasts, RUNX2 activates expression of the essential osteogenesis signaling factor: bone morphogenetic protein-2 (BMP-2) [50]. BMP-2 is a member of the transforming growth factor-β (TGF-β) superfamily of signaling molecules. It binds to TGF-β receptors, leading to activation of intracellular signaling via Smad-dependent and/or Smad-independent pathways ultimately resulting in further RUNX2 activation and promotion of a feed-forward loop. In MC3T3-E1 cells, BMP-2 was shown to enhance RUNX2 association with the promoter of Atf6 [51]. Atf6 is another transcription factor which is known to mediate osteoblast differentiation in a RUNX2 dependent manner [51]. Specifically, BMP-2-stimulated

**Table 3.1** Runx2 regulated genes

| Gene name | Gene symbol | Basic function | References |
|---|---|---|---|
| Bone morphogenetic protein-2 | BMP-2 | TGF-beta superfamily protein; cell signaling events; important for development of bone and cartilage | [50] |
| Snail1 | SNAI1 | Zinc finger transcription factor; promote and EMT through repression of E-Cadherin transcription | [99] |
| p27 (cyclin-dependent kinase inhibitor 1B) | CDKN1B | Controls G1 arrest; degradation is required to transition the cell from quiescence to proliferation | [80] |
| Bone sialoprotein 1 (osteopontin) | BSP1 (OPN) | Extracellular matrix component of the bone | [45, 59, 100] |
| Ribosomal ribonucleic acid | rRNA | RNA component of the ribosome; functions in protein synthesis | [57] |
| Matrix metalloproteinase 9 (gelatinase B) | MMP9 | Functions in extracellular matrix degradation | [36–38, 44, 100, 101] |
| Stromal cell-derived factor 1 | SDF1/CXCL12 | Cytokine/chemokine signaling factor functioning in migration and adhesion | [36, 38, 44] |
| Cystatin F | CST7 | Protease inhibitor; functions in immune regulation; promotes tumor metastasis | [38, 44] |
| Matrix metalloproteinase 13 | MMP13 | Functions in extracellular matrix degradation | [36, 37, 44, 52–55] |
| Prolactin-induced protein | PIP | Water transport; binds immunoglobulins; used as a biomarker in breast cancer; promotes cellular proliferation | [39, 44] |
| Vascular endothelial growth factor | VEGF | Promotes angiogenesis | [36, 37, 102, 103] |
| Insulin-like growth factor-binding protein 5 | IGFBP5 | Prolong the half-life of IGF's and alters IGF interactions with IGFR to promote cell survival, growth, and proliferation | [104] |
| Collagen type X alpha 1 | Col10a1 | Extracellular matrix component of the bone | [105] |
| Transcription factor Sox9 | Sox9 | Regulates transcription of genes important in osteogenesis | [36, 38] |
| Lipocalin-2 | LCN2 | Important for innate immunity; upregulated during inflammation; functions in apoptosis | [36] |
| Snail2 | SNAI2 | Zinc finger transcription factor; promotes oncogenic transformation of epithelial cells | [36, 38, 106] |
| Pepsinogen C (gastricsin) | PGC | Extracellular matrix degradation and invasiveness | [36] |
| Endothelin 2 | EDN2 | Important for angiogenesis | [36] |
| Receptor activator of nuclear factor kappa-b ligand | RANKL | Also called osteoclast differentiation factor; receptor that serves to activate osteoclasts; aids in metastasis to the bone | [36] |

| | | | |
|---|---|---|---|
| Parathyroid hormone-related protein | PTHrP | Regulator of cell growth, differentiation, migration, and survival; aids in bone metastasis | [36, 37, 64] |
| Interleukin 8 | IL8 | Cytokine/chemokine signaling; Potent activator of angiogenesis | [36, 37] |
| Sphingosine kinase 1 | SPHK1 | Regulates proliferation and survival; aids in metastasis to the bone | [36, 38] |
| Endothelial differentiation GPCR 3 | EDG3 | Receptor for sphingosine 1-phosphate; aids in metastasis to the bone | [36] |
| Kallikrein-2 | KLK2 | Cleaves PSA to active form; prognostic marker for prostate cancer risk | [36] |
| Ras small GTPase, Rab type | RAB3B | Modify membrane lipids and small GTPases of the Rab family; secretory pathway | [36] |
| Ras small GTPase, Rab type | RAB43 | Modify membrane lipids and small GTPases of the Rab family; secretory pathway | [36] |
| Ras small GTPase, Rab type | RAB35 | Modify membrane lipids and small GTPases of the Rab family; secretory pathway | [36] |
| Activating transcription factor 6 | ATF6 | Involved in the unfolded protein response; activates the transcription of ER chaperones | [51] |
| Bcl-2-associated X protein | Bax | Promotes apoptosis | [30, 107] |
| Exostosin-1 | Ext1 | Tumor suppressor; glycosyltransferase | [72] |
| Heparanase | HPSE | Helps degrade heparan sulfate; promotes tumor angiogenesis and metastasis | [72] |
| Bone morphogenetic protein - 3b | BMP-3B | Member of the TGF-beta superfamily of secreted molecules; aid in metastasis to the bone | [108] |
| Thyroglobulin | Tg | Used by thyroid to make T4 and T3; can be used as a biomarker in thyroid cancers | [109] |
| Cytochrome P450 11a1 | CYP11a1 | Converts cholesterol to pregnenolone; aids in cancer progression; sterol/steroid metabolism | [110] |
| Cytochrome P450 39a1 | CYP39a1 | Converts cholesterol to bile acids; aids in cancer progression; sterol/steroid metabolism | [110] |
| Cytochrome P450 51 | CYP51 | Converts lanosterol to cholesterol; aids in cancer progression; sterol/steroid metabolism | [110] |
| Lanosterol synthase | Lss | Functions in sterol/steroid metabolism | [110] |
| 7-dehydrocholesterol reductase | Dhcr7 | Functions in sterol/steroid metabolism | [110] |
| MicroRNA cluster | miR23a-27a-24-2 | MicroRNA cluster that promotes apoptosis; downregulated during tumor progression | [85] |
| Regulator of G-protein signaling 2 | Rgs2 | Functions in signaling; drug resistance; tumor angiogenesis | [111] |
| Regulator of G-protein signaling 4 | Rgs4 | Functions in signaling; drug resistance; tumor angiogenesis | [111] |
| Regulator of G-protein signaling 5 | Rgs5 | Functions in signaling; drug resistance; tumor angiogenesis | [111] |
| Regulator of G-protein signaling 16 | Rgs16 | Functions in signaling; drug resistance; tumor angiogenesis | [111] |

(continued)

**Table 3.1** (continued)

| Gene name | Gene symbol | Basic function | References |
|---|---|---|---|
| G-protein coupled receptor 23 | Gpr23 | Function in G-protein coupled signaling to activate cAMP signaling and PI3K to promote cell survival, growth, and proliferation | [111] |
| G-protein coupled receptor 30 | Gpr30 | Function in G-protein coupled signaling to activate cAMP signaling and PI3K to promote cell survival, growth, and proliferation | [111] |
| G-protein coupled receptor 54 | Gpr54 | Function in G-protein coupled signaling to activate cAMP signaling and PI3K to promote cell survival, growth, and proliferation | [111] |
| G-protein coupled receptor 64 | Gpr64 | Function in G-protein coupled signaling to activate cAMP signaling and PI3K to promote cell survival, growth, and proliferation | [111] |
| G-protein subunit alpha-13 | Gna13 | Function in G-protein coupled signaling to activate cAMP signaling and PI3K to promote cell survival, growth, and proliferation | [111] |
| Pannexin3 | PANX3 | Structural component of gap junctions | [112] |
| Survivin | BIRC5 | Member of the inhibitor of apoptosis family; functions in apoptosis; inhibit caspases | [113] |
| Galectin-3 | LGALS3 | Functions in cell adhesion, cell activation and chemoattraction, cell growth and differentiation, cell cycle, apoptosis | [114, 115] |
| Connective tissue growth factor | CTGF | Stimulates extracellular matrix production, chemotaxis, proliferation and integrin expression; angiogenesis | [56] |
| Lymphoid enhancer-binding factor 1 | Lef1 | Participates in Wnt signaling; promotes an EMT by repressing E-Cadherin | [116] |
| Dual specificity protein phosphatase 1 | DUSP1 | Negative regulator of cell proliferation | [38] |
| Osteoprotegerin | OPG | Cytokine receptor; decoy RANKL receptor; regulator of inflammation, innate immunity, and cell survival and differentiation | [114] |
| Aromatase | CYP19a1 | Functions in sterol/steroid metabolism | [100] |
| Osteocalcin | OC | Extracellular matrix component of the bone | [100, 114, 117, 118] |
| Type 1 collagen | COL1a2 | Extracellular matrix component | [100, 117] |
| Sclerostin | SOST | BMP antagonist; inhibits Wnt signaling; antagonizes bone formation | [42] |
| Interleukin 11 | IL-11 | Functions in cytokine/chemokine signaling; promotes bone cell proliferation and differentiation | [42] |
| Granulocyte-macrophage colony-stimulating factor | GM-CSF | Cytokine/chemokine signaling; regulates growth and differentiation | [36, 38, 44, 77] |

| | | | |
|---|---|---|---|
| Syndecan 1 | SDC1 | Integral cell membrane protein to participate in cell migration, proliferation and cell-matrix interactions | [72] |
| Syndecan 2 | SDC2 | Integral cell membrane protein to participate in cell migration, proliferation and cell-matrix interactions | [38, 72] |
| Syndecan 3 | SDC3 | Integral cell membrane protein to participate in cell migration, proliferation and cell-matrix interactions | [72] |
| Twinfilin | TWF | Regulations cytoskeletal dynamics via actin | [38] |
| SH3 and PX domain-containing protein 2A | SH3PXD2A | Involved in extracellular matrix degradation and cellular invasion | [38] |
| Alkaline phosphatase | ALP | Dephosphorylate proteins | [117] |
| Glypican | GPC | Modify signaling pathways and contribute to cell proliferation and tissue growth | [72] |
| Versican | VCAN | Functions in cell adhesion, migration, and proliferation | [72] |
| Axin2 | AXIN2 | Functions within the Wnt signaling pathway to maintain Beta-catenin | [119] |
| Smad 3 | SMAD3 | EMT and cell transformation | [38] |
| Dexamethasone-induced ras-related protein 1 | RASD1 | Plays a role in altered cell morphology, growth and extracellular matrix interactions | [38] |
| Fibroblast growth factor receptor 2 | FGFR2 | Receptor tyrosine kinase receptor that regulates cell signaling controlling cellular proliferation and differentiation | [72] |
| Fibroblast growth factor receptor 3 | FGFR3 | Receptor tyrosine kinase receptor that regulates cell signaling controlling cellular proliferation and differentiation | [72] |
| B-cell lymphoma-2 | Bcl-2 | Protects cells from apoptosis via maintaining mitochondrial membrane integrity; prevents cytochrome c release from the mitochondria | [8] |
| Prostate specific antigen | PSA | Protease to dissolve coagulum; serine protease; prostate cancer biomarker | [36] |
| Membrane type 1 matrix metalloproteinase | $MT_1MMP$ | Promotes tumor growth, tumor invasion, and angiogenesis; extracellular matrix protease | [1] |
| Cyclin D1 | CCND1 | Regulate CDK kinases during cell cycle; functions as a regulatory subunit for CDK4 and CDK6 during G1/S transition; contributes to tumorigenesis via altered cell cycle | [1] |
| Cyclin dependent kinase inhibitor 1A | p21 | Regulates cell cycle progression at G1 by inhibiting the activity of CDK4 and CDK2 | [1] |
| Cyclin dependent kinase inhibitor 2A | p14 | Induce cell cycle arrest at G1 and G2 phases; acts as a tumor suppressor | [1] |

RUNX2 activation of Atf6 enables the bone extracellular matrix, osteocalcin, gene to be expressed. A dominant negative construct of Atf6 was shown to inhibit RUNX2 activation of the osteocalcin promoter [51]. However, by restoring wild type Atf6, osteoblasts were able to differentiate and the expression of osteocalcin was restored.

During cancer progression, cancer cells express many bone specific proteins that mediate metastasis to the bone. Some of these proteins are regulated by RUNX2 transcriptional activity and mediate migration/motility and adhesion. Bone sialoprotein (BSP) and osteopontin (OPN) are two factors which mediate breast cancer metastasis to the bone and are activated by RUNX2 in breast cancer cells [45]. Using siRNA technology, Reufsteck et al. were able to demonstrate that targeting BSP and OPN drastically inhibits migration of MDA-MB-231 breast cancer cells when injected into athymic nude mice [45]. In prostate cancer, RUNX2 has been shown to upregulate genes not only associated with increased migration but genes associated with angiogenesis, epithelial-mesenchymal-transition (EMT), membrane trafficking/secretion, and osteolysis [36]. RUNX2 upregulates secretion factors including PTHrP, IL8, CSF2, and SDF-1 (See Table 3.1 for a description of protein function) [37, 38]. RUNX2 also upregulates MMP9, MMP13, VEGF, osteopontin, CST7, Sox9, SNAI2, Smad3, SDC2, Twinfilin, and SH3PXD2A [37, 38]. RUNX2 promoter occupancy in osteosarcoma was examined to determine potential RUNX2 target genes enabling progression of osteosarcoma. In SAOS-2 osteosarcoma cells, knockdown of RUNX2 resulted in an inhibition of motility [34]. RUNX2 upregulated matrix metalloproteinases during cancer progression to promote cell migration from primary tumor sites. MMP13 and MMP9 are two common MMP's upregulated by RUNX2 to mediate cancer cell invasion and metastasis [1, 52–55].

In addition to epithelial cells, RUNX2 has been shown to modulate gene expression in vascular cells. In human aortic smooth muscle cells and C3H10T1/2 cells RUNX2 inhibited the expression of connective tissue growth factor (CTGF) [56]. This is important for endothelial cells since CTGF has been shown to be a contributing factor to the development of atherosclerosis. Therefore, RUNX2 may protect the vasculature from development of atherosclerosis. Knockdown of RUNX2 enhanced CTGF expression [56] in a TGF-β dependent manner. Further analysis of the CTGF promoter revealed Smad binding elements that were able to interact with RUNX2/Smad3 heterodimers upon stimulation of the TGF-β receptors to inhibit transcription [56].

### *3.3.2 RUNX2 Cofactors*

Regulation of gene expression requires a transcriptional complex composed of RNA polymerase, transcription factors, and corepressors or coactivators. RUNX2 C-terminal domain contains many binding sites for both corepressors and coactivators (Table 3.2).

One group of proteins that function as strong corepressors of RUNX2 are histone deacetylases (HDACs). In osseous cells, HDAC1 has been shown to interact with

**Table 3.2** Runx2 cofactors

| Cofactor | References |
|---|---|
| CBP | [50] |
| p300 | [99] |
| HDAC1 | [57, 59] |
| G9a | [44] |
| HIF1α | [102] |
| p38 | [62] |
| SMAD | [56, 62, 63, 83, 111] |
| CBFβ | [25, 42, 120] |
| Rorβ | [121] |
| YAP | [21, 122] |
| TAZ | [91, 123] |
| C/EBPβ | [52] |
| Wip1 | [107] |
| WWP1 | [87] |
| ERα | [100, 129] |
| FOXO1 | [61] |
| AR | [39, 46, 124] |
| WWOX | [31, 125] |
| EWS-FLI | [126] |
| TLE1 | [58, 73] |
| CoAA | [119] |
| CBFβ-SMMHC | [49] |
| Gli2 | [64] |
| HDAC7 | [5, 60] |
| Suv39h1 | [108] |
| mSin3a | [122] |
| TLE2 | [122] |
| TLE3 | [122] |
| HDAC6 | [5, 122] |
| HDAC4 | [5, 122] |
| HDAC3 | [5, 122] |
| Smurf1 | [5] |
| Schnurri-3 | [5] |
| MOZ | [5] |
| MORF | [5] |
| HDAC5 | [5] |

RUNX2 to inhibit ribosomal RNA (rRNA) gene expression [57] thus inhibiting cellular proliferation and protein synthesis. Knockout of HDAC1 was shown to alleviate the RUNX2-mediated repression of rRNA expression resulting in an increase in cell proliferation and overall protein synthesis [57]. Transducin Like Enhancer-1 (TLE-1), functions to also promote RUNX2 inhibition of rRNA gene synthesis during mitosis [58]. In C3H10T1/2 cells HDAC1 bound to RUNX2 inhibited expression of osteopontin, thus inhibiting both proliferation and differentiation of the osteoblast cell [59]. HDAC7 was also shown to be a potent inhibitor of RUNX2 transcriptional activity [60] and HDAC5 was shown to repress RUNX2 expression [5].

In prostate cancer the forkhead box O (FoxO1) protein was found to be a corepressor of RUNX2 [61]. Inhibition of RUNX2 by upregulation of PTEN or FoxO1 protein inhibited prostate cancer cell migration and invasion. In prostate cancer specimens, immunohistochemistry revealed an inverse relationship between RUNX2 and FoxO1 nuclear localization [61]. PTEN inactivating mutations are often seen in prostate cancer [61] and therefore would potentiate RUNX2 activity to promote prostate cancer cell migration and invasion.

While some cofactors function as coactivators and corepressors a few are also able to function as both in a gene-dependent manner. One group used a doxycycline-regulated RUNX2 expression system in C4-2B prostate cancer cells to show that G9a (histone methyltransferase) is able to function as a corepressor for RUNX2 target genes MMP9, CSF2, SDF1, and CST7 [44]. However, G9a functions as a coactivator for RUNX2 transcription of MMP13 and PIP in the C4-2B prostate cancer cell line [44].

## 3.4 Transcriptional Regulation: Activation and Repression

### *3.4.1 Regulation of RUNX2 Activity*

Several intracellular signaling pathways have been shown to modulate RUNX2 activity (Fig. 3.3). When RUNX2 becomes activated it is able to bind DNA and either promote or inhibit the transcription of its target genes (Refer to Table 3.1). Mitogen-activated protein kinase (MAPK) and TGF-β intracellular signaling are both essential to RUNX2 activation to promote osteoblast differentiation or tumor progression. TGF-β/BMP signaling is a highly important signaling axis used by numerous cells to inhibit cell growth and proliferation under normal conditions. In cancer, tumor cells escape the inhibitory effects of TGF-β resulting in unrestricted proliferation. TGF-β/BMP intracellular signaling can occur via canonical Smad-dependent or non-canonical Smad-independent pathways. Upon stimulation, TGF-β receptors initiate intracellular signaling events leading to Smad activation (canonical TGF-β pathway), nuclear translocation, and Smad interaction with RUNX2 [6, 56, 62–64]. In osteogenesis, TGF-β/BMP signaling is responsible for activating RUNX2 to promote osteoblast differentiation [62, 65]. Conversely, in PC3 prostate cancer cells TGF-β cooperates with RUNX2 to promote cellular growth [46].

A potent regulator of RUNX2 activity is the extracellular signal-regulated kinase, ERK [66, 67, 138, 139]. ERK is a classical MAPK activated upstream in response to numerous extracellular factors including growth factors, cytokines, and G-protein coupled receptor ligands. MAPK signaling is essential for normal bone development. In many cancers where RUNX2 is aberrantly expressed, MAPK signaling is also altered generally through inactivating or activating mutations to components within the signaling cascade. In preosteoblast MC3T3-E1 cells it was demonstrated that S301 and S319 [66], two ERK phosphorylation sites on RUNX2, are

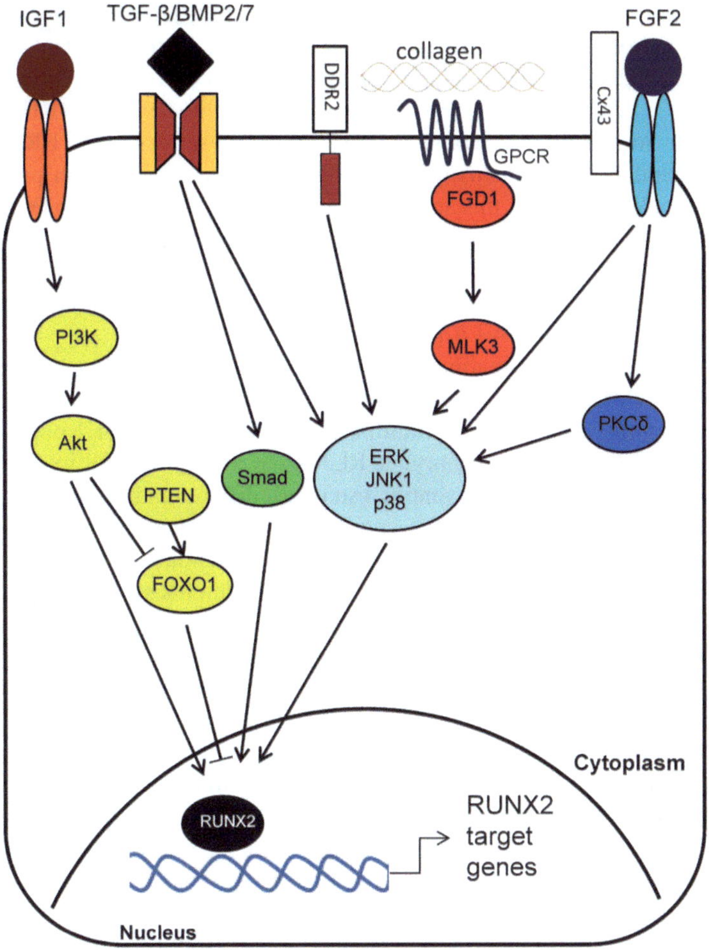


**Fig. 3.3 RUNX2 and signal transduction.** Many signaling pathways lead to activation or repression of RUNX2 transcriptional activity. These Pathways converge onto the MAPK pathway, the TGF-β/Smad pathway, and the PI3K/Akt pathway

phosphorylated upon nuclear translocation of ERK [67]. MAP3K mixed-lineage kinase 3 (MLK3) has been shown to activate ERK and p38 (another classical MAP kinase). Activation of ERK via MLK3 results in RUNX2 phosphorylation and subsequent osteoblast differentiation [68]. Insulin-like growth factor-1 (IGF-1) binding to IGF-1R (IGF-1 receptor tyrosine kinase) has been shown to be essential for normal skeletal development [69]. IGF-1 regulates the activation of RUNX2 via activation of ERK resulting in phosphorylation of RUNX2 [22, 69, 138]. p38, activated via several extracellular factors, has been implicated in modulating RUNX2 activity levels. p38 activation from TGF-β/BMP signaling (non-canonical TGF-β pathway) has been shown to mediate RUNX2 phosphorylation to promote osteoblast differentiation [62].

Fibroblast growth factor 2 (FGF2) is another extracellular growth factor implicated in promoting osteoblast differentiation through activation of RUNX2 [70–72]. In activating RUNX2, FGF2 functions as a double-edge sword. FGF2 activates protein kinase C (PKC), which in turn increases ERK activity, leading to RUNX2 phosphorylation in MC3T3 osteoblasts [71]. FGF2 also activates ERK through PKC-independent mechanisms (refer to Fig. 3.3). The gap junction protein, connexin-43 (Cx43), enhances this activation via stabilization of FGF2 to FGF2R (refer to Fig. 3.3). An inhibitor of gap junctions, 18β-glycyrrhetinic acid, attenuated the enhancement in RUNX2 transcriptional activity in MC3T3 osteoblasts [71]. In breast cancer MCF7 cells, FGF2 was able to increase BSP expression, which in turn upregulated RUNX2 mRNA [70].

Not all intracellular signaling pathways that lead to RUNX2 phosphorylation result in RUNX2 activation. The c-jun-N-terminal kinase (JNK) is activated by BMP-2 signaling in both C2C12 multipotent cells and MC3T3-E1 preosteoblastic cells [73]. JNK is able to phosphorylate RUNX2 at S104 [73] resulting in inhibition of RUNX2 activity and thus preventing osteoblast differentiation. Inhibition of JNK via a dominant negative JNK1, JNK knockdown, or treatment with a JNK inhibitor counteracted this inhibition to enable osteoblast differentiation.

RUNX2 has also been shown to play a role in endothelial signaling and cell cycle progression in response to physiological levels of glucose [22]. Euglycemic conditions were able to restore RUNX2 DNA binding through autocrine IGF1/IGFR signaling to promote endothelial cell migration, proliferation, and angiogenesis [22] indicating glucose has a function in modulating RUNX2 activity. Hyperglycemic conditions, however, inhibited RUNX2 activity through the aldose reductase polyol pathway [22]. Inhibition of RUNX2 by hyperglycemia inhibited endothelial cell migration, proliferation, and angiogenesis [20, 22]. Treatment with 2-deoxyglucose (inhibitor of glucose metabolism) under euglycemic growth conditions in endothelial cells resulted in a delayed exit from G1/S into G2 phases of the cell cycle with subsequent lower levels of RUNX2 DNA binding activity [20]. Similarly, nutrient and serum deprivation blocked endothelial cell exit from G1 [20]. Using shRNA lentiviral knockdown to reduce RUNX2 levels, exit from G1 and progression through the cell cycle was found to be dependent upon RUNX2 [20].

### *3.4.2 Regulation of RUNX2 Expression*

RUNX2 expression in osteoblasts promotes differentiation through tight control of RUNX2 activity to maintain the balance of bone formation and remodeling. Similarly, during tumorigenesis many of the negative regulators of RUNX2 are inhibited, thus allowing sustained expression and activity. In general, RUNX2 expression and DNA-binding activity depend on transcriptional mechanisms of activation versus repression and on post-translational protein modifications that include phosphorylation, acetylation, and ubiquitination [74].

#### 3.4.2.1 Transcriptional Activation and Repression

Osteoblast differentiation requires activation of RUNX2, but what regulates RUNX2 expression in pre-osteoblasts has remained an enigma for quite some time. Tu et al. determined that activation of RUNX2 expression is dependent upon Indian hedgehog (Ihh) signaling [75]. In Ihh null mice even the forced expression of RUNX2 failed to induce osteoblast differentiation [75] confirming that Ihh is not only important for activation of RUNX2 expression but also for RUNX2 activity. In osteoblasts, it was found that BMP-2 increased RUNX2 induction and expression [76]. Shu et al. generated BMP-2 knockout mice and found a dramatic reduction in RUNX2 expression resulting in severe chondrodysplasia [76]. BMP-2 was also able to upregulate PlexinA2 (PlxnA2) in pre-osteoblastic cells [77]. The upregulation of PlxnA2 was associated with increased RUNX2 expression, osteoblast differentiation, and bone mineralization [77]. This upregulation of RUNX2 was thought to be a result of PlxnA2 stabilization of BMP-2 binding to BMP-2 receptors.

TWIST, a transcription factor implicated in epithelial-mesenchymal-transition (EMT) and metastasis in many types of cancers is also an essential transcription factor in development. TWIST was shown to be an inhibitor of RUNX2 expression in bone marrow-derived mesenchymal stem cells [78]. Under low oxygen conditions (hypoxia), Hif1α upregulated expression of TWIST, which directly inhibited RUNX2 expression [78] and osteoblast differentiation. Conversely, studies in PC3 prostate cancer cells showed that TWIST enhances RUNX2 expression levels to promote metastasis to the bone [79]. Glucocorticoid receptor binding to the P2 promoter of RUNX2 was shown to inhibit RUNX2 expression [80]. This resulted in adipocyte differentiation [80] instead of osteoblast differentiation and reveals how signaling events can lead to cell fate determination by regulating RUNX2 expression.

RUNX2 expression is increased in breast tumors where it has been implicated in mediating metastasis to the bone. How RUNX2 expression is enhanced in breast cancer tissue is poorly understood. Using breast cancer cells, one group was able to show that serotonin induced parathyroid hormone related protein (PTHrP), which in turn increased transcription of RUNX2 [40]. Since PTHrP is also a RUNX2 target gene, it fuels a feed forward loop potentiating maximal RUNX2 expression to promote progression of breast cancer to a metastatic stage. In multiple myeloma (MM) it has been shown that Gfi1 is upregulated and represses RUNX2 expression [81]. This inhibition of RUNX2 then results in inhibition of osteoblast differentiation.

#### 3.4.2.2 Post-translational Regulation

Intracellular signaling is able to modulate RUNX2 post-translational modifications to regulate DNA-binding activity and RUNX2 levels. Using antibodies to detect phosphoserine sites on RUNX2 in endothelial cells, it was shown that RUNX2 is phosphorylated under euglycemic growth conditions by cyclin dependent kinase 4 (cdk4). This phosphorylation was abrogated by mutation of the cdk site S451 [20].

A RUNX2-S451A mutant showed inhibition of DNA binding in endothelial cells as well as a reduction in wound healing activity [20]. Inhibition of cdk4 produced similar results demonstrating that cdk4 can activate RUNX2 through phosphorylation of S451 in response to glucose. In addition to glucose modulation, RUNX2 activity was cell cycle regulated [136, 137]. RUNX2 was associated with DNA when cells were proliferating but was sequestered to subnuclear loci when cells were quiescent [20, 23, 136]. RUNX2 protein levels were maximal in endothelial cells in late G2 and M phases of the cell cycle [23]. Using RNA interference to knockdown RUNX2, endothelial cell exit from G2/M phases of the cell cycle was delayed [23] resulting in a decrease in cell proliferation. In vitro kinase assays showed S451 must be phosphorylated to allow RUNX2 to function in promoting progression through the cell cycle [23]. However, in osteoblasts, RUNX2 inhibited osteoblast proliferation and RUNX2 protein levels were maximal in G1 [24], suggesting that RUNX2 regulation may be cell type dependent.

BMP2 is able to regulate RUNX2 protein levels via inhibition of cdk4 [76]. This inhibition leads to protection from proteasomal degradation thus maintaining cellular protein levels. Recent research has focused on how micro-RNA's (miRNAs) regulate RUNX2 protein levels. MiRNAs modulate protein levels through an RNA interference pathway ultimately leading to mRNA degradation or reduced protein translation. MiRNAs are 18–25 nucleotide RNAs that repress translational activity of mRNAs [82]. Wu et al. have suggested that the miR-30 family of miRNAs may play an essential role in osteogenesis. Their data show that miR-30 was able to negatively regulate both Smad1 and RUNX2 [82]. Alkaline phosphatase (marker of osteoblast differentiation) was shown to be dramatically decreased after exogenous miR-30 expression [82]. Furthermore, miR-30 family miRNAs were able to bind to the 3′-untranslated region of both Smad1 and RUNX2 mRNA [82] thus inhibiting the effects of BMP-2-stimulated osteoblast differentiation pathways. MiR-203 is a known tumor suppressor miRNA which is downregulated in prostate cancer [83, 84]. It has been shown to bind RUNX2 mRNA resulting in a mesenchymal to epithelial transition (MET), inhibition of cell proliferation, and inhibition of cell migration and invasion [83, 84]. In addition to the miRNAs already described, it has been shown that miR-23a, miR-34 cluster, miR-133a, miR-135a, miR-137, miR-204, miR-205, miR-217, miR-218, and miR-338 all regulate RUNX2 expression [85, 86]. In addition to miRNA regulation of RUNX2, the proteasomal degradation pathway is implicated in regulating RUNX2 levels in the cell. For example, WWP1 (WW domain-containing E3 ubiquitin protein ligase 1) has been shown to function as the E3 ubiquitin ligase responsible for ubiquitinating RUNX2 and targeting it for proteasomal degradation [87].

## 3.5 RUNX2 as a Therapeutic Target

RUNX2 is a transcriptional regulator of gene expression. Mutations, amplification, or inappropriate expression of RUNX2 has the potential to amplify the expression or repression of a variety of target genes. This could regulate global changes in gene regulatory networks and lead to a process called transformation amplification. Therefore, RUNX2 may be a prime target for therapeutic intervention [88, 128] to treat disease because multiple transformation pathways could be inhibited. Many cancers develop resistance to therapies due to signaling pathway redundancy allowing signaling switches to occur. Being able to target transcription factors allows therapies to bypass the redundancy of signaling.

Prostate cancer bone metastases form osteolytic lesions before the development of osteoblastic lesions. Li et al. found that isoflavone and 3, 3′-diindolylmethone (BR-DIM) are able to multifunctionally inhibit these metastases from forming [88]. This combination therapy may inhibit not only osteoblast differentiation but also osteoclast differentiation. Their research showed that one of the ways in which this combination treatment functions is to inhibit signaling of RANKL [88]. However, further cellular analysis revealed isoflavone and BR-DIM combination therapy inhibited signaling from the Akt, AR (androgen receptor), PSA, and p27 signaling axis as well as blocking the RNA interference pathway by inhibiting miR-92a which is associated with RANKL signaling [88]. This is one example of inhibiting the effects of RUNX2 through targeting of its target genes.

Natural compounds have been shown to have potent medicinal benefits. Astragaloside II a compound from the plant, *Radix astragalus*, was tested on rat primary osteoblasts to determine its effects on viability, proliferation, differentiation and maturation [89]. Astragaloside II promoted proliferation, differentiation and mineralization of primary rat osteoblasts [89]. The effects of this drug on post-menopausal women could potentially prevent osteoporosis. One benefit in cancer patients could be prevention of bone fractures associated with osteolytic metastases by promoting osteoblast differentiation to prevent bone degradation caused by cancer cells. However, use of this drug for patients with metastatic cancers could increase the incidence of bone metastases and lead to early death by stimulating RUNX2 expression within tumor cells. A second compound, Neobavaisoflavone (NBIF), was isolated from the plant *Psoralea corylifolia L* and was shown to have a similar effect on RUNX2 in osteoblasts as Astragaloside II [90]. NBIF was shown to upregulate RUNX2 expression in MC3T3-E1 cells while also activating its gene regulatory functions [90]. NBIF was shown to upregulate osteocalcin, bone sialoprotein, and type 1 collagen [90]. While this drug may have pro-bone forming functions and could potentially be a way to restore bone loss as a result of bone degradative diseases, one must also take into consideration the dosing regimen that would make this drug specific for osteoblasts while not further stabilizing RUNX2 positive cancer cells.

An understanding of upstream signaling pathways that activate RUNX2 and how cofactors regulate RUNX2 activity in disease would improve the development of therapeutics. In MM it was shown that the Gfi1 targeted drug, Trichostatin-A, was able to block the inhibition that Gfi1 imposes on RUNX2 [81]. Bortezomib (Velcade) is a proteasomal inhibitor and is also used in the treatment of MM because it was shown to induce osteoblast differentiation [91]. Bortezomib inhibited FGF2 induced TAZ (a RUNX2 binding coactivator) protein degradation [91], thus allowing TAZ to interact with RUNX2 and promote osteoblast differentiation in MC3T3-E1 cells. Therefore, restoration of RUNX2 expression enabled osteoblast differentiation and disabled MM progression in the bone microenvironment.

Studies that show targeting upstream signaling pathways that activate RUNX2 or cofactors have been reported, but the research for direct RUNX2 inhibition is limited. Our laboratory showed for the first time that cholecalciferol (inactive Vitamin D3 precursor) directly modulates RUNX2 DNA-binding activity [25]. Cholecalciferol is produced in the skin as a result of exposure to UV or it can be absorbed in the digestive system through the diet [25]. Cholecalciferol is normally converted to active Vitamin D3 in the body to 1,25OH-D3, which interacts with Vitamin D Receptor (VDR) to promote calcium absorption in the gut and increase bone formation. However, Vitamin D3 also exhibits paracrine and autocrine activity by regulating epithelial cell differentiation and modulating immune system function [25]. Using a quantitative DNA binding assay (D-ELISA) it was shown that cholecalciferol was able to modulate RUNX2 DNA binding in a VDR-independent manner. Analysis of RUNX2-positive breast tumor cells (MCF7), endothelial cells (HBME), and osteosarcoma cells (SaOs2) showed that cholecalciferol was able to inhibit cellular proliferation [25], suggesting a RUNX2-specific function. Further research needs to be conducted to study the effects in animal models before this strategy could be modified for clinical trials in the treatment of RUNX2 positive tumors.

Direct targeting of epithelial or bone cell RUNX2 pathways and cofactors that modulate RUNX2 expression or activity is another therapeutic strategy. However, targeting the microenvironment is also another option. Angiogenesis is an essential process that must occur for many solid tumors to metastasize to bone. Endothelial cells express RUNX2, which mediates wound healing by stimulating new blood vessel formation. However, tumors use angiogenesis to vascularize and provide nutrients to tumors that also allow cancer cells to intravasate. Inhibition of angiogenesis using VEGF inhibitors has been a therapeutic strategy for many years. In theory if one starves tumors of their nutrient supply, it should lead to tumor regression or necrosis. Alternatively, by stabilizing blood vessels feeding the tumors then in theory chemotherapeutic agents would be able to get to sites of tumor growth more effectively. However, these therapies have been less successful in the clinic. LGD1069 is a selective retinoid X receptor ligand used to treat T-cell lymphoma but has also been shown to inhibit angiogenesis in lung cancer [92]. LGD1069 inhibited activation of the TGF-β/Smad pathway thus reducing both activation and expression of RUNX2 in human umbilical vein endothelial cells (HUVECs) [92]. However, endothelial cells are not the only cells that use the TGF-β/Smad pathway to activate the expression and activity of RUNX2. Therefore, LGD1069 activity in other cell lines (breast, prostate, and osteosarcoma) should be tested to determine whether it decreases RUNX2 protein or activity. In melanoma, it was shown that SD-208 (a TGF-β receptor I kinase inhibitor) blocked expression of RUNX2 through downregulation of TGF-β/Smad signaling [93], supporting the notion that targeting of the TGF-β/Smad axis could have therapeutic benefit in RUNX2 positive tumors by inhibiting expression within the tumor itself or by preventing angiogenesis.

## 3.6 Future Directions

The biological understanding of disease will be essential to creating new therapeutics to treat disease. Signaling pathways are difficult to ablate because of pathway redundancy. Therefore, it is essential to target upstream signaling as well as downstream signaling, especially of transcription factors. Understanding the role that RUNX2 plays in cancer progression will be essential to be able to use it as a therapeutic target to inhibit metastasis to the bone. RUNX2 has been shown to be highly upregulated in the cancer stem cell (CSC) populations of prostate cancer, breast cancer, and osteosarcoma [94–98]. Therefore, with the emergence of the CSC as a key contributor to cancer development, progression, and resistance to modern therapies, it will be important to understand the role that RUNX2 is playing in CSC regulation so that appropriate therapeutic strategies can be developed.

## References

1. Anglin I, Passaniti A (2004) Runx protein signaling in human cancers. Cancer Treat Res 119:189–215
2. Ito Y (2008) RUNX genes in development and cancer: regulation of viral gene expression and the discovery of RUNX family genes. Adv Cancer Res 99:33–76
3. Blyth K, Cameron ER, Neil JC (2005) The RUNX genes: gain or loss of function in cancer. Nat Rev Cancer 5(5):376–387
4. Ito Y (2004) Oncogenic potential of the RUNX gene family: 'overview'. Oncogene 23(24):4198–4208
5. Jonason JH, Xiao G, Zhang M, Xing L, Chen D (2009) Post-translational regulation of Runx2 in bone and cartilage. J Dent Res 88(8):693–703
6. Zaidi SK, Javed A, Pratap J, Schroeder TM, Westendorf JJ, Lian JB et al (2006) Alterations in intranuclear localization of Runx2 affect biological activity. J Cell Physiol 209(3):935–942
7. Soung DY, Talebian L, Matheny CJ, Guzzo R, Speck ME, Lieberman JR et al (2012) Runx1 dose-dependently regulates endochondral ossification during skeletal development and fracture healing. J Bone Miner Res 27(7):1585–1597
8. Pratap J, Lian JB, Stein GS (2011) Metastatic bone disease: role of transcription factors and future targets. Bone 48(1):30–36
9. Lou Y, Javed A, Hussain S, Colby J, Frederick D, Pratap J et al (2009) A Runx2 threshold for the cleidocranial dysplasia phenotype. Hum Mol Genet 18(3):556–568
10. Zhang X, Ting K, Pathmanathan D, Ko T, Chen W, Chen F et al (2012) Calvarial cleidocraniodysplasia-like defects with ENU-induced Nell-1 deficiency. J Craniofac Surg 23(1):61–66
11. Dalle Carbonare L, Innamorati G, Valenti MT (2011) Transcription factor Runx2 and its application to bone tissue engineering. Stem Cell Rev 8(3):891–897
12. Liu JC, Lengner CJ, Gaur T, Lou Y, Hussain S, Jones MD et al (2011) Runx2 protein expression utilizes the Runx2 P1 promoter to establish osteoprogenitor cell number for normal bone formation. J Biol Chem 286(34):30057–30070
13. Stein GS, Lian JB, van Wijnen AJ, Stein JL, Montecino M, Javed A et al (2004) Runx2 control of organization, assembly and activity of the regulatory machinery for skeletal gene expression. Oncogene 23(24):4315–4329
14. Zelzer E, Glotzer DJ, Hartmann C, Thomas D, Fukai N, Soker S et al (2001) Tissue specific regulation of VEGF expression during bone development requires Cbfa1/Runx2. Mech Dev 106(1–2):97–106

15. Bronckers AL, Sasaguri K, Cavender AC, D'Souza RN, Engelse MA (2005) Expression of Runx2/Cbfa1/Pebp2alphaA during angiogenesis in postnatal rodent and fetal human orofacial tissues. J Bone Miner Res 20(3):428–437
16. Barnes GL, Hebert KE, Kamal M, Javed A, Einhorn TA, Lian JB et al (2004) Fidelity of Runx2 activity in breast cancer cells is required for the generation of metastases-associated osteolytic disease. Cancer Res 64(13):4506–4513
17. Pratap J, Lian JB, Javed A, Barnes GL, van Wijnen AJ, Stein JL et al (2006) Regulatory roles of Runx2 in metastatic tumor and cancer cell interactions with bone. Cancer Metastasis Rev 25(4):589–600
18. Kwon TG, Zhao X, Yang Q, Li Y, Ge C, Zhao G et al (2011) Physical and functional interactions between Runx2 and HIF-1alpha induce vascular endothelial growth factor gene expression. J Cell Biochem 112(12):3582–3593
19. Sun L, Vitolo M, Passaniti A (2001) Runt-related gene 2 in endothelial cells: inducible expression and specific regulation of cell migration and invasion. Cancer Res 61(13):4994–5001
20. Pierce AD, Anglin IE, Vitolo MI, Mochin MT, Underwood KF, Goldblum SE et al (2012) Glucose-activated RUNX2 phosphorylation promotes endothelial cell proliferation and an angiogenic phenotype. J Cell Biochem 113(1):282–292
21. Vitolo MI, Anglin IE, Mahoney WM Jr, Renoud KJ, Gartenhaus RB, Bachman KE et al (2007) The RUNX2 transcription factor cooperates with the YES-associated protein, YAP65, to promote cell transformation. Cancer Biol Ther 6(6):856–863
22. D'Souza DR, Salib MM, Bennett J, Mochin-Peters M, Asrani K, Goldblum SE et al (2009) Hyperglycemia regulates RUNX2 activation and cellular wound healing through the aldose reductase polyol pathway. J Biol Chem 284(27):17947–17955
23. Qiao M, Shapiro P, Fosbrink M, Rus H, Kumar R, Passaniti A (2006) Cell cycle-dependent phosphorylation of the RUNX2 transcription factor by cdc2 regulates endothelial cell proliferation. J Biol Chem 281(11):7118–7128
24. San Martin IA, Varela N, Gaete M, Villegas K, Osorio M, Tapia JC et al (2009) Impaired cell cycle regulation of the osteoblast-related heterodimeric transcription factor Runx2-Cbfbeta in osteosarcoma cells. J Cell Physiol 221(3):560–571
25. Underwood KF, D'Souza DR, Mochin-Peters M, Pierce AD, Kommineni S, Choe M et al (2012) Regulation of RUNX2 transcription factor-DNA interactions and cell proliferation by vitamin D3 (cholecalciferol) prohormone activity. J Bone Miner Res 27(4):913–925
26. Fang CY, Xue JJ, Tan L, Jiang CH, Gao QP, Liang DS et al (2011) A novel single-base deletion mutation of the RUNX2 gene in a Chinese family with cleidocranial dysplasia. Genet Mol Res 10(4):3539–3544
27. Sugawara M, Kato N, Tsuchiya T, Motoyama T (2011) RUNX2 expression in developing human bones and various bone tumors. Pathol Int 61(10):565–571
28. Sase T, Suzuki T, Miura K, Shiiba K, Sato I, Nakamura Y et al (2012) Runt-related transcription factor 2 in human colon carcinoma: a potent prognostic factor associated with estrogen receptor. Int J Cancer 131(10):2284–2293
29. Slattery ML, Lundgreen A, Herrick JS, Caan BJ, Potter JD, Wolff RK (2011) Associations between genetic variation in RUNX1, RUNX2, RUNX3, MAPK1 and eIF4E and risk of colon and rectal cancer: additional support for a TGF-beta-signaling pathway. Carcinogenesis 32(3):318–326
30. Eliseev RA, Dong YF, Sampson E, Zuscik MJ, Schwarz EM, O'Keefe RJ et al (2008) Runx2-mediated activation of the Bax gene increases osteosarcoma cell sensitivity to apoptosis. Oncogene 27(25):3605–3614
31. Kurek KC, Del Mare S, Salah Z, Abdeen S, Sadiq H, Lee SH et al (2010) Frequent attenuation of the WWOX tumor suppressor in osteosarcoma is associated with increased tumorigenicity and aberrant RUNX2 expression. Cancer Res 70(13):5577–5586
32. Martin JW, Zielenska M, Stein GS, van Wijnen AJ, Squire JA (2011) The role of RUNX2 in osteosarcoma oncogenesis. Sarcoma 2011:282745

33. Sadikovic B, Thorner P, Chilton-Macneill S, Martin JW, Cervigne NK, Squire J et al (2010) Expression analysis of genes associated with human osteosarcoma tumors shows correlation of RUNX2 overexpression with poor response to chemotherapy. BMC Cancer 10:202
34. van der Deen M, Akech J, Lapointe D, Gupta S, Young DW, Montecino MA et al (2012) Genomic promoter occupancy of runt-related transcription factor RUNX2 in osteosarcoma cells identifies genes involved in cell adhesion and motility. J Biol Chem 287(7):4503–4517
35. Won KY, Park HR, Park YK (2009) Prognostic implication of immunohistochemical Runx2 expression in osteosarcoma. Tumori 95(3):311–316
36. Little GH, Noushmehr H, Baniwal SK, Berman BP, Coetzee GA, Frenkel B (2012) Genome-wide Runx2 occupancy in prostate cancer cells suggests a role in regulating secretion. Nucleic Acids Res 40(8):3538–3547
37. Akech J, Wixted JJ, Bedard K, van der Deen M, Hussain S, Guise TA et al (2010) Runx2 association with progression of prostate cancer in patients: mechanisms mediating bone osteolysis and osteoblastic metastatic lesions. Oncogene 29(6):811–821
38. Baniwal SK, Khalid O, Gabet Y, Shah RR, Purcell DJ, Mav D et al (2010) Runx2 transcriptome of prostate cancer cells: insights into invasiveness and bone metastasis. Mol Cancer 9:258
39. Baniwal SK, Little GH, Chimge NO, Frenkel B (2012) Runx2 controls a feed-forward loop between androgen and prolactin-induced protein (PIP) in stimulating T47D cell proliferation. J Cell Physiol 227(5):2276–2282
40. Hernandez LL, Gregerson KA, Horseman ND (2012) Mammary gland serotonin regulates parathyroid hormone-related protein and other bone-related signals. Am J Physiol Endocrinol Metab 302(8):E1009–E1015
41. Leong DT, Lim J, Goh X, Pratap J, Pereira BP, Kwok HS et al (2010) Cancer-related ectopic expression of the bone-related transcription factor RUNX2 in non-osseous metastatic tumor cells is linked to cell proliferation and motility. Breast Cancer Res 12(5):R89
42. Mendoza-Villanueva D, Zeef L, Shore P (2011) Metastatic breast cancer cells inhibit osteoblast differentiation through the Runx2/CBFbeta-dependent expression of the Wnt antagonist, sclerostin. Breast Cancer Res 13(5):R106
43. Onodera Y, Miki Y, Suzuki T, Takagi K, Akahira J, Sakyu T et al (2010) Runx2 in human breast carcinoma: its potential roles in cancer progression. Cancer Sci 101(12):2670–2675
44. Purcell DJ, Khalid O, Ou CY, Little GH, Frenkel B, Baniwal SK et al (2012) Recruitment of coregulator G9a by Runx2 for selective enhancement or suppression of transcription. J Cell Biochem 113(7):2406–2414
45. Reufsteck C, Lifshitz-Shovali R, Zepp M, Bauerle T, Kubler D, Golomb G et al (2012 Jun) Silencing of skeletal metastasis-associated genes impairs migration of breast cancer cells and reduces osteolytic bone lesions. Clin Exp Metastasis 29(5):441–456
46. van der Deen M, Akech J, Wang T, FitzGerald TJ, Altieri DC, Languino LR et al (2010) The cancer-related Runx2 protein enhances cell growth and responses to androgen and TGFbeta in prostate cancer cells. J Cell Biochem 109(4):828–837
47. Pratap J, Imbalzano KM, Underwood JM, Cohet N, Gokul K, Akech J et al (2009) Ectopic runx2 expression in mammary epithelial cells disrupts formation of normal acini structure: implications for breast cancer progression. Cancer Res 69(17):6807–6814
48. Chimge NO, Baniwal SK, Luo J, Coetzee S, Khalid O, Berman BP et al (2012) Opposing effects of Runx2 and estradiol on breast cancer cell proliferation: in vitro identification of reciprocally regulated gene signature related to clinical letrozole responsiveness. Clin Cancer Res 18(3):901–911
49. Kuo YH, Zaidi SK, Gornostaeva S, Komori T, Stein GS, Castilla LH (2009) Runx2 induces acute myeloid leukemia in cooperation with Cbfbeta-SMMHC in mice. Blood 113(14):3323–3332
50. Shim JH, Greenblatt MB, Singh A, Brady N, Hu D, Drapp R et al (2012) Administration of BMP2/7 in utero partially reverses Rubinstein-Taybi syndrome-like skeletal defects induced by Pdk1 or Cbp mutations in mice. J Clin Invest 122(1):91–106

51. Jang WG, Kim EJ, Kim DK, Ryoo HM, Lee KB, Kim SH et al (2012) BMP2 protein regulates osteocalcin expression via Runx2-mediated Atf6 gene transcription. J Biol Chem 287(2):905–915
52. Hirata M, Kugimiya F, Fukai A, Saito T, Yano F, Ikeda T et al (2012) C/EBPbeta and RUNX2 cooperate to degrade cartilage with MMP-13 as the target and HIF-2alpha as the inducer in chondrocytes. Hum Mol Genet 21(5):1111–1123
53. Mak IW, Cowan RW, Popovic S, Colterjohn N, Singh G, Ghert M (2009) Upregulation of MMP-13 via Runx2 in the stromal cell of giant cell tumor of bone. Bone 45(2):377–386
54. Singh S, Mak IW, Cowan RW, Turcotte R, Singh G, Ghert M (2011) The role of TWIST as a regulator in giant cell tumor of bone. J Cell Biochem 112(9):2287–2295
55. Takarada T, Yoneda Y (2009) Transactivation by Runt related factor-2 of matrix metalloproteinase-13 in astrocytes. Neurosci Lett 451(2):99–104
56. Ohyama Y, Tanaka T, Shimizu T, Matsui H, Sato H, Koitabashi N et al (2012) Runx2/Smad3 complex negatively regulates TGF-beta-induced connective tissue growth factor gene expression in vascular smooth muscle cells. J Atheroscler Thromb 19(1):23–35
57. Ali SA, Dobson JR, Lian JB, Stein JL, van Wijnen AJ, Zaidi SK et al (2012) A Runx2-HDAC1 co-repressor complex regulates rRNA gene expression by modulating UBF acetylation. J Cell Sci 125(Pt 11):2732–2739
58. Ali SA, Zaidi SK, Dobson JR, Shakoori AR, Lian JB, Stein JL et al (2010) Transcriptional corepressor TLE1 functions with Runx2 in epigenetic repression of ribosomal RNA genes. Proc Natl Acad Sci USA 107(9):4165–4169
59. Zhang Z, Deepak V, Meng L, Wang L, Li Y, Jiang Q et al (2012) Analysis of HDAC1-mediated regulation of Runx2-induced osteopontin gene expression in C3h10t1/2 cells. Biotechnol Lett 34(2):197–203
60. Jensen ED, Schroeder TM, Bailey J, Gopalakrishnan R, Westendorf JJ (2008) Histone deacetylase 7 associates with Runx2 and represses its activity during osteoblast maturation in a deacetylation-independent manner. J Bone Miner Res 23(3):361–372
61. Zhang H, Pan Y, Zheng L, Choe C, Lindgren B, Jensen ED et al (2011) FOXO1 inhibits Runx2 transcriptional activity and prostate cancer cell migration and invasion. Cancer Res 71(9):3257–3267
62. Chen G, Deng C, Li YP (2012) TGF-beta and BMP signaling in osteoblast differentiation and bone formation. Int J Biol Sci 8(2):272–288
63. Javed A, Bae JS, Afzal F, Gutierrez S, Pratap J, Zaidi SK et al (2008) Structural coupling of Smad and Runx2 for execution of the BMP2 osteogenic signal. J Biol Chem 283(13):8412–8422
64. Pratap J, Wixted JJ, Gaur T, Zaidi SK, Dobson J, Gokul KD et al (2008) Runx2 transcriptional activation of Indian Hedgehog and a downstream bone metastatic pathway in breast cancer cells. Cancer Res 68(19):7795–7802
65. Nishimura R, Hata K, Matsubara T, Wakabayashi M, Yoneda T (2012) Regulation of bone and cartilage development by network between BMP signalling and transcription factors. J Biochem 151(3):247–254
66. Ge C, Yang Q, Zhao G, Yu H, Kirkwood KL, Franceschi RT (2012) Interactions between extracellular signal-regulated kinase 1/2 and P38 map kinase pathways in the control of RUNX2 phosphorylation and transcriptional activity. J Bone Miner Res 27(3):538–551
67. Li Y, Ge C, Long JP, Begun DL, Rodriguez JA, Goldstein SA et al (2012) Biomechanical stimulation of osteoblast gene expression requires phosphorylation of the RUNX2 transcription factor. J Bone Miner Res 27(6):1263–1274
68. Zou W, Greenblatt MB, Shim JH, Kant S, Zhai B, Lotinun S et al (2011) MLK3 regulates bone development downstream of the faciogenital dysplasia protein FGD1 in mice. J Clin Invest 121(11):4383–4392
69. Chen J, Yuan K, Mao X, Miano JM, Wu H, Chen Y (2012) Serum response factor regulates bone formation via IGF-1 and Runx2 signals. J Bone Miner Res 27(8):1659–1668
70. Li Z, Wang Z, Yang L, Li X, Sasaki Y, Wang S et al (2010) Fibroblast growth factor 2 regulates bone sialoprotein gene transcription in human breast cancer cells. J Oral Sci 52(1):125–132

71. Niger C, Buo AM, Hebert C, Duggan BT, Williams MS, Stains JP (2012) ERK acts in parallel to PKCdelta to mediate the connexin43-dependent potentiation of Runx2 activity by FGF2 in MC3T3 osteoblasts. Am J Physiol Cell Physiol 302(7):C1035–C1044
72. Teplyuk NM, Haupt LM, Ling L, Dombrowski C, Mun FK, Nathan SS et al (2009) The osteogenic transcription factor Runx2 regulates components of the fibroblast growth factor/proteoglycan signaling axis in osteoblasts. J Cell Biochem 107(1):144–154
73. Huang YF, Lin JJ, Lin CH, Su Y, Hung SC (2012) c-Jun N-terminal kinase 1 negatively regulates osteoblastic differentiation induced by BMP-2 via phosphorylation of Runx2 at Ser104. J Bone Miner Res 27(5):1093–1105
74. Bae SC, Lee YH (2006) Phosphorylation, acetylation and ubiquitination: the molecular basis of RUNX regulation. Gene 366(1):58–66
75. Tu X, Joeng KS, Long F (2012) Indian hedgehog requires additional effectors besides Runx2 to induce osteoblast differentiation. Dev Biol 362(1):76–82
76. Shu B, Zhang M, Xie R, Wang M, Jin H, Hou W et al (2011) BMP2, but not BMP4, is crucial for chondrocyte proliferation and maturation during endochondral bone development. J Cell Sci 124(Pt 20):3428–3440
77. Oh JE, Kim HJ, Kim WS, Lee ZH, Ryoo HM, Hwang SJ et al (2012) PlexinA2 mediates osteoblast differentiation via regulation of Runx2. J Bone Miner Res 27(3):552–562
78. Yang DC, Yang MH, Tsai CC, Huang TF, Chen YH, Hung SC (2011) Hypoxia inhibits osteogenesis in human mesenchymal stem cells through direct regulation of RUNX2 by TWIST. PLoS One 6(9):e23965
79. Yuen HF, Kwok WK, Chan KK, Chua CW, Chan YP, Chu YY et al (2008) TWIST modulates prostate cancer cell-mediated bone cell activity and is upregulated by osteogenic induction. Carcinogenesis 29(8):1509–1518
80. Zhang YY, Li X, Qian SW, Guo L, Huang HY, He Q et al (2012) Down-regulation of type I Runx2 mediated by dexamethasone is required for 3T3-L1 adipogenesis. Mol Endocrinol 26(5):798–808
81. D'Souza S, del Prete D, Jin S, Sun Q, Huston AJ, Kostov FE et al (2011) Gfi1 expressed in bone marrow stromal cells is a novel osteoblast suppressor in patients with multiple myeloma bone disease. Blood 118(26):6871–6880
82. Wu T, Zhou H, Hong Y, Li J, Jiang X, Huang H (2012) miR-30 family members negatively regulate osteoblast differentiation. J Biol Chem 287(10):7503–7511
83. Saini S, Majid S, Yamamura S, Tabatabai L, Suh SO, Shahryari V et al (2011) Regulatory role of mir-203 in prostate cancer progression and metastasis. Clin Cancer Res 17(16):5287–5298
84. Viticchie G, Lena AM, Latina A, Formosa A, Gregersen LH, Lund AH et al (2011) MiR-203 controls proliferation, migration and invasive potential of prostate cancer cell lines. Cell Cycle 10(7):1121–1131
85. Hassan MQ, Gordon JA, Beloti MM, Croce CM, van Wijnen AJ, Stein JL et al (2010) A network connecting Runx2, SATB2, and the miR-23a~27a~24-2 cluster regulates the osteoblast differentiation program. Proc Natl Acad Sci USA 107(46):19879–19884
86. Zhang Y, Xie RL, Croce CM, Stein JL, Lian JB, van Wijnen AJ et al (2011) A program of microRNAs controls osteogenic lineage progression by targeting transcription factor Runx2. Proc Natl Acad Sci USA 108(24):9863–9868
87. Zhi X, Chen C (2012) WWP1: a versatile ubiquitin E3 ligase in signaling and diseases. Cell Mol Life Sci 69(9):1425–1434
88. Li Y, Kong D, Ahmad A, Bao B, Sarkar FH (2012) Targeting bone remodeling by isoflavone and 3,3'-diindolylmethane in the context of prostate cancer bone metastasis. PLoS One 7(3):e33011
89. Kong XH, Niu YB, Song XM, Zhao DD, Wang J, Wu XL et al (2012) Astragaloside II induces osteogenic activities of osteoblasts through the bone morphogenetic protein-2/MAPK and Smad1/5/8 pathways. Int J Mol Med 29(6):1090–1098
90. Don MJ, Lin LC, Chiou WF (2012) Neobavaisoflavone stimulates osteogenesis via p38-mediated up-regulation of transcription factors and osteoid genes expression in MC3T3-E1 cells. Phytomedicine 19(6):551–561

91. Eda H, Aoki K, Kato S, Okawa Y, Takada K, Tanaka T et al (2010) The proteasome inhibitor bortezomib inhibits FGF-2-induced reduction of TAZ levels in osteoblast-like cells. Eur J Haematol 85(1):68–75
92. Fu J, Wang W, Liu YH, Lu H, Luo Y (2011) In vitro anti-angiogenic properties of LGD1069, a selective retinoid X-receptor agonist through down-regulating Runx2 expression on human endothelial cells. BMC Cancer 11:227
93. Mohammad KS, Javelaud D, Fournier PG, Niewolna M, McKenna CR, Peng XH et al (2011) TGF-beta-RI kinase inhibitor SD-208 reduces the development and progression of melanoma bone metastases. Cancer Res 71(1):175–184
94. Dalle Carbonare L, Frigo A, Francia G, Davi MV, Donatelli L, Stranieri C et al (2012) Runx2 mRNA expression in the tissue, serum, and circulating non-hematopoietic cells of patients with thyroid cancer. J Clin Endocrinol Metab 97(7):E1249–E1256
95. Liao CP, Adisetiyo H, Liang M, Roy-Burman P (2010) Cancer stem cells and microenvironment in prostate cancer progression. Horm Cancer 1(6):297–305
96. Liao CP, Adisetiyo H, Liang M, Roy-Burman P (2010) Cancer-associated fibroblasts enhance the gland-forming capability of prostate cancer stem cells. Cancer Res 70(18):7294–7303
97. Wang L, Park P, Zhang H, La Marca F, Claeson A, Valdivia J et al (2011) BMP-2 inhibits the tumorigenicity of cancer stem cells in human osteosarcoma OS99-1 cell line. Cancer Biol Ther 11(5):457–463
98. Zhang Y, Xie RL, Gordon J, Leblanc K, Stein JL, Lian JB et al (2012) Control of mesenchymal lineage progression by microRNAs targeting the skeletal gene regulators Trps1 and Runx2. J Biol Chem 287(26):21926–21935
99. Hsu YL, Huang MS, Yang CJ, Hung JY, Wu LY, Kuo PL (2011) Lung tumor-associated osteoblast-derived bone morphogenetic protein-2 increased epithelial-to-mesenchymal transition of cancer by Runx2/Snail signaling pathway. J Biol Chem 286(43):37335–37346
100. Jeong JH, Choi JY (2011) Interrelationship of Runx2 and estrogen pathway in skeletal tissues. BMB Rep 44(10):613–618
101. Pratap J, Javed A, Languino LR, van Wijnen AJ, Stein JL, Stein GS et al (2005) The Runx2 osteogenic transcription factor regulates matrix metalloproteinase 9 in bone metastatic cancer cells and controls cell invasion. Mol Cell Biol 25(19):8581–8591
102. Lee SH, Che X, Jeong JH, Choi JY, Lee YJ, Lee YH et al (2012) Runx2 protein stabilizes hypoxia-inducible factor-1alpha through competition with von hippel-lindau protein (pVHL) and stimulates angiogenesis in growth plate hypertrophic chondrocytes. J Biol Chem 287(18):14760–14771
103. Sun X, Wei L, Chen Q, Terek RM (2009) HDAC4 represses vascular endothelial growth factor expression in chondrosarcoma by modulating RUNX2 activity. J Biol Chem 284(33):21881–21890
104. Peruzzi B, Cappariello A, Del Fattore A, Rucci N, De Benedetti F, Teti A (2012) c-Src and IL-6 inhibit osteoblast differentiation and integrate IGFBP5 signalling. Nat Commun 3:630
105. Ding M, Lu Y, Abbassi S, Li F, Li X, Song Y et al (2012) Targeting Runx2 expression in hypertrophic chondrocytes impairs endochondral ossification during early skeletal development. J Cell Physiol 227(10):3446–3456
106. Chimge NO, Baniwal SK, Little GH, Chen YB, Kahn M, Tripathy D et al (2011) Regulation of breast cancer metastasis by Runx2 and estrogen signaling: the role of SNAI2. Breast Cancer Res 13(6):R127
107. Goloudina AR, Tanoue K, Hammann A, Fourmaux E, Le Guezennec X, Bulavin DV et al (2012) Wip1 promotes RUNX2-dependent apoptosis in p53-negative tumors and protects normal tissues during treatment with anticancer agents. Proc Natl Acad Sci USA 109(2):E68–E75
108. Tandon M, Gokul K, Ali SA, Chen Z, Lian J, Stein GS et al (2012) Runx2 mediates epigenetic silencing of the bone morphogenetic protein-3B (BMP-3B/GDF10) in lung cancer cells. Mol Cancer 11(1):27
109. Endo T, Kobayashi T (2010) Runx2 deficiency in mice causes decreased thyroglobulin expression and hypothyroidism. Mol Endocrinol 24(6):1267–1273

110. Teplyuk NM, Zhang Y, Lou Y, Hawse JR, Hassan MQ, Teplyuk VI et al (2009) The osteogenic transcription factor runx2 controls genes involved in sterol/steroid metabolism, including CYP11A1 in osteoblasts. Mol Endocrinol 23(6):849–861
111. Teplyuk NM, Galindo M, Teplyuk VI, Pratap J, Young DW, Lapointe D et al (2008) Runx2 regulates G protein-coupled signaling pathways to control growth of osteoblast progenitors. J Biol Chem 283(41):27585–27597
112. Bond SR, Lau A, Penuela S, Sampaio AV, Underhill TM, Laird DW et al (2011) Pannexin 3 is a novel target for Runx2, expressed by osteoblasts and mature growth plate chondrocytes. J Bone Miner Res 26(12):2911–2922
113. Lim M, Zhong C, Yang S, Bell AM, Cohen MB, Roy-Burman P (2010) Runx2 regulates survivin expression in prostate cancer cells. Lab Invest 90(2):222–233
114. Vladimirova V, Waha A, Luckerath K, Pesheva P, Probstmeier R (2008) Runx2 is expressed in human glioma cells and mediates the expression of galectin-3. J Neurosci Res 86(11):2450–2461
115. Zhang HY, Jin L, Stilling GA, Ruebel KH, Coonse K, Tanizaki Y et al (2009) RUNX1 and RUNX2 upregulate Galectin-3 expression in human pituitary tumors. Endocrine 35(1):101–111
116. Hoeppner LH, Secreto F, Jensen ED, Li X, Kahler RA, Westendorf JJ (2009) Runx2 and bone morphogenic protein 2 regulate the expression of an alternative Lef1 transcript during osteoblast maturation. J Cell Physiol 221(2):480–489
117. Endo T, Ohta K, Kobayashi T (2008) Expression and function of Cbfa-1/Runx2 in thyroid papillary carcinoma cells. J Clin Endocrinol Metab 93(6):2409–2412
118. Makita N, Suzuki M, Asami S, Takahata R, Kohzaki D, Kobayashi S et al (2008) Two of four alternatively spliced isoforms of RUNX2 control osteocalcin gene expression in human osteoblast cells. Gene 413(1–2):8–17
119. Li X, Hoeppner LH, Jensen ED, Gopalakrishnan R, Westendorf JJ (2009) Co-activator activator (CoAA) prevents the transcriptional activity of Runt domain transcription factors. J Cell Biochem 108(2):378–387
120. Mendoza-Villanueva D, Deng W, Lopez-Camacho C, Shore P (2010) The Runx transcriptional co-activator, CBFbeta, is essential for invasion of breast cancer cells. Mol Cancer 9:171
121. Roforth MM, Liu G, Khosla S, Monroe DG (2012) Examination of nuclear receptor expression in osteoblasts reveals Rorbeta as an important regulator of osteogenesis. J Bone Miner Res 27(4):891–901
122. Westendorf JJ (2006) Transcriptional co-repressors of Runx2. J Cell Biochem 98(1):54–64
123. Byun MR, Jeong H, Bae SJ, Kim AR, Hwang ES, Hong JH (2012) TAZ is required for the osteogenic and anti-adipogenic activities of kaempferol. Bone 50(1):364–372
124. Baniwal SK, Khalid O, Sir D, Buchanan G, Coetzee GA, Frenkel B (2009) Repression of Runx2 by androgen receptor (AR) in osteoblasts and prostate cancer cells: AR binds Runx2 and abrogates its recruitment to DNA. Mol Endocrinol 23(8):1203–1214
125. Del Mare S, Kurek KC, Stein GS, Lian JB, Aqeilan RI (2011) Role of the WWOX tumor suppressor gene in bone homeostasis and the pathogenesis of osteosarcoma. Am J Cancer Res 1(5):585–594
126. Li X, McGee-Lawrence ME, Decker M, Westendorf JJ (2010) The Ewing's sarcoma fusion protein, EWS-FLI, binds Runx2 and blocks osteoblast differentiation. J Cell Biochem 111(4):933–943
127. Blyth K, Vaillant F, Hanlon L, Mackay N, Bell M, Jenkins A et al (2006) Runx2 and MYC collaborate in lymphoma development by suppressing apoptotic and growth arrest pathways in vivo. Cancer Res 66(4):2195–2201
128. Chua CW, Chiu YT, Yuen HF, Chan KW, Man K, Wang X et al (2009) Suppression of androgen-independent prostate cancer cell aggressiveness by FTY720: validating Runx2 as a potential antimetastatic drug screening platform. Clin Cancer Res 15(13):4322–4335
129. Das K, Leong DT, Gupta A, Shen L, Putti T, Stein GS et al (2009) Positive association between nuclear Runx2 and oestrogen-progesterone receptor gene expression characterises a biological subtype of breast cancer. Eur J Cancer 45(13):2239–2248

130. Dy P, Wang W, Bhattaram P, Wang Q, Wang L, Ballock RT et al (2012) Sox9 directs hypertrophic maturation and blocks osteoblast differentiation of growth plate chondrocytes. Dev Cell 22(3):597–609
131. Giuliani N, Mangoni M, Rizzoli V (2009) Osteogenic differentiation of mesenchymal stem cells in multiple myeloma: identification of potential therapeutic targets. Exp Hematol 37(8):879–886
132. Hassan MQ, Saini S, Gordon JA, van Wijnen AJ, Montecino M, Stein JL et al (2009) Molecular switches involving homeodomain proteins, HOXA10 and RUNX2 regulate osteoblastogenesis. Cells Tissues Organs 189(1–4):122–125
133. Jonsson S, Hjorth-Hansen H, Olsson B, Wadenvik H, Sundan A, Standal T (2012) Imatinib inhibits proliferation of human mesenchymal stem cells and promotes early but not late osteoblast differentiation in vitro. J Bone Miner Metab 30(1):119–123
134. Kawato Y, Hirao M, Ebina K, Tamai N, Shi K, Hashimoto J et al (2011) Nkx3.2-induced suppression of Runx2 is a crucial mediator of hypoxia-dependent maintenance of chondrocyte phenotypes. Biochem Biophys Res Commun 416(1–2):205–210
135. Kayed H, Jiang X, Keleg S, Jesnowski R, Giese T, Berger MR et al (2007) Regulation and functional role of the Runt-related transcription factor-2 in pancreatic cancer. Br J Cancer 97(8):1106–1115
136. Pockwinse SM, Kota KP, Quaresma AJ, Imbalzano AN, Lian JB, van Wijnen AJ et al (2011) Live cell imaging of the cancer-related transcription factor RUNX2 during mitotic progression. J Cell Physiol 226(5):1383–1389
137. Pockwinse SM, Rajgopal A, Young DW, Mujeeb KA, Nickerson J, Javed A et al (2006) Microtubule-dependent nuclear-cytoplasmic shuttling of Runx2. J Cell Physiol 206(2):354–362
138. Qiao M, Shapiro P, Kumar R, Passaniti A (2004) Insulin-like growth factor-1 regulates endogenous RUNX2 activity in endothelial cells through a phosphatidylinositol 3-kinase/ERK-dependent and Akt-independent signaling pathway. J Biol Chem 279(41):42709–42718
139. Zhang Y, Su J, Yu J, Bu X, Ren T, Liu X et al (2011) An essential role of discoidin domain receptor 2 (DDR2) in osteoblast differentiation and chondrocyte maturation via modulation of Runx2 activation. J Bone Miner Res 26(3):604–617

# Chapter 4
# Epigenetic Mechanisms of Cancer Metastasis

**Jing Liang and Yongfeng Shang**

**Abstract** Metastasis refers to the process that cancer cells leave their primary tumor mass, break into blood and lymphatic vessels, and travel to distant organ sites throughout the body where they may establish new colonies. Metastasis is responsible for 90 % of cancer mortality. Increasing evidence suggest that epigenetic mechanisms, such as DNA methylation and histone modifications, play an important role in mediating the invasion-metastasis cascade. Targeting deregulated epigenetic modification enzymes by small-molecule inhibitors is a promising therapeutic strategy for the treatment of metastatic cancers.

**Keywords** Epithelial-mesenchymal transition (EMT) • Cancer metastasis • DNA methylation • Histone modification • Epigenetic therapy

## Abbreviations

| | |
|---|---|
| DNMT | DNA methyltransferase |
| ECM | Extracellular matrix |
| EMT | Epithelial-mesenchymal transition |

J. Liang, Ph.D. (✉)
Department of Biochemistry and Molecular Biology, School of Basic Medical Sciences, Peking University Health Science Center, Beijing, People's Republic of China
e-mail: liang_jing@bjmu.edu.cn

Y. Shang, Ph.D. (✉)
Department of Biochemistry and Molecular Biology, School of Basic Medical Sciences, Peking University Health Science Center, Beijing, People's Republic of China

Department of Biochemistry and Molecular Biology, Tianjin Medical University, Tianjin, People's Republic of China
e-mail: yshang@hsc.pku.edu.cn

R. Kumar (ed.), *Nuclear Signaling Pathways and Targeting Transcription in Cancer*, Cancer Drug Discovery and Development, DOI 10.1007/978-1-4614-8039-6_4,

| | |
|---|---|
| HATs | Histone acetyltransferases |
| HDACi | HDAC inhibitors |
| HDACs | Histone deacetylases |
| MET | Mesenchymal-epithelial transition |
| MMPs | Matrix metalloproteinases |
| PTMs | Post-translational modifications |

Cancer is characterized by uncontrolled growth and spread of abnormal cells. Primary tumors do not necessarily cause obvious discomfort to the patient, simply because some organs are well expansible while maintaining their normal functions. A primary tumor in the breast will not cause much trouble to the patient's overall physiological function, unless breastfeeding is needed. Only 10 % of deaths from cancer are due to primary tumors, while metastasis is responsible for the remaining 90 % of cancer mortality. Metastasis refers to the process that cancer cells leave their primary tumor mass, break into blood and lymphatic vessels, and travel to distant organ sites throughout the body where they may establish new colonies. Metastases of breast cancer can be found in the brain, liver, bones, and lungs. Insidious growth of metastatic cancer cells in these sites is life-threatening because the physiological functions of such vital organs are greatly compromised.

While the molecular mechanisms underlying tumorigenesis have been studied in great detail, our understanding of cancer metastasis is limited, partly due to the difficulty to set up the in vitro and in vivo experimental models for this process. Currently, cancer metastasis is considered to be a cascade that consists of a series of interrelated steps, including (i) detachment of tumor cells from the primary tumor; (ii) invasion into the surrounding tissues; (iii) intravasation into the blood or lymphatic vessels; (iv) dissemination in the blood stream or the lymphatic system and, finally, (v) extravasation and colonization at a secondary site (Fig. 4.1) [1, 2]. While certain genetic lesions provide primary tumor cells good opportunity for successful metastasis, the interplay between tumor cells and the surrounding stroma eventually endow tumor cells the ability to fulfill these daunting tasks. Epigenetic mechanisms, including DNA methylation, histone modification, nucleosome remodeling, and RNA-mediated targeting, alter gene expression profile via changes in the chromatin states. The epigenome of cells is dynamic in response to extra- and intracellular signals, while under certain circumstances it can also be transmitted to the next generation of cells to maintain cell identity. Increasing evidence suggest that epigenetic mechanisms play important roles in tumor progression and metastasis. In this chapter, we first describe our current understanding of the major molecular events governing the invasion-metastasis cascade, and then discuss how epigenetic factors, particularly DNA methylation and histone modifications, contribute to the metastasis process.

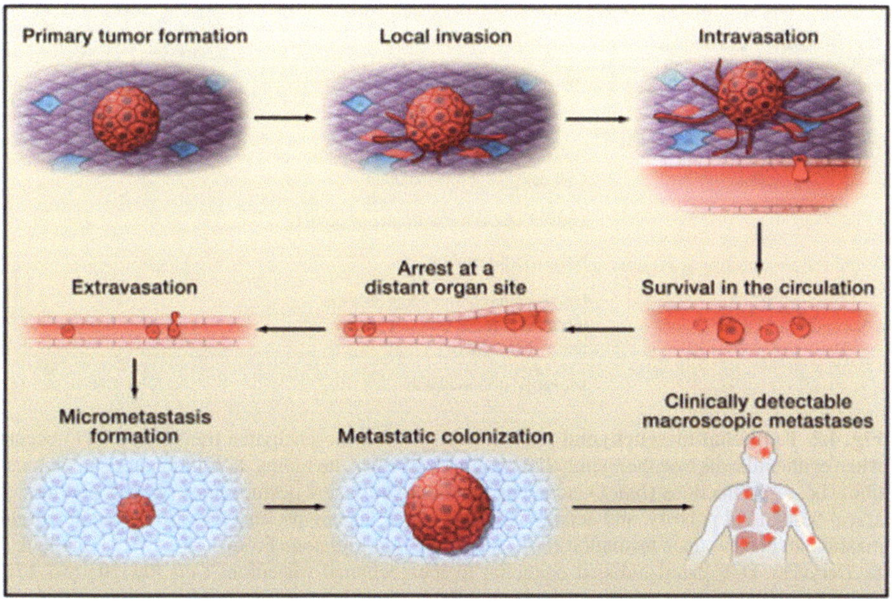


**Fig. 4.1** **The invasion-metastasis cascade**. During metastatic progression, tumor cells exit their primary sites of growth, invade into the extracellular matrix, and enter into the blood or lymphatic vessels (local invasion, intravasation). Circulating cancer cells translocate systemically (survival in the circulation, arrest at a distant organ site, extravasation), and adapt to survive and thrive in the foreign microenvironments of distant tissues (micrometastasis formation, metastatic colonization). Carcinoma cells are depicted in *red* (From S. Valastyan and R.A. Weinberg. Tumor Metastasis: Molecular insights and evolving paradigms. *Cell* 147: 275–292, 2011. Reprinted with kind permission from Elsevier Limited)

## 4.1 The Epithelial-Mesenchymal Transition Enables Carcinoma Cells to Become Invasive and Represents the Initial Step of Metastasis

Malignant cells arise from epithelial tissues form carcinomas, which contribute to the majority of life-threatening cancers. Early stage carcinoma cells retain the typical epithelial property, which includes lateral tight connection by specialized junction structures, aligned apical-basal polarity through association with the basement membrane, and lack of cell motility. Later on, many of these cancer cells acquire the ability to invade the nearby stroma, travel throughout the body via the lymphatic or hematogenous circulation, and form disseminated colonies in distant organs. To understand how epigenetic mechanisms play a role in cancer metastasis, we need to dissect the process into sequential biological steps, with the notion that these steps are interrelated to each other in reality.

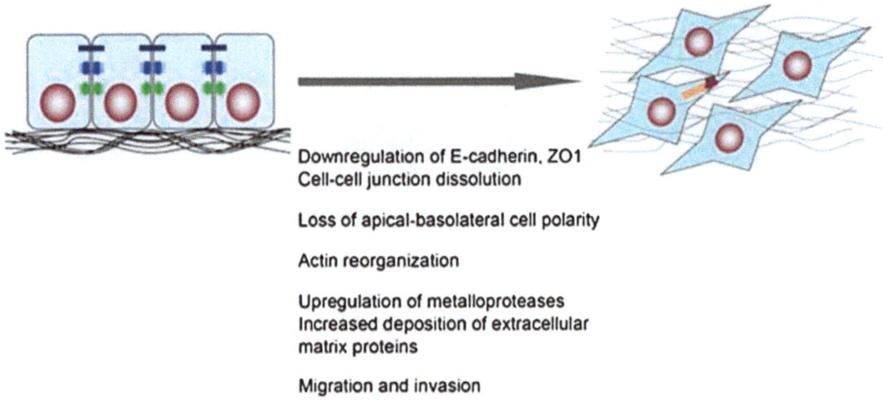


**Fig. 4.2 Epithelial-mesenchymal transition.** Epithelial-mesenchymal transition (EMT) occurs when epithelial cells lose their epithelial cell characteristics, including dissolution of cell-cell junctions, i.e. tight junctions (*black*), adherens junctions (*blue*) and desmosomes (*green*), and loss of apical-basolateral polarity, and acquire a mesenchymal phenotype, characterized by actin reorganization and stress fiber formation (*red*), migration and invasion (From J. Xu, S. Lamouille and R. Derynck. TGF-[beta]-induced epithelial to mesenchymal transition. *Cell Res* 19: 156–172, 2009. Reprinted with permission from Nature Publishing Group)

As long as tumor cells are separated from the surrounding tissues by an intact basement membrane, they are not truly "malignant". Once in situ carcinoma cells breach the basement membrane and invade into the stroma, the initial step of metastasis is manifested. Arise from epithelial tissues, original carcinoma cells has poor motility, and the lateral cell-cell connection and the apical-basal polarity do not allow these cells to invade through the underlying extracellular matrix (ECM). To initiate the first step of metastasis, carcinoma cells must transform someway to shed these epithelial properties that suppress invasion, and acquire mesenchymal phenotypes such as better motility, loose cell-cell contacts, and affinity to the stroma. The epithelial-mesenchymal transition (EMT) is now generally accepted to be the critical initial step of cancer metastasis (Fig. 4.2).

EMT was recognized as a feature of embryogenesis in the early 1980s. One example of EMT in embryonic development is the formation of the mesoderm, the precursor of mesenchymal tissues, during gastrulation in early embryogenesis. At this stage, ectoderm cells, which are located at outer side of the embryo and arrayed in an epithelial cell layer, migrate inward toward the center of the embryo to form the mesoderm, where fibroblasts and hematopoietic cells originate. EMT can also be witnessed in wound healing, in which process the epithelial cells at the edge of a wound acquire mesenchymal phenotype and become motile and invasive to fill in the gap in the epithelium created by the wounding. Therefore, EMT is not a unique nature of carcinoma cells; rather it is the activation of a reprogramming behavior that is usually confined to early embryogenesis and wound healing in adult.

Fundamental alterations of gene expression lead to reconstituted cellular machinery, and ultimately result in the phenotypic changes of EMT cells. Among all the molecules that influence epithelial versus mesenchymal phenotypes, the epithelial specific E-cadherin, encoded by *CDH1* gene, plays the dominant role. E-cadherin is a transmembrane glycoprotein, and the cytoplasmic domains of E-cadherin are tethered to the actin fibers of the cytoskeleton via a complex of α- and β-catenins. E-cadherin molecules are displayed on adjacent epithelial cells and tether the apposed plasma membranes to one another. Loss of E-cadherin is consistently observed at the sites of EMT. Epithelial originated cells acquire a mesenchymal morphology and increased motility when the expression of E-cadherin is suppressed, whereas re-expression of this protein in invasive cancer cells strongly suppressed their metastatic dissemination. Besides E-cadherin, loss of expression of other epithelial specific proteins, such as α-, β-, and γ-catenins, and gain of expression of mesenchymal markers, such as fibronectin, vimentin, N-cadherin, is constantly observed during EMT [3, 4].

## 4.2 The Interplay Between Cancer Cells and the Stroma Provide Signals for the Epithelial-Mesenchymal Transition

EMT is an induced, reversible process, as carcinoma cells often revert back to a more epithelial phenotype once these cells reach the distant organs, where EMT-inducing signals are no longer existed. Therefore, the metastases resemble the phenotype of the primary carcinomas. This process is often called mesenchymal-epithelial transition (MET). EMT can be induced or regulated by various growth and differentiation factors, including TGF-β (transforming growth factor-β), TNF-α (tumor necrosis factor- α), EGF (epidermal growth factor), HGF (hepatocyte growth factor), and IGF-1 (insulin-like growth factor-1). These heterotypic signals are mainly released by the stroma of primary carcinomas, and cells located at the outer edges of the neoplasm sense these signal and undergo EMT. Among these factors, the major inducer of EMT is TGF-β, which can also produced by cancer cells themselves to generate a positive feedback loop for EMT induction. Culturing EpRas tumor cells in the presence of TGF-β, these epithelial, cobblestone-like cells change to an elongated fibroblastic phenotype. The levels of tumor-associated TGF-β (often TGF-β1) were frequently found to be correlated with increasing degrees of tumor invasiveness. High levels of TGF-β in the blood are often predictive of poor prognosis for the cancer patient. EMT-inducing growth factors activate different intracellular signaling cascades in carcinoma cells, such as NF-κB and Wnt pathways, leading to the alteration of gene expression profiles that favor mesenchymal phenotypes.

In response to various inductive signals, EMT-inducing transcription factors can serve as major signaling mediators of the EMT program to promote metastasis. Ectopic expression of these transcription factors enables epithelial cells undergo

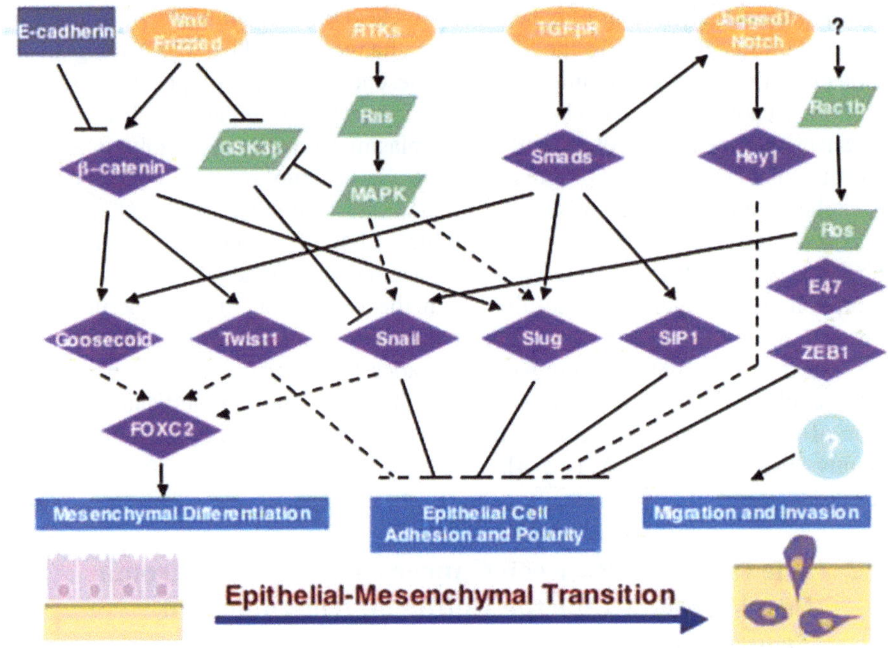


**Fig. 4.3 Signaling pathways and transcription factors that regulate the epithelial-mesenchymal transition in carcinoma cells.** In cancer cells, the TGF-β signaling pathway induces multiple EMT-inducing transcription factors, including Slug, SIP1, and Goosecoid, via activation of Smads. The Wnt pathway and loss of E-cadherin from adherens junctions activate β-catenin, which in turn induces several EMT-inducing transcription factors as well, such as Slug, Twist1, and Goosecoid. Multiple tyrosine kinase receptor (TKR) pathways, including FGFR, EGFR, PDGFR, and HGFR, can induce the expression of Snail and Slug through the Ras-MAPK pathway. Among all the EMT-inducing transcription factors, Snail, Slug, SIP1, and E47 directly suppress E-cadherin transcription, while Twist1, Goosecoid, and FOXC2 seem to function indirectly. FOXC2 is induced in tumor cells expressing Twist1, Snail, and Goosecoid and mediates mesenchymal differentiation. *Solid lines* indicate direct transcriptional or posttranscriptional regulations. *Dashed lines* indicate indirect regulation (From J. Yang and R.A. Weinberg. Epithelial-mesenchymal transition: At the crossroads of development and tumor metastasis. *Dev Cell* 14: 818–829, 2008. Reprinted with kind permission from Elsevier Limited)

EMT and acquire mesenchymal phenotype (Fig. 4.3). Examples of these EMT-inducing transcription factors include Twist, Snail, Slug, ZEB2/ZFXH1B/SIP1, and Goosecoid. These proteins are often highly expressed during embryogenesis, and their reactivation allows carcinoma cells to initiate the EMT program. Accumulating evidence associates these transcription factors with various malignancies, especially those with highly invasive behavior and poor prognosis. These transcription factors can regulate a panel of downstream target genes that collaboratively program an EMT process. For example, the zinc-finger transcription factors Snail, Slug, ZEB1 and ZEB2 are capable of directly repressing the transcription of E-cadherin and several polarity factors, including Crumbs3 and Lgl2 [5, 6].

## 4.3 Disseminated Cancer Cells Found Metastases in Distant Organs

Once primary tumor cells undergo EMT, they acquire abilities for further invasion. Before they enter the general circulation, EMT-transformed tumor cells first need to pave the way in the ECM by either dissolving the stroma or pushing aside any cells that stand in their path. Among the molecular events that govern this process, the best understood is the secretion of matrix metalloproteinases (MMPs), which is a class of secreted proteases that can degrade specific components of the ECM. MMPs are secreted by tumor-recruited stroma cells, such as macrophages, mast cells, and fibroblasts. Dissolving of the ECM by MMPs creates spaces for the tumor cells to move. MMPs also mobilize and activate some growth factors that have been tethered in the inactive form to the ECM or to the cell surface.

Motile tumor cells adhere to the vessels of blood or lymphatic systems in the stroma and enter the general circulation, a process that is often referred to intravasation. Before these cells reach their destination, they may experience much ordeal such as mechanical shearing forces of the blood stream and attack from immune cells in the circulation. The molecular mechanisms that govern these events are largely unknown. Theoretically, metastatic cancer cells can disseminate all over the body, but several organs, such as the lung, bone, liver, and brain are clearly more prone for metastases to grow. In addition, subtypes of tumor cells from different tissue origin have their own preference of metastatic organs. The vast majority of cells that end up forming small micrometastases (<2 mm diameter) never succeed in growing into macrometastases (>2 mm diameter) that are clinically relevant for the patients' health. In this sense, colonization is extremely inefficient, and often considered to be the rate-limiting step in the entire invasion-metastasis cascade. Although the detailed molecular mechanisms are yet to be clarified, it is important to understand that it is the complex interplay between metastatic cancer cells and the microenvironment of a foreign tissue that essentially determines whether colonization will be successful and metastatic diseases will ever develop. Dispersed cancer cells will only survive and found colonies in the environment with appropriate chemokines, trophic factors, and mitogens, while seeding cancer cells also release heterotypic signals to reshape the landed tissues. In this adaptation process, gene expression reprogramming in the metastatic cancer cells is inevitable.

Analogous to oncogenes and tumor suppressor genes that regulate tumorigenesis, there are many genes that function positively or negatively in the regulation of metastasis. Metastasis promoting regulators can be growth factors, growth factor receptors, transcription factors, or key components in signal transduction pathways. When genes encode these molecules are ectopically expressed in epithelial cells, they are able to elicit metastasis-prone phenotypic changes, such as increased cell motility and decreased cell polarity. On the other hand, metastasis suppressor genes encode proteins that specifically inhibit invasion and metastasis without affecting the growth of primary tumors. To date, over 20 metastasis suppressor genes have been identified, with the best characterized is *CDH1*, which encodes the epithelial-specific E-cadherin. Although the molecular functions of many metastasis suppressor genes are yet to be clarified, altered expression of these genes, whether due to genetic defects or epigenetic regulations, can significantly affect the overall invasion-metastasis process [7].

## 4.4 Promoter Methylation is an Important Epigenetic Mechanism to Regulate the Expression of Metastasis Related Genes

The long journey of invasion-metastasis consists of multiple steps that involve the interplay between primary tumor cells and the surrounding cells in the stroma, blood, and the distant foreign tissues. From the initial EMT to the final colonization, cancer cells constantly transduce heterotypic extracellular signals into changes of gene expression and cell behavior to better adapt themselves to metastasis. Unlike genetic defects such as gene amplification, deletion, or mutations, epigenetic regulations, such as chemical modifications of DNA sequences or histone proteins, are more flexible in response to various extracellular stimuli. In the mean time, certain epigenetic modifications can also be inherited by the next generation of cells to maintain relatively stable cell characteristics. In recent years, it is increasingly realized that epigenetic mechanisms play a profound role in the regulation of cancer metastasis. Epigenetics is commonly used to describe chromatin-based events that regulate DNA-templated processes. Direct DNA methylation at cytosine on CpG sequences, post-translational modifications (PTMs) of histones, the presence of histone variants, remodeling of nucleosomes, and non-coding RNA mediated targeting are the major epigenetic pathways that regulate many important biological processes [8]. In this chapter, we mainly focus on how DNA methylation and histone modifications play a role in the invasion-metastasis cascade.

Promoter cytosine methylation in CpG dinucleotides inactivates gene expression and can profoundly affect the metastasis cascade. DNA methylation is catalyzed by DNA methyltransferase (DNMT) enzymes. DNMT-1 is responsible for DNA methylation maintenance through its action on semimethylated CpG substrates. DNMT-3A and -3B are the *de novo* methyltransferases to newly methylate cytosine during early embryogenesis [9]. Promoter hypermethylation of certain metastatic suppressor genes is constantly observed in various invasive malignancies. Hypermethylation of the *CDH1* promoter in cancers leads to the loss of E-cadherin expression during cancer progression. A large CpG island in the 5′ proximal promoter region of the *CDH1* gene shows aberrant DNA methylation in many different human carcinomas and correlates with reduced E-cadherin protein expression. Exposure of cancer cells with demethylating agent 5′-aza-2′-deoxycytidine (5Aza-dC) reactivates E-cadherin expression in many cancer cell lines, leading to increased cell aggregation and reduced cell motility and invasiveness [6]. Kisspeptin (KISS-1) gene has been identified as a metastasis suppressor gene in various human

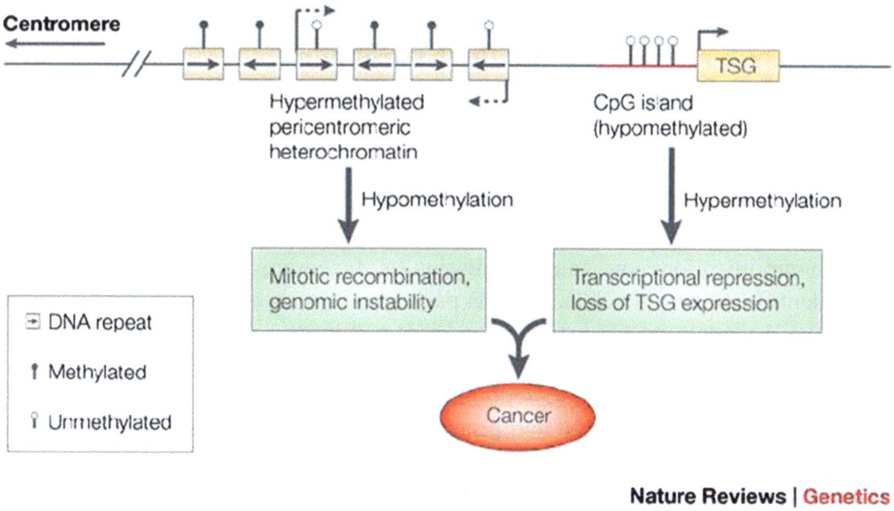


**Fig. 4.4 DNA methylation and cancer**. In normal cells, the repeat-rich, pericentromeric heterochromatin is hypermethylated and is transcriptionally silent. Actively transcribed tumour suppressor gene (*TSG*) is associated with a hypomethylated CpG island (indicated in *red*). In tumour cells, repeat-rich heterochromatin becomes hypomethylated and this contributes to genomic instability through increased mitotic recombination events. De novo methylation of CpG islands in the TSG or the metastasis suppressor gene can result in the transcriptional silencing of these genes, leading to unrestrained growth and spreading of the cancer cells (From K.D. Robertson. DNA methylation and human disease. *Nat Rev Genet* 6: 597–610, 2005. Reprinted with permission from Nature Publishing Group)

malignancies. KISS-1encodes a number of peptides (kp-54, kp-14, kp-13, kp-10), which are endogenous ligands to a G protein-coupled receptor called GPR54. The molecular basis of anti-metastatic activity of KISS-1 is not fully understood. Some evidences suggest that kisspeptin/GPR54 system negatively regulates MMP-9, a member of matrix metalloproteinases that degrade ECM. Secretion of KISS1 has also been demonstrated to be necessary to maintain dormancy in disseminated cancer cells, thus blocking metastatic colonization. Inactivation of KISS-1 gene by promoter hypermethylation has been observed in many invasive cancers such as melanoma, bladder cancer, and gastric carcinoma [10].

Genome-wide hypomethylation is a common feature of many malignant cells (Fig. 4.4). Hypomethylation of regulatory DNA sequences leads to overactivation of some oncogenes or EMT-inducing transcription factors, and general hypomethylation in heterochromatin regions is more deleterious as it results in genomic instability and subsequent multiple genetic defects [11]. Within a breast tumor mass, heterotypic cancer cells can often be sorted to two groups based on their phenotype and specific cell surface markers, namely stem cell-like, more invasive $CD44^+$ $CD24^-$cells and epithelial, more differentiated $CD44^-CD24^+$ cells. Promoter hypomethylation and the resultant overexpression of several EMT-inducing transcription factors have been observed in invasive $CD44^+$ $CD24^-$cells compare to their differentiated $CD44^-CD24^+$ counterparts [12].

## 4.5 Complex Post-Translational Histone Modifications Coordinately Regulate the Expression of Metastasis-Related Genes

In most cases, the role of promoter DNA methylation in gene regulation is relatively straightforward, in that hypermethylation inhibits gene expression whereas hypomethylation activates gene expression. By contrast, the influence of post-translational histone modifications on gene expression is far more complex and context-dependent. In eukaryotic cells, DNA is packaged into chromatin. The nucleosome is the fundamental unit of chromatin and it is composed of an octamer of the four core histones (H3, H4, H2A, H2B) around which 147 base pairs of DNA are wrapped. The core histones are predominantly globular except for their N-terminal "tails", which are unstructured and protrude from the globular core. Various post-translational chemical modifications, such as acetylation, phosphorylation, poly ADP-ribosylation, ubiquitination, and methylation, particularly those on the N-terminal tails, can alone or in combination affect gene transcription through direct remodeling of chromatin structure or recruitment of regulatory non-histone proteins (Fig. 4.5). For one particular histone modification, there are "writers" to catalyze the addition and "erasers" responsible for the removal of the mark. In addition, "readers" are those factors that bind to the specific modifications and respond to the "histone code" information conveyed by upstream signaling cascades. Aberrant histone modifications at either the candidate loci or genome-wide level, often caused by the deregulation of the writer/eraser/reader interplay, plays a significant role in mediating the invasion-metastasis cascade [13–15].

### *4.5.1 Histone Acetylation*

Histone lysine acetylation neutralizes lysine's positive charge and consequently weakens the electrostatic interaction between histones and negatively charged DNA, leading to an "open" chromatin conformation in favor of transcriptional activation. In addition, acetylated lysine residues may serve as a "docking" site for non-histone proteins, which can carry their enzymatic activities and further modify chromatin. Histone acetylation is catalyzed by histone acetyltransferases (HATs) and is removed by histone deacetylases (HDACs). Usually, the catalytic activity of HATs and HDACs is not restricted to one particular lysine residue. Most of the identified human HATs, which mainly include three families, GNAT, MYST, and CBP/p300, function as transcriptional co-activators, and are recruited to chromatin by interacting with sequence-specific DNA-binding proteins [16]. For example, nuclear protein p300 acetylates multiple lysine residues on H3, H4, H2A, and H2B. p300 binds to transcription factor HNF3, and these two factors coordinately activate the transcription of *CDH1* gene, which encodes the metastasis suppressor E-cadherin. Ectopic expression of p300 in certain cancer cells can restore E-cadherin expression and repress their metastatic potential [17].

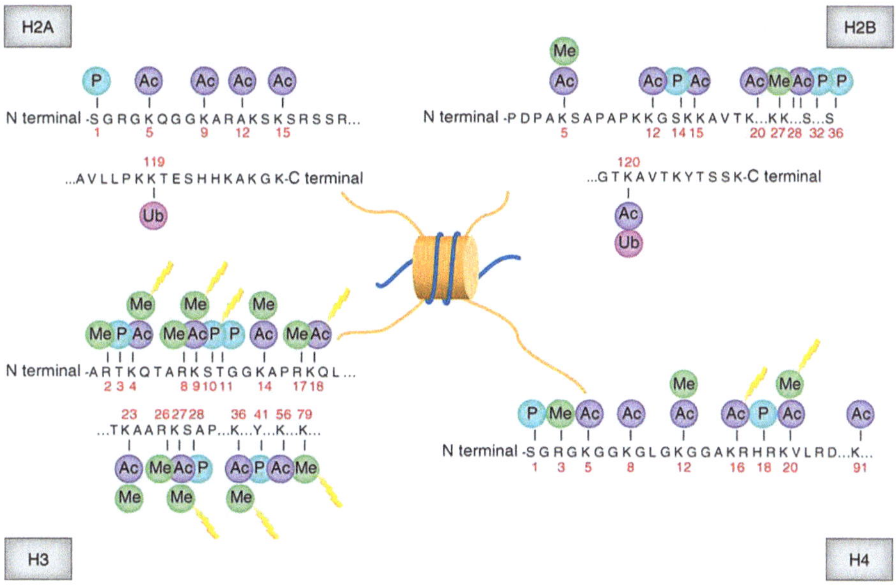


**Fig. 4.5 Histone modification patterns in normal and cancer cells.** Histones can undergo diverse post-translational modifications, especially on their protruding N-terminal tails. In the right combination and translated by the appropriate effectors, these modifications contribute to establishing the global and local condensed or decondensed chromatin states that eventually determine gene expression. This figure depicts the main modifications of the four core histones in normal cells (type and position in the amino acid sequence). Histone modifications typically associated with cancer have also been highlighted. *Ac* acetylation, *Me* methylation, *P* phosphorylation, *Ub* ubiquitination (From M. Rodríguez-Paredes and M. Esteller. Cancer epigenetics reaches mainstream oncology. *Nat Med* 17: 330–339, 2011. Reprinted with permission from Nature Publishing Group)

## 4.5.2 *Histone Deacetylation*

Deacetylation on lysine residues of histone proteins by HDACs are believed to cause chromosomal condensation and gene repression. There are three distinct families of histone deacetylases: the class I and class II histone deacetylases and the class III NAD-dependant enzymes of the Sir family. Class I HDACs include HDAC1, HDAC2, HDAC3, and HDAC8, and they are mainly localized to the nucleus. Class II HDACs include HDAC4-7, HDAC9, and HDAC10, and these proteins can shuttle between the cytoplasm and the nucleus. Class III HDACs are homologs of Sir2, a yeast transcriptional repressor that requires the cofactor $NAD^+$ for its deacetylase activity. Most of classes I HDACs are subunits of multiprotein nuclear complexes that are crucial for transcriptional repression [18]. At the E-cadherin promoter, EMT-inducing transcription factor Snail1 binds to the specific sequence called E-box elements, and recruits a repressive complex consisting of HDAC1, HDAC2 and SIN3A. Other EMT-inducing transcription factors Snail2/Slug, ZEB1, and ZEB2 similarly repress E-cadherin transcription by recruiting repressive protein complexes containing different class I HDACs [19].

### 4.5.3 Histone Methylation

Histone methylation occurs on all basic residues: arginines, lysines and histidines. The best-characterized sites of histone methylation are those that occur on lysine residues. The histone lysine methyltransferases and demethylases are responsible for addition or removal of methyl groups from different lysine residues on histones. These enzymes are highly specific, in that each enzyme regulates mono-, di-, or trimethylation of a single or a few lysine residues on histones. Because lysine methylation does not usually alter the charge of histone proteins, this modification per se has little influence on the interaction between histones and DNA. Instead, the location of the methyl-lysine residue on a histone tail and the degree of methylation (whether mono-, di-, or trimethylation) serve as the docking signal to recruit various reader proteins containing methyl-binding domains. To date, many methyl-binding domains have been identified. The classic methyl-binding domains are the Royal superfamily, including chromodomains, double chromodomains, chromobarrels, Tudor domains, double or tandem Tudor domains and the malignant brain tumor (MBT) repeats. Proteins with other domains, such as PHD finger, WD40 repeats, CW domains, PWWP domains, ankyrin repeats, have also been reported to recognize and bind to methylated lysine residues. Functionally, the combinations of the recruited regulatory proteins ultimately determine whether the transcription of local genes is activated or repressed [20].

The most extensively studied histone methylation sites include histone H3 lysine 4 (H3K4), H3K9, H3K27, H3K36, H3K79 and H4K20. The effects of methylation on gene transcription are diverse and context-dependent. In general, H3K4me3 is associated with active transcription or with genes that are poised for activation, whereas H3K27me3 is associated with repressed chromatin. H3K4me1 is often associated with enhancer function, whereas H3K4me3 is linked to promoter activity [21].

Aberrant histone methylation plays a role in cancer progression. Changes in global levels of certain histone methylation events are correlated with increased cancer recurrence and poor survival [20]. Mutations in or altered expression of histone methyltransferases correlate with various invasive cancers. These enzymes are often critical to the transcriptional regulation of key metastasis-related genes.

The SET-domain containing proteins and DOT1-like proteins have been shown to methylate lysine residues in histone and non-histone substrates. Enhancer of zeste homolog 2 (EZH2) contains a SET domain and is the catalytic component of the PRC2 complex, which is primarily responsible for catalyzing the trimethylation of histone H3 lysine 27 (H3K27me3). EZH2 has been found to be overexpressed in metastatic prostate cancer, and the expression level of EZH2 directly correlates with the aggressiveness of breast cancer. Ectopic expression of EZH2 in immortalized human mammary epithelial cell lines promotes anchorage-independent growth and cell invasion. In the molecular level, EZH2 usually acts as a transcriptional repressor through its H3K27me3 methyltransferase activity. Presumably, EZH2 promotes cancer metastasis by transcriptional inhibition of anti-metastasis genes. EZH2 is recruited to the promoter region of *CDH1* and represses E-cadherin expression in an invasive prostate cancer cell line DU145 [22]. Interestingly, loss-of-function mutations in EZH2 gene confer a poor prognosis in certain hematopoietic malignancies, suggesting a tumor-suppressive role for EZH2 in these cell lineages [23, 24]. To date, the precise role of gain and loss of EZH2 activity in cancers and the underlying molecular mechanisms are an area of active investigation.

G9a (also known as KMT1C or EHMT2) catalyzes dimethylation of histone lysine 9 (H3K9me2). In TGF-β induced EMT cell line models, it has been demonstrated that transcription factor Snail recruits G9a and DNA methyltransferases to the *CDH1* promoter, leading to the inhibition of E-cadherin expression and the induction of EMT. Knockdown of G9a restores E-cadherin expression, inhibits cell migration and invasion, and reduces lung colonization of breast cancer metastasis [25].

SET8 (also known as PR-Set7/9, SETD8, KMT5A), a member of the SET domain-containing methyltransferase family, catalyzes monomethylation of H4K20. Recent studies indicate that SET8 and the transcription factor Twist are functionally interdependent to promote EMT and enhance the invasiveness of breast cancer cells in vitro and in vivo. SET8 is recruited to the *CDH1* promoter by Twist and repress E-cadherin expression; interestingly, SET8 and Twist are also present at the N-cadherin promoter, where they enhance the transcription of the gene. Together, SET8 and Twist coordinately promote EMT and cell invasiveness, and the dual function of SET8 at different promoters reinforce the notion the effect of histone methylation on gene transcription is context-dependent, relying on the combinations of histone modifications nearby and the distinct sets of regulatory proteins recruited [26].

### 4.5.4 Histone Demethylation

Two families of demethylases, the amine oxidases and jumonji C (JmjC)-domain-containing, iron-dependent dioxygenases, have been identified thus far to demethylate methyl-lysines [27]. Lysine-specific demethylase 1 (LSD1) is an amine oxidase that catalyzes the removal of mono- and di-methylation from histone H3 lysine 4 (H3K4). LSD1 regulates several intracellular signaling pathways including that of TGFβ1, which plays a critical role in cancer metastasis as we mentioned earlier in this chapter. In breast cancer tissue samples, the level of LSD1 is negatively correlated with that of TGFβ1. LSD1 downregulates TGFβ1 and inhibits the invasiveness and metastasis of breast cancer cells in vitro and in vivo [28]. Responsible for the removal of H3K4 trimethylation, JARID1B/PLU-1 belongs to the JmjC family of demethylases. Through its enzymatic activity, JARID1B removes the active H3K4me3 marks at the promoter regions of various genes, including CCL14, an epithelial derived chemokine. JARID1B inhibits the expression of CCL14 and suppresses the angiogenic and metastatic potential of breast cancer cells [29].

In many cases, multiple histone and DNA modification enzymes collaborate to regulate target gene expression. These factors may form a stable protein complex; alternatively they can assemble and dissemble dynamically at different chromatin locations. H3K27me3 methyltransferase EZH2 physically interacts with DNA methyltransferases at the promoters of certain target genes, leading to the silence of these genes. The transcriptional repressive protein complex NuRD contains multiple components with different chromatin-related activities, and these components include chromatin remodeling factors (metastasis tumor antigen, MTA), histone deacetylases (HDAC1 and HDAC2), histone binding proteins (RbAp46 and RbAp48), methyl CpG-binding proteins (MBD2 and MBD3), DNA helicase/ATPase (Mi-2α/β), and in certain circumstances histone demethylases (LSD1 and JARID1B). Except for the core components, multiple NuRD subunits could be dynamically incorporated into the big protein complex at different chromatin locations, allowing more specific and fine-tuned regulation of different target genes [30, 31].

## 4.6 Epigenetic Therapy for Cancer Metastasis

Unlike genetic defects, epigenetic aberrations that lead to cancer metastasis are theoretically reversible, making them the ideal drug targets for cancer therapeutics. By inhibition of the particular modification enzymes that are responsible for the addition or removal of the chemical groups from DNA or histones, we may adjust the aberrant epigenome and the abnormal gene expression in cancer cells (Fig. 4.6). Although effective inhibitors of many DNA or histone modification enzymes have been developed, these epigenetic drugs have certain embedded problems that may limit their applications until we get better understanding of the molecular mechanisms of cancer epigenetics. To what degree an inhibitor is specific to a particular enzyme? To what degree inhibition of this enzyme is specific to regulate growth- and metastasis-associated genes in cancer cells? Does inhibition of epigenetic modification enzymes affect the biological function of non-cancer cells? These problems directly affect the efficacy and toxicity of the potential epigenetic anti-cancer drugs. Despite the above concerns and the lack of detailed molecular mechanisms, there are several epigenetic drugs that have been shown to effectively reverse metastatic phenotype in different cancer cell lines, and many clinical studies have evaluated specific enzyme inhibitors in the treatment of cancer metastasis [6, 32, 33].

As we have repeatedly stated earlier, the E-cadherin encoding gene *CDH1* is a critical metastasis suppression gene. Transcription of *CDH1* gene is silenced or reduced in many cancers with high metastasis potential. Reactivation of the *CDH1* gene expression is an important anti-metastasis therapeutic strategy. It has been demonstrated that inhibition of the activity of either DNMTs or histone modification enzymes that remove certain repressive histone modifications can effectively upregulate E-cadherin expression. Treatment of several invasive cancer cell lines with the DNMT inhibitor 5-aza-2′-deoxycytidine leads to restoration of E-cadherin expression and reversion of these cells to the epithelial phenotype. DNMT inhibitors 5-azacytidine and 2′-deoxy-5-azacytidine have been approved for the treatment of myelodysplastic syndrome, a pre-leukemic bone marrow disorder. It is important

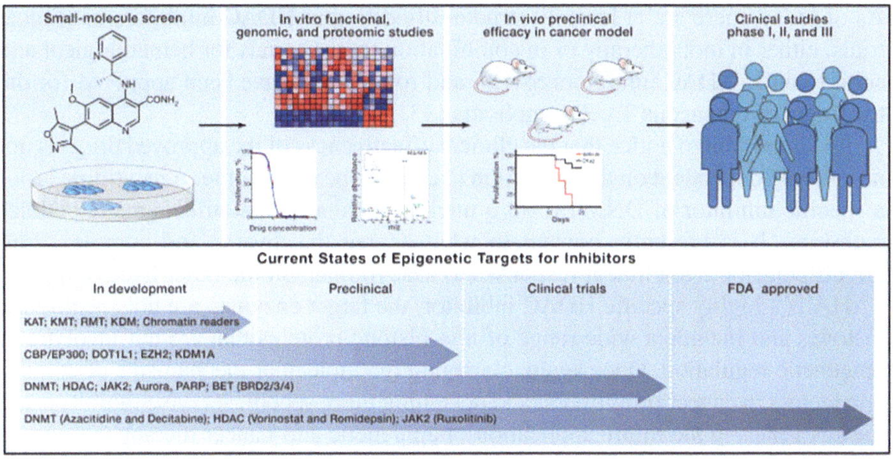


**Fig. 4.6 The process to develop epigenetic drugs and the current status of various epigenetic therapies in cancer.** Candidate small-molecule inhibitors are first tested in vitro in malignant cell lines for specificity and phenotypic response. These may, in the first instance, assess the inhibition of proliferation, induction of apoptosis, or cell-cycle arrest. These phenotypic assays are often coupled to genomic and proteomic methods to identify potential molecular mechanisms for the observed response. Potentially effective inhibitors are then tested in vivo in animal models of cancer to ascertain whether they may provide therapeutic benefit in terms of survival. Animal studies also provide valuable information regarding the toxicity and pharmacokinetic properties of the drug. Based on these preclinical studies, candidate molecules may be taken forward into the clinical setting. *KAT* histone lysine acetyltransferase, *KMT* histone lysine methyltransferase, *RMT* histone arginine methyltransferase, *PARP* poly ADP ribose polymerase (From M.A. Dawson and T. Kouzarides. Cancer epigenetics: From mechanism to therapy. *Cell* 150: 12–27, 2012. Reprinted with kind permission from Elsevier Limited)

to bear in mind the exact role of the DNMT inhibitors in cancer progression is not fully understood, and in some cases these drugs can elicit opposite response in different cancer cell lines. Recent data showed that treatment of MCF-7 breast cancer cells, which are largely epithelial, with 5-aza-2′-deoxycytidine may increase their ability for invasion and metastasis, concomitant with the upregulation of several pro-invasive genes. These observations raise concerns about the potential use of DNA methyltransferase inhibitors for the treatment of breast cancer [6].

HDAC inhibitors (HDACi) have been shown in preclinical studies to selectively target cancer cells with high specificity. The effects of HDACi include the induction of apoptosis and cell cycle arrest, and suppression of angiogenesis and tumor cell invasion. The anti-metastasis activity of HDACi is at least partially dependent on their potent capacity to upregulate E-cadherin expression. Butyrate, the first HDACi to be identified, induces cell cycle arrest and enhances cell-cell adhesion in two breast cancer cell lines, and these phenotypic changes can be inhibited by the addition of E-cadherin antibodies. Subsequently, butyrate was found to upregulate E-cadherin expression in colon cancer cells and in endometrial carcinoma cells.

As of today, there are at least 20 structurally different HDAC inhibitors in clinical trials, either in monotherapy or in combination therapy trials for hematological and solid tumors. HDAC inhibitors SAHA and romidepsin have been approved for the treatment of cutaneous T cell lymphoma [33, 34].

It is important to notice that the clinical effectiveness of the approved drugs is not necessarily dependent on their function to change the epigenome. Azacytidine is not a specific inhibitor of DNMTs, but a nucleoside analog that affects many cellular pathways. It is presently unclear to what degree the diverse and complex drug effects contribute to clinical responses to azacytidine. On the other hand, although SAHA is a highly specific HDAC inhibitor, the target enzymes are not specific for histones and include a wide range of non-histone proteins that are not involved in epigenetic regulation. Once again, clarifying the molecular mechanisms of epigenetic factors in cancer progression and designing more specific enzyme inhibitors are the key issues in the future application of epigenetic anti-cancer therapy.

## 4.7 Concluding Remarks and Future Perspective

Metastasis is life-threatening and accounts for 90 % of cancer mortality. Along the invasion-metastasis cascade, the initial commitment for carcinoma cells to move out is the EMT process. Extensive investigations have identified key signaling pathways and transcription factors that mediate EMT. Most of our current knowledge of metastasis epigenetics is essentially about the deregulated epigenome leading to abnormal expression of genes that induce or inhibit EMT. By contrast, while colonization in and adaptation to the distant foreign tissues is the final rate-limiting step for metastasis, our understanding to the molecular mechanisms of colonization is still limited, partly due to the difficulty to establish experimental models of this step. In fact, many patients may already have occult micrometastases at the time of primary cancer diagnosis; therefore targeting the final colonization step is a more appropriate therapeutic strategy to treat metastatic cancers. It is no doubt we should expect more studies to elucidate the molecular mechanisms governing metastatic colonization and the underlying epigenetic factors contribute to this process.

In recent years, we have witnessed the rapid development of many genome-wide based technologies. Global alteration of epigenomics is constantly observed in various types of invasive cancers. The integration of these data with the information coming from genomics and transcriptomics will exponentially expand understanding of cancer metastasis and yield better epigenetic biomarkers for detection, prognosis and therapy prediction. Considering that the effects of most of the epigenetic drugs are still nonspecific and may cause undesirable side effects, it will be necessary to design new agents against specific enzymes of the epigenetic machinery involved in specific types of cancer. At that time, epigenetics will truly enter the center stage of cancer research.

## References

1. Eccles SA, Welch DR (2007) Metastasis: recent discoveries and novel treatment strategies. Lancet 369(9574):1742–1757
2. Valastyan S, Weinberg RA (2011) Tumor metastasis: molecular insights and evolving paradigms. Cell 147(2):275–292
3. Weinberg RA (2007) The biology of cancer: garland Science. Taylor & Francis, New York
4. Polyak K, Weinberg RA (2009) Transitions between epithelial and mesenchymal states: acquisition of malignant and stem cell traits. Nat Rev Cancer 9(4):265–273
5. Yang J, Weinberg RA (2008) Epithelial-mesenchymal transition: at the crossroads of development and tumor metastasis. Dev Cell 14(6):818–829
6. Wang Y, Shang Y (2013) Epigenetic control of epithelial-to-mesenchymal transition and cancer metastasis. Exp Cell Res 319(2):160–169
7. Smith SC, Theodorescu D (2009) Learning therapeutic lessons from metastasis suppressor proteins. Nat Rev Cancer 9(4):253–264
8. Allis CD, Jenuwein T, Reinberg D (2007) Epigenetics. Cold Spring Harbor Laboratory Press, New York
9. Jones PA (2012) Functions of DNA methylation: islands, start sites, gene bodies and beyond. Nat Rev Genet 13(7):484–492
10. Makri A, Pissimissis N, Lembessis P, Polychronakos C, Koutsilieris M (2008) The kisspeptin (KiSS-1)/GPR54 system in cancer biology. Cancer Treat Rev 34(8):682–692
11. Robertson KD (2005) DNA methylation and human disease. Nat Rev Genet 6(8):597–610
12. Bloushtain-Qimron N, Yao J, Snyder EL, Shipitsin M, Campbell LL, Mani SA et al (2008) Cell type-specific DNA methylation patterns in the human breast. Proc Natl Acad Sci USA 105(37):14076–14081
13. Wu CY, Tsai YP, Wu MZ, Teng SC, Wu KJ (2012) Epigenetic reprogramming and post-transcriptional regulation during the epithelial-mesenchymal transition. Trends Genet 28(9):454–463
14. Kouzarides T (2007) Chromatin modifications and their function. Cell 128(4):693–705
15. Strahl BD, Allis CD (2000) The language of covalent histone modifications. Nature 403(6765):41–45
16. Chen H, Tini M, Evans RM (2001) HATs on and beyond chromatin. Curr Opin Cell Biol 13(2):218–224
17. Liu YN, Lee WW, Wang CY, Chao TH, Chen Y, Chen JH (2005) Regulatory mechanisms controlling human E-cadherin gene expression. Oncogene 24(56):8277–8290
18. Narlikar GJ, Fan HY, Kingston RE (2002) Cooperation between complexes that regulate chromatin structure and transcription. Cell 108(4):475–487
19. Peinado H, Olmeda D, Cano A (2007) Snail, Zeb and bHLH factors in tumour progression: an alliance against the epithelial phenotype? Nat Rev Cancer 7(6):415–428
20. Greer EL, Shi Y (2012) Histone methylation: a dynamic mark in health, disease and inheritance. Nat Rev Genet 13(5):343–357
21. Chi P, Allis CD, Wang GG (2010) Covalent histone modifications–miswritten, misinterpreted and mis-erased in human cancers. Nat Rev Cancer 10(7):457–469
22. Cao Q, Yu J, Dhanasekaran SM, Kim JH, Mani RS, Tomlins SA et al (2008) Repression of E-cadherin by the polycomb group protein EZH2 in cancer. Oncogene 27(58):7274–7284
23. Nikoloski G, Langemeijer SM, Kuiper RP, Knops R, Massop M, Tonnissen ER et al (2010) Somatic mutations of the histone methyltransferase gene EZH2 in myelodysplastic syndromes. Nat Genet 42(8):665–667
24. Ernst T, Chase AJ, Score J, Hidalgo-Curtis CE, Bryant C, Jones AV et al (2010) Inactivating mutations of the histone methyltransferase gene EZH2 in myeloid disorders. Nat Genet 42(8):722–726
25. Dong C, Wu Y, Yao J, Wang Y, Yu Y, Rychahou PG et al (2012) G9a interacts with Snail and is critical for Snail-mediated E-cadherin repression in human breast cancer. J Clin Invest 122(4):1469–1486

26. Yang F, Sun L, Li Q, Han X, Lei L, Zhang H et al (2011) SET8 promotes epithelial-mesenchymal transition and confers TWIST dual transcriptional activities. EMBO J 31(1):110–123
27. Mosammaparast N, Shi Y (2010) Reversal of histone methylation: biochemical and molecular mechanisms of histone demethylases. Annu Rev Biochem 79:155–179
28. Wang Y, Zhang H, Chen Y, Sun Y, Yang F, Yu W et al (2009) LSD1 is a subunit of the NuRD complex and targets the metastasis programs in breast cancer. Cell 138(4):660–672
29. Li Q, Shi L, Gui B, Yu W, Wang J, Zhang D et al (2011) Binding of the JmjC demethylase JARID1B to LSD1/NuRD suppresses angiogenesis and metastasis in breast cancer cells by repressing chemokine CCL14. Cancer Res 71(21):6899–6908
30. Lai AY, Wade PA (2011) Cancer biology and NuRD: a multifaceted chromatin remodelling complex. Nat Rev Cancer 11(8):588–596
31. Li DQ, Pakala SB, Nair SS, Eswaran J, Kumar R (2012) Metastasis-associated protein 1/nucleosome remodeling and histone deacetylase complex in cancer. Cancer Res 72(2):387–394
32. Mund C, Lyko F (2010) Epigenetic cancer therapy: proof of concept and remaining challenges. Bioessays 32(11):949–957
33. Dawson MA, Kouzarides T (2012) Cancer epigenetics: from mechanism to therapy. Cell 150(1):12–27
34. Minucci S, Pelicci PG (2006) Histone deacetylase inhibitors and the promise of epigenetic (and more) treatments for cancer. Nat Rev Cancer 6(1):38–51
35. Xu J, Lamouille S, Derynck R (2009) TGF-[beta]-induced epithelial to mesenchymal transition. Cell Res 19:156–172
36. Rodríguez-Paredes M, Esteller M (2011) Cancer epigenetics reaches mainstream oncology. Nat Med 17:330–339

# Part II
# Signaling Pathway and Cancer

# Chapter 5
# Regulatory Effects of Arsenic on Cellular Signaling Pathways: Biological Effects and Therapeutic Implications

**Elspeth M. Beauchamp, Ruth Serrano, and Leonidas C. Platanias**

**Abstract** Arsenic compounds exert important biological effects and arsenic trioxide has been approved by the Food and Drug Administration (FDA) for the treatment of patients with acute promyelocytic leukemia (APL). Much of arsenic's actions in cells reflect its ability to bind thiol groups in cellular proteins or to affect the production of reactive oxygen species (ROS), leading to the engagement and regulation of several cellular signaling pathways. Arsenic has been also shown to degrade abnormal fusion proteins found in myeloid leukemias. It has also been shown to effect NFκB, MAPK, mTOR and Hedgehog pathways which can modulate the viability of cancer cells. Many clinical trials have been performed to examine the clinical efficacy of arsenic trioxide alone or in combination with other agents in the treatment of various hematological malignancies. The continuous advances in basic and translational research and the better understanding of the mechanisms of action of arsenic should lead to more effective combinations with other agents that could result in better clinical outcomes.

**Keywords** Arsenic • Leukemia • Cancer • Cell signaling

## 5.1 Clinical Uses of Arsenic Trioxide

Arsenic has been used empirically for centuries, for the treatment of countless diseases, including syphilis, cancer, malaria, and ulcers [1]. It was first described as a drug to treat leukemia in 1878 [2]. In the modern medical era, one of the compounds of arsenic, arsenic trioxide, has shown significant clinical activity in certain malignant diseases, as discussed below.

E.M. Beauchamp, Ph.D. • R. Serrano, M.D. • L.C. Platanias, M.D., Ph.D. (✉)
Robert H. Lurie Comprehensive Cancer Center, Northwestern University, Chicago, IL, USA
e-mail: e-beauchamp@northwestern.edu; ruth_csp@hotmail.com; l-platanias@northwestern.edu

R. Kumar (ed.), *Nuclear Signaling Pathways and Targeting Transcription in Cancer*, Cancer Drug Discovery and Development, DOI 10.1007/978-1-4614-8039-6_5,

### 5.1.1 Acute Promyelocytic Leukemia (APL)

Over the last two decades there has been extensive evidence accumulated indicating that arsenic trioxide (ATO) has major clinical activity in the treatment of one form of acute myeloid leukemia (AML), acute promyelocytic leukemia (APL) and ATO was approved for the treatment of relapsed APL by the Food and Drug Administration (FDA) of the United States in 2001 [3]. This relatively rare variant of AML is associated with the reciprocal chromosomal translocation t(15;17) that brings together the promyelocytic leukemia (PML) gene on chromosome 15 and the retinoic acid receptor (RAR)α gene on chromosome 17 [4]. The resultant chimeric protein (PML–RARα) causes a maturation block of myeloid cells at the promyelocytic stage, resulting in the accumulation of abnormal promyelocytes in the bone marrow [4]. Historically, APL has been associated with a severe bleeding dysfunction associated with disseminated intravascular coagulation (DIC) and a fatal course of only weeks [5]. With the implementation of chemotherapy, a complete remission (CR) rate of 75–80 % in newly diagnosed patients was achieved, however the median duration of remission ranged from 11 to 25 months, with only 35–45 % of the patients being cured [4]. The introduction of a regimen consistent of *all-trans* retinoic acid (ATRA), which targets the RAR moiety of the fusion transcript, together with anthracycline-based chemotherapy dramatically raised the remission rate up to 90–95 % and the 5-year disease free survival (DFS) to 74 % [6]. Since the early 1990s, ATO was introduced for the treatment of relapsed APL, and has shown major clinical activity [7]. Since ATO is less toxic than chemotherapy, its role as a single agent in newly diagnosed patients is currently being researched with the aim to minimize the use of cytotoxic chemotherapy in this condition, especially for those with a compromised cardiac function and/or for older patients [8, 9].

### 5.1.2 Clinical Trials of ATO in Multiple Myeloma

In vitro studies have shown that ATO induces apoptosis in myeloma cells [10–13], therefore investigators have evaluated its potential in the treatment of refractory and relapsed multiple myeloma (MM) [14]. Some clinical activity was seen in a phase II study performed in 14 patients with refractory or relapsed MM [15]. In another trial using a higher but not as frequent dose of ATO, reduction of M-protein in serum of more than 25 % was obtained in eight patients (33 %), while six patients had stable disease, with a median duration response time of 130 days [16]. Investigators have also developed combination studies using ATO together other agents previously known to be useful for the treatment of this condition. Berenson et al. administered a combination of melphalan, ATO and ascorbic acid to 65 patients with MM who had failed more than two previous regimens [17]. This combination (also known as MAC regimen) produced objective responses in 31 patients (48 %), ranging from CR in two patients to minor responses in 14 of them [17]. More recently, the combination of MAC regimen plus bortezomib was evaluated in a different randomized trial and was found to be safe and well tolerated by patients [18]. Other combination regimens including ATO have also demonstrated efficacy in patients with relapsed or refractory MM [19].

### *5.1.3 Myelodysplastic Syndromes*

There has been also evidence for some clinical activity of ATO in the treatment of myelodysplastic syndromes (MDS). Hematologic improvement was obtained in MDS with the use of single agent ATO in two different trials [20, 21]. In other studies, thalidomide was used in combination with ATO in 28 patients with transfusion dependent MDS, accomplishing a response in 25 % of them, including one CR and responses in three of five patients with high baseline levels of EVI1, which is a known poor prognostic marker [22]. More recently, the combination of thalidomide, ATO, dexamethasone, and ascorbic acid (TADA regimen) was used in patients with myelodysplastic/myeloproliferative neoplasms (MDS/MPN) or primary myelofibrosis (PMF), achieving a response in 29 % of patients [23].

## 5.2 Effects of Arsenic on Cellular Signaling Pathways in Malignant Cells

### *5.2.1 Arsenic Compounds*

Arsenic is found is two different oxidative states, As (III) or trivalent arsenic and As(V) or pentavalent arsenic. Pentavalent arsenic can substitute for phosphate and cause hydrolysis of compounds such as ATP [24]. Trivalent arsenic can bind to thiol groups in the cysteines of proteins in cells and alter their structure resulting in the modulation of protein stability, folding, and function, thus affecting cellular signaling pathways [24–26]. For instance, arsenic can bind to tubulin and other cytoskeletal proteins and affect polymerization and mitosis [27–30]. Arsenic can also affect signaling pathways through the production of reactive oxygen species (ROS) and there is evidence that it increases ROS in cells in two ways. First, arsenic can inhibit the activity of enzymes, such as thioredoxin reductase by its ability to bind via cysteine groups, which are involved in regulating the cellular redox state [31]. Second, methylation of arsenic during its cellular metabolism also leads to the production of ROS [32, 33].

### *5.2.2 Effects on Fusion Proteins in Leukemia*

Arsenic trioxide has been shown to cause the degradation of multiple fusion proteins found in leukemia by various mechanisms. ATO's proposed mechanism of action in acute promeylocytic leukemia is via degradation of the PML-RAR fusion protein [34]. In APL, the fusion protein alters the localization of PML from nuclear bodies, which contributes to aberrant cell growth [35, 36]. Arsenic trioxide targets both PML and PML-RAR to nuclear bodies in APL cells and leads to its subsequent degradation [37]. Targeting PML protein expression with arsenic has also been shown in quiescent leukemia initiating stem cells in CML [38]. A recent publication demonstrated that arsenic specifically binds to the PML zinc finger domain at cysteine residues displacing the zinc and causing a shift in secondary structure as well as aggregation that leads to increased sumolyation and degradation [39, 40]. Another recent publication showed that autophagy induction by ATO and ATRA also contributes to the degradation of the PML-RAR fusion protein [41].

Besides APL, arsenic has shown cytotoxicity in chronic myleogenous leukemia (CML), as well. It is of particular interest that historically, arsenic was used to treat CML in the nineteenth and twentieth centuries [1]. Imatinib combined with arsenic sulfide showed enhanced anti-leukemic effects over either agent alone in a mouse model of CML [42]. Recent evidence has shown that arsenic is cytotoxic in Ph + leukemia cells by degradation of the BCR-ABL fusion protein by the autophagic machinery, where p62 binds to BCR-ABL in the autophagosome [43]. Arsenic trioxide has been also shown to degrade another fusion protein, AML1/MDS1/EVI1, via targeting of the MDS1/EVI1 portion of the fusion protein [44]. The EVI1 portion contains two zinc finger DNA binding domains therefore similar to PML, arsenic could be binding to the cysteine residues in zinc finger domains in EVI1 and lead to the degradation of the fusion protein [40, 44].

### *5.2.3 mTOR Pathway*

Arsenic has been shown to activate the mTOR pathway although the precise mechanism of such engagement is unknown (Fig. 5.1) [45]. Treatment with rapamycin or the dual PI3K/mTOR inhibitor, PI-103, was shown to enhance the antileukemic effects of arsenic, indicating that activation of mTOR occurs in a negative feedback manner in order to suppress the cytotoxic effects of arsenic [45, 46]. Therefore combining arsenic with mTOR pathway inhibitors could conceivably enhance its antileukemic effects in vivo and this needs to be examined in future work.

### *5.2.4 MAPK Pathways*

Arsenic has been shown to affect the various MAPK pathways such as p38 MAPK, JNK and ERK. JNK activation has been shown to be important for the anti-leukemic effects of arsenic (Fig. 5.1) [47–49]. ATO-resistant APL cell lines showed little activation of JNK due to upregulation of glutathione (GSH) [47]. Treating cells with compounds that deplete GSH in cells enhance ATO's cytotoxic effects [47, 50]. Increased GSH levels in leukemia cells has been correlated with a decrease in

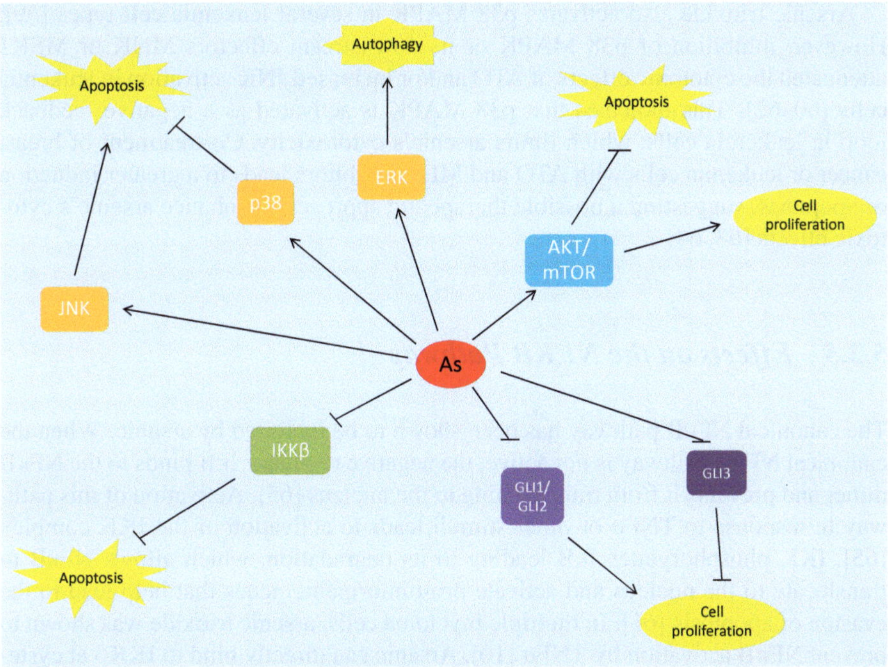


**Fig. 5.1** Arsenic's positive and negative effects on cell viability and proliferation. Arsenic can affect MAPK pathways, by activating the MEK/ERK branch leading to the induction of autophagy. At the same time it can either activate p38 or JNK leading to the inhibition, or induction of apoptosis. Additionally, arsenic can activate the PI3K/mTOR pathway by activation of AKT signaling or mTOR signaling which leads to the inhibition of apoptosis and increase in cellular proliferation. Arsenic can inhibit GLI1 and GLI2 which leads to an inhibition of cellular proliferation. Arsenic's inhibition of GLI3, however, can lead to activation of cellular proliferation in some cellular contexts

sensitivity to arsenic, which could affect sensitivity by either GSH decreasing the amount of ROS in cells directly, or binding arsenic leading to its metabolism and subsequent excretion [51–53]. Ascorbic acid has been shown to synergize with arsenic in multiple myeloma and myeloid leukemia cells by decreasing GSH levels and increasing ROS levels [52, 54, 55]. In chronic lymphocytic leukemia (CLL), JNK activation was an early event leading to the upregulation of PTEN, which results in PI3K, AKT, NFκB inhibition, and an increase in ROS production [56]. In addition, combining arsenic with PI3K inhibitors was shown to enhance arsenic's cytotoxic effects on CLL cells [56].

Other studies have shown that arsenic modulates ERK activity. The induction of autophagy by arsenic trioxide was shown to be important for its antileukemic properties and the ERK pathway is required for induction of the autophagic state in this context [57]. ATO-dependent ERK2-mediated phosphorylation of PML has also been shown to lead to increased sumoylation/degradation of the PML protein and ultimately resulting in induction of apoptosis [58].

Arsenic trioxide also activates p38 MAPK in several leukemia cell types [59]. However, inhibition of p38 MAPK or its downstream effectors MNK or MSK1 attenuated the cytotoxic effects of ATO and/or increased JNK activation in leukemia cells [60–62]. This indicates that p38 MAPK is activated as a negative feedback loop in leukemia cells, which limits arsenic's cytotoxicity. Co-treatment of breast cancer or leukemia cells with ATO and MEK inhibitors leads to a greater induction of apoptosis, suggesting a possible therapeutic approach to enhance arsenic's cytotoxic effects [63, 64].

### 5.2.5 Effects on the NFKB Pathway

The canonical NFκB pathway has been shown to be inhibited by arsenic. When the canonical NFκB pathway is not active, the negative regulator IκB binds to the NFκB dimer and prevents it from translocating to the nucleus [65]. Activation of this pathway in response to TNFα or other stimuli leads to activation of the IKK complex [65]. IKK phosphorylates IκB leading to its degradation, which allows NFκB to translocate to the nucleus and activate pro-tumorigenic genes that help lead to the evasion of apoptosis [66]. In multiple myeloma cells, arsenic trioxide was shown to prevent NFκB activation by TNFα [10]. Arsenic can directly bind to IKKβ at cysteine residue 179 in the activation loop of the catalytic subunit of IKKβ and inhibit its activity, to engage the NFκB canonical signaling (Fig. 5.1) [67]. Since IKKβ can have effects independently of NFκB such as by regulating MAPK and mTOR pathways [66], the inhibition of IKKβ by arsenic can also conceivably effect those pathways in addition to NFκB.

### 5.2.6 Hedgehog Pathway

Recent work has shown that arsenic can inhibit the hedgehog pathway by inhibiting GLI1/2 (Fig. 5.1) [68, 69]. Such inhibition was shown to be at the level of GLI1/2 because ATO was found to inhibit hedgehog signaling when GLI1/2 was overexpressed or in SUFU$^{-/-}$ MEFs, in contrast to upstream pathway inhibitors that cannot inhibit Hh signaling in this context [68, 69]. Notably, some tumors activate the pathway by overexpression of ligand, patched inactivation or mutations that activate Smoothened [70–76]. Other cancers, however, can activate the pathway at the level of GLI, independent of Smoothened or Patched, either by mutations in negative regulators SUFU or REN, chromosomal amplification of GLI, chromosomal translocations that involve GLI, an increase in GLI protein stability or activation via non-canonical mechanisms involving other pathways [77–89]. Arsenic is able to inhibit the growth of both upstream activated medulloblastoma cancer cell lines as well as Ewing sarcoma cells lines which have activation of GLI1 independently of SMO [68].

The exact mechanisms by which arsenic inhibits GLI1/2 still need further investigation. Since one of the studies demonstrated that arsenic can directly bind to GLI1 [68] and given prior evidence of arsenic's ability to bind to cysteines in the zinc finger domain of PML, it is highly plausible that ATO binds to the zinc finger domains in GLI1. However, this remains to be directly addressed in future studies and the overall mechanisms by which arsenic affects GLI function necessitates further investigation.

Another study showed that arsenic activates Hedgehog signaling [90]. The authors of that study found that arsenic activated GLI1/Hedgehog signaling in these cells by inhibiting the GLI3 repressor. However, in this study sodium arsenite was used, whereas the other two studies used arsenic trioxide. It is possible that sodium arsenite has preferential binding to GLI3 over GLI1 and GLI2 and thus activates signaling instead of repressing it. Notably, sodium arsenite has been previously shown to have opposing effects to the ones of arsenic trioxide in other cancer models. For instance, arsenic trioxide promotes apoptosis in breast cancer cell lines [91, 92], while sodium arsenite binds to the estrogen receptor-α (ER-α) and increases the proliferation of MCF-7 cells [93]. Arsenic trioxide and sodium arsenite have been also shown to exhibit differential effects when combined with radiation [94].

The precise mechanisms of how arsenic induces autophagy are not known, other than the requirement for MEK/ERK signaling [54]. Recent evidence suggests that the hedgehog pathway antagonizes autophagy through inhibition of autophagosome synthesis most likely through repression of genes required for autophagy [95]. Thus, the inhibition of the hedgehog pathway by arsenic could mechanistically contribute to its ability to induce autophagy and this hypothesis remains to be examined in future studies.

### *5.2.7 Effects on Nuclear Receptor Pathways*

Arsenic has been shown to alter multiple nuclear receptor pathways. Notably, it has been shown to directly bind and inhibit the glucocorticoid receptor [96]. Nuclear receptor function has been shown to be inhibited by arsenic trioxide though JNK activation and phosphorylation of the retinoid X receptor (RXR) [97]. Arsenic's effects on the estrogen receptor are controversial as multiple groups have shown differential effects. As mentioned previously, sodium arsenite can bind to the ligand pocket of ER-α and activate it, leading to proliferation of MCF-7 cells [93]. Arsenic trioxide was shown to lead to a decrease in expression of ER-α in ER-positive breast cancer cell lines, resulting in suppression of cellular proliferation [98, 99]. More recently arsenic trioxide treatment was found to result in increased expression of ER-α in ER-negative breast cancer cells by promoting demethylation of the promoter, leading to re-sensitization to endocrine therapy [100, 101]. The differences in effects may be due to differences in cell contexts (ER-positive vs. ER-negative cells) as well as mentioned previously the differential effects of sodium arsenite and arsenic trioxide.

## References


1. Waxman S, Anderson KC (2001) History of the development of arsenic derivatives in cancer therapy. Oncologist 6(Suppl 2):3–10
2. Forkner CE, Scott TF (1931) Arsenic as a therapeutic agent in chronic myelogenous leukemia: preliminary report. J Am Med Assoc 97(1):3–5
3. Cohen MH, Hirschfeld S, Flamm Honig S, Ibrahim A, Johnson JR, O'Leary JJ et al (2001) Drug approval summaries: arsenic trioxide, tamoxifen citrate, anastrazole, paclitaxel, bexarotene. Oncologist 6(1):4–11
4. Tallman MS (2008) What is the role of arsenic in newly diagnosed APL? Best Pract Res Clin Haematol 21(4):659–666
5. Rodeghiero F, Castaman G (1994) The pathophysiology and treatment of hemorrhagic syndrome of acute promyelocytic leukemia. Leukemia 8(Suppl 2):S20–S26
6. Soignet S, Fleischauer A, Polyak T, Heller G, Warrell RP Jr (1997) All-trans retinoic acid significantly increases 5-year survival in patients with acute promyelocytic leukemia: long-term follow-up of the New York study. Cancer Chemother Pharmacol 40(Suppl):S25–S29
7. Wang ZY, Chen Z (2008) Acute promyelocytic leukemia: from highly fatal to highly curable. Blood 111(5):2505–2515
8. Ghavamzadeh A, Alimoghaddam K, Ghaffari SH, Rostami S, Jahani M, Hosseini R et al (2006) Treatment of acute promyelocytic leukemia with arsenic trioxide without ATRA and/or chemotherapy. Ann Oncol 17(1):131–134
9. Mathews V, Chendamarai E, George B, Viswabandya A, Srivastava A (2011) Treatment of acute promyelocytic leukemia with single-agent arsenic trioxide. Mediterranean J Hematol Infect Dis 3(1):e2011056
10. Hayashi T, Hideshima T, Akiyama M, Richardson P, Schlossman RL, Chauhan D et al (2002) Arsenic trioxide inhibits growth of human multiple myeloma cells in the bone marrow microenvironment. Mol Cancer Ther 1(10):851–860
11. Liu Q, Hilsenbeck S, Gazitt Y (2003) Arsenic trioxide-induced apoptosis in myeloma cells: p53-dependent G1 or G2/M cell cycle arrest, activation of caspase-8 or caspase-9, and synergy with APO2/TRAIL. Blood 101(10):4078–4087
12. Park WH, Seol JG, Kim ES, Hyun JM, Jung CW, Lee CC et al (2000) Arsenic trioxide-mediated growth inhibition in MC/CAR myeloma cells via cell cycle arrest in association with induction of cyclin-dependent kinase inhibitor, p21, and apoptosis. Cancer Res 60(11):3065–3071
13. Rousselot P, Labaume S, Marolleau JP, Larghero J, Noguera MH, Brouet JC et al (1999) Arsenic trioxide and melarsoprol induce apoptosis in plasma cell lines and in plasma cells from myeloma patients. Cancer Res 59(5):1041–1048
14. Rousselot P, Larghero J, Arnulf B, Poupon J, Royer B, Tibi A et al (2004) A clinical and pharmacological study of arsenic trioxide in advanced multiple myeloma patients. Leukemia 18(9):1518–1521
15. Munshi NC, Tricot G, Desikan R, Badros A, Zangari M, Toor A et al (2002) Clinical activity of arsenic trioxide for the treatment of multiple myeloma. Leukemia 16(9):1835–1837
16. Hussein MA, Saleh M, Ravandi F, Mason J, Rifkin RM, Ellison R (2004) Phase 2 study of arsenic trioxide in patients with relapsed or refractory multiple myeloma. Br J Haematol 125(4):470–476
17. Berenson JR, Boccia R, Siegel D, Bozdech M, Bessudo A, Stadtmauer E et al (2006) Efficacy and safety of melphalan, arsenic trioxide and ascorbic acid combination therapy in patients with relapsed or refractory multiple myeloma: a prospective, multicentre, phase II, single-arm study. Br J Haematol 135(2):174–183
18. Sharma M, Khan H, Thall PF, Orlowski RZ, Bassett RL Jr, Shah N et al (2012) A randomized phase 2 trial of a preparative regimen of bortezomib, high-dose melphalan, arsenic trioxide, and ascorbic acid. Cancer 118(9):2507–2515

19. Abou-Jawde RM, Reed J, Kelly M, Walker E, Andresen S, Baz R et al (2006) Efficacy and safety results with the combination therapy of arsenic trioxide, dexamethasone, and ascorbic acid in multiple myeloma patients: a phase 2 trial. Med Oncol 23(2):263–272
20. Schiller GJ, Slack J, Hainsworth JD, Mason J, Saleh M, Rizzieri D et al (2006) Phase II multicenter study of arsenic trioxide in patients with myelodysplastic syndromes. J Clin Oncol 24(16):2456–2464
21. Vey N, Bosly A, Guerci A, Feremans W, Dombret H, Dreyfus F et al (2006) Arsenic trioxide in patients with myelodysplastic syndromes: a phase II multicenter study. J Clin Oncol 24(16):2465–2471
22. Raza A, Buonamici S, Lisak L, Tahir S, Li D, Imran M et al (2004) Arsenic trioxide and thalidomide combination produces multi-lineage hematological responses in myelodysplastic syndromes patients, particularly in those with high pre-therapy EVI1 expression. Leuk Res 28(8):791–803
23. Bejanyan N, Tiu RV, Raza A, Jankowska A, Kalaycio M, Advani A et al (2012) A phase 2 trial of combination therapy with thalidomide, arsenic trioxide, dexamethasone, and ascorbic acid (TADA) in patients with overlap myelodysplastic/myeloproliferative neoplasms (MDS/MPN) or primary myelofibrosis (PMF). Cancer 118(16):3968–3976
24. Dilda PJ, Hogg PJ (2007) Arsenical-based cancer drugs. Cancer Treat Rev 33(6):542–564
25. Jiang G, Gong Z, Li XF, Cullen WR, Le XC (2003) Interaction of trivalent arsenicals with metallothionein. Chem Res Toxicol 16(7):873–880
26. Ramadan D, Rancy PC, Nagarkar RP, Schneider JP, Thorpe C (2009) Arsenic(III) species inhibit oxidative protein folding in vitro. Biochemistry 48(2):424–432
27. Li YM, Broome JD (1999) Arsenic targets tubulins to induce apoptosis in myeloid leukemia cells. Cancer Res 59(4):776–780
28. Ramirez P, Eastmond DA, Laclette JP, Ostrosky-Wegman P (1997) Disruption of microtubule assembly and spindle formation as a mechanism for the induction of aneuploid cells by sodium arsenite and vanadium pentoxide. Mutat Res 386(3):291–298
29. Zhang X, Yang F, Shim JY, Kirk KL, Anderson DE, Chen X (2007) Identification of arsenic-binding proteins in human breast cancer cells. Cancer Lett 255(1):95–106
30. Ling YH, Jiang JD, Holland JF, Perez-Soler R (2002) Arsenic trioxide produces polymerization of microtubules and mitotic arrest before apoptosis in human tumor cell lines. Mol Pharmacol 62(3):529–538
31. Lu J, Chew EH, Holmgren A (2007) Targeting thioredoxin Reductase is a basis for cancer therapy by arsenic trioxide. Proc Natl Acad Sci USA 104(30):12288–12293
32. Kojima C, Ramirez DC, Tokar EJ, Himeno S, Drobna Z, Styblo M et al (2009) Requirement of arsenic biomethylation for oxidative DNA damage. J Natl Cancer Inst 101(24):1670–1681
33. Tezuka M, Hanioka K, Yamanaka K, Okada S (1993) Gene damage induced in human alveolar type II (L-132) cells by exposure to dimethylarsinic acid. Biochem Biophys Res Commun 191(3):1178–1183
34. Nasr R, Guillemin MC, Ferhi O, Soilihi H, Peres L, Berthier C et al (2008) Eradication of acute promyelocytic leukemia-initiating cells through PML-RARA degradation. Nat Med 14(12):1333–1342
35. Weis K, Rambaud S, Lavau C, Jansen J, Carvalho T, Carmo-Fonseca M et al (1994) Retinoic acid regulates aberrant nuclear localization of PML-RAR alpha in acute promyelocytic leukemia cells. Cell 76(2):345–356
36. Dyck JA, Maul GG, Miller WH Jr, Chen JD, Kakizuka A, Evans RM (1994) A novel macromolecular structure is a target of the promyelocyte-retinoic acid receptor oncoprotein. Cell 76(2):333–343
37. Zhu J, Koken MH, Quignon F, Chelbi-Alix MK, Degos L, Wang ZY et al (1997) Arsenic-induced PML targeting onto nuclear bodies: implications for the treatment of acute promyelocytic leukemia. Proc Natl Acad Sci USA 94(8):3978–3983
38. Ito K, Bernardi R, Morotti A, Matsuoka S, Saglio G, Ikeda Y et al (2008) PML targeting eradicates quiescent leukaemia-initiating cells. Nature 453(7198):1072–1078

39. Jeanne M, Lallemand-Breitenbach V, Ferhi O, Koken M, Le Bras M, Duffort S et al (2010) PML/RARA oxidation and arsenic binding initiate the antileukemia response of As2O3. Cancer Cell 18(1):88–98
40. Zhang XW, Yan XJ, Zhou ZR, Yang FF, Wu ZY, Sun HB et al (2010) Arsenic trioxide controls the fate of the PML-RARalpha oncoprotein by directly binding PML. Science 328(5975):240–243
41. Isakson P, Bjørås M, Bøe SO (2010) Simonsen a. Autophagy contributes to therapy-induced degradation of the PML/RARA oncoprotein. Blood 116(13):2324–2331
42. Zhang QY, Mao JH, Liu P, Huang QH, Lu J, Xie YY et al (2009) A systems biology understanding of the synergistic effects of arsenic sulfide and Imatinib in BCR/ABL-associated leukemia. Proc Natl Acad Sci USA 106(9):3378–3383
43. Goussetis DJ, Gounaris E, Wu EJ, Vakana E, Sharma B, Bogyo M et al (2012) Autophagic degradation of the BCR-ABL oncoprotein and generation of antileukemic responses by arsenic trioxide. Blood 120(17):3555–3562
44. Shackelford D, Kenific C, Blusztajn A, Waxman S, Ren R (2006) Targeted degradation of the AML1/MDS1/EVI1 oncoprotein by arsenic trioxide. Cancer Res 66(23):11360–11369
45. Altman JK, Yoon P, Katsoulidis E, Kroczynska B, Sassano A, Redig AJ et al (2008) Regulatory effects of mammalian target of rapamycin-mediated signals in the generation of arsenic trioxide responses. J Biol Chem 283(4):1992–2001
46. Hong Z, Xiao M, Yang Y, Han Z, Cao Y, Li C et al (2011) Arsenic disulfide synergizes with the phosphoinositide 3-kinase inhibitor PI-103 to eradicate acute myeloid leukemia stem cells by inducing differentiation. Carcinogenesis 32(10):1550–1558
47. Davison K, Mann KK, Waxman S, Miller WH Jr (2004) JNK activation is a mediator of arsenic trioxide-induced apoptosis in acute promyelocytic leukemia cells. Blood 103(9):3496–3502
48. Huang C, Ma WY, Li J, Dong Z (1999) Arsenic induces apoptosis through a c-Jun NH2-terminal kinase-dependent, p53-independent pathway. Cancer Res 59(13):3053–3058
49. Mann KK, Davison K, Colombo M, Colosimo AL, Diaz Z, Padovani AM et al (2006) Antimony trioxide-induced apoptosis is dependent on SEK1/JNK signaling. Toxicol Lett 160(2):158–170
50. Chen D, Chan R, Waxman S, Jing Y (2006) Buthionine sulfoximine enhancement of arsenic trioxide-induced apoptosis in leukemia and lymphoma cells is mediated via activation of c-Jun NH2-terminal kinase and up-regulation of death receptors. Cancer Res 66(23):11416–11423
51. Jing Y, Dai J, Chalmers-Redman RM, Tatton WG, Waxman S (1999) Arsenic trioxide selectively induces acute promyelocytic leukemia cell apoptosis via a hydrogen peroxide-dependent pathway. Blood 94(6):2102–2111
52. Dai J, Weinberg RS, Waxman S, Jing Y (1999) Malignant cells can be sensitized to undergo growth inhibition and apoptosis by arsenic trioxide through modulation of the glutathione redox system. Blood 93(1):268–277
53. Davison K, Cote S, Mader S, Miller WH (2003) Glutathione depletion overcomes resistance to arsenic trioxide in arsenic-resistant cell lines. Leukemia 17(5):931–940
54. Bachleitner-Hofmann T, Gisslinger B, Grumbeck E, Gisslinger H (2001) Arsenic trioxide and ascorbic acid: synergy with potential implications for the treatment of acute myeloid leukaemia? Br J Haematol 112(3):783–786
55. Grad JM, Bahlis NJ, Reis I, Oshiro MM, Dalton WS, Boise LH (2001) Ascorbic acid enhances arsenic trioxide-induced cytotoxicity in multiple myeloma cells. Blood 98(3):805–813
56. Redondo-Muñoz J, Escobar-Díaz E, Hernández Del Cerro M, Pandiella A, Terol MJ, García-Marco JA et al (2010) Induction of B-chronic lymphocytic leukemia cell apoptosis by arsenic trioxide involves suppression of the phosphoinositide 3-kinase/Akt survival pathway via c-jun-NH2 terminal kinase activation and PTEN upregulation. Clin Cancer Res 16(17):4382–4391
57. Goussetis DJ, Altman JK, Glaser H, McNeer JL, Tallman MS, Platanias LC (2010) Autophagy is a critical mechanism for the induction of the antileukemic effects of arsenic trioxide. J Biol Chem 285(39):29989–29997

58. Hayakawa F, Privalsky ML (2004) Phosphorylation of PML by mitogen-activated protein kinases plays a key role in arsenic trioxide-mediated apoptosis. Cancer Cell 5(4):389–401
59. Giafis N, Katsoulidis E, Sassano A, Tallman MS, Higgins LS, Nebreda AR et al (2006) Role of the p38 mitogen-activated protein kinase pathway in the generation of arsenic trioxide-dependent cellular responses. Cancer Res 66(13):6763–6771
60. Dolniak B, Katsoulidis E, Carayol N, Altman JK, Redig AJ, Tallman MS et al (2008) Regulation of arsenic trioxide-induced cellular responses by Mnk1 and Mnk2. J Biol Chem 283(18):12034–12042
61. Kannan-Thulasiraman P, Katsoulidis E, Tallman MS, Arthur JS, Platanias LC (2006) Activation of the mitogen- and stress-activated kinase 1 by arsenic trioxide. J Biol Chem 281(32):22446–22452
62. Verma A, Mohindru M, Deb DK, Sassano A, Kambhampati S, Ravandi F et al (2002) Activation of Rac1 and the p38 mitogen-activated protein kinase pathway in response to arsenic trioxide. J Biol Chem 277(47):44988–44995
63. Lunghi P, Tabilio A, Lo-Coco F, Pelicci PG, Bonati A (2005) Arsenic trioxide (ATO) and MEK1 inhibition synergize to induce apoptosis in acute promyelocytic leukemia cells. Leukemia 19(2):234–244
64. Ye J, Li A, Liu Q, Wang X, Zhou J (2005) Inhibition of mitogen-activated protein kinase kinase enhances apoptosis induced by arsenic trioxide in human breast cancer MCF-7 cells. Clin Exp Pharmacol Physiol 32(12):1042–1048
65. Gilmore TD (2006) Introduction to NF-kappaB: players, pathways, perspectives. Oncogene 25(51):6680–6684
66. Oeckinghaus A, Hayden MS, Ghosh S (2011) Crosstalk in NF-kappaB signaling pathways. Nat Immunol 12(8):695–708
67. Kapahi P, Takahashi T, Natoli G, Adams SR, Chen Y, Tsien RY et al (2000) Inhibition of NF-kappa B activation by arsenite through reaction with a critical cysteine in the activation loop of ikappa B kinase. J Biol Chem 275(46):36062–36066
68. Beauchamp EM, Ringer L, Bulut G, Sajwan KP, Hall MD, Lee YC et al (2011) Arsenic trioxide inhibits human cancer cell growth and tumor development in mice by blocking hedgehog/GLI pathway. J Clin Invest 121(1):148–160
69. Kim J, Lee JJ, Kim J, Gardner D, Beachy PA (2010) Arsenic antagonizes the hedgehog pathway by preventing ciliary accumulation and reducing stability of the Gli2 transcriptional effector. Proc Natl Acad Sci USA 107(30):13432–13437
70. Hahn H, Wicking C, Zaphiropoulous PG, Gailani MR, Shanley S, Chidambaram A et al (1996) Mutations of the human homolog of drosophila patched in the nevoid basal cell carcinoma syndrome. Cell 85(6):841–851
71. Johnson RL, Rothman AL, Xie J, Goodrich LV, Bare JW, Bonifas JM et al (1996) Human homolog of patched, a candidate gene for the basal cell nevus syndrome. Science 272(5268):1668–1671
72. Cowan R, Hoban P, Kelsey A, Birch JM, Gattamaneni R, Evans DG (1997) The gene for the naevoid basal cell carcinoma syndrome acts as a tumour-suppressor gene in medulloblastoma. Br J Cancer 76(2):141–145
73. Pietsch T, Waha A, Koch A, Kraus J, Albrecht S, Tonn J et al (1997) Medulloblastomas of the desmoplastic variant carry mutations of the human homologue of drosophila patched. Cancer Res 57(11):2085–2088
74. Reifenberger J, Wolter M, Weber RG, Megahed M, Ruzicka T, Lichter P et al (1998) Missense mutations in SMOH in sporadic basal cell carcinomas of the skin and primitive neuroectodermal tumors of the central nervous system. Cancer Res 58(9):1798–1803
75. Taipale J, Chen JK, Cooper MK, Wang B, Mann RK, Milenkovic L et al (2000) Effects of oncogenic mutations in smoothened and patched can be reversed by cyclopamine. Nature 406(6799):1005–1009
76. Berman DM, Karhadkar SS, Maitra A, Montes De Oca R, Gerstenblith MR, Briggs K et al (2003) Widespread requirement for Hedgehog ligand stimulation in growth of digestive tract tumours. Nature 425(6960):846–851

77. Beauchamp E, Bulut G, Abaan O, Chen K, Merchant A, Matsui W et al (2009) GLI1 is a direct transcriptional target of EWS-FLI1 oncoprotein. J Biol Chem 284(14):9074–9082
78. Dahlén A, Fletcher CD, Mertens F, Fletcher JA, Perez-Atayde AR, Hicks MJ et al (2004) Activation of the GLI oncogene through fusion with the beta-actin gene (ACTB) in a group of distinctive pericytic neoplasms: pericytoma with t(7;12). Am J Pathol 164(5):1645–1653
79. Di Marcotullio L, Ferretti E, De Smaele E, Argenti B, Mincione C, Zazzeroni F et al (2004) REN(KCTD11) is a suppressor of hedgehog signaling and is deleted in human medulloblastoma. Proc Natl Acad Sci USA 101(29):10833–10838
80. Kasper M, Schnidar H, Neill GW, Hanneder M, Klingler S, Blaas L et al (2006) Selective modulation of hedgehog/GLI target gene expression by epidermal growth factor signaling in human keratinocytes. Mol Cell Biol 26(16):6283–6298
81. Nolan-Stevaux O, Lau J, Truitt ML, Chu GC, Hebrok M, Fernández-Zapico ME et al (2009) GLI1 Is regulated through smoothened-independent mechanisms in neoplastic pancreatic ducts and mediates PDAC cell survival and transformation. Genes Dev 23(1):24–36
82. Pasca DI, Magliano M, Sekine S, Ermilov A, Ferris J, Dlugosz AA, Hebrok M (2006) Hedgehog/Ras interactions regulate early stages of pancreatic cancer. Genes Dev 20(22):3161–3173
83. Riobó NA, Lu K, Ai X, Haines GM, Emerson CP Jr (2006) Phosphoinositide 3-kinase and Akt are essential for sonic hedgehog signaling. Proc Natl Acad Sci USA 103(12):4505–4510
84. Roberts WM, Douglass EC, Peiper SC, Houghton PJ, Look AT (1989) Amplification of the gli gene in childhood sarcomas. Cancer Res 49(19):5407–5413
85. Sheng T, Li C, Zhang X, Chi S, He N, Chen K et al (2004) Activation of the hedgehog pathway in advanced prostate cancer. Mol Cancer 13:3–29
86. Stecca B, Mas C, Clement V, Zbinden M, Correa R, Piguet V et al (2007) Melanomas require HEDGEHOG-GLI signaling regulated by interactions between GLI1 and the RAS-MEK/AKT pathways. Proc Natl Acad Sci USA 104(14):5895–5900
87. Taylor MD, Liu L, Raffel C, Hui CC, Mainprize TG, Zhang X et al (2002) Mutations in SUFU predispose to medulloblastoma. Nat Genet 31(3):306–310
88. Zwerner JP, Joo J, Warner KL, Christensen L, Hu-Lieskovan S, Triche TJ et al (2008) The EWS/FLI1 oncogenic transcription factor deregulates GLI1. Oncogene 27(23):3282–3291
89. Wang Y, Ding Q, Yen CJ, Xia W, Izzo JG, Lang JY et al (2012) The crosstalk of mTOR/S6K1 and hedgehog pathways. Cancer Cell 21(3):374–387
90. Fei DL, Li H, Kozul CD, Black KE, Singh S, Gosse JA et al (2010) Activation of hedgehog signaling by the environmental toxicant arsenic may contribute to the etiology of arsenic-induced tumors. Cancer Res 70(5):1981–1988
91. Baj G, Arnulfo A, Deaglio S, Mallone R, Vigone A, De Cesaris MG et al (2002) Arsenic trioxide and breast cancer: analysis of the apoptotic, differentiative and immunomodulatory effects. Breast Cancer Res Treat 73(1):61–73
92. Chow SK, Chan JY, Fung KP (2004) Inhibition of cell proliferation and the action mechanisms of arsenic trioxide (As2O3) on human breast cancer cells. J Cell Biochem 93(1):173–187
93. Stoica A, Pentecost E, Martin MB (2000) Effects of arsenite on estrogen receptor-alpha expression and activity in MCF-7 breast cancer cells. Endocrinology 141(10):3595–3602
94. Hornhardt S, Gomolka M, Walsh L, Jung T (2006) Comparative investigations of sodium arsenite, arsenic trioxide and cadmium sulphate in combination with gamma-radiation on apoptosis, micronuclei induction and DNA damage in a human lymphoblastoid cell line. Mutat Res 600(1–2):165–176
95. Jimenez-Sanchez M, Menzies FM, Chang YY, Simecek N, Neufeld TP, Rubinsztein DC (2012) The hedgehog signalling pathway regulates autophagy. Nat Commun 3:1200
96. Stancato LF, Hutchison KA, Chakraborti PK, Simons SS Jr, Pratt WB (1993) Differential effects of the reversible thiol-reactive agents arsenite and methyl methanethiosulfonate on steroid binding by the glucocorticoid receptor. Biochemistry 32(14):3729–3736
97. Mann KK, Padovani AM, Guo Q, Colosimo AL, Lee HY, Kurie JM et al (2005) Arsenic trioxide inhibits nuclear receptor function via SEK1/JNK-mediated RXRalpha phosphorylation. J Clin Invest 115(10):2924–2933

98. Chen GC, Guan LS, Hu WL, Wang ZY (2002) Functional repression of estrogen receptor a by arsenic trioxide in human breast cancer cells. Anticancer Res 22(2A):633–638
99. Chow SK, Chan JY, Fung KP (2004) Suppression of cell proliferation and regulation of estrogen receptor alpha signaling pathway by arsenic trioxide on human breast cancer MCF-7 cells. J Endocrinol 182(2):325–337
100. Du J, Zhou N, Liu H, Jiang F, Wang Y, Hu C et al (2012) Arsenic induces functional re-expression of estrogen receptor alpha by demethylation of DNA in estrogen receptor-negative human breast cancer. PLoS One 7(4):e35957
101. Zhang W, Wang L, Fan Q, Wu X, Wang F, Wang R et al (2011) Arsenic trioxide re-sensitizes ERalpha-negative breast cancer cells to endocrine therapy by restoring ERalpha expression in vitro and in vivo. Oncol Rep 26(3):621–628

# Chapter 6
# Nuclear Factors Linking Cancer and Inflammation

**Kunzang Chosdol, Mohita Bhagat, Bhawana Dikshit, Evanka Madan, Parthaprasad Chattopadhyay, and Subrata Sinha**

**Abstract** An association between inflammation and cancer has long-been known, but the past decade has witnessed a spurt in the research linking the two processes. On the one hand chronic inflammation predisposes to cancer, on the other, neoplastic transformation predisposes towards an intrinsic pro-inflammatory microenvironment, which further promotes the progression of the malignancy. Irrespective of the stimulus, whether extrinsic (bacteria, viruses, non-healing wounds, irritants etc.) or intrinsic (oncogenes, protein kinases etc.), all signals that trigger the inflammatory microenvironment in tumor cells converge in the nucleus and coordinate inflammatory transcriptional activity by activating various nuclear/transcription factors. This forms a vicious cycle, further promoting inflammation, facilitating tumor progression, proliferation, survival and angiogenesis. This chapter focuses on transcriptional mediators intrinsic to tumour cells that enhance the inflammatory processes.

The well known nuclear/transcription factors important for this linkage are NFkB, HIF1α, AP-1 and STAT-3. The cross talk between these factors results in a complex web of signalling processes that have a marked influence on the cellular phenotype, cell-cell interactions and the interaction of the tumour with the microenvironment. This chapter focuses on the above mentioned factors and their effects, while also looking at the possibilities of utilising the same for therapeutic intervention.

---

K. Chosdol, MD, Ph.D. • M. Bhagat M.Sc, Persuing Ph.D. • B. Dikshit, Ph.D.
• E. Madan, M.Sc, Persuing Ph.D. student • P. Chattopadhyay, M.D., Ph.D.
Department of Biochemistry, All India Institute of Medical Sciences,
New Delhi, Delhi 110029, India
e-mail: kunzangchosdol@yahoo.com; bhawana1112@gmail.com;
mohita444@gmail.com; evanka.madan@gmail.com; parthoaiims@hotmail.com

S. Sinha, M.D., Ph.D. (✉)
National Brain Research Centre (NBRC), Near NSG Campus, Nainwal Mode,
Manesar, Haryana 122050, India
e-mail: subrata.sinha@nbrc.ac

R. Kumar (ed.), *Nuclear Signaling Pathways and Targeting Transcription in Cancer*, Cancer Drug Discovery and Development, DOI 10.1007/978-1-4614-8039-6_6,

**Keywords** Transcription factors • NFkB • AP-1 • STAT3 • HIF1α • Cancer • Inflammation

## Abbreviations

| | |
|---|---|
| AOM | Azoxymethane |
| AP-1 | Activator protein 1 |
| ARP 2/3 | Actin-related protein 2/3 complex subunit 1B |
| ATF | Activating transcription factor |
| BAK | BCL2-antagonist/killer |
| BAX | BCL2-associated X protein |
| Bcl-2 | B-cell CLL/lymphoma *2* |
| Bcl-xl | B-cell lymphoma-extra large |
| CapG | Capping protein (actin filament) |
| cIAP2 | Baculoviral IAP repeat-containing protein 3 |
| c-Rel | Proto-oncogene *c-Rel* |
| COX-2 | Cyclooxygenase-2 |
| CRE | cAMP response element |
| CXCL | Chemokine (C-X-C motif) ligand |
| DEN | Diethylnitrosamine |
| DSS | Dextran-sulphate sodium |
| EGF | Epidermal growth factor |
| EGFR | Epidermal growth factor |
| EMT | Epithelial-Mesenchymal Transition |
| ERK | Extracellular signal-regulated kinase |
| FIH | Factor inhibiting HIF |
| Fos | FBJ murine osteosarcoma viral oncogene homolog |
| Fra | Fos–gene family member |
| GBM | Glioblastoma multifome |
| GM-CSF | Granulocyte-macrophage colony-stimulating factor |
| HIF1α | Hypoxia-inducible factor 1 α |
| IFN-γ | Interferon-gamma |
| IGSF4 | Immunoglobulin (Ig)-like intercellular adhesion molecule |
| IκB | Inhibitor of kappa light polypeptide gene enhancer in B-cells |
| IKK | IκB kinase |
| IL-1β | Interleukin 1-beta |
| Inos | Inducible Nitric acid synthase |
| JNK | c-Jun N-terminal kinase |
| JUN | *jun* proto-oncogene |
| KC | Keratinocyte chemokine |
| LPS | Bacterial lipopolysaccharides |
| MAF | Musculoaponeurotic fibrosarcoma |
| MAPK | Mitogen-activated protein kinase |

| | |
|---|---|
| Mcl-1 | Myeloid cell leukemia sequence 1 |
| MIP | Major intrinsic protein of lens fiber |
| MMP | Matrix metalloproteinase |
| MUC1 | Mucin 1, cell surface associated |
| MYC | v-*myc* myelocytomatosis viral oncogene homolog |
| NEMO | NF-kappa B essential modulator also known as IKKγ |
| NF-Kb | Nuclear factor kappa-light-chain-enhancer of activated B cells |
| NIK | NF-κB inducing kinase |
| NSAID | Nonsteroidal anti-inflammatory drug |
| PDCD4 | Programmed cell death protein 4 |
| PHDs | Proline hydroxylases |
| PIM-1 | Proto-oncogene serine/threonine-protein kinase |
| PTEN | Phosphatase and tensin homolog |
| RAF1 | v-*raf*-1 murine leukemia viral oncogene homolog |
| RAS | *RAS* oncogene |
| RelA | v-rel reticuloendotheliosis viral oncogene homolog A |
| RelB | v-rel reticuloendotheliosis viral oncogene homolog B |
| RHD | Rel homology domain |
| RNS | Reactive nitrogen species |
| ROS | Reactive oxygen species |
| SPARC | Secreted protein acidic and rich in cysteines |
| STAT3 | Signal transducer and activator of transcription 3 |
| STAT6 | Signal transducer and activator of transcription 6 |
| TAMs | Tumor associated macrophages |
| TGF-β | Transforming growth factor beta |
| TNF | Tumor necrosis factor |
| TPA | 12-0-tetradecanoylphorbol 13-acetate |
| TRAMP | Transgenic adenocarcinoma of mouse prostrate |
| TRE | TPA response element |
| TWIST | Twist homolog (Drosophila) |
| VEGF | Vascular endothelial growth factor |
| VHL | Von Hipple Lindau |
| ZEB | Zinc finger E-box binding homeobox |

## 6.1 Introduction

Inflammation has long been known as a protective measure to tissue from injury, infection or irritation. While acute inflammation occurs as a part of immediate defense mechanism, chronic inflammation due to persistent infections can lead to several diseases including arthritis, cardiovascular, neurological disorders, cancer etc. [1].

Association between inflammation and cancer has long been known. Initial information linking inflammation with carcinogenesis dates back to 1863 when Rudolf Virchow noted malignant neoplasm arise at regions of chronic inflammation and reasoned that various "irritants" cause tissue injury, inflammation and increased cell proliferation [2, 3]. Studies way back in early 1960s have pointed out the role of inflammation in the promotion of established cancers [4]. In the recent years, it has been proven that the inflammatory microenvironment driven by the presence of cytokines in solid tumors have major role in tumor progression and invasion [5, 6].

Epidemiological studies have shown that around 1.2 million cases of infection related malignancies per year are caused due to chronic inflammation as a result of bacterial and viral infections [6]. Viruses such as human papillomavirus, hepatitis C virus and hepatitis B virus, in addition to inhibiting tumor suppressor proteins or promoting tumor promoter proteins, can cause malignancy through inflammation related mechanisms [7–9]. Along with gastrointestinal tract, other organs with high susceptibility to tumor development following a chronic inflammation are the lungs, bladder, liver, pancreas and oesophagus etc. [3].

Inflammation is an indispensable participant in neoplastic progression [6, 7] and suppression of pro-inflammatory pathways demonstrates potential for the prevention and treatment of cancer [10]. The importance of inflammation in tumor initiation and malignant progression has become one of the major focus and researchers are analyzing the mechanism linking the two.

The trigger(s) for inflammatory microenvironment in tumors can be extrinsic for example by inflammatory and infectious conditions [6, 8] or intrinsic stimulus through oncogenes and cytokines [5, 11]. Persistence of chronic inflammation is one of the major extrinsic factors involved in neoplastic transformation of the cells and higher risk of developing many age related malignancies [12]. A lot of evidence has established that inflammation regulates different stages of tumor development, such as initiation, promotion, invasion and metastasis [5], whereby various pro-inflammatory mediators triggered by inflammation aid tumor progression by regulating cascades of cytokines, chemokines, adhesion and pro-angiogenic activities [6].

In addition, intrinsic factors like oncogenes triggering the pro-cancer inflammatory microenvironment in solid tumors are equally important. The role of protein tyrosine kinase cascade [13–15] was the first evidence on the role of intracellular molecule in triggering inflammatory program. Similarly, RAS-RAF signaling pathway [16–18], MYC [19], TGFβ etc. are known to promote inflammatory microenvironment by upregulating the expression of various cytokines and inflammatory modulators like COX-2 etc.

Irrespective of the stimulus, whether extrinsic or intrinsic, all signals that trigger the inflammatory microenvironment in the tumor cells converge in nucleus and coordinate inflammatory transcriptional activities by activating various nuclear/transcription factors [20]. It forms a vicious cycle further promoting the inflammatory microenvironment of the tumor cells facilitating tumor progression, proliferation, survival and angiogenesis.

The well known nuclear/transcription factors activated by these proinflammatory signaling pathways, upregulating the inflammatory microenvironment and cancer progression are NFkB, HIF1α, AP-1 and STAT-3. This chapter focuses on the role of these nuclear factors in triggering inflammatory microenvironment in cancers, helping tumor cells to survive, proliferate and promote tumor aggressiveness.

## 6.2 NF-κB

### 6.2.1 Introduction

NF-κB (nuclear factor kappa light chain enhancer of activated B cells) is a protein complex that controls transcription of a number of genes involved in inflammation, cell proliferation and immune response. Five members of NF-κB family have been identified in mammals: RelA/p65, c-Rel, RelB, NF-κB1/p50 and NF-κB2/p52. All proteins of the NF-κB share a Rel homology domain(RHD) in their N-terminus, which is required for DNA binding, homo and heterodimerization, nuclear localization and inhibitor (IκB) binding [9]. Only RelA, Rel B and c-Rel contain Transactivation Domains (TDs- required to act as transcriptional activators) at their C-terminal regions. p50 and p52,which do not possess these TDs, are produced as p105 and p100 precursors respectively. p105 and p100 are then processed by ubiquitin proteosome pathway that cleaves the C-terminal ankyrin repeats(IκB like portion) containing region to respectively generate p50 and p52 [21].

Two distinct pathways lead to activation of different NF-kB transcription factors (Fig. 6.1). The classical or the cannonical pathway triggered by pro-inflammatory cytokines in bacterial and viral infections led to activation of the IκB kinase (IKK). IKK complex, which has two catalytic subunits, IKK- α and IKK-β and a regulatory subunit, IKK-γ (NEMO), phosphorylates NFκB-bound IκBs leading to release of

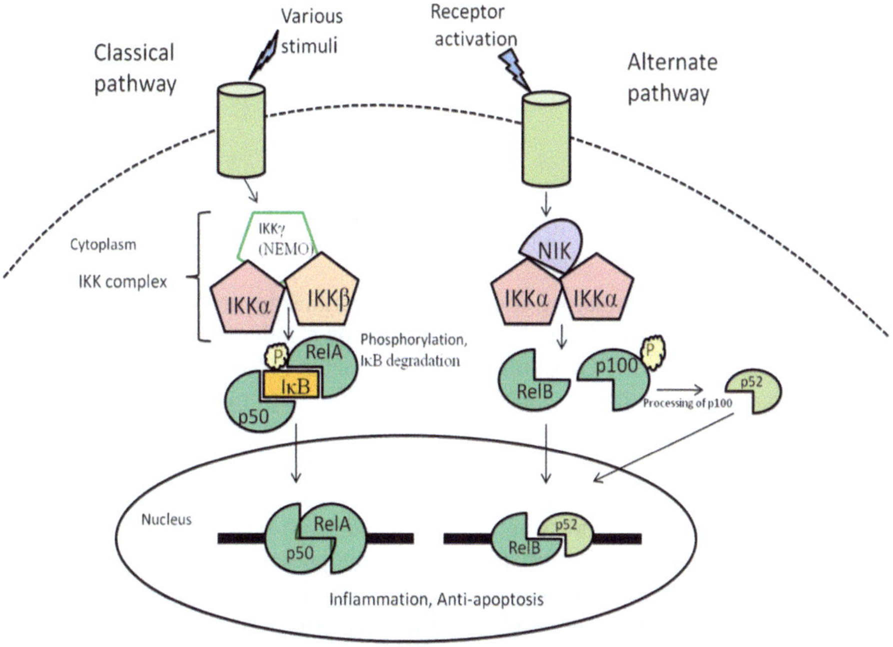


**Fig. 6.1** NF-κB signaling via classical and alternative pathways

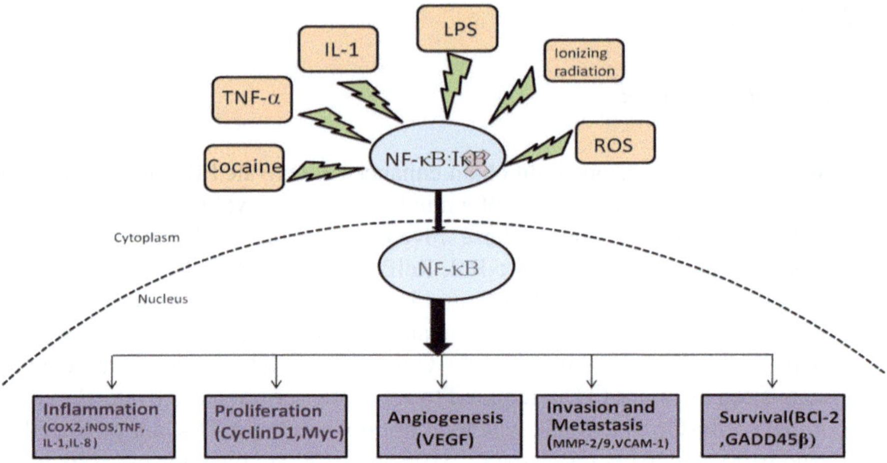


**Fig. 6.2** NF-κB activation and its functional effects on cells

NF-κB. The free NF-κB is the activated form which then translocates to the nucleus where it can 'turn on' the expression of specific genes that have DNA-binding sites for NF-κB. The phosphorylated IκB undergoes proteosomal degradation [21].

The non-cannonical pathway involves activation of NF-κB inducing kinase (NIK) in response to certain members of TNF family, which activates IKK-α (independent of IKK-β and IKK- γ) leading to phosphorylation and processing of p100 to p52 [22]. Both the pathways mediate different immune responses by activating different sets of genes.

### *6.2.2 NF-κB Activation and Inflammatory Networking in Cancer*

A wide variety of inflammatory stimuli (e.g. infectious agents, cytokines and carcinogens) are known to trigger NF-κB [23]. Known inducers of NF-κB include reactive oxygen species(ROS), tumor necrosis factor (TNFα), interleukin 1-beta (IL-1β), bacterial lipopolysaccharides (LPS), ionizing radiation and cocaine (Fig. 6.2) [1, 24, 25].

Aberrant NF-κB activation leads to induction of a diverse array of genes involved in tumorogenesis. These include the expression of anti-apoptotic genes promoting cancer cell survival, the expression of cyclins and proto-oncogenes leading to cell proliferation, the expression of matrix metalloproteinases and cell adhesion genes promoting metastasis and induction of genes responsible for new blood vessel growth stimulating angiogenesis (Table 6.1) [26]. All these actions are also very important in defense mechanisms of the body towards infection and injury but in a tumor microenvironment, these inappropriately regulated set of genes may provide an advantage to the malignant cells to proliferate and survive.

**Table 6.1** NF-κB target genes involved in oncogenesis

| Gene | Function | References |
|---|---|---|
| Bcl-2 | Pro-survival factor | Catz and Johnson 2001 |
| IL-6 | Interleukin-6, inflammatory cytokine | Son et al. 2008 |
| IL-8 | Interleukin-8, alpha-chemokine | Kang et al. 2007 |
| ICAM-1 | Intercellular adhesion molecule-1 | Bunting et al. 2007 |
| Epidermal Growth Factor Receptor | Receptor for EGF | Thornburg and Raab-Traub 2007 |
| Bax | Pro-apoptotic Bcl-2 homologue | Grimm et al. 2004 |
| Fas-Ligand | Inducer of apoptosis | Singh et al. 2006 |
| TRAF-1 | TNF-receptor associated factor | Schwenzer et al. 1999 |
| Angiopoietin | Tie-2 receptor ligand | Scott et al. 2005 |
| GM-CSF | Granulocyte Macrophage Colony Stimulating Factor | Bunting et al. 2007 |
| EPO | Erythropoietin | Figueroa et al. 2002 |
| A20 | TNF-inducible zinc finger | Krikos et al. 1992 |
| ABIN-3 | NF-kB inhibitor | Verstrepen et al. 2007 |
| E2F3a | Cell cycle regulator | Cheng et al. 2003 |
| HIF-1alpha | Hypoxia-inducible factor | Bonello et al. 2007 |
| MMP-3, matrix metalloproteinaase-3 | Secreted collagenase involved in metastasis | Borghaei et al. 2004 |
| MMP-9, matrix metalloproteinaase-9 | Secreted collagenase involved in metastasis | Yan et al. 2004 |
| Cyclin D1 | Cell-cycle regulation | Toualbi-Abed et al. 2008 |
| Cyclin D2 | Cell-cycle regulation | Iwanga et al. 2008 |
| Gadd45beta | DNA repair/cell cycle | Qiu et al. 2004 |

### *6.2.3 NF-κB: A Defensive Route to Malignancy*

Since its discovery, many studies have shown NF-κB to be a key regulator of inflammation [27]. Indeed, promoters of many genes encoding for cytokines and chemokines contain NF-κB binding sites [28]. The first hint implicating an association between NF-κB and cancer was the observation that it shares sequence homology with viral oncoprotein v-Rel [29]. This, together with the findings of activated NF-κB in cancers hypothesized a link between cancer and inflammation. Recently many studies using mice models have been carried out establishing a central role of NF-κB in cancer. NF-κB is found to have role in all three stages of carcinogenesis (Tumor initiation, tumor promotion and tumor progression) [30].

#### 6.2.3.1 NF-κB–A Tumor Initiator

Many studies have established that infection with *Helicobacter pyroli* in gastric epithelium leads to NF-κB activation [30]. The first molecular evidence linking inflammation to tumor initiation was shown by Houghton et al. in 2004, in which a mouse model of Helicobacter induced gastric cancer was used. It was shown that chronic inflammation caused by bacterial inflammation can lead to gastric mucosa erosion followed by repopulation of bone marrow derived cells at the site of erosion. These repopulated cell differentiated into gastric epithelial phenotype giving rise to gastric carcinoma [31]. NF-κB is also involved in the induction of iNOS (inducible Nitric acid synthase). iNOS has been shown to be involved in cellular changes like transformation of normal cells, triggering angiogenesis and metastasis of malignant cells [32].

#### 6.2.3.2 NF-κB–A Tumor Promoter

Recent work in mouse studies has provided strong genetic evidence demonstrating a crucial role of NF-κB in tumor promotion. Studies on Mdr2 knockout mice have found spontaneous development of hepatocellular carcinoma (HCC) within 8–10 months of birth due to inflammation of the bile duct followed by cholangitis and cancer [5, 33]. TNF-α, upregulated in the endothelial and inflammatory cell had led to activation of NF-κB in the adjacent hepatocytes. Suppressing NF-κB through expression of a non-degradable IκBα variant (IκB super-repressor) or by using neutralizing antibodies for TNF-α resulted in apoptosis of premalignant hepatocytes and failure to progress to hepatocellular carcinoma [33]. Activation of NF-κB through TLR signaling was found in mouse model of colitis-associated cancer on exposure to mutagen azoxymethane (AOM) followed by administration of dextran-sulphate sodium (DSS) salt led to development of cancer. Deletion of IKK-β in the intestinal epithelial cells resulted in decrease in tumor incidence and an increase in expression levels of pro-apoptotic BAX and BAK proteins indicating a role of NF-κB in suppressing apotosis in colon cancer [5].

#### 6.2.3.3 NF-κB: Role in Tumor Progression

Once a tumor is established, the role of NF-κB does not end and it continues to be a central signaling molecule in tumor progression. NF-κB is known to regulate many genes which are involved in tumor invasion and metastasis such as serine proteases, adhesion molecules, MMPs, and chemokines [6, 34]. Also, NF-κB induces iNOS leading to increased production of reactive nitrogen species (RNS) which might result in generation and accumulation of additional DNA mutations that drive tumor progression [32]. In addition to inducing the expression of inflammatory cytokines, NF-κB is also seen in polarization of tumor associated macrophages or TAMs [35]. In response to different stimuli, macrophages undergo different activation programmes. M1 macrophages (which produce IL-12, GM-CSF and TNF when exposed to IFN-γ) are proposed to be tumoricidal, whereas M2 macrophages (producing IL-10, VEGF, iNOS when exposed to IL-4) are proposed to be tumor promoting [5]. It was shown that inhibition of NF-κB in macrophages converts them from M2 tumor promoting phenotype to M1 cytotoxic phenotype, resulting in tumor regression [36].

NF-κB pathway is extensively studied in both solid tumors and hematological malignancies (Fig. 6.3). Constitutive activation of NF-κB is seen in GBM tumors and inhibition of NF-κB activity induces apoptosis in glioma [37]. Deletion of

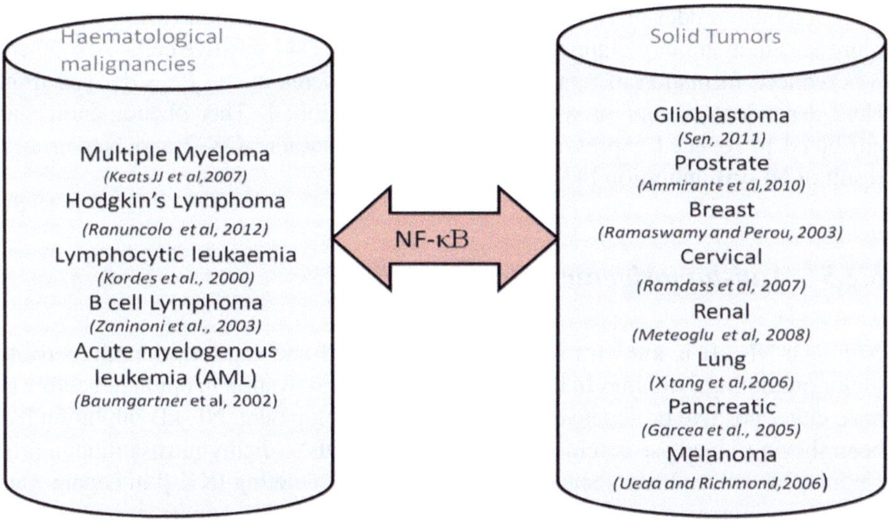


**Fig. 6.3** Human cancers (heamatological and solid) with activated NF-κB expression

NFKBIA, a gene coding for IκBα is a frequent oncogenic mutation seen in glioblastoma tumors and is mutually exclusive with EGFR amplification. It was shown that deletion of NFKBIA (which normally inhibits EGFR signaling), and thus activation of NF-κB, had an effect that was similar to the effect of EGFR amplification in the pathogenesis of glioblastoma and was associated with comparatively short survival [38]. NF-κB is found to be activated in castration resistant prostate cancer [29]. Using a TRAMP (Transgenic adenocarcinoma of mouse prostrate) model, it was shown that deletion of IKKβ in epithelial cells did not have any effect on tumor development and progression [39]. Furthermore, IKKα and not IKKβ enhanced metastatic progression in these mice [40]. Constitutive NF-κB activation has been shown in primary myeloma cells and cell lines. Furthermore, inhibitors of NF-κB signaling such as proteosome inhibitors, IKK inhibitors and IκB phosphorylation inhibitors have induced apoptosis and inhibited growth of multiple myeloma cells [29]. Most of the genetic alterations in multiple myeloma activate both classical and alternate pathways. These studies also showed stabilization of NIK in multiple myeloma (which is normally degraded in normal cells) which is required to activate both the pathways, thus leading to B cell survival [41].

### *6.2.4 NF-κB and Its Role in Cancer Stem Cells*

The connection between NF-κβ in cancer stem cells has been well explored in breast cancers. A membrane bound receptor tyrosine kinase Her2, which controls NF-κB through the canonical pathway is known to be overexpressed in 30 % of breast cancers. It has also been shown to play a role in regulating cancer stem cell population [42]

In a mouse model of Her2 breast cancer, NF-κB suppression (using IκB super repressor) in mammary gland was seen to decrease CD44-positive cells. Also, there was reduced formation of non-adherent mammospheres in cell lines derived from Her2-dependent tumors in which NF-κB was inhibited. This phenomenon was attributed to reduced expression of stem cell regulators SOX-2 and Nanog as a result of NF-κB inhibition [43].

### *6.2.5 Anti-tumorigenic Effects of NF-κB*

Although NF-κB is known to promote cell survival and transcriptionally activate tumor promoting cytokines in inflammatory cells, its activation in epithelial cells can have either positive or negative effect on tumor development. NF-κB inhibition has been shown to increase carcinogenesis in mice. In DEN (diethylnitrosamine -a pro-carcinogen) induced hepatocarcinoma mouse model, deleting IKK-β in hepatocytes resulted in increase in tumor growth and size [44]. In skin cancer mouse model, tumor was initiated by administration of 7,12-dimethylbenzanthracene (DMBA-a carcinogen) followed by phorbol esters administration. Inhibition of NF-κB in the epidermis in this model promoted development of squamous cell carcinoma [45, 46]. TNF-deficient and TNFR1-deficient mice were, however, found to be resistant [47].

### *6.2.6 NF-κB: A Target in Therapeutics*

The findings that NF-κB is seen to be activated in many malignancies provide a strong rationale for the use of NF-κB inhibitors as therapeutics in treating cancer. Efforts have been made by the pharmaceutical companies to develop IKKβ and NF-κB inhibitors [48]. Such an inhibitor, should specifically prevent NF-κB activation in malignant cells without showing any side effects on normal cells. Also, the dose should be such that so as to minimize systemic toxicity and avoid broad suppression of innate immunity. Many reports have shown the role of these NF-κB/IKKβ inhibitors as anti-tumor compounds in experimental cancer models [49, 50]. However, such a drug is unlikely to be effective alone in tumor regression and can be used as adjuvants for existing chemotherapeutic drugs or radiation therapy. However, excessive and prolonged NF-κB inhibition can be detrimental owing to its important role in innate immunity. Due to these factors, NF-κB inhibitors have not been clinically approved till date.

Also, selective interference in upstream activation or downstream targets of NF-κB such as, TNF-α, IL-6 or targeting individual cytokines can also prove effective [30]. Targeting proteosome by using inhibitors that can block NF-κB activation can also be developed as a good therapeutic strategy, as is the case with bortezomib, which has reached phase III clinical trial in treating multiple myeloma [51, 52].

Curcumin, also has been perpetually a promising molecule for NF-κB inhibition. It has been shown to inhibit NF-κB activation and cell proliferation in many cancers

[51, 53]. Curcumin blocks the IKK β mediated phosphorylation and degradation of IκBα, thus NF-κB remains bound to IκBα in the cytoplasm and is not able to enter the nucleus to activate transcription [53]. Recently, curcumin has been tested for phase II clinical trial against pancreatic cancer in which curcumin downregulated NF-κB expression [54].

## 6.3 HIF-1$\alpha$: A Synergistic Link Between Cancer and Inflammation

Solid tumors are characterized by intratumoral hypoxia that result from dysregulated cell proliferation. Physiological responses triggered by hypoxia impact on all critical aspects of cancer progression, including immortalization, transformation, differentiation, genetic instability, angiogenesis, metabolic adaptation, autocrine growth factor signaling, invasion, metastasis, and resistance to therapy [55].

Hypoxic conditions in in-vitro as well as in-vivo result in induction of HIF-1α (Hypoxia Inducible Factor), whereas the beta subunit is not sensitive to oxygen. Oxygen sensitivity of HIF-1α is conferred by a family of oxygen-dependent hydroxylase enzymes including three proline hydroxylases (PHDs) and one asparigine hydroxylase known as the factor inhibiting HIF (FIH). These hydroxylases promote oxygen-dependent hydroxylation of HIF-1α leading to its ubiquitylation by the Von Hipple Lindau (VHL) E3 ligase and subsequent degradation in the proteasome. This effect is reversed in hypoxia leading to HIF-1α stabilization and activation [56].

HIF-1α, on activation is known to stimulate the transcription of a number of genes linking inflammation and tumor survival such as vascular endothelial growth factor (VEGF), Carbonic anhydrase IX, iNOS, COX-2 and several glycolytic enzymes. HIF-1α can also be activated under normoxia in response to pro-inflammatory cytokines IL-1β and TNF-α. The IL-1β induced HIF-1α activity is mediated by NF-κB [37].

In innate immunity, HIF-1α is shown to regulate phagocyte immune functions by inhibiting apoptosis, releasing antimicrobial peptides, inducing iNOS and promoting the expression of NF-κB regulated cytokines [56]. HIF-1α is also shown to mediate NF-κB activation in neutrophils under anoxic conditions [57, 58]. Hypoxia is also hypothesized to down regulate adaptive immune response by inhibiting IFN-γ production and T cell activation [57]. HIF-1α, therefore promotes tumorigenesis by curtailing adaptive immune system and enhancing innate immune system (which during chronic inflammation can lead to cancer).

### *6.3.1 HIF1$\alpha$ -Hypoxia- NF-$\kappa$B: Who Regulates Whom?*

Studies linking NF-κB signaling pathway to hypoxia dates back to 1994 when it was shown by Koong et al. (1994) that exposure of cells to hypoxia (0.02 % O2)

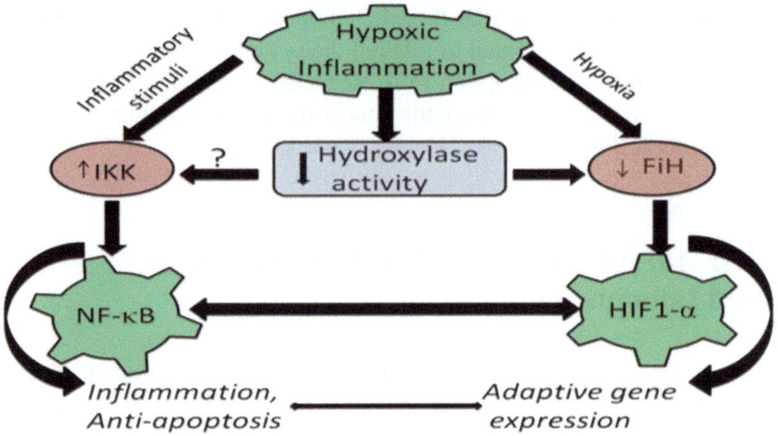


**Fig. 6.4** Crosstalk between NF-κB and HIF1 in hypoxic inflammation

resulted in IκBα degradation and increased NF-κB DNA binding activity . This link between hypoxia and NF-κB was then supported by several reports demonstrating activation of NF-κB dependent genes such as TNF-α , IL-6 and COX-2 in response to hypoxia [59]. Furthermore, it has been proved recently, that hypoxia activates NF-κB through a pathway involving activation of IKKβ, which in turn phosphorylates IκBα, liberating it from NF-κB (Fig. 6.4) [60].

The presence of conserved motif in activation loop of IKKβ and IKKα for hydroxylation by prolyl hydroxylase (PHD-which also inhibits HIF-1α activation) has been identified. Site directed mutagenesis of the proline residue in these putative IKKβ hydroxylation site did not induce IKKβ in hypoxia. Moreover, it has been shown that knockdown of PHD in HeLa cells led to activation of NF-κB. These findings hypothesize that hypoxia releases NF-κB repression through decreased PHD-dependent hydroxylation of IKKβ, in a manner, similar to that of HIF-1α activation in hypoxia [61].

### *6.3.2 HIF-1α: NF-κB Crosstalk*

A plethora of evidence now exists that hypoxia activates both HIF-1α and NF-κB, which act as transcriptional activators for a number of genes involved in tumor development [56, 59, 61]. However, recent studies have also exhibited an interdependent role of HIF-1 and NF-κB in regulating the expression of each other. HIF-1α, apart from being induced by hypoxia, is also shown to be activated in response to a number of non-hypoxic stimuli including bacterial LPS, TNF-α, ROS and IL-18 in NF-κB dependent manner by increasing HIF-1α mRNA levels [62]. It was shown that expression of the NF-κB subunits p50 and p65 enhances HIF-1α mRNA levels, whereas blocking NF-κB by an inhibitor attenuated HIF-1α mRNA induction by hypoxia [63]. Furthermore, presence of an NF-κB binding site at position within the HIF-1α promoter has been identified and mutation of this site abolished induction by hypoxia [64].

Interestingly, mice lacking IKKβ in different cell types resulted in defective induction of various HIF-1α target genes including VEGF [[60]. The role of HIF-1α in regulating NF-κB signaling has also been reported. Mice overexpressing HIF-1α in keratinocytes showed increased NF-κB activity and increased expression of pro-inflammatory targets including macrophage inflammatory protein-2 (MIP-2/CXCL2/3), keratinocyte chemokine (KC/CXCL1), and TNF-α, leading to hyper-responsiveness to inflammatory stimuli. HIF-1α -induced NF-κB activation was due to phosphorylation of IκB (leading to its degradation) and Ser276 on p65, thus enhancing its nuclear localization and transcriptional activity [65]. Therefore, there is an interdependency between HIF-1α and NF-κB signaling leading to their synergistic role in tumor development and progression.

## 6.4 AP-1

### *6.4.1 Introduction*

AP-1 (**Activator Protein 1**), one of the most extensively studied transcription factors, is involved in regulating the expression of a variety of genes by forming either a homodimer or heterodimers . The dimers are formed between the proteins belonging to different families like Jun, Fos, Fra, MAF (musculoaponeurotic fibrosarcoma) and ATF (activating transcription factor) [66–69]. There are multiple proteins included in each protein family of AP1 complex, like in Jun family, there are c-Jun, JunB and JunD; in Fos family- c-Fos and FosB; in Fra family - Fra1 and Fra2; in ATF family - ATFa, ATF2 and ATF3; and in Maf family - v-Maf, c-Maf, Nrl, MafB, MafF, MafG and MafK [67, 70–73].

The Jun and ATF family of proteins form homodimers and Jun also forms heterodimers with Fos, ATF and Maf [66, 68, 69]. The Fos proteins forms stable heterodimers with Jun but cannot form a stable homodimer [68, 69]. The cMaf and Nrl form heterodimers with Jun and Fos but other Maf proteins like MafB, MafF, MafG and MafK forms heterodimers only with Fos [74].

Among Jun proteins, cJun exhibits the highest transcriptional activity and the cJun-cFos dimer forms the most stable and active AP-1 complex [68, 69]. There are considerable variations in the activity as well as the expression of AP-1 components, exhibiting cell type or tissue specific activity [75]. Some protein combinations of the AP-1 complexes are inhibitory e.g. JunB and JunD suppress the transformation ability of the cells by forming homodimers or heterodimers with cFos [76–79] and JunB acts as an inhibitor of c-Jun [80].

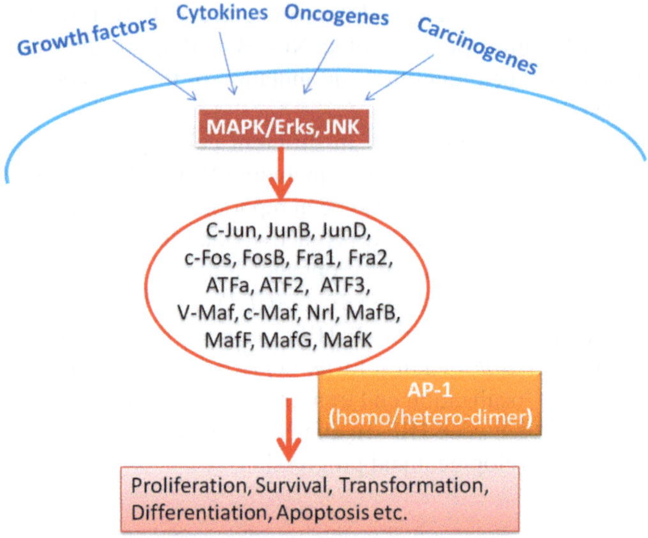


**Fig. 6.5** AP-1 signaling: Depending on the stimulus, either homo- or hetero-dimers of different protein components are formed and affect the cellular consequences

## *6.4.2 AP-1 Activation*

AP-1 acts as a mediator to convert vast number of extracellular stimulus by cytokines, growth factors, carcinogens etc. signaling through conserved pathways like transforming growth factor beta, mitogen-activated protein kinases and PI3K into the expression of specific target genes regulating cell proliferation, survival, transformation, invasion, apoptosis etc. (Fig. 6.5). The signaling pathways that predominantly mediate AP-1 activation are the MAPK (mitogen-activated protein kinase) cascade [81, 82] through multiple mechanisms including ERK (extracellularsignal-regulated kinase), JNK (c-Jun N-terminal kinase) and p38 MAPK signal transduction pathways [83]. The precise regulation of AP-1 is critical for maintaining normal functioning of the cell as AP-1 controls both basal and induced expression of genes. AP-1 is stimulated in response to TPA (12-0-tetradecanoylphorbol 13-acetate) and EGF (Epidermal growth factor) and is thus required for TPA or EGF induced transformation [84]. Jun is known to be positively autoregulated by AP-1 [85]. AP-1 dependent transcription is reported to be suppressed by sequestration of c-Fos at the nuclear envelope through interaction with A-type lamins [86]. The phosphatase PTEN is also known to down-regulate AP-1 activity in human glioma cells [23].

Upon activation, the AP-1 protein components dimerize through lucine zipper domains producing a contiguous DNA contact interface that, depending on the dimer (homo/hetero) complexes formed, binds to either TRE (TPA response element) or cAMP response element (CRE) within the promoter region of target genes [87, 88].

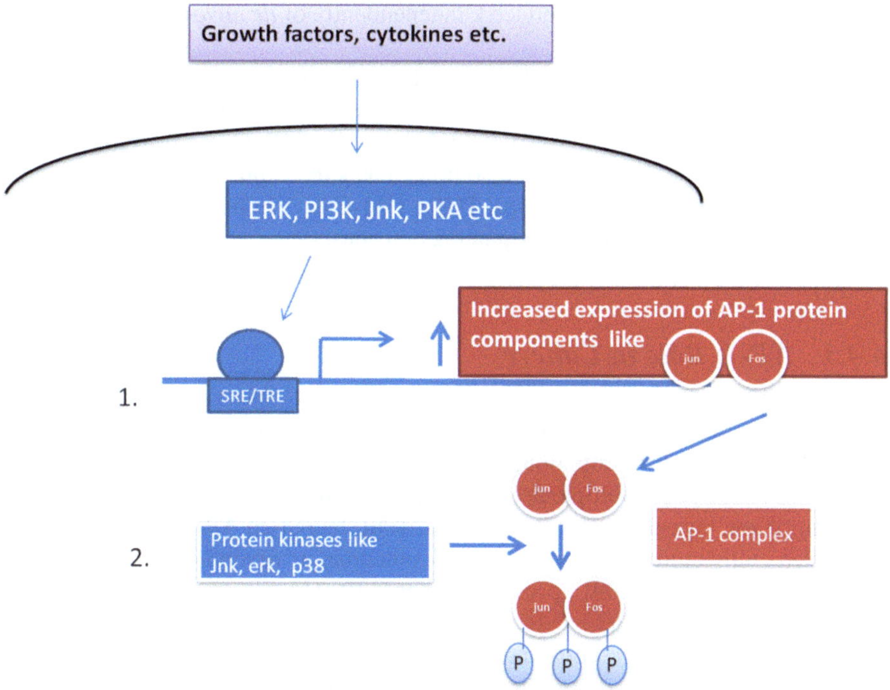


**Fig. 6.6** Regulation of AP-1 activation by transcriptional and post-translational modifications

The different dimer combinations of AP-1 regulate the expression of target genes positively or negatively by recognizing different sequence elements in the promoters and enhancers of the target genes and differentially affect the cellular consequences [80, 89].

AP-1 activity may be induced by several mechanisms which may either increase the abundance of AP-1 components or stimulate their activity. AP1 also acts through interactions with protein kinases which causes phosphorylation of the protein complexes and activate them and also interact with a variety of transcriptional coactivators [72]. MAPKs are known to be the major regulators of AP-1 activity which transduce the extracellular signals to activate AP-1 mediated transcription [90].

The ability of AP-1 to transduce diverse signals relies on two features: (1) the overall protein composition either forming homodimer or heterodimer among different proteins of the AP-1 complex [91–94] and (2) the post-translational modification of the protein complex by various kinases (Fig. 6.6) [90, 95, 96]. Phosphorylation at specific sites enhances the transactivating potential of several AP-1 proteins, including c-Jun and c-Fos, independent of DNA binding activities [97].

Tight regulation of the function of c-Jun/c-Fos and c-Jun/ATF2 is a must for preventing the overexpression of the target genes which might lead to pathological changes such as sustained inflammation as well as cell proliferation [72]. The genes encoding AP-1 components mostly behave as “immediate-early” genes, i.e. genes whose transcription is rapidly induced, independently of *de novo* protein synthesis, following cell stimulation [90].

Target genes known to be regulated by AP-1 are those involved in regulating cell motility and invasion e.g. proteases like MMPs, SPARC (secreted protein acidic and rich in cysteine)), autotoxin, STAT6, IGSF4, ARP 2/3 p41B, CapG etc. [95, 98–103], cell transformation [95, 104]. AP1 also upregulates the expression of various interleukins (IL-1, IL-2, IL-4, IL-5, IL-8), TNFa, GM-CSF, VEGF, COX-2 etc. [5, 105] involved in various inflammatory signaling cascades.

### 6.4.3 AP-1: Role in Cancer

AP-1 transcription factor is known to play significant roles in tumorigenesis, inflammation, apoptosis, differentiation, and developmental processes [33]. The AP-1 complex is involved in multiple processes linked to tumorigenesis such as proliferation, migration, invasion and metastasis [95, 106]. Increased AP-1 activity has been associated with more aggressive clinical outcome in several cancers including breast cancer [92, 107–109] , prostate cancer [110, 111], keratinocytes [112, 113] and colon cancer [114] etc.

The role for AP-1 in controlling cell proliferation has been proposed based on observations that AP-1 activity is induced upon mitogenic stimulation [104, 115] as well as the reversal of *ras* induced transformed phenotype of keratinocytes by dominant negative constructs of c-*fos* and c-*jun* [82, 113]. The transformation and growth regulatory activity of AP-1 was evident from the experiments studying the effects of overexpression of the components of AP-1 complex in immortalized rat fibroblast [116], Chick embryo fibroblast [117, 118], as well as by blocking the expression of the AP-1 components by antisense/si-RNA or by antibody microinjection of AP-1. AP-1 proteins of c-Fos-JunB complex have also been shown to have tumor-suppressor activity [76, 78, 79].

### 6.4.4 Role of AP-1 in Linking Cancer and Inflammation

Basal levels of AP-1 activity are important for normal cellular processes like cell proliferation and survival, but over activation of the AP-1 transcriptional activity by extracellular stimulus may increase the expression of AP-1 dependent genes leading to inflammation, angiogenesis, cell proliferation, invasion etc. that drive normal cell into tumorigenesis. AP-1 is one of the major transcription factors involved in mediating inflammatory response and plays an important role in cancer mediated inflammation [82].

The evidence that AP-1 proteins are involved in the inflammatory response of cancer cells is demonstrated in epidermal cells where loss of AP-1 is shown to control cytokine expression through transcriptional and post-transcriptional pathways

[20]. The Jun proteins are also known to control the expression of cytokines and chemokines such as granulocyte colony-stimulating factor, IL-6 and tumour necrosis factor alpha that are involved in inflammatory skin diseases such as psoriasis [75]. Recently the JNK/p38 pathway has been shown to be activated by pro-inflammatory cytokines like TNF-α and IL-1 and may also have a role in inflammatory responses [51]. The promoters of many inflammatory response genes, especially those encoding cytokines and chemokines, have AP-1 binding sites suggesting a possible role of AP-1/JNK pathway in their regulation. JNK is also thought to be involved in the induction of cyclo-oxygenase 2/prostaglandin synthase 2 which plays an important part in the inflammatory response by catalysing the production of prostaglandins [119].

Targeted inhibition of AP-1 activity suggests a pivotal role for AP-1 in oncogenic transformation and progression [106]. Inflammatory stimuli are known to cause activation of AP-1, which in turn induces cytokine expression and cellular immune responses [120, 121]. Hasselblatt et al. [122] had proposed AP-1 as an essential mediator of oncogenic β-catenin signaling in the intestine. Inflammation induced activation of AP-1 pathway has also been implicated in ovarian carcinogenesis [123]. The production of tumor-promoting cytokines by inflammatory cells are known to activate transcription factors such as AP-1 in premalignant cells to induce genes that stimulate cell proliferation and survival [5, 29].

AP-1 also interacts with NF-κB, and a dominant-negative Jun has been reported to inhibit both AP-1 and NF-κB activity in HPV-immortalized human keratinocytes [124]. AP-1 and NF-κB transcription factors are known to be constitutively active in head and neck squamous cell carcinoma cell lines that express pro-inflammatory and pro-angiogenic cytokines [125]. Moreover, the dominant negative constructs of c-*fos* and c-*jun* have been shown to reverse the transformed phenotype induced by activated Ras and also inhibit the invasiveness and tumorigenesis of keratinocytes [113]. The interleukins have been linked to inflammation and subsequent cancer development. Various ILs like IL-1α, IL-1β, IL-6, IL-8 and IL-17 are known to activate transcription factors like AP-1 and NF-κB. IL -1alpha is known to promote NF-κB and AP-1-induced IL-8 expression, cell survival, and proliferation in head and neck squamous cell carcinomas [126]. IL17A is known to induce AP-1 activity and causing neoplastic transformation of JB6 Cl41 cells. Knockdown of AP-1 inhibited the tumorigenesis of MCF7 cell line, pointing to the critical role of inflammatory microenvironment in carcinogenesis [123].

Adding to the complexity of AP1 regulation, a study showed that the knockdown of a trans-membrane protein, FAT1, leads to inhibition of AP-1 mediated transcription and hence pro-inflammatory modulators like COX-2, IL6 and IL1β in glial tumour cell lines [105]. This is apparently mediated via up-regulation of tumour suppressor PDCD4 (Programmed Cell Death4). PDCD4 is a known tumor-suppressor whose expression is decreased under inflammatory tumor promoting conditions. PDCD4 expression is known to be inversely correlated with AP-1 mediated transcription [37, 127–129]. There are several factors responsible for regulating PDCD4 expression including COX-2 inhibitors, miR-21, pAkt etc. PDCD4 levels are known to be induced by cytokines such as IL-12 (interleukin 12), but

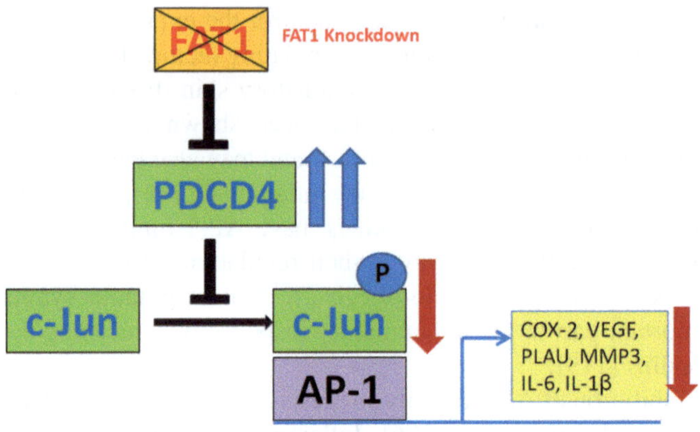


**Fig. 6.7** Knockdown of FAT1 led to increased PDCD4 expression, decreased c-Jun expression as well as c-Jun phosphorylation and decreased expression of AP-1 regulated transcripts. From Dikshit et al. [105] reprinted with permission from Kunzang Chosdol and Nature Publishing Group

down-regulated by IL-2 and IL-15 [130]. FAT1, a trans-membrane protein is identified as a novel regulator of PDCD4 expression and consequently AP-1 mediated transcription (Fig. 6.7).

A study by Zhang et al. [131] had identified FAT1 as a differentially expressed gene in human colon carcinoma cells treated with a selective COX-2 inhibitor. Also, AP-1 is known to regulate COX-2 expression via binding to cAMP response element (CRE) [132, 133]. Cyclooxygenase-2 (COX-2) and the prostaglandins resulting from its enzymatic activity have been shown to play a major role in modulating cell growth and development of human neoplasia [134]. COX-2 contributes to tumorigenesis and the malignant phenotype of tumor cells by different mechanisms, including: (1) inhibition of apoptosis; (2) increased angiogenesis; (3) increased invasiveness; (4) modulation of inflammation/immuno-suppression; and (5) conversion of procarcinogens to carcinogens [135].

Thus, AP-1 family of proteins acts as a master regulator of gene expression in response to inflammatory stimuli and oncogenic signal transduction cascades in a wide variety of tumor cell and animal models. It may be considered as an important target for novel anti-cancer therapies. However the ubiquitous nature of its effects may affect the specificity of the response.

### 6.4.5 *Other Members of Activated Proteins (APs)*

In addition to AP-1, the other members of activated Proteins (APs) family include AP-2, AP-3, AP-4 and AP-5. The DNA binding motif of all the APs is conserved and binds to 5′-TGANTCA-3′ consensus sequence of the target DNA. Among all activated proteins, AP-1 is the most extensively studied one. AP-2 transcription factors represent a family of five closely related proteins, AP-2alpha, -beta, –gamma, -delta and -epsilon [136] and are involved in physiological processes, such as morphogenesis and in pathological processes such as tumorigenesis and genetic disease [137]. Recent evidence suggests that the oncogenic [138, 139] as well as tumor suppressive role [140–144] of AP-2 in different cancers, indicating the context dependent effect of AP-2 depending on the ratio and dimerization of AP-2 isoforms present as well as AP-2 modulating factors.

## 6.5 STAT3

### *6.5.1 Introduction*

STATs (signal transducer and activator of transcription) are members of seven related proteins (STAT1, STAT2, STAT3, STAT4, STAT5A, STAT5B and STAT6) containing SH2 domain which helps in homo- or hetero-dimerization during activation [145, 146]. STAT3 (92-kDa protein) is 770 amino acids long with sequential N terminal coiled-coil domain, DNA-binding domain, a linker, SH2 domain, and C-terminal transactivation domain [147]. STAT3 is predominantly present in the cytoplasm and on activation by growth factors and cytokines [148, 149] gets phosphorylated followed by dimerization and nuclear localization where it acts as transcription factor mediating expression of target genes [150–153]. STAT3 is a point of convergence for numerous oncogenic signaling pathways [23] known to upregulate genes involved in cell proliferation, invasion, angiogenesis and inflammation [23, 153, 154]). STAT3 has a role in inducing procarcinogenic inflammatory microenvironment leading to malignant transformation and cancer progression [5].

### *6.5.2 Mode of Action*

STAT3 transcription is triggered by many cytokines and growth factors, including epidermal growth factor [34], platelet-derived growth factor [155], and IL-6 [146] as well as by oncogenic proteins, such as Src5 [115] and Ras6 [156] as well as by carcinogens like cigarette smoke [157] and by tumor promoters [158, 159].

Activation of STAT3 is regulated by various kinases that phosphorylate at tyrosine 705 residue [151, 160–163] and serine 727 residue [164–170] of STAT3 within the C-terminal region. Along with the phosphorylation, acetylation at lysine 685 residue [171] is also critical for stabilization of the STAT3 dimer state and enhancing its transcriptional activity. On phosphorylation and dimerization, STAT3 gets translocated to nucleus, binds to DNA and upregulates the transcription of various genes known to promote cell proliferation, angiogenesis, survival, migration and inflammation (Fig. 6.8) [145, 169, 170].

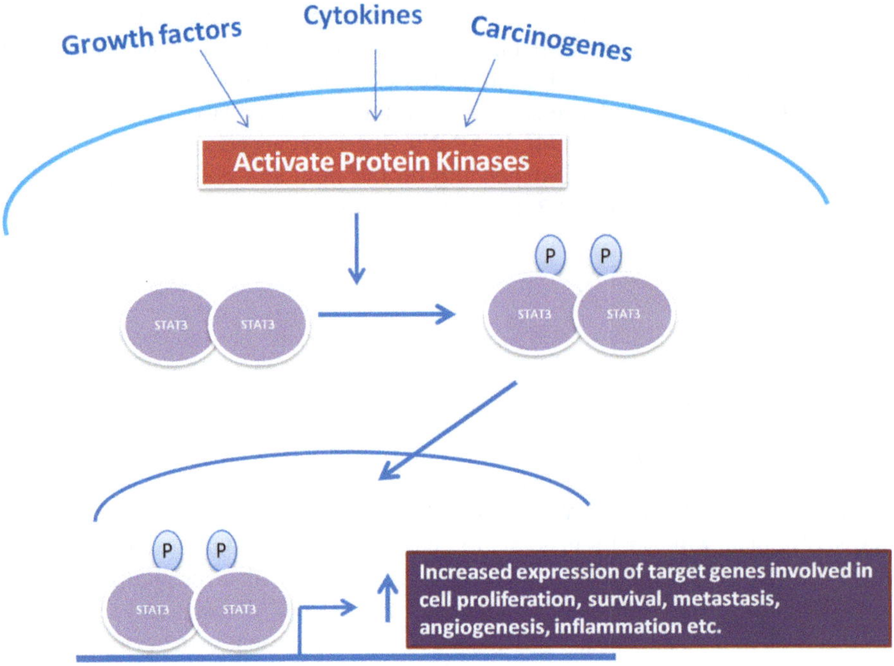


**Fig. 6.8** STAT (Signal transducer and activator of transcription)-3 activation and the functional effect in cancer

### *6.5.3 STAT3 in Linking Inflammation and Cancer*

STAT-3 was initially discovered as an acute-phase response protein and a known mediator of inflammation [172]. Most proinflammatory molecules including IL-6, IL-10, IL-17, IL-23 etc., were found to mediate their effects through the activation of the STAT-3 pathway [149]. In some cell types, IL-6-induced STAT-3 activation has been shown to be dependent on cyclooxygenase 2 (COX-2) [173]. Oncogenic transformation in inflammation-associated gastric epithelial cell mediated by cytokine IL-11 and its glycoprotein 130 (gp130) receptor was found to be dependent on increased activation of STAT-3 [174]. Similarly, tumor promoters, lipopolysaccharides, and cigarette smoke also activate STAT-3 pathway [157, 175].

Aside from its role in inflammation, STAT-3 activation by various stimuli are known for transformation of cells [176]. The activation of STAT-3 by src protein kinase [115, 154, 177], by v-Fps; by polyoma virus middle T antigen and v-Sis [161], by human T-cell lymphotropic virus I [162] and by Hepatitis C virus core protein [118] were all found to transform cells. STAT-3 signaling is also required for hepatocyte growth factor/scatter factor-Met-mediated tumorigenesis [178]. STAT-3 is considered an oncogene [154] due to its capability to transform cells and its over activity detected in different tumors.

Constitutive activation of STAT3 promotes tumor cells proliferation and survival [179] and are found to be activated in wide variety of cancers including gliomas [5, 23, 170, 180, 181]. STAT3 mediate expression of various cell survival gene products like bcl-xl [182, 183], bcl-2 [184], survivin [185], Mcl-1 [186] and cIAP2 [187], cyclin D1 [188], VEGF, c-Myc [189]) and pim-1 [190]. Additionally, activated STAT3 is known to suppress apoptosis by suppressing expression of FAS protein [191, 192]). So, Stat3 in many human cancers functions as a critical mediator of oncogenic signaling through transcriptional activation of genes encoding apoptosis inhibitors (e.g. Bcl-x(L), Mcl-1 and survivin), cell-cycle regulators (e.g. cyclin D1 and c-Myc) and inducers of angiogenesis (e.g. vascular endothelial growth factor).

STAT-3 activation plays a major role in tumor cell invasion by regulating the expression of matrix metalloproteinase (MMP)-2 and MMP-1 [119, 193] as well as the expression of the *MUC1* gene [194]. Overexpression of phosphorylated STAT-3 in cutaneous squamous cell carcinoma showed increased invasion and metastasis [182].

STAT3 is found to upregulate the transcription of genes promoting angiogenesis in chick chorioallantoic membrane [137]. Constitutive STAT-3 activity upregulates the expression of VEGF [119, 195, 196] and TWIST [197], promoting tumor angiogenesis and metastasis. STAT3 is also known to regulate EMT through activation of cytokines, including IL-1, IL-6, TNFα, TGFβ [114, 198] and by regulating EMT modulators such as Snail, Twist and ZEB [114, 199].

Overall findings from various studies show that STAT-3 has a central role in regulating cancer-associated inflammation by controlling the expression of various inflammatory modulators and oncogenes, making tumors more aggressive with increased metastatic potential. Role of STAT3 in linking inflammation with cancer is well established as evident from various studies over the past decade in gastric carcinoma, hepatocellular carcinoma, lung cancer, neuroblastoma, multiple myloma etc. [42, 123, 174, 200–206]. So targeting STAT3 in cancers with activated STAT3 form may attenuate tumor progression and invasion as well as it may regulate inflammatory microenvironment of tumors thus making tumors more sensitive to conventional therapy.

## 6.6 Conclusions

Even with the enormous knowledge and advancement in the field of cancer related inflammation, many questions remain unanswered. Activation of signaling pathways involved in regulating the pro-inflammatory microenvironment in cancers, by the extracellular triggers (either extrinsic or intrinsic) converges onto the common intracellular signaling molecules which eventually stimulate well known transcription factors upregulating the expression of pro-inflammatory cytokines and modulators. As discussed above, the transcription factors NF-κB, HIF1α, AP-1 and STAT3 play enormous role in regulating the expression of genes which are involved in linking cancer and inflammation. In addition to their canonical effect, these transcription factors also cross talk with each other in further maintaining and aggravating the pro-inflammatory microenvironment in cancers (Fig. 6.9) [47, 178, 207].

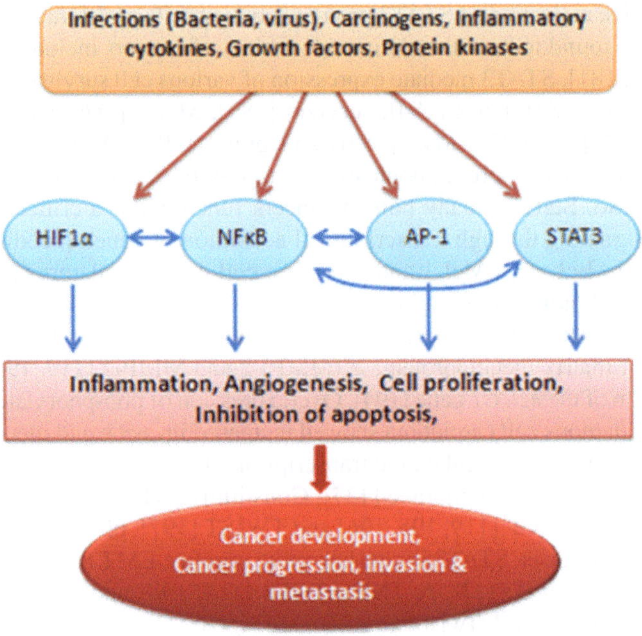


**Fig. 6.9** Stimulation of transcription factors having pro-inflammatory and pro-cancerous activities by various stimuli in cancer. Blue arrows depict direct effect of the individual transcription factors, as well as crosstalk among different transcription factors

These transcription factors are the major contributors in regulating the expression of cytokines and inflammatory modulators, and targeting them for molecular therapeutics are considered to be promising with many such drugs undergoing trials. With the identification of these transcription factors in tumor progression and invasion, it has become a major goal for a number of laboratories and companies to develop novel therapeutic molecules so as to target transcription factors for cancer therapy.

Moreover, the mechanisms of action of several old drugs have been re-evaluated and some of them e.g. NSAID were discovered to act, at least, partially through some of these transcription factors. The antitumor activity by NSAID was due to the inhibition of cyclooxygenase (COX) enzyme activity as well as by inhibiting the LPS-induced NFkB-dependent transcription by preventing IkBa degradation [9, 208]. Similarly sulfasalazine was shown to inhibit NF-kB activation [209, 210]. The cardiac glycosides drugs like digitoxin and oleandrin also found to have potential implications in inflammation and tumorigenesis by inhibiting TNF-α-induced activation of NF-kB, JNK and AP-1 [34, 211–214].

Curcumin is another promising anticancer drug, with wide range of effect on cancer and inflammation. Curcumin is known to affect many signaling pathways and found to interfere with the activity of NFkB, STAT3 as well as JNK [196, 215–217]. Recently, in in-vitro and preclinical studies, it has been shown to suppress STAT3 phosphorylation in lung cancer with concomitant reduction in cell proliferation and suggested to be a promising chemopreventive agent in high-risk populations such as smokers [218]. Curcumin as well as other natural chemopreventive agents like retinoid, resveratrol phenethyl isothiocyanate and sulforaphane were found to inhibit AP1 activity [219].

Many novel agents have been identified since then against these transcription factors and are under study at different levels of clinical trials. One such molecule is the bortezomib (proteosome inhibitor) was found to induce apoptosis in a number of otherwise resistant tumor cells, and sensitize such cancer cells to conventional chemotherapies and radiation therapy [158, 220–224]. Other well-known drugs used in cancer inflammation are avastin against VEGF, Rituxan to represses B cells in rheumatoid arthritis & B cell lymphoma and antibodies targeting COX-2 and cytokines like IL-6, IL-1β, IL-8, CCL2 etc.

Along with these, there are drugs targeted exclusively against specific transcription factors which may be used as adjuvant therapeutic molecules in cancer and are under study. These are IKK inhibitor PS1145 and BMS-345541 [225–227], HIF1α inhibitors e.g. echinomycin [3, 55, 228], AP1 inhibitors [94] and STAT3 inhibitors BP-1–102 and WP1066 [229, 230].

The expression of target genes by different protein-complexes of a transcription factor may vary and may not always be same, and cross talk among different transcription factors may also stimulate different subset of target genes. With the result, each tumor type could demonstrate distinct clinical presentations and activities depending on the genes being expressed by these transcription factors. Therefore, for each tumor type, the therapeutic molecules would ideally have to be carefully chosen on the basis of the genetic/epigenetic profile of a patient and the combination of transcription factors being activated. Hopefully, with the explosion in our knowledge and understanding of the biology of neoplasia, personalized therapy is soon likely to be a reality for cancer patients.

**Acknowledgement** Research work of the authors cited in the text is funded by Defence Research and Development Organization (DRDO), India and Department of Biotechnology (DBT), India. Authors are grateful to National Brain Research Centre (NBRC), Manesar, Gurgaon, India for the facilities and co-support.

## References

1. Aggarwal BB, Shishodia S, Sandur SK, Pandey MK, Sethi G (2006) Inflammation and cancer: how hot is the link? Biochem Pharmacol 72(11):1605–1621
2. Kuper H, Adami HO, Trichopoulos D (2000) Infections as a major preventable cause of human cancer. J Intern Med 248(3):171–183
3. Wang R, Zhou SLi S (2011) Cancer therapeutic agents targeting hypoxia-inducible factor-1. Curr Med Chem 18(21):3168–3189
4. Shabad LM (1962) Some experimental data regarding the relationship between inflammation and cancer. Vopr Onkol 8(6):74–80

5. Grivennikov SI, Greten FR, Karin M (2010) Immunity, inflammation, and cancer. Cell 140(6):883–899
6. Coussens LM, Werb Z (2002) Inflammation and cancer. Nature 420(6917):860–867
7. Balkwill F, Mantovani A (2001) Inflammation and cancer: back to Virchow? Lancet 357(9255):539–545
8. de Martel C, Ferlay J, Franceschi S, Vignat J, Bray F, Forman D et al (2012) Global burden of cancers attributable to infections in 2008: a review and synthetic analysis. Lancet Oncol 13(6):607–615
9. Kopp E, Ghosh S (1994) Inhibition of NF-kappa B by sodium salicylate and aspirin. Science 265(5174):956–959
10. Aggarwal BB, Gehlot P (2009) Inflammation and cancer: how friendly is the relationship for cancer patients? Curr Opin Pharmacol 9(4):351–369
11. Borrello MG, Degl'Innocenti D, Pierotti MA (2008) Inflammation and cancer: the oncogene-driven connection. Cancer Lett 267(2):262–270
12. Schwartsburd PM (2004) Age-promoted creation of a pro-cancer microenvironment by inflammation: pathogenesis of dyscoordinated feedback control. Mech Ageing Dev 125(9):581–590
13. Borrello MG, Alberti L, Fischer A, Degl'innocenti D, Ferrario C, Gariboldi M et al (2005) Induction of a proinflammatory program in normal human thyrocytes by the RET/PTC1 oncogene. Proc Natl Acad Sci USA 102(41):14825–14830
14. De Falco V, Guarino V, Avilla E, Castellone MD, Salerno P, Salvatore G et al (2007) Biological role and potential therapeutic targeting of the chemokine receptor CXCR4 in undifferentiated thyroid cancer. Cancer Res 67(24):11821–11829
15. Xu K, Shu HK (2007) EGFR activation results in enhanced cyclooxygenase-2 expression through p38 mitogen-activated protein kinase-dependent activation of the Sp1/Sp3 transcription factors in human gliomas. Cancer Res 67(13):6121–6129
16. Guerra C, Schuhmacher AJ, Canamero M, Grippo PJ, Verdaguer L, Perez-Gallego L et al (2007) Chronic pancreatitis is essential for induction of pancreatic ductal adenocarcinoma by K-Ras oncogenes in adult mice. Cancer Cell 11(3):291–302
17. Sparmann A, Bar-Sagi D (2004) Ras-induced interleukin-8 expression plays a critical role in tumor growth and angiogenesis. Cancer Cell 6(5):447–458
18. Sumimoto H, Imabayashi F, Iwata T, Kawakami Y (2006) The BRAF-MAPK signaling pathway is essential for cancer-immune evasion in human melanoma cells. J Exp Med 203(7):1651–1656
19. Shchors K, Shchors E, Rostker F, Lawlor ER, Brown-Swigart L, Evan GI (2006) The Myc-dependent angiogenic switch in tumors is mediated by interleukin 1beta. Genes Dev 20(18):2527–2538
20. Schonthaler HB, Guinea-Viniegra J, Wagner EF (2011) Targeting inflammation by modulating the Jun/AP-1 pathway. Ann Rheum Dis 70(Suppl 1):i109–i112
21. Bonizzi G, Karin M (2004) The two NF-kappaB activation pathways and their role in innate and adaptive immunity. Trends Immunol 25(6):280–288
22. Perkins ND (2012) The diverse and complex roles of NF-kappaB subunits in cancer. Nat Rev Cancer 12(2):121–132
23. Koul D, Shen R, Shishodia S, Takada Y, Bhat KP, Reddy SA et al (2007) PTEN down regulates AP-1 and targets c-fos in human glioma cells via PI3-kinase/Akt pathway. Mol Cell Biochem 300(1–2):77–87
24. Qin H, Wilson CA, Lee SJ, Zhao X, Benveniste EN (2005) LPS induces CD40 gene expression through the activation of NF-kappaB and STAT-1alpha in macrophages and microglia. Blood 106(9):3114–3122
25. Fitzgerald DC, Meade KG, McEvoy AN, Lillis L, Murphy EP, MacHugh DE et al (2007) Tumour necrosis factor-alpha (TNF-alpha) increases nuclear factor kappaB (NFkappaB) activity in and interleukin-8 (IL-8) release from bovine mammary epithelial cells. Vet Immunol Immunopathol 116(1–2):59–68

26. Fan Y, Dutta J, Gupta N, Fan G, Gelinas C (2008) Regulation of programmed cell death by NF-kappaB and its role in tumorigenesis and therapy. Adv Exp Med Biol 615:223–250
27. Ben-Neriah Y, Karin M (2011) Inflammation meets cancer, with NF-kappaB as the matchmaker. Nat Immunol 12(8):715–723
28. Smale ST (2011) Hierarchies of NF-kappaB target-gene regulation. Nat Immunol 12(8):689–694
29. Baud V, Karin M (2009) Is NF-kappaB a good target for cancer therapy? hopes and pitfalls. Nat Rev Drug Discov 8(1):33–40
30. Karin M, Greten FR (2005) NF-kappaB: linking inflammation and immunity to cancer development and progression. Nat Rev Immunol 5(10):749–759
31. Houghton J, Stoicov C, Nomura S, Rogers AB, Carlson J, Li H et al (2004) Gastric cancer originating from bone marrow-derived cells. Science 306(5701):1568–1571
32. Bogdan C (2001) Nitric oxide and the immune response. Nat Immunol 2(10):907–916
33. Bitton-Worms K, Pikarsky E, Aronheim A (2010) The AP-1 repressor protein, JDP2, potentiates hepatocellular carcinoma in mice. Mol Cancer 9:54
34. Yang Q, Huang W, Jozwik C, Lin Y, Glasman M, Caohuy H et al (2005) Cardiac glycosides inhibit TNF-alpha/NF-kappaB signaling by blocking recruitment of TNF receptor-associated death domain to the TNF receptor. Proc Natl Acad Sci USA 102(27):9631–9636
35. Mantovani A, Sica A (2010) Macrophages, innate immunity and cancer: balance, tolerance, and diversity. Curr Opin Immunol 22(2):231–237
36. Hagemann T, Lawrence T, McNeish I, Charles KA, Kulbe H, Thompson RG et al (2008) Re-educating tumor-associated macrophages by targeting NF-kappaB. J Exp Med 205(6):1261–1268
37. Yang HS, Jansen AP, Nair R, Shibahara K, Verma AK, Cmarik JL et al (2001) A novel transformation suppressor, Pdcd4, inhibits AP-1 transactivation but not NF-kappaB or ODC transactivation. Oncogene 20(6):669–676
38. Bredel M, Scholtens DM, Yadav AK, Alvarez AA, Renfrow JJ, Chandler JP et al (2011) NFKBIA deletion in glioblastomas. N Engl J Med 364(7):627–637
39. Ammirante M, Luo JL, Grivennikov S, Nedospasov S, Karin M (2010) B-cell-derived lymphotoxin promotes castration-resistant prostate cancer. Nature 464(7286):302–305
40. Luo JL, Tan W, Ricono JM, Korchynskyi O, Zhang M, Gonias SL et al (2007) Nuclear cytokine-activated IKKalpha controls prostate cancer metastasis by repressing Maspin. Nature 446(7136):690–694
41. Keats JJ, Fonseca R, Chesi M, Schop R, Baker A, Chng WJ et al (2007) Promiscuous mutations activate the noncanonical NF-kappaB pathway in multiple myeloma. Cancer Cell 12(2):131–144
42. Charafe-Jauffret E, Ginestier C, Iovino F, Wicinski J, Cervera N, Finetti P et al (2009) Breast cancer cell lines contain functional cancer stem cells with metastatic capacity and a distinct molecular signature. Cancer Res 69(4):1302–1313
43. Merkhofer EC, Cogswell P, Baldwin AS (2010) Her2 Activates NF-kappaB and induces invasion through the canonical pathway involving IKKalpha. Oncogene 29(8):1238–1248
44. Maeda S, Kamata H, Luo JL, Leffert H, Karin M (2005) IKKbeta couples hepatocyte death to cytokine-driven compensatory proliferation that promotes chemical hepatocarcinogenesis. Cell 121(7):977–990
45. Arnott CH, Scott KA, Moore RJ, Robinson SC, Thompson RG, Balkwill FR (2004) Expression of both TNF-alpha receptor subtypes is essential for optimal skin tumour development. Oncogene 23(10):1902–1910
46. Dajee M, Lazarov M, Zhang JY, Cai T, Green CL, Russell AJ et al (2003) NF-kappaB blockade and oncogenic Ras trigger invasive human epidermal neoplasia. Nature 421(6923): 639–643
47. Lind MH, Rozell B, Wallin RP, van Hogerlinden M, Ljunggren HG, Toftgard R et al (2004) Tumor necrosis factor receptor 1-mediated signaling is required for skin cancer development induced by NF-kappaB inhibition. Proc Natl Acad Sci USA 101(14):4972–4977

48. Korkaya H, Paulson A, Iovino F, Wicha MS (2008) HER2 Regulates the mammary stem/progenitor cell population driving tumorigenesis and invasion. Oncogene 27(47):6120–6130
49. Liu M, Sakamaki T, Casimiro MC, Willmarth NE, Quong AA, Ju X et al (2010) The canonical NF-kappaB pathway governs mammary tumorigenesis in transgenic mice and tumor stem cell expansion. Cancer Res 70(24):10464–10473
50. Karin M, Yamamoto Y, Wang QM (2004) The IKK NF-kappa B system: a treasure trove for drug development. Nat Rev Drug Discov 3(1):17–26
51. Beyaert R, Cuenda A, Vanden Berghe W, Plaisance S, Lee JC, Haegeman G et al (1996) The p38/RK mitogen-activated protein kinase pathway regulates interleukin-6 synthesis response to tumor necrosis factor. EMBO J 15(8):1914–1923
52. Amschler K, Schon MP, Pletz N, Wallbrecht K, Erpenbeck L, Schon M (2010) NF-kappaB inhibition through proteasome inhibition or IKKbeta blockade increases the susceptibility of melanoma cells to cytostatic treatment through distinct pathways. J Invest Dermatol 130(4):1073–1086
53. Wilken R, Veena MS, Wang MB, Srivatsan ES (2011) Curcumin: a review of anti-cancer properties and therapeutic activity in head and neck squamous cell carcinoma. Mol Cancer 10:12
54. Dhillon N, Aggarwal BB, Newman RA, Wolff RA, Kunnumakkara AB, Abbruzzese JL et al (2008) Phase II trial of curcumin in patients with advanced pancreatic cancer. Clin Cancer Res 14(14):4491–4499
55. Semenza GL (2009) HIF-1 inhibitors for cancer therapy: from gene expression to drug discovery. Curr Pharm Des 15(33):3839–3843
56. Nizet V, Johnson RS (2009) Interdependence of hypoxic and innate immune responses. Nat Rev Immunol 9(9):609–617
57. Melillo G (2011) Hypoxia: jump-starting inflammation. Blood 117(9):2561–2562
58. Walmsley SR, Print C, Farahi N, Peyssonnaux C, Johnson RS, Cramer T et al (2005) Hypoxia-induced neutrophil survival is mediated by HIF-1alpha-dependent NF-kappaB activity. J Exp Med 201(1):105–115
59. Taylor CT, Cummins EP (2009) The role of NF-kappaB in hypoxia-induced gene expression. Ann N Y Acad Sci 1177:178–184
60. Rius J, Guma M, Schachtrup C, Akassoglou K, Zinkernagel AS, Nizet V et al (2008) NF-kappaB links innate immunity to the hypoxic response through transcriptional regulation of HIF-1alpha. Nature 453(7196):807–811
61. Cummins EP, Berra E, Comerford KM, Ginouves A, Fitzgerald KT, Seeballuck F et al (2006) Prolyl hydroxylase-1 negatively regulates IkappaB kinase-beta, giving insight into hypoxia-induced NFkappaB activity. Proc Natl Acad Sci USA 103(48):18154–18159
62. Taylor CT (2008) Interdependent roles for hypoxia inducible factor and nuclear factor-kappaB in hypoxic inflammation. J Physiol 586(Pt 17):4055–4059
63. Belaiba RS, Bonello S, Zahringer C, Schmidt S, Hess J, Kietzmann T et al (2007) Hypoxia up-regulates hypoxia-inducible factor-1alpha transcription by involving phosphatidylinositol 3-kinase and nuclear factor kappaB in pulmonary artery smooth muscle cells. Mol Biol Cell 18(12):4691–4697
64. van Uden P, Kenneth NS, Rocha S (2008) Regulation of hypoxia-inducible factor-1alpha by NF-kappaB. Biochem J 412(3):477–484
65. Scortegagna M, Cataisson C, Martin RJ, Hicklin DJ, Schreiber RD, Yuspa SH et al (2008) HIF-1alpha regulates epithelial inflammation by cell autonomous NFkappaB activation and paracrine stromal remodeling. Blood 111(7):3343–3354
66. Angel P, Imagawa M, Chiu R, Stein B, Imbra RJ, Rahmsdorf HJ et al (1987) Phorbol ester-inducible genes contain a common cis element recognized by a TPA-modulated trans-acting factor. Cell 49(6):729–739
67. Angel P, Karin M (1991) The role of Jun, Fos and the AP-1 complex in cell-proliferation and transformation. Biochim Biophys Acta 1072(2–3):129–157
68. Halazonetis TD, Georgopoulos K, Greenberg ME, Leder P (1988) c-Jun dimerizes with itself and with c-Fos, forming complexes of different DNA binding affinities. Cell 55(5):917–924

69. Kouzarides T, Ziff E (1988) The role of the leucine zipper in the fos-jun interaction. Nature 336(6200):646–651
70. Kataoka K, Noda M, Nishizawa M (1994) Maf nuclear oncoprotein recognizes sequences related to an AP-1 site and forms heterodimers with both Fos and Jun. Mol Cell Biol 14(1):700–712
71. Wisdom R (1999) AP-1: one switch for many signals. Exp Cell Res 253(1):180–185
72. Karin M, Liu Z, Zandi E (1997) AP-1 function and regulation. Curr Opin Cell Biol 9(2):240–246
73. Fujiwara KT, Kataoka K, Nishizawa M (1993) Two new members of the maf oncogene family, mafK and mafF, encode nuclear b-Zip proteins lacking putative trans-activator domain. Oncogene 8(9):2371–2380
74. Landschulz WH, Johnson PF, McKnight SL (1988) The leucine zipper: a hypothetical structure common to a new class of DNA binding proteins. Science 240(4860):1759–1764
75. Wagner EF, Schonthaler HB, Guinea-Viniegra J, Tschachler E (2010) Psoriasis: what we have learned from mouse models. Nat Rev Rheumatol 6(12):704–714
76. Schutte J, Minna JD, Birrer MJ (1989) Deregulated expression of human c-jun transforms primary rat embryo cells in cooperation with an activated c-Ha-ras gene and transforms rat-1a cells as a single gene. Proc Natl Acad Sci USA 86(7):2257–2261
77. Schutte J, Viallet J, Nau M, Segal S, Fedorko J, Minna J (1989) Jun-B inhibits and c-fos stimulates the transforming and trans-activating activities of c-jun. Cell 59(6):987–997
78. Pfarr CM, Mechta F, Spyrou G, Lallemand D, Carillo S, Yaniv M (1994) Mouse JunD negatively regulates fibroblast growth and antagonizes transformation by ras. Cell 76(4):747–760
79. Passegue E, Wagner EF (2000) JunB suppresses cell proliferation by transcriptional activation of p16 (INK4a) expression. EMBO J 19(12):2969–2979
80. Chiu R, Angel P, Karin M (1989) Jun-B differs in its biological properties from, and is a negative regulator of, c-Jun. Cell 59(6):979–986
81. Chang L, Karin M (2001) Mammalian MAP kinase signalling cascades. Nature 410(6824):37–40
82. Matthews CP, Colburn NH, Young MR (2007) AP-1 a target for cancer prevention. Curr Cancer Drug Targets 7(4):317–324
83. Whitmarsh AJ, Davis RJ (1996) Transcription factor AP-1 regulation by mitogen-activated protein kinase signal transduction pathways. J Mol Med (Berl) 74(10):589–607
84. Dong Z, Birrer MJ, Watts RG, Matrisian LM, Colburn NH (1994) Blocking of tumor promoter-induced AP-1 activity inhibits induced transformation in JB6 mouse epidermal cells. Proc Natl Acad Sci USA 91(2):609–613
85. Angel P, Hattori K, Smeal T, Karin M (1988) The jun proto-oncogene is positively autoregulated by its product, Jun/AP-1. Cell 55(5):875–885
86. Gonzalez JM, Navarro-Puche A, Casar B, Crespo P, Andres V (2008) Fast regulation of AP-1 activity through interaction of lamin A/C, ERK1/2, and c-Fos at the nuclear envelope. J Cell Biol 183(4):653–666
87. Ellenberger TE, Brandl CJ, Struhl K, Harrison SC (1992) The GCN4 basic region leucine zipper binds DNA as a dimer of uninterrupted alpha helices: crystal structure of the protein-DNA complex. Cell 71(7):1223–1237
88. Schumacher MA, Goodman RH, Brennan RG (2000) The structure of a CREB bZIP. Somatostatin CRE complex reveals the basis for selective dimerization and divalent cation-enhanced DNA binding. J Biol Chem 275(45):35242–35247
89. Zenz R, Eferl R, Scheinecker C, Redlich K, Smolen J, Schonthaler HB et al (2008) Activator protein 1 (Fos/Jun) functions in inflammatory bone and skin disease. Arthritis Res Ther 10(1):201
90. Karin M (1995) The regulation of AP-1 activity by mitogen-activated protein kinases. J Biol Chem 270(28):16483–16486
91. van Dam H, Castellazzi M (2001) Distinct roles of Jun : Fos and Jun : ATF dimers in oncogenesis. Oncogene 20(19):2453–2464
92. Baan B, Pardali E, ten Dijke P, van Dam H (2010) In situ proximity ligation detection of c-Jun/AP-1 dimers reveals increased levels of c-Jun/Fra1 complexes in aggressive breast cancer cell lines in vitro and in vivo. Mol Cell Proteomics 9(9):1982–1990

93. Ryseck RP, Bravo R (1991) c-JUN, JUN B, and JUN D differ in their binding affinities to AP-1 and CRE consensus sequences: effect of FOS proteins. Oncogene 6(4):533–542
94. Leaner VD, Donninger H, Birrer MJ (2007) Transcription factors as targets for cancer therapy: AP-1 a potential therapeutic target, in Current Cancer Therapy Reviews. Bentham Science Publishers, pp 1–6
95. Eferl R, Wagner EF (2003) AP-1: a double-edged sword in tumorigenesis. Nat Rev Cancer 3(11):859–868
96. Karin M, Delhase M (1998) JNK or IKK, AP-1 or NF-kappaB, which are the targets for MEK kinase 1 action? Proc Natl Acad Sci USA 95(16):9067–9069
97. Smeal T, Binetruy B, Mercola D, Grover-Bardwick A, Heidecker G, Rapp UR et al (1992) Oncoprotein-mediated signalling cascade stimulates c-Jun activity by phosphorylation of serines 63 and 73. Mol Cell Biol 12(8):3507–3513
98. Black EJ, Clair T, Delrow J, Neiman P, Gillespie DA (2004) Microarray analysis identifies autotaxin, a tumour cell motility and angiogenic factor with lysophospholipase D activity, as a specific target of cell transformation by v-Jun. Oncogene 23(13):2357–2366
99. Bos TJ, Margiotta P, Bush L, Wasilenko W (1999) Enhanced cell motility and invasion of chicken embryo fibroblasts in response to Jun over-expression. Int J Cancer 81(3):404–410
100. Briggs J, Chamboredon S, Castellazzi M, Kerry JA, Bos TJ (2002) Transcriptional upregulation of SPARC, in response to c-Jun overexpression, contributes to increased motility and invasion of MCF7 breast cancer cells. Oncogene 21(46):7077–7091
101. el Bahassi M, Karyala S, Tomlinson CR, Sartor MA, Medvedovic M, Hennigan RF (2004) Critical regulation of genes for tumor cell migration by AP-1. Clin Exp Metastasis 21(4):293–304
102. Cohen SB, Waha A, Gelman IH, Vogt PK (2001) Expression of a down-regulated target, SSeCKS, reverses v-Jun-induced transformation of 10 T1/2 murine fibroblasts. Oncogene 20(2):141–146
103. McGarry LC, Winnie JN, Ozanne BW (2004) Invasion of v-Fos(FBR)-transformed cells is dependent upon histone deacetylase activity and suppression of histone deacetylase regulated genes. Oncogene 23(31):5284–5292
104. Young MR, Li JJ, Rincon M, Flavell RA, Sathyanarayana BK, Hunziker R et al (1999) Transgenic mice demonstrate AP-1 (activator protein-1) transactivation is required for tumor promotion. Proc Natl Acad Sci USA 96(17):9827–9832
105. Dikshit B, Irshad K, Madan E, Aggarwal N, Sarkar C, Chandra PS et al (2012) FAT1 acts as an upstream regulator of oncogenic and inflammatory pathways, via PDCD4, in glioma cells. Oncogene
106. Shaulian E (2010) AP-1–the Jun proteins: oncogenes or tumor suppressors in disguise? Cell Signal 22(6):894–899
107. Dahlman-Wright K, Qiao Y, Jonsson P, Gustafsson JA, Williams C, Zhao C (2012) Interplay between AP-1 and estrogen receptor alpha in regulating gene expression and proliferation networks in breast cancer cells. Carcinogenesis 33(9):1684–1691
108. Milde-Langosch K, Janke S, Wagner I, Schroder C, Streichert T, Bamberger AM et al (2008) Role of Fra-2 in breast cancer: influence on tumor cell invasion and motility. Breast Cancer Res Treat 107(3):337–347
109. Milde-Langosch K, Roder H, Andritzky B, Aslan B, Hemminger G, Brinkmann A et al (2004) The role of the AP-1 transcription factors c-Fos, FosB, Fra-1 and Fra-2 in the invasion process of mammary carcinomas. Breast Cancer Res Treat 86(2):139–152
110. Kajanne R, Miettinen P, Tenhunen M, Leppa S (2009) Transcription factor AP-1 promotes growth and radioresistance in prostate cancer cells. Int J Oncol 35(5):1175–1182
111. Parra E, Ferreira J, Ortega A (2011) Overexpression of EGR-1 modulates the activity of NF-kappaB and AP-1 in prostate carcinoma PC-3 and LNCaP cell lines. Int J Oncol 39(2):345–352
112. Domann FE, Levy JP, Birrer MJ, Bowden GT (1994) Stable expression of a c-JUN deletion mutant in two malignant mouse epidermal cell lines blocks tumor formation in nude mice. Cell Growth Differ 5(1):9–16

113. Dong Z, Crawford HC, Lavrovsky V, Taub D, Watts R, Matrisian LM et al (1997) A dominant negative mutant of jun blocking 12-O-tetradecanoylphorbol-13-acetate-induced invasion in mouse keratinocytes. Mol Carcinog 19(3):204–212
114. Zhang HS, Yan B, Li XB, Fan L, Zhang YF, Wu GH, et al (2012) Pax2 induces expression of cyclin D1 through activating AP-1 and promotes proliferation of colon cancer cells. J Biol Chem 287(53):44164–44172
115. Saez E, Rutberg SE, Mueller E, Oppenheim H, Smoluk J, Yuspa SH et al (1995) c-fos is required for malignant progression of skin tumors. Cell 82(5):721–732
116. Miller AD, Curran T, Verma IM (1984) c-fos protein can induce cellular transformation: a novel mechanism of activation of a cellular oncogene. Cell 36(1):51–60
117. Bos TJ, Monteclaro FS, Mitsunobu F, Ball AR Jr, Chang CH, Nishimura T et al (1990) Efficient transformation of chicken embryo fibroblasts by c-Jun requires structural modification in coding and noncoding sequences. Genes Dev 4(10):1677–1687
118. Yoshida T, Shindo Y, Ohta K, Iba H (1989) Identification of a small region of the v-fos gene product that is sufficient for transforming potential and growth-stimulating activity. Oncogene Res 5(2):79–89
119. Xie W, Herschman HR (1995) v-src induces prostaglandin synthase 2 gene expression by activation of the c-Jun N-terminal kinase and the c-Jun transcription factor. J Biol Chem 270(46):27622–27628
120. Meixner A, Karreth F, Kenner L, Wagner EF (2004) JunD regulates lymphocyte proliferation and T helper cell cytokine expression. EMBO J 23(6):1325–1335
121. Wagner EF, Eferl R (2005) Fos/AP-1 proteins in bone and the immune system. Immunol Rev 208:126–140
122. Hasselblatt P, Gresh L, Kudo H, Guinea-Viniegra J, Wagner EF (2008) The role of the transcription factor AP-1 in colitis-associated and beta-catenin-dependent intestinal tumorigenesis in mice. Oncogene 27(47):6102–6109
123. Kim MK, Kim K, Han JY, Lim JM, Song YS (2011) Modulation of inflammatory signaling pathways by phytochemicals in ovarian cancer. Genes Nutr 6(2):109–115
124. Li JJ, Rhim JS, Schlegel R, Vousden KH, Colburn NH (1998) Expression of dominant negative Jun inhibits elevated AP-1 and NF-kappaB transactivation and suppresses anchorage independent growth of HPV immortalized human keratinocytes. Oncogene 16(21):2711–2721
125. Ondrey FG, Dong G, Sunwoo J, Chen Z, Wolf JS, Crowl-Bancroft CV et al (1999) Constitutive activation of transcription factors NF-(kappa)B, AP-1, and NF-IL6 in human head and neck squamous cell carcinoma cell lines that express pro-inflammatory and pro-angiogenic cytokines. Mol Carcinog 26(2):119–129
126. Wolf JS, Chen Z, Dong G, Sunwoo JB, Bancroft CC, Capo DE et al (2001) IL (interleukin)-1alpha promotes nuclear factor-kappaB and AP-1-induced IL-8 expression, cell survival, and proliferation in head and neck squamous cell carcinomas. Clin Cancer Res 7(6):1812–1820
127. Wang Q, Zhang Y, Yang HS (2012) Pdcd4 knockdown up-regulates MAP4K1 expression and activation of AP-1 dependent transcription through c-Myc. Biochim Biophys Acta 1823(10):1807–1814
128. Yang HS, Knies JL, Stark C, Colburn NH (2003) Pdcd4 suppresses tumor phenotype in JB6 cells by inhibiting AP-1 transactivation. Oncogene 22(24):3712–3720
129. Yang HS, Matthews CP, Clair T, Wang Q, Baker AR, Li CC et al (2006) Tumorigenesis suppressor Pdcd4 down-regulates mitogen-activated protein kinase kinase kinase kinase 1 expression to suppress colon carcinoma cell invasion. Mol Cell Biol 26(4):1297–1306
130. Azzoni L, Zatsepina O, Abebe B, Bennett IM, Kanakaraj P, Perussia B (1998) Differential transcriptional regulation of CD161 and a novel gene, 197/15a, by IL-2, IL-15, and IL-12 in NK and T cells. J Immunol 161(7):3493–3500
131. Zhang Z, DuBois RN (2001) Detection of differentially expressed genes in human colon carcinoma cells treated with a selective COX-2 inhibitor. Oncogene 20(33):4450–4456
132. Subbaramaiah K, Cole PA, Dannenberg AJ (2002) Retinoids and carnosol suppress cyclooxygenase-2 transcription by CREB-binding protein/p300-dependent and -independent mechanisms. Cancer Res 62(9):2522–2530

133. Subbaramaiah K, Norton L, Gerald W, Dannenberg AJ (2002) Cyclooxygenase-2 is overexpressed in HER-2/neu-positive breast cancer: evidence for involvement of AP-1 and PEA3. J Biol Chem 277(21):18649–18657
134. Trifan OC, Hla T (2003) Cyclooxygenase-2 modulates cellular growth and promotes tumorigenesis. J Cell Mol Med 7(3):207–222
135. Dempke W, Rie C, Grothey A, Schmoll HJ (2001) Cyclooxygenase-2: a novel target for cancer chemotherapy? J Cancer Res Clin Oncol 127(7):411–417
136. Hilger-Eversheim K, Moser M, Schorle H, Buettner R (2000) Regulatory roles of AP-2 transcription factors in vertebrate development, apoptosis and cell-cycle control. Gene 260(1–2):1–12
137. Orso F, Penna E, Cimino D, Astanina E, Maione F, Valdembri D et al (2008) AP-2alpha and AP-2gamma regulate tumor progression via specific genetic programs. FASEB J 22(8):2702–2714
138. Bosher JM, Williams T, Hurst HC (1995) The developmentally regulated transcription factor AP-2 is involved in c-erbB-2 overexpression in human mammary carcinoma. Proc Natl Acad Sci USA 92(3):744–747
139. Gilbertson RJ, Perry RH, Kelly PJ, Pearson AD, Lunec J (1997) Prognostic significance of HER2 and HER4 coexpression in childhood medulloblastoma. Cancer Res 57(15):3272–3280
140. Gee JM, Robertson JF, Ellis IO, Nicholson RI, Hurst HC (1999) Immunohistochemical analysis reveals a tumour suppressor-like role for the transcription factor AP-2 in invasive breast cancer. J Pathol 189(4):514–520
141. Karjalainen JM, Kellokoski JK, Eskelinen MJ, Alhava EM, Kosma VM (1998) Downregulation of transcription factor AP-2 predicts poor survival in stage I cutaneous malignant melanoma. J Clin Oncol 16(11):3584–3591
142. Pellikainen J, Kataja V, Ropponen K, Kellokoski J, Pietilainen T, Bohm J et al (2002) Reduced nuclear expression of transcription factor AP-2 associates with aggressive breast cancer. Clin Cancer Res 8(11):3487–3495
143. Ropponen KM, Kellokoski JK, Pirinen RT, Moisio KI, Eskelinen MJ, Alhava EM et al (2001) Expression of transcription factor AP-2 in colorectal adenomas and adenocarcinomas; comparison of immunohistochemistry and in situ hybridisation. J Clin Pathol 54(7):533–538
144. Ruiz M, Troncoso P, Bruns C, Bar-Eli M (2001) Activator protein 2alpha transcription factor expression is associated with luminal differentiation and is lost in prostate cancer. Clin Cancer Res 7(12):4086–4095
145. Darnell JE Jr, Kerr IM, Stark GR (1994) Jak-STAT pathways and transcriptional activation in response to IFNs and other extracellular signaling proteins. Science 264(5164):1415–1421
146. Akira S, Nishio Y, Inoue M, Wang XJ, Wei S, Matsusaka T et al (1994) Molecular cloning of APRF, a novel IFN-stimulated gene factor 3 p91-related transcription factor involved in the gp130-mediated signaling pathway. Cell 77(1):63–71
147. Zhang T, Kee WH, Seow KT, Fung W, Cao X (2000) The coiled-coil domain of Stat3 is essential for its SH2 domain-mediated receptor binding and subsequent activation induced by epidermal growth factor and interleukin-6. Mol Cell Biol 20(19):7132–7139
148. Sadowski HB, Shuai K, Darnell JE, Gilman MZ Jr (1993) A common nuclear signal transduction pathway activated by growth factor and cytokine receptors. Science 261(5129):1739–1744
149. Zhong Z, Wen Z, Darnell JE Jr (1994) Stat3: a STAT family member activated by tyrosine phosphorylation in response to epidermal growth factor and interleukin-6. Science 264(5155):95–98
150. Stahl N, Boulton TG, Farruggella T, Ip NY, Davis S, Witthuhn BA et al (1994) Association and activation of Jak-Tyk kinases by CNTF-LIF-OSM-IL-6 beta receptor components. Science 263(5143):92–95
151. Lutticken C, Wegenka UM, Yuan J, Buschmann J, Schindler C, Ziemiecki A et al (1994) Association of transcription factor APRF and protein kinase Jak1 with the interleukin-6 signal transducer gp130. Science 263(5143):89–92

152. Shuai K, Horvath CM, Huang LH, Qureshi SA, Cowburn D, Darnell JE Jr (1994) Interferon activation of the transcription factor Stat91 involves dimerization through SH2-phosphotyrosyl peptide interactions. Cell 76(5):821–828
153. Shain KH, Yarde DN, Meads MB, Huang M, Jove R, Hazlehurst LA et al (2009) Beta1 integrin adhesion enhances IL-6-mediated STAT3 signaling in myeloma cells: implications for microenvironment influence on tumor survival and proliferation. Cancer Res 69(3):1009–1015
154. Bromberg JF, Wrzeszczynska MH, Devgan G, Zhao Y, Pestell RG, Albanese C et al (1999) Stat3 as an oncogene. Cell 98(3):295–303
155. Vignais ML, Sadowski HB, Watling D, Rogers NC, Gilman M (1996) Platelet-derived growth factor induces phosphorylation of multiple JAK family kinases and STAT proteins. Mol Cell Biol 16(4):1759–1769
156. Giordano V, De Falco G, Chiari R, Quinto I, Pelicci PG, Bartholomew L et al (1997) Shc mediates IL-6 signaling by interacting with gp130 and Jak2 kinase. J Immunol 158(9):4097–4103
157. Arredondo J, Chernyavsky AI, Jolkovsky DL, Pinkerton KE, Grando SA (2006) Receptor-mediated tobacco toxicity: cooperation of the Ras/Raf-1/MEK1/ERK and JAK-2/STAT-3 pathways downstream of alpha7 nicotinic receptor in oral keratinocytes. FASEB J 20(12):2093–2101
158. Chauhan D, Li G, Podar K, Hideshima T, Neri P, He D et al (2005) A novel carbohydrate-based therapeutic GCS-100 overcomes bortezomib resistance and enhances dexamethasone-induced apoptosis in multiple myeloma cells. Cancer Res 65(18):8350–8358
159. Tharappel JC, Lee EY, Robertson LW, Spear BT, Glauert HP (2002) Regulation of cell proliferation, apoptosis, and transcription factor activities during the promotion of liver carcinogenesis by polychlorinated biphenyls. Toxicol Appl Pharmacol 179(3):172–184
160. Megeney LA, Perry RL, LeCouter JE, Rudnicki MA (1996) bFGF and LIF signaling activates STAT3 in proliferating myoblasts. Dev Genet 19(2):139–145
161. Garcia R, Yu CL, Hudnall A, Catlett R, Nelson KL, Smithgall T et al (1997) Constitutive activation of Stat3 in fibroblasts transformed by diverse oncoproteins and in breast carcinoma cells. Cell Growth Differ 8(12):1267–1276
162. Migone TS, Lin JX, Cereseto A, Mulloy JC, O'Shea JJ, Franchini G et al (1995) Constitutively activated Jak-STAT pathway in T cells transformed with HTLV-I. Science 269(5220):79–81
163. Tian SS, Tapley P, Sincich C, Stein RB, Rosen J, Lamb P (1996) Multiple signaling pathways induced by granulocyte colony-stimulating factor involving activation of JAKs, STAT5, and/or STAT3 are required for regulation of three distinct classes of immediate early genes. Blood 88(12):4435–4444
164. Fu AK, Fu WY, Ng AK, Chien WW, Ng YP, Wang JH et al (2004) Cyclin-dependent kinase 5 phosphorylates signal transducer and activator of transcription 3 and regulates its transcriptional activity. Proc Natl Acad Sci USA 101(17):6728–6733
165. Wen Z, Darnell JE Jr (1997) Mapping of Stat3 serine phosphorylation to a single residue (727) and evidence that serine phosphorylation has no influence on DNA binding of Stat1 and Stat3. Nucleic Acids Res 25(11):2062–2067
166. Jain N, Zhang T, Kee WH, Li W, Cao X (1999) Protein kinase C delta associates with and phosphorylates Stat3 in an interleukin-6-dependent manner. J Biol Chem 274(34):24392–24400
167. Aziz MH, Manoharan HT, Church DR, Dreckschmidt NE, Zhong W, Oberley TD et al (2007) Protein kinase Cepsilon interacts with signal transducers and activators of transcription 3 (Stat3), phosphorylates Stat3Ser727, and regulates its constitutive activation in prostate cancer. Cancer Res 67(18):8828–8838
168. Yokogami K, Wakisaka S, Avruch J, Reeves SA (2000) Serine phosphorylation and maximal activation of STAT3 during CNTF signaling is mediated by the rapamycin target mTOR. Curr Biol 10(1):47–50
169. Brantley EC, Benveniste EN (2008) Signal transducer and activator of transcription-3: a molecular hub for signaling pathways in gliomas. Mol Cancer Res 6(5):675–684

170. Yu H, Pardoll D, Jove R (2009) STATs in cancer inflammation and immunity: a leading role for STAT3. Nat Rev Cancer 9(11):798–809
171. Yuan ZL, Guan YJ, Chatterjee D, Chin YE (2005) Stat3 dimerization regulated by reversible acetylation of a single lysine residue. Science 307(5707):269–273
172. Pfitzner E, Kliem S, Baus D, Litterst CM (2004) The role of STATs in inflammation and inflammatory diseases. Curr Pharm Des 10(23):2839–2850
173. Dalwadi H, Krysan K, Heuze-Vourc'h N, Dohadwala M, Elashoff D, Sharma S et al (2005) Cyclooxygenase-2-dependent activation of signal transducer and activator of transcription 3 by interleukin-6 in non-small cell lung cancer. Clin Cancer Res 11(21):7674–7682
174. Ernst M, Najdovska M, Grail D, Lundgren-May T, Buchert M, Tye H et al (2008) STAT3 and STAT1 mediate IL-11-dependent and inflammation-associated gastric tumorigenesis in gp130 receptor mutant mice. J Clin Invest 118(5):1727–1738
175. Kobierski LA, Srivastava S, Borsook D (2000) Systemic lipopolysaccharide and interleukin-1beta activate the interleukin 6: STAT intracellular signaling pathway in neurons of mouse trigeminal ganglion. Neurosci Lett 281(1):61–64
176. Frank DA (2007) STAT3 as a central mediator of neoplastic cellular transformation. Cancer Lett 251(2):199–210
177. Bromberg JF, Horvath CM, Besser D, Lathem WW Jr, Darnell JE Jr (1998) Stat3 activation is required for cellular transformation by v-src. Mol Cell Biol 18(5):2553–2558
178. Yu Z, Zhang W, Kone BC (2002) Signal transducers and activators of transcription 3 (STAT3) inhibits transcription of the inducible nitric oxide synthase gene by interacting with nuclear factor kappaB. Biochem J 367(Pt 1):97–105
179. Rahaman SO, Harbor PC, Chernova O, Barnett GH, Vogelbaum MA, Haque SJ (2002) Inhibition of constitutively active Stat3 suppresses proliferation and induces apoptosis in glioblastoma multiforme cells. Oncogene 21(55):8404–8413
180. Atkinson GP, Nozell SE, Benveniste ET (2010) NF-kappaB and STAT3 signaling in glioma: targets for future therapies. Expert Rev Neurother 10(4):575–586
181. Yu H, Jove R (2004) The STATs of cancer–new molecular targets come of age. Nat Rev Cancer 4(2):97–105
182. Catlett-Falcone R, Landowski TH, Oshiro MM, Turkson J, Levitzki A, Savino R et al (1999) Constitutive activation of Stat3 signaling confers resistance to apoptosis in human U266 myeloma cells. Immunity 10(1):105–115
183. Karni R, Jove R, Levitzki A (1999) Inhibition of pp 60c-Src reduces Bcl-XL expression and reverses the transformed phenotype of cells overexpressing EGF and HER-2 receptors. Oncogene 18(33):4654–4662
184. Zushi S, Shinomura Y, Kiyohara T, Miyazaki Y, Kondo S, Sugimachi M et al (1998) STAT3 mediates the survival signal in oncogenic ras-transfected intestinal epithelial cells. Int J Cancer 78(3):326–330
185. Mahboubi K, Li F, Plescia J, Kirkiles-Smith NC, Mesri M, Du Y et al (2001) Interleukin-11 up-regulates survivin expression in endothelial cells through a signal transducer and activator of transcription-3 pathway. Lab Invest 81(3):327–334
186. Liu H, Ma Y, Cole SM, Zander C, Chen KH, Karras J et al (2003) Serine phosphorylation of STAT3 is essential for Mcl-1 expression and macrophage survival. Blood 102(1):344–352
187. Bhattacharya S, Schindler C (2003) Regulation of Stat3 nuclear export. J Clin Invest 111(4):553–559
188. Masuda M, Suzui M, Yasumatu R, Nakashima T, Kuratomi Y, Azuma K et al (2002) Constitutive activation of signal transducers and activators of transcription 3 correlates with cyclin D1 overexpression and may provide a novel prognostic marker in head and neck squamous cell carcinoma. Cancer Res 62(12):3351–3355
189. Kiuchi N, Nakajima K, Ichiba M, Fukada T, Narimatsu M, Mizuno K et al (1999) STAT3 is required for the gp130-mediated full activation of the c-myc gene. J Exp Med 189(1): 63–73
190. Shirogane T, Fukada T, Muller JM, Shima DT, Hibi M, Hirano T (1999) Synergistic roles for Pim-1 and c-Myc in STAT3-mediated cell cycle progression and antiapoptosis. Immunity 11(6):709–719

191. Konnikova L, Kotecki M, Kruger MM, Cochran BH (2003) Knockdown of STAT3 expression by RNAi induces apoptosis in astrocytoma cells. BMC Cancer 3:23
192. Ivanov VN, Bhoumik A, Krasilnikov M, Raz R, Owen-Schaub LB, Levy D et al (2001) Cooperation between STAT3 and c-jun suppresses Fas transcription. Mol Cell 7(3):517–528
193. Itoh M, Murata T, Suzuki T, Shindoh M, Nakajima K, Imai K et al (2006) Requirement of STAT3 activation for maximal collagenase-1 (MMP-1) induction by epidermal growth factor and malignant characteristics in T24 bladder cancer cells. Oncogene 25(8):1195–1204
194. Gaemers IC, Vos HL, Volders HH, van der Valk SW, Hilkens J (2001) A stat-responsive element in the promoter of the episialin/MUC1 gene is involved in its overexpression in carcinoma cells. J Biol Chem 276(9):6191–6199
195. Niu G, Wright KL, Huang M, Song L, Haura E, Turkson J et al (2002) Constitutive Stat3 activity up-regulates VEGF expression and tumor angiogenesis. Oncogene 21(13):2000–2008
196. Senft C, Polacin M, Priester M, Seifert V, Kogel D, Weissenberger J (2010) The nontoxic natural compound Curcumin exerts anti-proliferative, anti-migratory, and anti-invasive properties against malignant gliomas. BMC Cancer 10:491
197. Cheng GZ, Zhang WZ, Sun M, Wang Q, Coppola D, Mansour M et al (2008) Twist is transcriptionally induced by activation of STAT3 and mediates STAT3 oncogenic function. J Biol Chem 283(21):14665–14673
198. Wang Q, Sun Z, Yang HS (2008) Downregulation of tumor suppressor Pdcd4 promotes invasion and activates both beta-catenin/Tcf and AP-1-dependent transcription in colon carcinoma cells. Oncogene 27(11):1527–1535
199. Sullivan NJ, Sasser AK, Axel AE, Vesuna F, Raman V, Ramirez N et al (2009) Interleukin-6 induces an epithelial-mesenchymal transition phenotype in human breast cancer cells. Oncogene 28(33):2940–2947
200. Nagpal JK, Mishra R, Das BR (2002) Activation of Stat-3 as one of the early events in tobacco chewing-mediated oral carcinogenesis. Cancer 94(9):2393–2400
201. Boehm AL, Sen M, Seethala R, Gooding WE, Freilino M, Wong SM et al (2008) Combined targeting of epidermal growth factor receptor, signal transducer and activator of transcription-3, and Bcl-X(L) enhances antitumor effects in squamous cell carcinoma of the head and neck. Mol Pharmacol 73(6):1632–1642
202. Rebouissou S, Amessou M, Couchy G, Poussin K, Imbeaud S, Pilati C et al (2009) Frequent in-frame somatic deletions activate gp130 in inflammatory hepatocellular tumours. Nature 457(7226):200–204
203. Kim G, Khanal P, Lim SC, Yun HJ, Ahn SG, Ki SH et al (2013) Interleukin-17 induces AP-1 activity and cellular transformation via upregulation of tumor progression locus 2 activity. Carcinogenesis 34(2):341–350
204. Gao SP, Mark KG, Leslie K, Pao W, Motoi N, Gerald WL et al (2007) Mutations in the EGFR kinase domain mediate STAT3 activation via IL-6 production in human lung adenocarcinomas. J Clin Invest 117(12):3846–3856
205. Bharti AC, Shishodia S, Reuben JM, Weber D, Alexanian R, Raj-Vadhan S et al (2004) Nuclear factor-kappaB and STAT3 are constitutively active in CD138+ cells derived from multiple myeloma patients, and suppression of these transcription factors leads to apoptosis. Blood 103(8):3175–3184
206. Riehle KJ, Campbell JS, McMahan RS, Johnson MM, Beyer RP, Bammler TK et al (2008) Regulation of liver regeneration and hepatocarcinogenesis by suppressor of cytokine signaling 3. J Exp Med 205(1):91–103
207. von Knethen A, Callsen D, Brune B (1999) NF-kappaB and AP-1 activation by nitric oxide attenuated apoptotic cell death in RAW 264.7 macrophages. Mol Biol Cell 10(2):361–372
208. Yin MJ, Yamamoto Y, Gaynor RB (1998) The anti-inflammatory agents aspirin and salicylate inhibit the activity of I(kappa)B kinase-beta. Nature 396(6706):77–80
209. Wahl C, Liptay S, Adler G, Schmid RM (1998) Sulfasalazine: a potent and specific inhibitor of nuclear factor kappa B. J Clin Invest 101(5):1163–1174
210. Weber CK, Liptay S, Wirth T, Adler G, Schmid RM (2000) Suppression of NF-kappaB activity by sulfasalazine is mediated by direct inhibition of IkappaB kinases alpha and beta. Gastroenterology 119(5):1209–1218

211. Egan LJ, Mays DC, Huntoon CJ, Bell MP, Pike MG, Sandborn WJ et al (1999) Inhibition of interleukin-1-stimulated NF-kappaB RelA/p65 phosphorylation by mesalamine is accompanied by decreased transcriptional activity. J Biol Chem 274(37):26448–26453
212. Manna SK, Sah NK, Newman RA, Cisneros A, Aggarwal BB (2000) Oleandrin suppresses activation of nuclear transcription factor-kappaB, activator protein-1, and c-Jun NH2-terminal kinase. Cancer Res 60(14):3838–3847
213. Mijatovic T, Op De Beeck A, Van Quaquebeke E, Dewelle J, Darro F, de Launoi Y et al (2006) The cardenolide UNBS1450 is able to deactivate nuclear factor kappaB-mediated cytoprotective effects in human non-small cell lung cancer cells. Mol Cancer Ther 5(2):391–399
214. Sreenivasan Y, Sarkar A, Manna SK (2003) Oleandrin suppresses activation of nuclear transcription factor-kappa B and activator protein-1 and potentiates apoptosis induced by ceramide. Biochem Pharmacol 66(11):2223–2239
215. Aggarwal BB, Shishodia S (2004) Suppression of the nuclear factor-kappaB activation pathway by spice-derived phytochemicals: reasoning for seasoning. Ann N Y Acad Sci 1030:434–441
216. Bachmeier BE, Mohrenz IV, Mirisola V, Schleicher E, Romeo F, Hohneke C et al (2008) Curcumin downregulates the inflammatory cytokines CXCL1 and -2 in breast cancer cells via NFkappaB. Carcinogenesis 29(4):779–789
217. Ravindran J, Prasad S, Aggarwal BB (2009) Curcumin and cancer cells: how many ways can curry kill tumor cells selectively? AAPS J 11(3):495–510
218. Alexandrow MG, Song LJ, Altiok S, Gray J, Haura EB, Kumar NB (2012) Curcumin: a novel Stat3 pathway inhibitor for chemoprevention of lung cancer. Eur J Cancer Prev 21(5):407–412
219. Jeong WS, Kim IW, Hu R, Kong AN (2004) Modulation of AP-1 by natural chemopreventive compounds in human colon HT-29 cancer cell line. Pharm Res 21(4):649–660
220. Hideshima T, Richardson P, Chauhan D, Palombella VJ, Elliott PJ, Adams J et al (2001) The proteasome inhibitor PS-341 inhibits growth, induces apoptosis, and overcomes drug resistance in human multiple myeloma cells. Cancer Res 61(7):3071–3076
221. Tummala R, Romano RA, Fuchs E, Sinha S (2003) Molecular cloning and characterization of AP-2 epsilon, a fifth member of the AP-2 family. Gene 321:93–102
222. Mitsiades N, Mitsiades CS, Richardson PG, Poulaki V, Tai YT, Chauhan D et al (2003) The proteasome inhibitor PS-341 potentiates sensitivity of multiple myeloma cells to conventional chemotherapeutic agents: therapeutic applications. Blood 101(6):2377–2380
223. Nencioni A, Wille L, Dal Bello G, Boy D, Cirmena G, Wesselborg S et al (2005) Cooperative cytotoxicity of proteasome inhibitors and tumor necrosis factor-related apoptosis-inducing ligand in chemoresistant Bcl-2-overexpressing cells. Clin Cancer Res 11(11):4259–4265
224. Yanamandra N, Colaco NM, Parquet NA, Buzzeo RW, Boulware D, Wright G et al (2006) Tipifarnib and bortezomib are synergistic and overcome cell adhesion-mediated drug resistance in multiple myeloma and acute myeloid leukemia. Clin Cancer Res 12(2):591–599
225. Burke JR, Pattoli MA, Gregor KR, Brassil PJ, MacMaster JF, McIntyre KW et al (2003) BMS-345541 is a highly selective inhibitor of I kappa B kinase that binds at an allosteric site of the enzyme and blocks NF-kappa B-dependent transcription in mice. J Biol Chem 278(3):1450–1456
226. Olivier S, Robe P, Bours V (2006) Can NF-kappaB be a target for novel and efficient anticancer agents? Biochem Pharmacol 72(9):1054–1068
227. Shao L, Wu L, Zhou D (2012) Sensitization of tumor cells to cancer therapy by molecularly targeted inhibition of the inhibitor of nuclear factor kappaB kinase. Transl Cancer Res 1(2):100–108
228. Xia Y, Choi HK, Lee K (2012) Recent advances in hypoxia-inducible factor (HIF)-1 inhibitors. Eur J Med Chem 49:24–40
229. Horiguchi A, Asano T, Kuroda K, Sato A, Asakuma J, Ito K et al (2010) STAT3 inhibitor WP1066 as a novel therapeutic agent for renal cell carcinoma. Br J Cancer 102(11): 1592–1599
230. Zhang X, Yue P, Page BD, Li T, Zhao W, Namanja AT et al (2012) Orally bioavailable small-molecule inhibitor of transcription factor Stat3 regresses human breast and lung cancer xenografts. Proc Natl Acad Sci USA 109(24):9623–9628

# Chapter 7
# Regulation of the Jak/STATs Pathways by Histone Deacetylases

**David M. Woods, Maritza Lienlaf-Moreno, Eduardo Sotomayor, Edward Seto, and Alejandro Villagra**

**Abstract** The regulation of the Jak/STAT pathways involves multiple mechanisms, some of them exclusive to individual cascades, and other shared amongst them. Additionally, there is increasing evidence to support the interconnection of these cascades with other intracellular signaling pathways, which are regulated by independent factors. Thus, the understanding of the Jak/STAT pathway regulation is still growing and recently being revisited to account for the participation of non-conventional regulators. Acetylation of proteins is an emerging regulatory mechanism with demonstrated involvement in the regulation of several intracellular pathways. Specifically, the participation of histone deacetylases (HDACs) is garnering attention due to the feasibility of inhibition of their activity and the subsequent control over the cellular processes being modulated by them. This chapter

---

D.M. Woods, M.Sc.
Pathology, University of South Florida College of Medicine, Tampa, FL, USA
e-mail: dwoods1@health.usf.edu

M. Lienlaf-Moreno, B.Sc. (Biochemistry)
Immunology, H. Lee Moffitt Cancer Center and Research Institute, Tampa, FL, USA
e-mail: maritza.lienlafmoreno@moffitt.org

E. Sotomayor, M.D.
Department of Malignant Hematology, H. Lee Moffitt Cancer Center, Oncologic Sciences, University of South Florida, Tampa, FL, USA
e-mail: Eduardo.sotomayor@moffitt.org

E. Seto, Ph.D. (✉)
Department of Molecular Oncology, Moffitt Cancer Center and Research Institute, Tampa, FL, USA
e-mail: ed.seto@moffitt.org

A. Villagra, Ph.D. (✉)
Immunology, H. Lee Moffitt Cancer Center and Research Institute, Tampa, FL, USA

Oncologic Sciences, University of South Florida, Tampa, FL, USA
e-mail: alejandro.villagra@moffitt.org

R. Kumar (ed.), *Nuclear Signaling Pathways and Targeting Transcription in Cancer*, Cancer Drug Discovery and Development, DOI 10.1007/978-1-4614-8039-6_7,

will explore and present the current knowledge about the role of HDACs and their inhibitors in the control of members of the Jak/STAT pathways and direct activators/modulators of their signaling.

**Keywords** HDACs • STATs • Cytokine regulation • HDAC inhibitors

## 7.1 Introduction

Cellular responses to the environment are mediated by numerous signaling pathways. The proper reception and translation of these signals allows integration of single cells into a multicellular system where every member has its own function. The Janus Kinase/signal transducers and activators of the transcription (JAK/STAT) signaling pathway is one of the most well studied yet enigmatic integrated signaling systems. This family of pathways' major responsibility is allowing for cellular adaptation to an ever-changing microenvironment. Despite the name, the JAK/STAT pathway contains multiple signaling cascades with a plethora of proteins in addition to the Jaks and STATs.

The initial step in decoding the Jak/STAT pathway was the early finding of inducible genes by IFN treatment achieved almost simultaneously by the Darnell and Stark groups [1, 2]. In the following years these two groups consolidated their findings, identifying and characterizing several components of the Jak/STAT pathway, including several STATs and Jaks [3–5]. By 1994, these reports as well as findings from different groups, lead to the description of the Jak/STATs pathways. Further details about this fascinating race can be found in the article "The JAK-STAT Pathway at Twenty" [6].

The evolutionary conserved Jak/STAT pathway comprises several combinatorial subsets of functional units activated differentially by dozens of cytokines and growth factors. The final cellular responses triggered by these pathways depend on the signal, tissue, and cellular context, and include proliferation, differentiation, migration, apoptosis, and cell survival. Moreover, the JAK/STAT signaling is essential for numerous developmental and homeostatic processes, including hematopoiesis, immune cell development, stem cell maintenance, organism growth, and mammary gland development [7–9].

Some of the regulatory mechanisms affecting the Jak/STAT pathway functionality are those involving post-translational modifications, such as phosphorylation, acetylation, methylation and sumoylation [6]. In particular, this chapter will focus on the current knowledge of protein acetylation as a major modulator of the Jak/STAT pathway. A special emphasis will be assigned to the role of histone deacetylases (HDACs) and their inhibitors in this emerging and expanding field. The protagonists of these pathways will be introduced, and then the participation of HDACs at different levels during the Jak/STAT pathway activation, including their direct participation in formal Jak/STAT pathway members. In addition, their indirect role as regulators of activators of these pathways will be discussed.

## 7.2 Jak and STATs at a Glance

In a general description, activation of the Jak/STAT pathway begins with the binding of cytokines and growth factors to their corresponding receptors. This, in turn activates JAK proteins, which then introduce specific phosphorylation to either, receptors or STATs proteins on specific residues. The next step involves the homo- or heterodimerization of STATs accordingly with the respective signal sensed by the membrane receptors. This dimerization allows the translocation of STATs to the nucleus, binding to the consensus DNA sequence of 5'-TT($N_{4-6}$)AA-3' and initiates the transcription of specific target genes [10].

The mammalian JAK family of kinases is composed by four members: JAK1, JAK2, JAK3 and tyrosine kinase 2 (TYK2) [11]. Each of these proteins contains a conserved kinase domain (JH1) and a pseudo-kinase domain at the carboxyl terminus catalytically inactivated (JH2), which has been proposed as a regulator of the former kinase domain [12]. JAKs also have a Src homology 2 (SH2) domain and an N-terminal domain (FERM), which allows its association with cytokine receptors. Also, all members of this family distribute five blocks of sequence similarity throughout the amino-terminal region [13]. JAK1, JAK2, and TYK2 appear to be ubiquitously expressed, while JAK3 expression is normally limited to lymphoid cells.

There are seven mammalian STATs: STAT1, STAT2, STAT3, STAT4, STAT5A, STAT5B and STAT6. These proteins are highly homologous in several regions, including a N-terminal domain (NTD), a coiled-coil domain (CC), a beta-barrel DNA binding domain (DBD), a linker domain (LD), a SRC homology 2 (SH2) domain, and a C-terminal transactivation domain (TAD), which is located at the carboxyl terminus (Fig. 7.1) [12, 14]. However, the amino acid sequence diversity and their tissue-specific distributions account for the diverse roles of STATs in response to extracellular cytokines.

Abnormal activation of JAK-STAT pathways has been described in various cancers and immune disorders [8, 15]. These oncogenic processes have been largely described for either Jaks or STATs, and most of them involve constitutive activation. For example, the identification of an activating point mutation in the JAK2 kinase, JAK2V617F, in a vast majority of patients with BCR-ABL myeloproliferative neoplasms (MPNs), provided a significant advance in the knowledge of the molecular pathogenesis of these disorders [16]. This JAK2 mutant activates downstream signaling through the STATs, RAS/mitogen-activated protein kinase (MAPK), and phosphatidyl-inositol 3 (PI3). Noteworthy, treatment with the pan-HDACi LBH589, also known as panobinostat, seems to counteract the constitutive activation of JAK2V617F by inhibition of its autophosphorylation and promotion of its proteasomal degradation. Additionally, panobinostat has shown inhibition in other tumor cell lines such as the human erythroleukemia HEL92.1.7 and Ba/F3-JAK2V617F cells [17].

In most malignancies, it has been found that oncogenic mutations in STATs 1, 3, and 5 lead to persistent tyrosine phosphorylation [18]. Furthermore, abnormal activation of STAT3 has been detected in many cancers, including glioma, melanoma, non-small cell lung carcinoma, head and neck, breast, cervical, ovarian and prostate cancers [19]. The oncogenic property of STAT3 is mainly due to its function as a

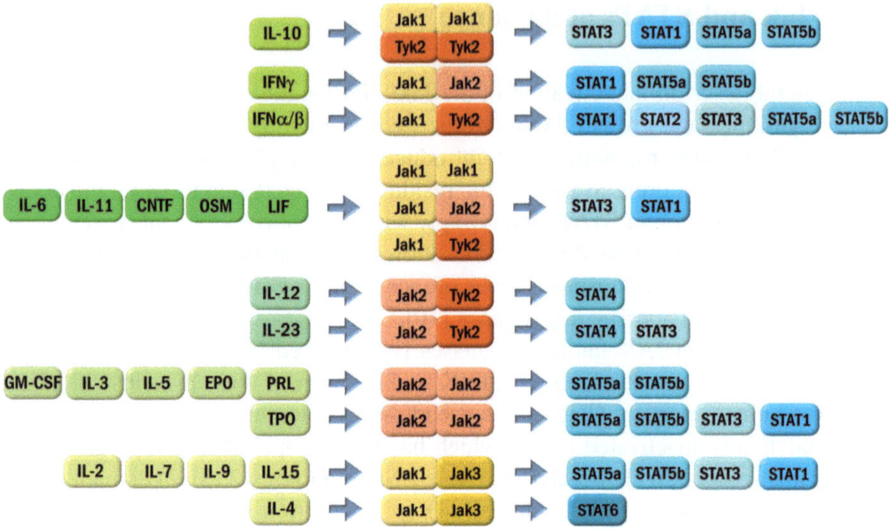


**Fig. 7.1** Activation of Jak/STAT pathways by cytokines

modulator of cellular processes such as cell cycle, cell proliferation, and apoptosis by regulation of cyclin D1, Bcl2 and Bcl-xl, genes that ultimately control tumor growth and metastasis [20]. Interestingly, STAT3 can also be abnormally induced by uncontrolled activation of its pathway. For example, paracrine sources of IL-6 can induce autocrine production of IL-6 and pStat3 expression in tumor cells, leading to heterogeneous levels of pStat3 [18].

## 7.3 Histone Deacetylases and Their Inhibitors

One of the most studied posttranslational protein modifications is the acetylation of lysine amino acids. Initially, these modifications were found at the N-terminal end of histones as a transcriptional regulatory mechanism. Briefly, in a non-modified steady-state, the highly positive N-terminal ends of histones wrap around histones, generating an obstacle for the binding of transcription factors and the recruitment of other proteins that need to read the "writing pattern" on nucleosomes to exert their transcriptional functions. In this context, acetylation of histones neutralizes these positive charges promoting a relaxed nucleosome conformation, and allowing the binding of transcription factors and "writer" proteins. Acetyl modifications are introduced by a heterogeneous group of proteins named histone acetyltransferases (HATs), most of them forming multi-protein complexes that can be selectively recruited to DNA sequences upon exogenous or endogenous cellular stimuli [21]. In opposition, these acetyl modifications can be removed by another group of proteins, HDACs. The 18 HDACs identified in humans are subdivided in two families; the classical HDAC family of zinc dependent metalloproteins, composed by Classes I, II and IV, and the Class III

NAD+-dependent proteins belonging to the sirtuin family of HDACs. The Class I HDACs (HDAC1, 2, 3 and 8) are most closely related to the yeast deacetylase RPD3, and the Class II HDACs are subdivided into Class IIa (HDAC4, 5, 7 and 9) and Class IIb (HDAC6 and 10), both subclasses sharing homology with the yeast deacetylase HDA1 [22]. Finally, the newest HDAC discovered, HDAC11, comprises its own Class IV, and does not share homology with either RPD3 or HDA1 yeast deacetylases. HDAC knock-out mice have severe malfunctions at multiple cellular processes and are in some cases embryonically lethal (HDAC1, 3 and 7) or lethal at the perinatal stage (HDAC2, 4 and 8) [23].

HDACs, originally described as histone modifiers have more recently been demonstrated to modify a variety of other proteins involved in diverse cellular processes non-related to the chromatin environment. This includes deacetylation of multiple non-histone targets, including several proteins involved in cell cycle/apoptosis and immune regulation [23, 24]. The expanded role opens the possibility that the effects of HDACs and HDACi may affect non-epigenetic regulatory pathways, including the Jak/STAT pathway. Indeed, as will be shown throughout this chapter, multiple studies show that this is the case.

A logical approach to elucidating the precise mechanisms of action and functions of HDACs is the use of specific inhibitors that block HDAC enzymatic activity. Some of the compounds that inhibit HDAC activity include butyrates, TSA, trapoxin, SAHA, apicidin, depsipeptide, depudecin, CBHA, MS-275, CI-994, oxamflatin, pyroxamide, scriptaid, CHAP, and valproic acid [25, 26]. Early works indicated that extremely low concentrations of an HDAC inhibitor, TSA, effectively induced cell differentiation and G1/G2 cell cycle arrest. Many studies followed to show that cells cultured with an HDAC inhibitor result in a change in gene expression patterns. A large number of studies have shown that HDAC inhibitors can effectively arrest and counteract transformation of some cells and inhibit the growth of cancers in tumor-bearing animals. In fact, two HDAC inhibitors, Vorinostat and Romidepsin, are FDA approved for treatment of cutaneous T cell lymphoma, and many others HDACi are being pursued in clinical trials at different stages.

## 7.4 Signal Transducer and Activator of Transcription

### *7.4.1 STAT1*

STAT1 is the principal transcriptional mediator of interferon (IFN) signaling and plays a central role in the regulation of innate and adaptive immune responses. In addition to IFNs, many other stimuli are able to activate STAT1, including cytokines belonging to the gp130 family such as IL-6, LIF, and OSM and growth factors such as EGF and PDGF (Fig. 7.2) [27]. All of these stimuli can also lead to STAT1 phosphorylation in association with other STATs (STAT3 and STAT5) accordingly, with the availability of other STATs and the combination of stimuli received from the environment. Probably the most well documented immunological function of

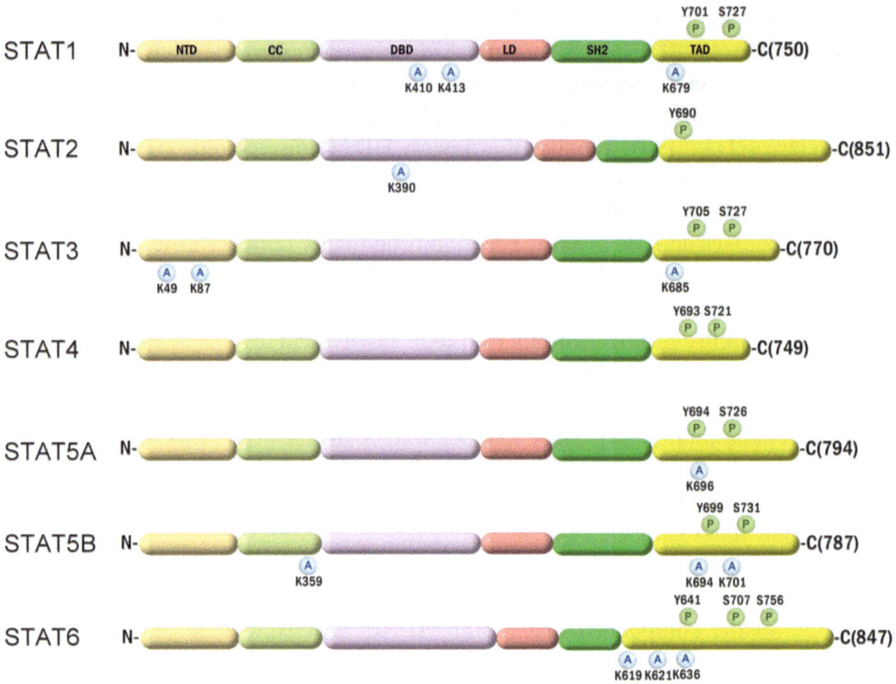


**Fig. 7.2** Conserved domains of STATs: Distribution of phosphorylated and acetylated residues

STAT1 is its participation in T cell responses. It is important to mention its positive role in the induction of Th1 differentiation of T-cells, while repressing regulatory T-cells [28]. Additionally, STAT1 is considered as an anti-oncogenic protein. The mechanism by which STAT1 modulates cell death appears to involve both transcription-dependent and independent processes. At the transcriptional level, STAT1 has been described as master regulator of genes involved in the cell cycle and apoptosis such as Bcl-xL [29], p21 [30], iNOS [31], caspases [32], and death receptors [32]. Among its indirect non-transcriptional functions, its association with several pro-apoptotic mediators such as TRADD [33] and p53 [34] are well characterized. An example of the participation of STAT1 in cell death occurs in breast carcinomas, where the p21 activation by STAT1 also involves BRCA1, the most important susceptibility gene known in hereditary breast carcinoma. In fact, BRCA1 and STAT1 synergize to activate p21/waf1 transcription by means of an interaction between the BRCA1 and STAT1 TAD domain. This domain contains serine 727, whose phosphorylation is crucial for the transcriptional activity of STAT1 and which is directly implicated in the recruitment of STAT1 transcriptional coactivators; its mutation causes defective STAT1-BRCA1 binding [35].

The binding of IFN-γ allows the dimerization of its receptors and the reciprocal phosphorylation of Jaks. Once phosphorylated, Jaks will in turn phosphorylate the IFN receptors and provide a suitable STAT1 docking site. While associated with the receptors, STAT1 proteins are phosphorylated on conserved tyrosine 701 by JAKs, allowing the final dimerization of STAT1. There are two well described phosphorylation sites in STAT1; tyrosine 701 (Y701) and serine 727(S727), both occurring at the TAD domain and triggered by IFN stimulation (Fig. 7.1) [32]. The Y701 phosphorylation within the TAD is required for STAT1 dimerization, nuclear translocation, and DNA binding [36]. The second phosphorylation, S727, seems to be crucial for its transactivational activity, since STAT1 protein mutation at serine 727 has its transcriptional activity reduced by 20 % [37] and fails to recruit transcriptional coactivators [38]. The C-terminal domain TAD has been described as essential for its interaction with other regulatory proteins such as BRCA1 [38], CREB-binding protein (CBP)/p300 family of histone acetyl transferases [39], and MCM5 [40]. The same domain is also a target for the acetylation mediated by CBP, which also has been demonstrated as a STAT1 partner [41].

The magnitude and duration of the events downstream of STAT1 activation are influenced by several intracellular mechanisms and posttranscriptional modifications. In addition to the regulation mediated by phosphorylation, STAT1 can be regulated by acetylation. The first evidence of this mechanism came from early studies in melanoma cells, demonstrating that STAT1 was acetylated in the presence of HDACi [42]. This posttranslational modification was demonstrated to have a negative effect over STAT1 phosphorylation and its subsequent transcriptional activity [43]. The same group also demonstrated that cycles of phosphorylation/acetylation occur after IFN, in addition to other stimuli, treatment. A further hypothesis based on these observations proposes a "pulsing" mechanism of control. Essentially, the sequence of acetylation/phosphorylation observed in activated cells generates a brake for the action of STAT1 over its gene targets. Thus, the acetylation of STAT1 allows its interaction with the phosphatase TCP45 and its subsequent dephosphorylation (Fig. 7.3). This regulatory mechanism promotes a discrete action for STAT1, allowing the cell to escape from death [43].

Several acetylated residues have been described for STAT1 (Fig. 7.1) [44]; however, only a few sites appear to have an impact over STAT1 function. The use of STAT1 mutants has helped to identify two specific residues undergoing acetylation, Lys410 and Lys413 [42]. Interestingly, both residues are located at the DNA-binding domain, suggesting a potential influence in the DNA-binding capability of STAT1 mediated by acetylation/deacetylation.

The first attempts to identify the role of acetylation over the functionality of STAT1 included the identification of the acetyltransferase CBP as responsible for this covalent modification. As well, experiments using HDACi and specific RNAi against HDAC1, 2, 3 helped to identify these HDACs as the counteracting enzymes in this process (Fig. 7.3) [41, 45, 46]. Additionally, a recent report indicates that HDAC4 is also involved in the deacetylation of STAT1 in ovarian cancer cells [47]. The well documented differential tissue distribution of HDAC4 [48] is opening a new perspective about the protagonists in the deacetylation/acetylation of STAT1. It is possible that cell/tissues with low levels of a specific HDAC are less susceptible to the effect of HDACi against STAT1 activity, a hypothesis that is amenable to testing in the near future.

Other groups have demonstrated the role of HDACs over the STAT1 function. For example, Klampfer et al. showed that HDACs are required for signaling by STAT1, and that HDAC inhibitors, as well as small interfering RNA specific for

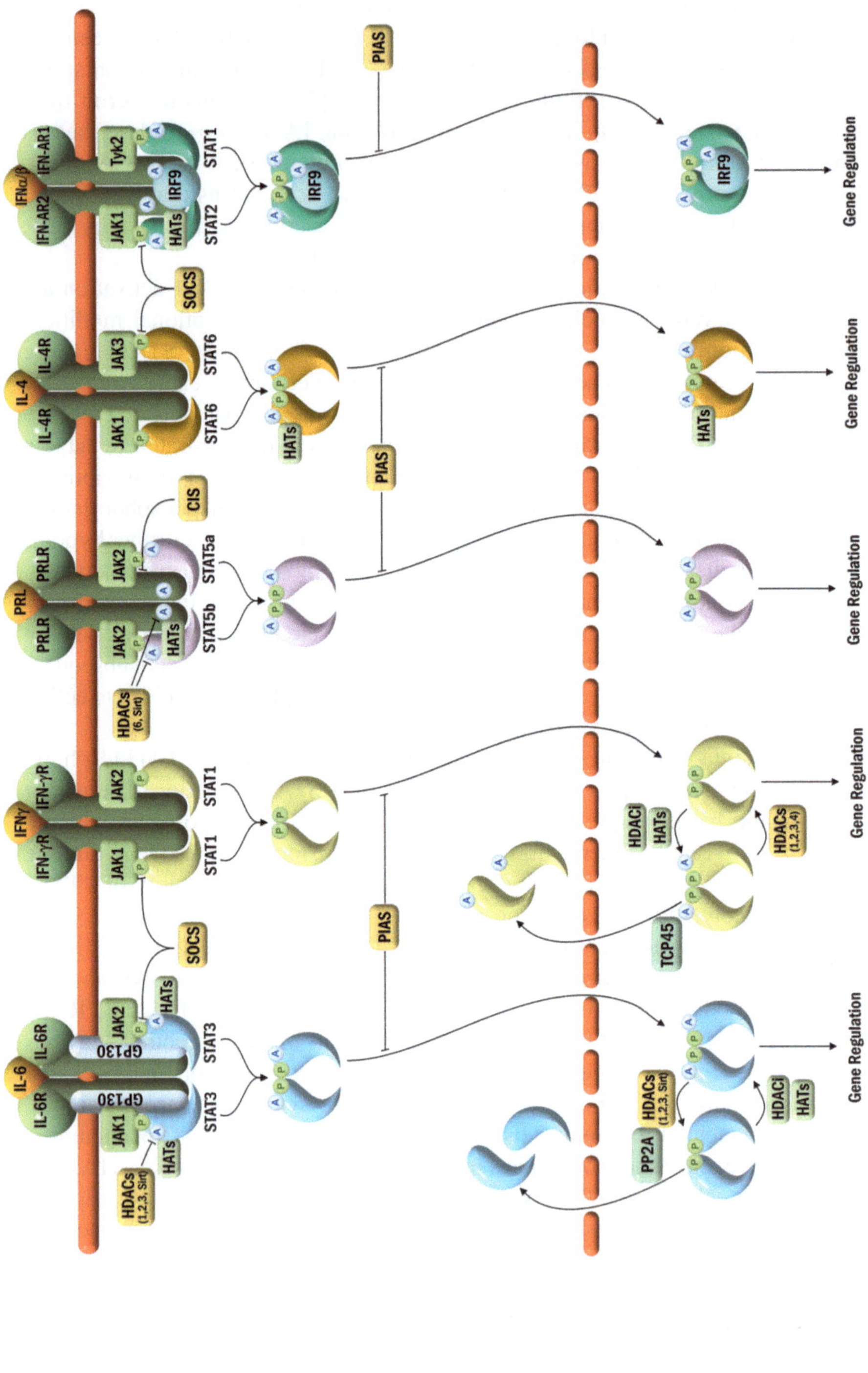


**Fig. 7.3** Points of regulation in five classical Jak/STAT pathways

HDAC1, 2 and 3, are able to prevent IFN-γ induced JAK-1 activation, STAT1 phosphorylation and STAT1 dependent gene activation in a colorectal carcinoma cell line [46]. Similar outcomes were observed by TSA treatment in several myeloid cell and tumor cells [49]. Additionally, in colon cancer cells lines treated with the HDACi butyrate, TSA and SAHA, a preferential induction of apoptosis in transformed cells bearing k-RAS mutations was observed [50].

A more detailed functional role of HDACs over the STAT1 functionality and preference for DNA targets has been provided in a recent report [45]. Briefly, the treatment of different cell lines with HDACi counteracts the phosphorylation of STAT1 mediated by IFN. Interestingly, a mutant form of STAT1 replacing both lysines 410 and 413 for glutamine, but not individual mutations, mimicked the effect of a constitutive acetylated STAT1 and the subsequent inactivation of STAT1, suggesting that the number of deacetylated residues controls the functionality of STAT1. The same work also demonstrated that cells treated with IFNα can rescue the negative effect of the double mutant STAT1. A plausible explanation proposed is that IFNα promotes the phosphorylation of other STATs, such as STAT3, that in turn will heterodimerize with STAT1 and allows its recruitment to target DNAs [45].

### *7.4.2 STAT3*

STAT3 modulates the expression of important genes involved in the regulation of a variety of physiological and non-physiological cellular functions, including cell cycle control, cellular differentiation, apoptosis, angiogenesis, metastasis, and innate and acquired immune responses [51, 52]. These genes include IL-17, IL-23, Bcl-xL, Bcl-2, MCL1, CCDN1, VEGF, c-Myc, p53, in addition to others [15, 27]. The Stat3 pathway is activated in response to a wide variety of cytokines such as the IL-6 and IL-10 family of cytokines, GCSF, leptin, IL-21, IL-23, etc. Moreover, STAT3 can be activated by other receptors and signals, including growth factors like PDGFR and EGFR, among others (Fig. 7.2) [15].

In the conventional STAT3 activation pathway (Fig. 7.1), the activation of cell surface receptors by growth factors or cytokines induces the phosphorylation of receptor tyrosine residues allowing the interaction of the STAT3 SH2 domain with the receptor and its subsequent phosphorylation. Phosphorylation of STAT3 can be mediated by the intrinsic tyrosine kinase activity of the activated growth factor receptor or by the Janus kinase (JAK) that associates with activated cytokine receptors, including Gp130. The final activation of STAT3 attained by its phosphorylation at the Tyrosine 705 (Y705) allows its dimerization (hetero- or homo-) and the subsequent translocation to the nucleus to modulate the expression of target genes. Additionally, a second phosphorylation at serine 727 (S727) in the C-terminal transactivation domain allows for the maximal activation of STAT3 target genes [51]. However, some evidence gathered from melanoma models suggests that the phosphorylation of Ser727 is not merely a secondary event and has a role in the survival activity and nuclear translocation of STAT3 in melanocytic cells, and its mechanism of action is independent from the Tyr705 phosphorylation [53].

It is accepted that constitutive activation of STAT3 mediates tumor-promoting inflammation. STAT3 has a dual effect in tumor inflammation and immunity by enhancing pro-oncogenic inflammatory pathways, including nuclear factor-kB (NF-KB) and interleukin-6 (IL-6)-GP130-JAK pathways, and by adding a brake to the STAT1 anti-tumor immune responses, mainly mediated by T cells. Thus, STAT3 has emerged as an attractive target for controlling oncogenic processes due to its critical role in tumor cell survival, proliferation, angiogenesis, metastasis, and immune modulation.

The important role of STAT3 in malignant processes was established after initial studies showed that STAT3 was constitutively activated during v-Src transformation [54], and that its downstream pathway was required for oncogenic transformation by v-Src [55]. Later, other transforming tyrosine kinases, such as v-Eyk, v-Ros, v-Fps, Etk/BMX, and Lck were found to constitutively activate STAT3 in the context of oncogenesis [56]. Moreover, early studies show that the constitutive activation of STAT3 in human breast cancer cells correlates with EGF receptors family kinase signaling and also with aberrant JAK and s-SRC activity [57]. These findings laid the groundwork to understanding the heterogeneous network involved in the activation of STAT3, leading to the discovery of other autocrine and paracrine stimuli influencing the aberrant functionality of STAT3. One of the most well understood signals triggering the oncogenic properties of STAT3 is the cytokine IL-6, which is particularly relevant in multiple myeloma and prostate cancer as IL-6-mediated activation of STAT3 enhances the production of key regulators of cell cycle and apoptosis such as Bcl-2, MCL-1, cyclin D1 and c-myc, which in turn prevents apoptosis and stimulates growth in these tumors [58]. In this context, in a recent report it was demonstrated that STAT3 plays a key role in G1-to S-phase cell-cycle transition through the up-regulation of cyclins D2, D3, A, and cdc25A and the concomitant down regulation of p21 and p27. Thus, constitutive Stat3 activation may lead to a growth advantage of the malignant counterpart [59].

Additionally, STAT3 tumorigenic activation has also been linked to other non-cytokine stimuli, such as EGFR, an event particularly important in lung cancer [60]. Noteworthy, it has been found that STAT3 is a common factor in many oncogenic signaling pathways, and is constitutively activated at a frequency of 50–90 % in diverse human cancers, including multiple-myeloma cells, chronic lymphocytic leukemia, head and neck cancer and many solid tumors such a breast, lung, prostate, ovarian cancers and malignant melanomas [15, 19, 61, 62]. The STAT3 activity is necessary for proliferation and/or survival of many different types of established or primary tumor cells bearing constitutive STAT3 activity, and its inhibition impairs tumor growth in vivo in different types of tumor [56]. Moreover, in murine models it has been reported that STAT3 represses p53 expression by directly binding to its promoter, inhibiting p53 pro-apoptotic activity and contributing to cell survival [63]. In a mouse model of melanoma, it was found that the alternative splice variant STAT3β with dominant negative properties suppressed tumor growth of B16 melanoma cells, supporting the role of STAT3 in these malignancies.

In addition to the well-characterized regulation mediated by phosphorylation, STAT3 can be regulated by acetylation. Interestingly, this mechanism seems to operate in the opposite direction compared to that observed in STAT1 (Fig. 7.3). Thus, like tyrosine phosphorylation, acetylation is necessary for the STAT3 activation [27]. In this context, persistent acetylation of STAT3 is generally observed in diverse human cancers. For this reason both phosphorylation and acetylation of STAT3 are crucial for STAT3-mediated up-regulation of oncogenic genes [64]. Cytokine-dependent acetylation of STAT3 at the C-terminal lysine-685 (K685) was reported by two independent groups [64, 65]. Both showed that STAT3 undergoes acetylation in various adherent cancer-derived cell lines treated with the related cytokines IL6 or OSM and with the Class I interferon IFNα. In addition to the opposite effect of acetylation observed between STAT1 and STAT3, several other discrepancies have been described. For example, STAT1 has been demonstrated to be acetylated only by CBP [42], and STAT3 can be acetylated by either CBP or p300 [65]. Moreover, the acetylated residues in both STATs are located in different structural domains; STAT1 is mainly acetylated at the residues K410 and K413 located at the DNA binding domain, and STAT3 is acetylated at the K685 located at the transactivation domain. These divergences suggest that the regulatory mechanisms behind these covalent modifications do not affect the same functions.

Early studies showed that HDAC1, HDAC2 and HDAC3 associate with STAT3 and the phosphatase PP2A. Among them; HDAC3 showed the highest deacetylase activity over STAT3. The mechanism proposed points to a facilitation of dephosphorylation of STAT3 mediated by the recruitment of PP2A through the interaction with HDACs [66]. This is further supported by the finding of hyperacetylation of STAT3 in cells treated with the Class I selective HDACi MS-275 [67], and the persistent phosphorylation of STAT3 in cells treated with HDACi [66]. Another body of evidence relating acetylation with the transcriptional regulation mediated by STAT3 involves a potential role in the control of the methylation status of gene promoters. It has been demonstrated that STAT3 interacts with the DNA methyltransferase 1 (DNMT1) [68]. In this context, a recent report showed that the impairment of the K685 acetylation is accompanied by demethylation and reactivation of several tumor-suppressor genes [69]. The same report also identified the potential role of sirtuins in this process through observations that the treatment of cells with the sirtuin activator resveratrol is directly linked to the acetylation status of STAT3 in melanoma cells, and this can reverse aberrant CpG island methylation in melanoma and several other malignancies. The participation of sirtuins in the deacetylation of STAT3 has been also reported by other groups [70, 71], positioning these deacetylases as potential targets in the modulation of the STAT3 activity.

### *7.4.3 STAT5*

Two highly related STAT5 proteins exist: STAT5a and STAT5b. These proteins are encoded by two distinct but chromosomally linked genes with high similarity, although there are functional differences between them, including the DNA binding affinities [72]. STAT5a and STAT5b are activated in response to a variety of cytokines as well as tyrosine kinase receptors [73]. These include prolactin (PRL), growth hormone (GH), erythropoietin (Epo), trombopoietin (Tpo), granulocyte-macrophage colony-stimulating factor (GM-CSF), IL-2, IL-12, IL-3, IL-5, IL-7, IL-9, and IL-15, which are involved in a great many functions regarding cell growth regulation (Fig. 7.2) [74]. The canonical JAK2/STAT5 pathway is one of the most widely studied cellular signaling cascades and is critical for normal hematopoiesis [75]. Moreover, STAT5 has been reported to have oncogenic properties, mainly through its promotion of cell survival and proliferation by tightly controlling the expression of genes involved in cell cycle progression and survival, such as c-myc, bcl-2 and bcl-XL, cyclin D1, thus intervening in the growth control of these cells [76]. Since STAT5 regulates the transcription of cyclin D1/D2 and c-myc in some cell types [77], the constitutive activation of STAT5a/b, probably promotes tumorigenesis by deregulating the cyclin complexes D/CDK4-6, which is responsible for the control progression from the G1 to the S-phase of the cell cycle [56]. Thus, the constitutive activation of STAT5 is a hallmark of hematopoietic malignancies, chronic myelogenous leukemia, erythro-leukemia, acute lymphocytic leukemia, myeloproliferative neoplasms such as polycythemia vera, essential thrombocytopenia [78] and other types of cancer, including breast cancer [79], prostate cancer [80], squamous cell carcinoma of the head and neck, melanoma and hepatocellular carcinoma [81]. Its constitutive activation may also be triggered off by the expression of fusion proteins causing persistent PTK activation, such as JAK2, PDGF-R or ABL [82].

A widely pursued question is whether the STAT5a/b proteins are critical for cellular transformation, as suggested by the presence of activated STATs in a variety of tumor types and the observation that over expression of various forms of dominant-negative STAT5 proteins can partially suppress cell growth. For example, in chronic myelogenous leukemia (CML) the presence of activated STAT5a/b, in the presence of dominant negative can suppress the transformed phenotype [83]. Additionally, considerable attention has been assigned to STAT5a/b as transcriptional regulators of anti-apoptotic genes such as Bcl-X, a characteristic shared with other STATs such as STAT1. Particularly, STAT5a/b dominant negatives have a considerable reduction in Bcl-X levels, which leads to the suppression of cell growth [73].

One of the most studied roles of STAT5 is its participation in the prolactic receptor (PRLR) pathway. The PRLR is an essential type I cytokine receptor involved in mammary gland development during pregnancy and lactation. JAK2 phosphorylates PRLR on multiple tyrosine sites in a cytoplasmatic loop, participating also in the phosphorylation of STAT5a and STAT5b on a conserved tyrosine residue within the C-terminal SH2-dimerization domain. The phosphorylation of STAT5 triggers its dissociation from the receptor and the subsequent dimerization, which activates STAT5 and facilitates its translocation into the nucleus where it regulates gene expression associated with the functions of the ligand prolactin (PRL) [84].

As with other STATs, the activity of STAT5 can be regulated by phosphorylation. Additionally, recent studies in human breast cancer TD47 cells, reported that CBP induces acetylation of STAT5 and, simultaneously, PRLR. The dimerization and subsequent activation of the STAT5 pathway is also enhanced by the pan-HDACi

TSA and the sirtuin inhibitor nicotinamide. These observations were further validated by inhibition of STAT5 after over expression of SIRT2 or HDAC6 [84]. This study also reported the existence of specific sites for acetylation in STAT5b, K359, K694, and K701, as reflected from mass spec analysis and site directed mutagenesis [84]. Interestingly, the residues K694 and K701 are located in close proximity to the phosphorylation site of STAT5, however, their acetylation by CBP was found to occur independently of the phosphorylation on Y699.

In a recent study it was reported that the selective class HDAC6 inhibitors NQN-1 decrease levels of constitutively active STAT5 and attenuated Erk phosphorylation by the acetylation of Hsp90 in human acute myeloid leukemia cell line MV4-11. This demonstrated that inhibition of HDAC6 and the subsequent interference with the Hsp90 chaperone function resulting in the degradation of critical proteins like STAT5, and attenuation of signaling cascades promoting leukemic cell growth [85]. Apparently, STAT5 seems to interact with a variety of HDACs. Thus, another recent report using co-immunoprecipitation and ChIP assays demonstrated that STAT5 is responsible for the recruitment of HDAC1 to the Id-1 gene through direct recruitment to the pro-B-cell enhancer (PBE) regulatory region. This recruitment induces deacetylation of the promoter region of the Id-1 gene as well as deacetylation of the transcription factor C/EBPbeta, whose acetylation diminishes its DNA-binding activity. Therefore, this new function for STAT5 enhances the transcriptional activity of Id-1 [86].

### *7.4.4 STAT2*

STAT2 is mainly phosphorylated and activated by stimuli mediated by IFNα/β, cytokines that activate the ISGF3 complex, which consists of STAT1, STAT2 and IRF9 (Fig. 7.3) [73]. The tyrosine residue Tyr690 of STAT2, required for SH-phosphotyrosine interaction and thus STAT activation, is located near the SH2 domains. Contrarily to other STATs, STAT2 is the only one that does not act as a homodimer, and is the only member of the STAT family that does not bind GAS elements as homodimers [87].

IFNα/β stimulation induces the dimerization of the IFNARs and the subsequent cross-phosphorylation of Tyk2 and of JAK1. Once activated, JAKs phosphorylate the IFNAR2 receptor subunit to generate a docking site for STAT2 and promote its phosphorylation. Nevertheless, IFNAR2 phosphorylation only generates the docking site for STAT2; the subsequent phosphorylation of STAT2 in tyrosine 690, once again mediated by JAKs, will favor binding with STAT1. The binding of STAT1 to IFNAR2 depends on STAT2, but not vice versa [35]. The recruitment of STAT1 to the receptor allows its phosphorylation by JAKs at the tyrosine 701, this permits the release of the heterodimer STAT1/STAT2, which associates with the IRF9 nuclear factor and forms the ISGF3 factor [88], that in turn activates specific target genes within the nucleus recognizing promoter sequences called IFN-stimulated response elements (IRSE) [89, 90]. STAT2 deficient mice are viable and develop normally. However similar to STAT1 deficient mice, STAT2 null mice are susceptible to viral infections, and cells from these mice are unresponsive to IFNα/β, supporting the essential role that Stat2 plays in the ISGF-3 complex induced by IFNα. In addition, the absence of reduced activation of STAT1 is consistent with the concept that STAT2 facilitates recruitment to, or activation by, the IFNα/β receptor complex [73].

Like many eukaryotic transcription factors, the carboxy-terminal TADs of STAT1, STAT2, STAT3, STAT5 and STAT6 interact with co-activators histone acetyl transferases (HATs) especially p300/CBP. In this context, it has been reported that STAT2 recruits the acetyltransferase protein GCN5 (general control non-repressed), which lead to acetylation of histones in the promoters of IFN-α–regulated genes [13].

All subunits of the ISG3 complex, STAT1, IRF9 and STAT2, are highly acetylated, a process mediated mainly by CBP in IFNα stimulated Hela and 293T cells [87]. The Lys390 within the STAT2 DBD is poorly conserved among other STATs. Although it is unclear whether Lys390 is an evolutionary mistake requiring acetylation for correction or it has a function evolved specially for human cells, STAT2 acetylation on this site may warrant a more flexible interaction between STAT2 and STAT1, allowing STAT1 to interact with IRF9 and DNA within the ISGF3 complex [87]. This potential mechanism has been shown for other transcription factors such as p53, where the acetylation within DBD mediates regulatory interaction with other cofactors without affecting its DNA-binding activity [91]. Whereas acetylation of STAT2 at K390 appears to be required to allow association of STAT2 with IRF9, $STAT2^{K390R}$ constitutively binds STAT1 but fails to recruit IRF9 in IFNα treated cells [44]. The importance of the acetylation of STAT2 was also demonstrated by experiments using the HDACi TSA, which prevent its association with IRF9 and the subsequent recruitment of RNA polymerase II to the ISG54 promoter. Interestingly, ectopic expression of IRF9 reverses the inhibitory actions of TSA, suggesting that IRF9 functions to recruit RNA polymerase II to the promoter of interferon-stimulated genes [92].

### *7.4.5 STAT6*

STAT6 is activated by IL-4 and IL-13 and regulates Th2 differentiation of lymphocytes [73], playing important roles in asthma and other inflammatory lung disease [93]. The absence of STAT6 blocks the differentiation of TH2 cells, and lack of STAT4 impairs IFNγ production by T cells and development of natural killer cells during bacterial and viral infections [13]. STAT6 has been shown to be persistently activated in various hematopoietic malignancies, such as lymphomas and leukemias [94], and upregulates genes important for hematopoietic tumor survival and proliferation when persistently activated in tumor cells [15]. As expected, STAT6 deficient mice lack most of the physiological functions associated with IL-4. In particular, the ability of IL-4 to induce the in vitro differentiation of Th2 cell is lost. This phenotype has provided support for the hypothesis that IL-4, through the activation of STAT6, induces the transcription of the IgE constant region of the heavy chain locus and thereby makes it accessible for the recombinase system [73].

A recent study showed that in several breast cell lines, STAT6 may also be involved in oncogenesis, like STAT1, STAT3 and STAT5. The inhibition of cell growth and induction of apoptosis, reported by other authors in human breast cancer cells after treatment with IL-4 seems to be mediated by STAT6 activation [95]. STAT6 is phosphorylated at low levels on the S707 and 756 serine residues in non-stimulated cells, and IL-4 stimulation triggers higher levels of phosphorylation. These serine phosphorylations occur independently of the Y641 tyrosine phosphorylation in STAT6, which is required for the STAT2 dimerization and subsequent nuclear translocation [35].

In addition to serine phosphorylation, STAT6 can be regulated by acetylation. In fact, STAT6 was the first member of the STAT family to be recognized as being acetylated [96]. However, the real contribution of its acetylation is not completely understood [44].

### 7.4.6 STAT4

STAT4 is the only STAT that has not been reported to be regulated by acetylation/deacetylation. STAT4 expression is restricted to myeloid cells [97] and is predominantly activated by IL-12, a cytokine produced mainly by APCs that is involved in the differentiation of naive T cells into Th1 cells. In fact, STAT4 and STAT6 are therefore involved in maintaining the equilibrium of the TH1 and TH2 responses, alterations in these proteins cause severe immune disorders, like autoimmune diseases for TH1 and allergic diseases for TH2 [98]. IL-12 is also important in the activation and growth of T cells. It stimulates the production of IFN-γ TNF-α from T and NK cells [99]. The IFNα/β can also activate STAT4 [100] but in a lesser magnitude.

### 7.4.7 STAT1 and STAT3 Acetylation; Opposing Roles in Survival and Immunogenicity

Although STAT3 and STAT1 control several common targets and share some characteristics, the majority of the cellular events unchained by their activation are opposite; STAT1 inhibits proliferation and enhances both, innate and adaptive immune responses. In the other hand, STAT3 promotes proliferation, survival, and immune tolerance (Fig. 7.4). The activation of these two pathways by cytokine/growth factors is indeed finely tuned by regulatory mechanisms present on each specific cell type, evidencing the intrinsic differences between cells to read and translate signals from the environment. Therefore, the fate of the final cellular outcome will depend on the composition, magnitude and duration during the activation of these two pathways, having a tremendous importance in pathological and pathophysiological conditions such as cancer, autoimmune diseases and immune response against foreign insults.

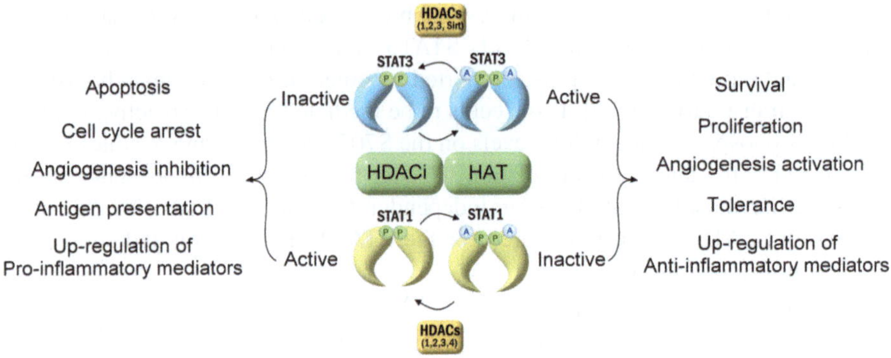


**Fig. 7.4** Opposite roles of STAT1 and STAT3 in survival and immunogenicity

As detailed before, the activation of STAT1 is primarily mediated by IFNs. The main function of STAT1 is triggering the expression of pro-apoptotic and anti-proliferative genes [101], and is considered a major barrier in neoplastic proliferation and expansion [27]. In addition to this function, STAT1 is considered a pro-inflammatory mediator due to its capacity to promote the recruitment and activation of immune cells during inflammatory processes. This function is in large part mediated for its capacity to promote antigen presentation by increasing the expression of MHC I and MHC II [27] and its positive regulatory role in the production of chemokines and adhesion molecules such as Mig [102], ICAM-1 [103], CXCL10 [104] and VCAM-1 [102]. A direct proof of the participation of STAT1 in antigenic processes came from early studies using STAT1 -/- and IRF-1 -/- hepatic cells, where the expression of all these mediators was diminished compared with wild type hepatic cells [102]. Additionally, STAT1 modulates the production of COX2 and nitric oxide, important mediators of vasodilatation during inflammatory processes [105].

In the opposite direction, STAT3 is a major regulator of anti-inflammatory events. This role is mainly attributed to its capacity to enhance the production of the anti-inflammatory cytokine IL-10, which in turn will counteract the inflammatory response. Additionally, by recent genome-wide analysis reports, several other target genes potentially responsible for the anti-inflammatory role of STAT3 [106, 107], including Bcl3, Il4ra and Socs3 [9, 10] have been identified. Stat3 can also indirectly ameliorate inflammation by down-regulation of STAT1 target genes, with function principally mediated by IL-10 which interferes with STAT1 phosphorylation [108]. STAT3 also promotes proliferation and inhibits apoptosis by direct regulation of several genes, including the anti-apoptotic factors Bcl-xL [27, 58], Survivin [109], MCL1 [110], and the cell cycle regulators Cyclin D1 [111] and MYC [77]. Additionally, STAT3 facilitates angiogenic events by induction of VEGF [112], HGF [113] and bFGF [114]. Interestingly, STAT1 seems to counteract the effect of STAT3 in angiogenesis by inhibiting the action of VEGF and bFGF [104].

The opposite effects of STAT1 and STAT3 in proliferation and immunogenicity have been explored in different contexts, and outstanding advances have been made by studies using selective inhibitors and direct genetic abrogation or silencing of their respective expressions [27]. However, new approaches are pointing to dual strategies to control differentially both pathways subsets, for example by manipulation of SOCS regulators [115].

As presented in this review, acetylation/deacetylation is emerging as a regulatory mechanism to control the activation of STATs. Particularly for STAT1 and STAT3 there is evidence indicating that their acetylation status has opposite outcomes in their functions; while STAT3 acetylation is necessary for its activation [64], this covalent modification impairs the transcriptional activation mediated by STAT1 [41]. This duality in the effect obtained by their acetylation status is diagrammed in Fig. 7.4, indicating the opposite outcomes in survival and immunogenicity attained upon acetylation of either, STAT1 and STAT3. This diagram also suggests that finding a common outcome in the function of STAT1 and STAT3 after either acetylation or deacetylation should be expected. Specifically, acetylation of both STATs will tip the balance toward the survival and anti-inflammatory phenotype, and their deacetylation will promote cell cycle arrest/apoptosis and stronger inflammatory responses (Fig. 7.4). This scenario is suggesting that pan-HDACi will favor survival and tolerance; however, in practice, these compounds are promoting exactly the opposite in several malignancies, indicating that the effect of pan-HDACi over other intracellular regulatory processes predominates in the final outcome. Another explanation could be that transformed cells are resistant to this putative regulatory mechanism. This resistance is supported by the fact that several malignancies bear aberrant activation of STATs [8, 15], and perhaps the regulatory mechanisms triggered by acetylation/deacetylation of STATs is bypassed by the hyper-activation of Jak/STATs pathways. Additionally, it is well documented that HDACi are able to decrease proliferation and modify other signaling pathways preferentially in oncogenic cells rather than non-transformed cells [25].

As described before, these two STATs can be acetylated by CBP [44]. However, only STAT3 can be acetylated by p300 [65]. In spite of the well characterized role of HAT complexes in the acetylation of STAT1 and STAT3, only a few HDACs have been tested for the removal of acetyl groups on STAT1 and STAT3. Interestingly, it seems like class I HDACs are able to deacetylate both proteins equally, and only a few reports describe the participation of other HDACs in this process. Noteworthy is the finding that Sirt1 deacetylates STAT3 to inhibit its function [70], and HDAC4 has the same effect over STAT1 [47]. These differences in substrate preferences by HDACs suggests that the differential inhibition of them could be used to differentially inhibit or activate STAT1 and STAT3, a possibility that must be explored in the future.

## 7.5 Regulators of Cytokine Signal Transduction

As with all biological processes, cytokine signal transduction must be tightly regulated. This regulation in the JAK/STAT signaling pathways is accomplished through multiple families of proteins, which include SH2-containing phosphatases (SHP), protein inhibitors of activated STATs (PIAS), and suppressors of cytokine signaling (SOCS). An overview of these three families of proteins, their role in regulating the JAK/STAT pathways, and the known interactions of HDACs and their inhibitors with these negative regulatory pathways is given below.

### *7.5.1 Negative Regulation of JAK/STAT Pathway by SOCs Proteins*

The SOCS family of proteins includes SOCS1 through SOCS7 as well as the cytokine-inducible SH2-containing protein (CIS). In addition to the SOCS family proteins listed above, several other protein subgroups that contain the SOCS box, but lack the SH2 domain are sometimes categorizes as SOCS proteins. These include the subgroups: ankryin-repeat-containing proteins, WD-40 repeat containing proteins, SPRY domain-containing proteins, and the RAR-like GTPases. All members of the SOCS family contain a conserved sequence of 40 amino acids referred to as the SOCS box [116]. In unstimulated cells, SOCS proteins are generally expressed at low levels, and become rapidly induced by cytokines, thereby inhibiting JAK-STAT signaling, forming a classic negative-feedback loop [12]. The SOCS proteins are known to achieve signaling suppression through multiple mechanisms. SOC2, SOCS3 and CIS can bind to phosphotyrosine residues on cytokine receptors, thereby competing with STATs for binding sites. SOCS1 binds to phosphotyrosine residues on JAKs, directly inhibiting JAK activity. As well, SOCS can target bound proteins for proteasomal degradation by formation of an E3 ubiquitin ligase complex and subsequent ubiquitination of the bound protein.

SOCS members appear to have overlaps in function, thereby making exact determination of the individual roles difficult. However, knockout studies of SOCS1 mice show increased STAT1 activation and sensitivity to IFN-γ [117]. The SOCS1 promoter is also known to contain an STAT1 binding region; however, it also contains binding sites for STAT3 and STAT6 [116]. In regards to other SOCS: SOCS2 knockout mice have increased signaling by insulin-like growth factor 1, while SOCS3 knockout mice die due to placental insufficiency, a phenotype also seen in STAT3 knockout mice. In macrophages with SOCS3 knockdown, prolonged activation of STAT1 and STAT3 by IL-6 is seen. In addition, an induction of genes associated with IFN-γ signaling is observed [117].

An increase in SOCS expression in cells is dependent on tissue type and the type of cytokine or growth factor stimulation. For example IL-6 stimulation increases mRNA levels of SOCS1, SOCS2, SOCS3, and CIS in liver tissue, while growth hormone stimulation induces expression of only SOCS3 and CIS. In mammary tissue, SOCS2 and CIS are instead induced in response to growth hormone stimulation. As well as tissue and stimulus influencing SOCS expression, expression dynamics are variable. While both SOCS1 and SOCS3 expression liver levels are apparent as little as 20 min after stimulation with IL-6, SOCS1 expression returns to basal levels in 4 h, while SOCS3 take about 8 h. As well, SOCS2 and CIS levels last for 24 h [116].

Currently, little is known about the roles that histone deacetylases play in regulating SHPs, PIASs and SOCs. Indeed, only a few studies to date have been published evaluating the role of HDACi or individual HDACs on these families of negative regulators of cytokine signal transduction. Of these few studies, all of them address SOCS.

The HDACi TSA has been shown to suppress JAK2/STAT3 signaling through its effects on SOCS. In colorectal cancer cells, treatment with TSA led to hyperacetylation of the promoter regions of both SOCS1 and SOCS3, with no effect on the SHP1 promoter. This led to an increase in SOCS1 and SOCS3 expression and a decrease in JAK2/STAT3 signaling. Ultimately, TSA treatment lead to cell cycle arrest and apoptosis in colorectal cancer cells through downstream targets of JAK2/STAT3 signaling such as BCL-2, survivin, and p16 [118]. Treatment with TSA has also been reported to increase SOCS3 expression in N-1 neurons. A similar increase in SOCS3 expression is seen with valproic acid treatment. In contrast, the same study showed that TSA suppressed SOCS3 expression in 3T3-L1 adipocytes [119].

In RAW264.7 macrophages, LPS induces the phosphorylation of the transcription factor ATF-2 as well as the expression of SOCS3. However, when ATF-2 is silenced by siRNA, SOCS3 expression is significantly reduced. Further investigation showed that HDAC1 was found to interact with ATF2 after, but not before, LPS treatment. When treated with TSA prior to LPS stimulation, SOCS3 expression was inhibited. These results indicate that HDAC1 positively regulates AFT2 expression, and thereby positively regulates the expression of SOCS3 [120].

### 7.5.2 Regulation of STATs by PTPs and PIAS

Several protein tyrosine phosphatases (PTP) have been indicated to regulate JAKs, including SHP1, SHP2, CD45, PTP1B and T-cell PTP (TCPTP). SHP1 and SHP2 are SH2-domain-containing PTPs. Genetic studies indicate that SHP2 is involved in the negative regulation of JAK1 tyrosine phosphorylation after IFN-gamma stimulation is increased [121]. STATs can be negatively regulated by PTPs (such as PTP1B and TCPTP) in the cytoplasm and the nucleus [12], for example the protein tyrosine phosphatase TC45 is responsible for the dephosphorylation of STAT1 in the nucleus [122].

The PIAS family consists of PIAS1, PIAS3, PIASx, and PIASy. As well, there are two known splice variants of PIASx, the alpha and beta forms. PIAS proteins interact with STATs in response to cytokine stimulation, inhibiting the transcriptional activity of STATs. The mechanisms of STAT inhibition vary between the PIAS family members. PIAS1 and PIAS3 are known to directly bind STAT1 and STAT3 respectively, preventing association with DNA. As well, PIAS1, PIAS3, and PIASx can sumoylate STAT1 at Lys-703, close to the site of JAK phosphorylation. However, the impact of sumoylation on STAT1 activity is not understood. Interestingly, PIASx mediated STAT4 repression can be disrupted by HDACi [117]. PIAS proteins have also been implicated in various processes that have no apparent connection to STAT proteins, including induction of apoptosis, modulation of ion channels, interaction with androgen receptors and interaction with RNA Helicase [13].

The SHP family contains two members, SHP-1 and SHP-2. Both of these proteins contain two N-terminal SH2 domains in addition to a C-terminal protein-tyrosine phosphatase domain. These proteins are constitutively expressed and act by dephosphorylating signaling molecules. To achieve this, the SH2 domain of SHPs binds to phosphotyrosine residues on various cytokine receptors or other target molecules. SHP-1 is known to desphosphorylate the IL-4 receptor, c-kit, erythropoietin receptor and JAK2. Conversely, SHP-2 appears to act as a positive regulator of signaling. As well, activation of SHP2 through gp130 requires Jak1. The proliferation depends on STAT3 and SHP2 activation. In addition, SHP2 might be involved in the induction of serine phosphorylation of STAT1 and STAT3 via activation of the MAPK pathway [123]. Interestingly, upwards of 25 % of sporadic juvenile myelomonocytic leukemia patients have mutations in SHP-2 [124].

## 7.6 Cytokine Regulation by Histone Deacetylases

As cytokine signaling is the initiator of, and cytokine production is often the consequence of the JAK/STAT signaling pathways, discussing the role of HDACs in modulating cytokine expression is essential to understanding their effects on signaling pathways. Histone deacetylases are known to influence expression of a wide variety of cytokines. This knowledge came from several reports utilizing manipulation of specific HDACs and/or pharmacological inhibition of them. A list of the current effects by individual HDACs and HDACi in specific cell types is detailed in Table 7.1. Generally, inhibition of HDACs results in repression of pro-inflammatory cytokines, making histone deacetylase inhibitors (HDACi) attractive therapeutic agents in inflammatory diseases. However, in some cell types and contexts, HDAC inhibition can instead favor inflammatory cytokine production while repressing anti-inflammatory cytokine production [24, 125]. This is the result of the dynamic nature of HDACs with regards to cell type, stages of differentiation and disease context. Most studies to date conducted on histone deacetylase control of cytokine production address the effects of one or more HDACi. The majority of HDACi currently utilized target multiple HDACs, making determination of the role of individual HDACs difficult. However, using recently developed isotype specific HDACi or other experimental approaches to manipulating expression of individual HDACs, a small, but increasing number of studies have identified the role of individual HDACs in the regulation of cytokine production.

### *7.6.1 Innate Immune Cells*

A large number of studies over the past decade have addressed the effect of HDAC inhibitors on leukocyte cytokine production. In a 2002 study, it was demonstrated that administration of the pan-HDACi suberoylanilide hydroamic acid (SAHA), also

**Table 7.1** Effect of HDACs and HDACi over the production of relevant cytokines involved in the activation of Jak/STAT pathways

| Cytokine | Effect | Condition | Cell type | References |
|---|---|---|---|---|
| IL-10 | ↑ | HDAC11 | Macrophages | [138] |
| | ↑ | HDAC1 | T cells | [149] |
| | ↑ | Valproic Acid | Dendritic cells | [141] |
| | ↑ | LBH589 | Dendritic cells | [156] |
| | ↑ | TSA | Macrophages | [130] |
| | ↑ | SAHA | Splenocytes | [157] |
| | ↑ | LAQ824 | Macrophages | [125] |
| IL-12 | ↑ | HDAC3 | Macrophages | [140] |
| | ↑ | HDAC1 | Macrophages | [158] |
| | ↑ | Valproic acid | Dendritic cells | [141] |
| | ↑ | Valproic acid | Macrophages | [132] |
| | ↑ | LBH589 | Dendritic cells | [156] |
| | ↑ | SAHA | Splenocytes | [157] |
| | ↑ | TSA | Macrophages | [159] |
| | ↑ | Apicidin | Dendritic cells | [133] |
| | ↑ | Butyrate | Dendritic cells | [134] |
| | ↑ | SAHA | Dendritic cells | [131] |
| | ↑ | LAQ824 | Macrophages | [125] |
| IL-6 | ↑ | HDAC3 | PBMCs | [160] |
| | ↑ | LBH589 | Dendritic cells | [156] |
| | ↑ | TSA | Macrophages | [130] |
| | ↑ | SAHA | Splenocytes | [157] |
| | ↑ | Apicidin | Dendritic cells | [133] |
| | ↑ | Butyrate | Dendritic cells | [134] |
| | ↑ | MS-275 | Monocytes | [161] |
| IFNγ | ↑ | HDAC1 and HDAC2 | T Cells | [162] |
| | ↑ | ITF2357 | Macrophages | [153] |
| | ↑ | TSA | PBMCs | [145] |
| | ↑ | TSA | T cells | [146] |
| | ↑ | TSA, VPA, NaB | NK cells | [147] |
| IL-1B | ↑ | TSA | Macrophages | [130] |
| | ↑ | MS-275 | Monocytes | [161] |
| IFNα | ↑ | TSA | Dendritic cells | [137] |
| IFNβ | ↑ | HDAC1 and HDAC8 | Various cell lines | [163] |
| | ↑ | HDAC6 | Various cell lines | [163] |
| IL-4 | ↑ | TSA | T cells | [164] |
| | ↑ | TSA | T cells | [142] |
| IL-2 | ↑ | TSA | PBMCs | [145] |
| IL-5 | ↑ | TSA | PBMCs | [145] |
| IL-23 | ↑ | LBH589 | Dendritic cells | [156] |
| | ↑ | TSA | Macrophages | [136] |

known as vorinostat, to mice results in a reduction of TNF, IL-1β, IL-6 and IFN-γ serum levels after induction by lipopolysaccharide (LPS). Similarly, when human monocytes are stimulated with LPS in the presence of SAHA, there is a pronounced reduction in production of TNF, IL-1β, IL-12 and IFN-γ. As well, IL-18 plus IL-12 stimulation of these cells in the presence of SAHA results in drastic reduction in

IFN-γ production. Additionally, IFN-γ and TNF mRNA levels are reduced by SAHA administration, but IL-8 and IL-1β levels are unaffected. Intriguingly, IFN-γ production by CD3 stimulation is not affected by SAHA treatment [126].

In dendritic cells, pretreatment with SAHA results in a dose dependent reduction in expression of TNF, IL-12 and IL-6 after LPS stimulation [127]. These results are similar with the use of another HDACi, ITF 2,357, also known as givinostat. Furthermore, these reductions are the result of increased STAT3 acetylation [128]. Other studies in dendritic cells show that both SAHA and TSA are also able to modulate the expression of IL-12 and IL-23 [129].

Trichostatin-A (TSA), an HDACi with activity against Class I and II HDACs, has been demonstrated to increase expression of IL-10 in bone marrow derived macrophages. TSA treatment also reduces levels of TNF, IL-6 and IL-1β production in these cells [130]. A separate study demonstrated in both human and murine cells, that treatment of dendritic cells with TSA or SAHA also inhibits production of IL-12p40 post TLR stimulation. Intriguingly, despite decreased expression of protein, HDACi treatment results in increased acetylation of the IL-12p40 locus [131].

In LPS stimulated macrophages, treatment with sodium valproate represses expression of IL-12 and TNF-α while increasing IL-10 expression [132]. As well, in bone marrow derived dendritic cells, apicidin, a fungal metabolite with HDACi properties, inhibits production of IL-12, along with IL-6 and TNF production. Resultantly, treatment of dendritic cells abrogates their ability to produce a Th1 response in T-cells [133]. In agreement, a separate study shows that treatment of LPS stimulated human monocyte derived dendritic cells with butyrate results in a reduction in IL-12p40 as well as IL-6 production. As with apicidin treatment, this impairment results in a deficiency in generating Th1 responses by T-cells [134]. Treatment of LPS stimulated human dendritic cells with the pan-HDACi LBH589 also reduces expression of a variety of cytokines including IL-10, IL-12p70, IL-6, IL-23 and TNF-alpha. Again, the functional consequence of which is impaired activation of T-cells [135].

Interleukin 23 is a heterodimeric cytokine produced by both dendritic cells as well as macrophages. The p40 subunit of IL-23 is shared with IL-12, and like IL-12, the IL-23 receptor signals through the STAT4 pathway. Functionally, IL-23 along with other cytokines is responsible for directing Th17 CD4+ T-cell differentiation. Its role as a pro-inflammatory cytokine and involvement in autoimmunity is becoming increasingly clear. As with various studies in IL-12, IL-23 production is abrogated with HDACi treatment. A study using TSA and SAHA showed human and murine dendritic cells have reduced production of IL-12 and IL-23 when treated with HDACi. Intriguingly, the results are similar when HDACi was given concomitantly with LPS or IFN-γ stimulation as well as administered after stimulation [136].

The Type I interferons are composed of IFN-α and IFN-β. IFN-α is a pleiotropic protein with over twenty known variants. It is mainly produced by innate immune cells in the context of viral infection. There is also evidence of IFN-α having anti-tumor effects. Like IFN-α, IFN-β is generally associated with viral infection and anti-tumor effects. IFN-β upregulates TRAIL expression on CD8+ T-cells. TRAIL upregulation is associated with both anti-viral and anti-tumor responses.

Plasmacytoid dendritic cells are the main producers of type I interferons. In a recent study, a reduction in type I interferons, TRAIL, IL-6 and TNFα was observed after TSA treatment of these cells [137].

While many studies have demonstrated the effects of pan-HDACi or class specific HDACi on cytokine production, more recently the role of individual HDACs in regulating cytokine production has become subject to experimentation. In 2008, HDAC11, the newest discovered HDAC, was discovered to negatively regulate the production of IL-10 in both human and mouse antigen presenting cells. An increase in both IL-10 and IL-12 production by LPS stimulated HDAC11-knockout macrophages is seen, while a decrease in both cytokines is seen in HDAC11-overexpressing macrophages. Interaction of HDAC11 with the proximal promoter region of the IL-10 gene, resulting in deacetylation activity, and thus transcriptional regulation was determined as the mechanism for this regulation [138].

The results of one study show that treatment of mouse macrophages with the pan-HDACi LAQ824 induces chromatin changes at the IL-10 promoter leading to repressed expression of IL-10. Importantly, this study also reveals that LAQ824 treatment results in enhanced recruitment of HDAC11, a negative regulator of IL-10 production, as well as the transcriptional repressor PU.1 to the IL-10 promoter region. Furthermore, an increase in IL-12 production by treated macrophages concomitant to the decrease in IL-10 production is seen. Functionally, in contrast to the described results with other HDACi, this results in increased Th1 activation of T-cells. Additionally, LAQ824 treated macrophages are able to restore function to anergized T-cells [139].

Another study shows that in intestinal macrophages, IL-10 regulates IL12p40 through HDAC3. In these cells, inhibition of HDAC3 by MS275, which target only Class I HDACs, results in increased histone 4 acetylation at the IL-12p40 promoter after IL-10 induction by LPS. As well, shRNA silencing of HDAC3 results in a diminished IL-12p40 repression by LPS induced IL-10 [140]. As in studies described above, another study demonstrated downregulation of IL-12p40 mRNA levels in mouse macrophages after treatment with TSA or SAHA. Additionally, when these cells are treated with the HDACi MS-275, this downregulation is not seen. However, when treated with the HDAC6 specific inhibitor referred to as 17a, a similar reduction in IL-12p40 expression is seen as when treated with TSA [132]. In dendritic cells, it has been recently demonstrated that treatment with valproic acid prior to LPS stimulation decreases the production of IL-10 as well as IL-12p70. These differences are also seen under IFN-γ stimulation. However, the roles of HDAC3 or other specific HDACs were not addressed in this study. Additionally, another study has also demonstrated that HDAC1 mediates repression of IL-12 expression in macrophages [141].

### *7.6.2 T-Cells*

While the above studies have investigated the effects of various HDACi and roles HDACs in leukocytes, several reports have also addressed the direct effects of

HDACi, or the role of specific HDACs, in T-cell cytokine production. In T-cells, as with leukocytes, HDACi treatment often, but not always, decreases expression of pro-inflammatory cytokines.

Results from a study in murine T-cells show that treatment with low dose of TSA and other HDACi results in marked reduction in the production of IL-2. However, sodium butylate treatment does not result in changes in IL-2 production. In addition to its effects on IL-2 production, TSA treatment also results in dramatic decreases in IL-4, IL-13 and IFN-γ production by T-cells. Treatment with both scriptaid and sodium butylate results in changes in only IFN-γ, and to a lesser degree than those seen with TSA. As well, the hyperacetylation associated with HDACi treatment is pronouncedly longer lasting with TSA treatment [142]. Conversely, a study published near the same time reported a TSA induced increase in IL-4 production by T-cells in a murine, collagen induced, rheumatoid arthritis model. This difference may be partially explained by an increase in Th1 cell apoptosis. However, Th2 cells also display an increase in IL-4 gene expression and an associated increase in histone 4 acetylation following TSA treatment [142]. As well, in regulatory T-cells TSA drastically reduces the formation of IL-17 producing cells [143].

Treatment of Jurkat cells with TSA or sodium butyrate elevates production of IL-5, a cytokine responsible for stimulating B-cell growth and antibody secretion as well as the maturation and differentiation of eosinophils. The increase in IL-5 production is associated with increased IL-5 promoter activity and hyperacetylation of both histone 3 and histone 4 at the IL-5 promoter region [144].

A study of whole human peripheral blood mononuclear cells (PBMC) showed ex vivo treatment with TSA results in decreased levels of IFN-γ and IL-2 production concomitant with increases in IL-4 and IL-13 production. When these cells are stimulated with phytohemagglutinin, an activator of T-cells, the cytokine differences resultant from TSA treatment are exacerbated [145]. A separate study also showed that Th1 T-cells when treated with various HDACi including TSA and butyrate have reduced levels of both IL-2 and IFNγ production. Resultantly, treatment with TSA is able to induce tolerance in OT-II ovalbumin-specific naïve CD4+ T-cells [146].

In addition to inhibiting IFNγ production in T-cells and PBMCs, HDACi treatment inhibits production in NK cells. When treated with TSA, valproic acid, or sodium butyrate, NK cells produce lower levels of IFNγ, even upon stimulation with exogenous cytokines such as IL-2 [147].

Class I HDACs are known to complex with Switch independent (SIN3) or Nucleosome-Remodeling protein (NuRD) in T-cells. In proliferating CD4+ T-cells, Sin3-HDAC complexes are recruited to the IFN-γ locus. This complex is displaced by the transcription factor T-bet in Th1 polarization of these cells. Therefore, treatment with HDACi is capable of augmenting IFNγ production by preventing the deacetylation of the IFNγ promoter by the Sin3-HDAC complex. However, as shown above, several HDACi reduce both IFNγ and IL-2 levels. This reduction of IL-2 has been reported to be characteristic of HDACi induced anergy in Th1 T-cells. Furthermore, T-bet has also been reported to indirectly interact with HDAC3 and HDAC5 in respect to GATA-3 binding, a transcription factor driving Th2 polarization [129].

Another study demonstrated that HDAC1 and HDAC2 in complex with Sin3A are displaced from IFN-γ locus in T-cells in Th1 differentiation. When non-stimulated, Th0 CD4+ cells were treated with TSA, there was an acquisition of histone 4 acetylation at the IFN-γ locus not seen in non-treated cells. When these cells were polarized to Th1 cells, no differences in histone 4 acetylation are seen. This increase in histone acetylation was shown to be independent of T-bet and Stat4. Sin3A\HDAC complex is found to associate with several areas of the IFNγ locus, including a known binding site of T-bet, a positive transcriptional regulator of IFNγ, in Th0 cells. When polarized to a Th1 phenotype, association of this complex at the promoter region of IFNγ is lost. However, in both Stat4 and T-bet deficient mice, when T-cells were polarized in Th1 conditions, no loss in Sin3A\HDAC complex at the IFNγ promoter is seen, indicating that dissociation of the complex is dependent on T-bet and STAT4 [148].

A separate group has also reported HDAC1 as a regulator of IL-10 production though E26 transformation-specific 1 (Ets-1). Th1 cells with Ets-1 deficiency had lower levels of HDAC1 recruitment to the IL-10 gene regulatory region and increased IL-10 production. A physical interaction between Ets-1 and HDAC1 was also demonstrated, indicating an Ets-1/HDAC1 synergism in IL-10 transcriptional repression [149].

### 7.6.3 *Disease State Cells*

As HDACi are potent modulators of cytokine production, and have a proclivity to down-regulate inflammatory cytokine responses, their use in autoimmune disease contexts is under continued investigation. Highlighting the complex and dynamic roles of HDACs, HDACi are utilized in the seemingly opposite setting as cancer therapeutics, due to their anti-proliferative and sometimes apoptotic effects. Indeed, two HDACi, SAHA and romidepsin, are FDA approved for the treatment of cutaneous T-cell lymphoma. Given this dynamic nature of HDACi, the effects on cytokine production in diseased cells can vary from those seen in healthy cells.

Rheumatoid arthritis is a chronic autoimmune disease with progressive destruction of the joints. Recently, a group has demonstrated that peripheral blood mononuclear cells (PBMCs) from rheumatoid arthritis patients have significantly increased HDAC activity when compared to healthy PBMCs. As well, treatment with TSA reduces levels of TNF and IL-6 production in both healthy and rheumatoid arthritis PBMCs, while treatment with an HDAC3 specific inhibitor, MI192, reduces production of IL-6 in rheumatoid arthritis PBMCs and having no effect on healthy PBMCs [150]. In a separate study, E11 cells, a cell line derived from human rheumatoid arthritis synovial fibroblasts, shows a dose-dependent reduction in IL-6, IL-18 and VEGF production when treated with either MS-275 or SAHA. In addition, treatment with either HDACi results in dose-dependent reduction in IL-1β, IL-6, IL-18 and TNF- α production by an LPS stimulated human monocyte cell line

[151]. A similar study demonstrated that PBMCs or lamina propria cells from Crohn's disease patients produced less TNF mRNA as well as protein when cultured with butyrate and stimulated with LPS. This reduction results from decreased NFkB activation [152].

In a Type 1 diabetes mouse model, the HDACi ITF2357, now known as givinostat, reduces production of IFNγ as well as TNF in peritoneal macrophages and spleenocytes. As well, when insulin-producing cells are challenged with IL-1β plus IFNγ, cells treated with givinostat have a marked reduction in apoptosis [153].

In systemic lupus erythematous, CD4+ T-cells produce high levels of IL-10 and reduced levels of IFNγ. When these cells are treated with the HDACi trichostatin A (TSA), a recovery of cytokine phenotype is seen [154]. In spleenocytes from MRL-*lpr/lpr* mice, a systemic lupus erythematous model, treatment with either TSA or SAHA downregulates expression of IL-10, IL-12, IFNγ, and IL-6. In vivo, TSA treatment of these mice results in a reduction in disease associated symptoms (e.g. proteinuria) [135].

Treatment of the human fibrosarcoma cell line 2fTGH with TSA reduces IFNβ production in response to viral challenge. Using a luciferase report assay and RNA interference, TSA treatment was found to interfere with the transcription factor interferon regulatory factor 3 (IRF3). Utilizing siRNA against individual HDACs, it was found that HDAC1 and HDAC8 both repressed IFNβ expression. However, these results did not explain the reduction in IFNβ production with TSA treatment. Further investigation revealed that HDAC6 enhances IFNβ expression by way of the IRF3 complex [155].

## 7.7 Conclusions

The Jak/STAT pathways represent one of the most well studied cellular signaling systems. From the initial description of IFN inducible genes, to the describing of the Jak/STAT pathways, the knowledge of these pathways has grown by a staggering amount in the last two decades. Indeed, recently regulation of the Jak/STAT pathway by mechanism beyond phosphorylation, including acetylation, has become evident. At the same time, it is becoming increasingly clear that the Jak/STAT pathways are highly intricate and that much work remains to be done. Additionally, it is also becoming increasingly evident that deregulation of the Jak/STAT pathways plays an important role in disease states, particularly cancer. In the disease context, understanding of not only the Jak/STAT pathways, but also its regulation by cytokines, HDACs, and the influence of HDAC inhibitors are essential to developing more effective and precise therapies. Overall, the Jak/STAT pathways represent a complex family of signaling pathways with equally complex regulation. Future studies will no doubt further unravel these complexities.

# References

1. Larner AC et al (1984) Transcriptional induction of two genes in human cells by beta interferon. Proc Natl Acad Sci USA 81:6733–6737
2. Friedman RL, Manly SP, McMahon M, Kerr IM, Stark GR (1984) Transcriptional and posttranscriptional regulation of interferon-induced gene expression in human cells. Cell 38:745–755
3. Shuai K, Schindler C, Prezioso VR, Darnell JE Jr (1992) Activation of transcription by IFN-gamma: tyrosine phosphorylation of a 91-kD DNA binding protein. Science 258: 1808–1812
4. Schindler C, Shuai K, Prezioso VR, Darnell JE Jr (1992) Interferon-dependent tyrosine phosphorylation of a latent cytoplasmic transcription factor. Science 257:809–813
5. Velazquez L, Fellous M, Stark GR, Pellegrini S (1992) A protein tyrosine kinase in the interferon alpha/beta signaling pathway. Cell 70:313–322
6. Stark, George R, Darnell Jr, James E (2012) The JAK-STAT pathway at twenty. Immunity 36:503–514
7. Ghoreschi K, Laurence A, O'Shea JJ (2009) Janus kinases in immune cell signaling. Immunol Rev 228:273–287
8. O'Shea JJ, Plenge R (2012) JAK and STAT signaling molecules in immunoregulation and immune-mediated disease. Immunity 36:542–550
9. Aaronson DS, Horvath CM (2002) A road map for those who don't know JAK-STAT. Science 296:1653–1655
10. Murray PJ (2007) The JAK-STAT signaling pathway: input and output integration. J Immunol 178:2623–2629
11. Stark GR, Kerr IM, Williams BR, Silverman RH, Schreiber RD (1998) How cells respond to interferons. Ann Rev Biochem 67:227–264
12. Shuai K, Liu B (2003) Regulation of JAK-STAT signalling in the immune system. Nat Rev Immunol 3:900–911
13. Levy DE, Darnell JE (2002) STATs: transcriptional control and biological impact. Nat Rev Mol Cell Biol 3:651–662
14. Darnell JE Jr (1997) STATs and gene regulation. Science 277:1630–1635
15. Yu H, Pardoll D, Jove R (2009) STATs in cancer inflammation and immunity: a leading role for STAT3. Nat Rev Cancer 9:798–809
16. Pahl HL (2012) A JAK-in-the-cell cycle. Blood 119:1096–1097
17. Wang Y et al (2009) Cotreatment with panobinostat and JAK2 inhibitor TG101209 attenuates JAK2V617F levels and signaling and exerts synergistic cytotoxic effects against human myeloproliferative neoplastic cells. Blood 114:5024–5033
18. Sansone P, Bromberg J (2012) Targeting the interleukin-6/Jak/stat pathway in human malignancies. J Clin Oncol 30:1005–1014
19. Darnell JE (2005) Validating Stat3 in cancer therapy. Nat Med 11:595–596
20. Levy DE, Inghirami G (2006) STAT3: a multifaceted oncogene. Proc Natl Acad Sci USA 103:10151–10152
21. Lee KK, Workman JL (2007) Histone acetyltransferase complexes: one size doesn't fit all. Nat Rev Mol Cell Biol 8:284–295
22. Yang XJ, Seto E (2008) The Rpd3/Hda1 family of lysine deacetylases: from bacteria and yeast to mice and men. Nat Rev Mol Cell Biol 9:206–218
23. Woan KV, Sahakian E, Sotomayor EM, Seto E, Villagra A (2012) Modulation of antigen-presenting cells by HDAC inhibitors: implications in autoimmunity and cancer. Immunol Cell Biol 90:55–65
24. Villagra A, Sotomayor EM, Seto E (2010) Histone deacetylases and the immunological network: implications in cancer and inflammation. Oncogene 29:157–173
25. Marks PA (2010) The clinical development of histone deacetylase inhibitors as targeted anticancer drugs. Expert Opin Investig Drugs 19:1049–1066

26. Philippe B (2010) Inside HDAC with HDAC inhibitors. Eur J Med Chem 45:2095–2116
27. Regis G, Pensa S, Boselli D, Novelli F, Poli V (2008) Ups and downs: the STAT1: STAT3 seesaw of Interferon and gp130 receptor signalling. Semin Cell Dev Biol 19:351–359
28. Caretto D, Katzman SD, Villarino AV, Gallo E, Abbas AK (2010) Cutting edge: the Th1 response inhibits the generation of peripheral regulatory T cells. J Immunol 184:30–34
29. Stephanou A, Brar BK, Knight RA, Latchman DS (2000) Opposing actions of STAT-1 and STAT-3 on the Bcl-2 and Bcl-x promoters. Cell Death Differ 7:329–330
30. Agrawal S, Agarwal ML, Chatterjee-Kishore M, Stark GR, Chisolm GM (2002) Stat1-dependent, p53-independent expression of p21(waf1) modulates oxysterol-induced apoptosis. Mol Cell Biol 22:1981–1992
31. Ohmori Y, Hamilton TA (2001) Requirement for STAT1 in LPS-induced gene expression in macrophages. J Leukoc Biol 69:598–604
32. Kim HS, Lee M-S (2007) STAT1 as a key modulator of cell death. Cell Signal 19:454–465
33. Wang Y, Wu TR, Cai S, Welte T, Chin YE (2000) Stat1 as a component of tumor necrosis factor alpha receptor 1-TRADD signaling complex to inhibit NF-kappaB activation. Mol Cell Biol 20:4505–4512
34. Townsend PA et al (2004) STAT-1 interacts with p53 to enhance DNA damage-induced apoptosis. J Biol Chem 279:5811–5820
35. Calò V et al (2003) STAT proteins: from normal control of cellular events to tumorigenesis. J Cell Physiol 197:157–168
36. O'Shea JJ, Gadina M, Schreiber RD (2002) Cytokine signaling in 2002: new surprises in the Jak/Stat pathway. Cell 109(Suppl):S121–S131
37. Imada K, Leonard WJ (2000) The Jak-STAT pathway. Mol Immunol 37:1–11
38. Ouchi T, Lee SW, Ouchi M, Aaronson SA, Horvath CM (2000) Collaboration of signal transducer and activator of transcription 1 (STAT1) and BRCA1 in differential regulation of IFN-gamma target genes. Proc Natl Acad Sci USA 97:5208–5213
39. Zhang JJ et al (1996) Two contact regions between Stat1 and CBP/p300 in interferon gamma signaling. Proc Natl Acad Sci USA 93:15092–15096
40. Zhang JJ et al (1998) Ser727-dependent recruitment of MCM5 by Stat1alpha in IFN-gamma-induced transcriptional activation. EMBO J 17:6963–6971
41. KrÃmer OH et al (2009) A phosphorylation-acetylation switch regulates STAT1 signaling. Genes Dev 23:223–235
42. Kramer OH et al (2006) Acetylation of Stat1 modulates NF-kappaB activity. Genes Dev 20:473–485
43. Krämer OH, Heinzel T (2010) Phosphorylation–acetylation switch in the regulation of STAT1 signaling. Mol Cell Endocrinol 315:40–48
44. Wieczorek M, Ginter T, Brand P, Heinzel T, Kramer OH (2012) Acetylation modulates the STAT signaling code. Cytokine Growth Factor Rev 23(6):293–305
45. Ginter T et al (2012) Histone deacetylase inhibitors block IFNgamma-induced STAT1 phosphorylation. Cell Signal 24:1453–1460
46. Klampfer L, Huang J, Swaby L-A, Augenlicht L (2004) Requirement of histone deacetylase activity for signaling by STAT1. J Biol Chem 279:30358–30368
47. Stronach EA et al (2011) HDAC4-regulated STAT1 activation mediates platinum resistance in ovarian cancer. Cancer Res 71:4412–4422
48. Paul AM (2010) Histone deacetylase inhibitors: a chemical genetics approach to understanding cellular functions. Biochim Biophys Acta (BBA) Gene Regu Mech 1799:717–725
49. Tomasi TB, Magner WJ, Khan AN (2006) Epigenetic regulation of immune escape genes in cancer. Cancer Immunol Immunother CII 55:1159–1184
50. Klampfer L, Huang J, Shirasawa S, Sasazuki T, Augenlicht L (2007) Histone deacetylase inhibitors induce cell death selectively in cells that harbor activated kRasV12: the role of signal transducers and activators of transcription 1 and p21. Cancer Res 67:8477–8485
51. Frank DA (2007) STAT3 as a central mediator of neoplastic cellular transformation. Cancer Lett 251:199–210

52. Johnston PA, Grandis JR (2011) STAT3 signaling: anticancer strategies and challenges. Mol Interv 11:18–26
53. Sakaguchi M, Oka M, Iwasaki T, Fukami Y, Nishigori C (2012) Role and regulation of STAT3 phosphorylation at Ser727 in melanocytes and melanoma Cells. J Invest Dermatol 132:1877–1885
54. Yu CL et al (1995) Enhanced DNA-binding activity of a Stat3-related protein in cells transformed by the Src oncoprotein. Science 269:81–83
55. Bromberg JF, Horvath CM, Besser D, Lathem WW, Darnell JE Jr (1998) Stat3 activation is required for cellular transformation by v-src. Mol Cell Biol 18:2553–2558
56. Bowman T, Garcia R, Turkson J, Jove R (2000) STATs in oncogenesis. Oncogene 19:2474–2488
57. Garcia R et al (1997) Constitutive activation of Stat3 in fibroblasts transformed by diverse oncoproteins and in breast carcinoma cells. Cell Growth Differ 8:1267–1276
58. Catlett-Falcone R et al (1999) Constitutive activation of Stat3 signaling confers resistance to apoptosis in human U266 myeloma cells. Immunity 10:105–115
59. Schuringa J-J, Wierenga ATJ, Kruijer W, Vellenga E (2000) Constitutive Stat3, Tyr705, and Ser727 phosphorylation in acute myeloid leukemia cells caused by the autocrine secretion of interleukin-6. Blood 95:3765–3770
60. Gao SP et al (2007) Mutations in the EGFR kinase domain mediate STAT3 activation via IL-6 production in human lung adenocarcinomas. J Clin Invest 117:3846–3856
61. Messina JL et al (2008) Activated stat-3 in melanoma. Cancer Control 15:196–201
62. Kortylewski M, Jove R, Yu H (2005) Targeting STAT3 affects melanoma on multiple fronts. Cancer Metastasis Rev 24:315–327
63. Gritsko T et al (2006) Persistent activation of stat3 signaling induces survivin gene expression and confers resistance to apoptosis in human breast cancer cells. Clin Cancer Res 12:11–19
64. Yuan Z-L, Guan Y-J, Chatterjee D, Chin YE (2005) Stat3 dimerization regulated by reversible acetylation of a single lysine residue. Science 307:269–273
65. Wang R, Cherukuri P, Luo J (2005) Activation of stat3 sequence-specific DNA binding and transcription by p300/CREB-binding protein-mediated acetylation. J Biol Chem 280: 11528–11534
66. Togi S et al (2009) HDAC3 influences phosphorylation of STAT3 at serine 727 by interacting with PP2A. Biochem Biophys Res Commun 379:616–620
67. Shen L et al (2012) Class I histone deacetylase inhibitor entinostat suppresses regulatory T cells and enhances immunotherapies in renal and prostate cancer models. PLoS One 7:e30815
68. Zhang Q et al (2005) STAT3- and DNA methyltransferase 1-mediated epigenetic silencing of SHP-1 tyrosine phosphatase tumor suppressor gene in malignant T lymphocytes. Proc Natl Acad Sci USA 102:6948–6953
69. Lee H et al (2012) Acetylated STAT3 is crucial for methylation of tumor-suppressor gene promoters and inhibition by resveratrol results in demethylation. Proc Natl Acad Sci USA 109:7765–7769
70. Nie Y et al (2009) STAT3 inhibition of gluconeogenesis is downregulated by SirT1. Nat Cell Biol 11:492–500
71. Sestito R et al (2011) STAT3-dependent effects of IL-22 in human keratinocytes are counter-regulated by sirtuin 1 through a direct inhibition of STAT3 acetylation. FASEB J 25:916–927
72. Boucheron C et al (1998) A single amino acid in the DNA binding regions of STAT5A and STAT5B confers distinct DNA binding specificities. J Biol Chem 273:33936–33941
73. Ihle JN (2001) The Stat family in cytokine signaling. Curr Opin Cell Biol 13:211–217
74. Takeda K, Akira S (2000) STAT family of transcription factors in cytokine-mediated biological responses. Cytokine Growth Factor Rev 11:199–207
75. Dawson, Mark A et al (2012) Three distinct patterns of histone H3Y41 phosphorylation mark active genes. Cell Reports 2:470–477

76. Matsumura I et al (1999) Transcriptional regulation of the cyclin D1 promoter by STAT5: its involvement in cytokine-dependent growth of hematopoietic cells. EMBO J 18:1367–1377
77. Bowman T et al (2001) Stat3-mediated Myc expression is required for Src transformation and PDGF-induced mitogenesis. Proc Natl Acad Sci USA 98:7319–7324
78. Buettner R, Mora LB, Jove R (2002) Activated STAT signaling in human tumors provides novel molecular targets for therapeutic intervention. Clin Cancer Res 8:945–954
79. Walker SR et al (2009) Reciprocal effects of STAT5 and STAT3 in breast cancer. Mol Cancer Res 7:966–976
80. Li H et al (2005) Activation of signal transducer and activator of transcription-5 in prostate cancer predicts early recurrence. Clin Cancer Res 11:5863–5868
81. Leong PL et al (2002) Differential function of STAT5 isoforms in head and neck cancer growth control. Oncogene 21:2846–2853
82. Lin JX, Leonard WJ (2000) The role of Stat5a and Stat5b in signaling by IL-2 family cytokines. Oncogene 19:2566–2576
83. Danial NN, Rothman P (2000) JAK-STAT signaling activated by Abl oncogenes. Oncogene 19:2523–2531
84. Ma L et al (2010) Acetylation modulates prolactin receptor dimerization. Proc Natl Acad Sci USA 107:19314–19319
85. Inks ES, Josey BJ, Jesinkey SR, Chou CJ (2012) A novel class of small molecule inhibitors of HDAC6. ACS Chem Biol 7:331–339
86. Xu M, Nie L, Kim SH, Sun XH (2003) STAT5-induced Id-1 transcription involves recruitment of HDAC1 and deacetylation of C/EBPbeta. EMBO J 22:893–904
87. Tang X et al (2007) Acetylation-dependent signal transduction for type I interferon receptor. Cell 131:93–105
88. Horvath CM, Stark GR, Kerr IM, Darnell JE Jr (1996) Interactions between STAT and non-STAT proteins in the interferon-stimulated gene factor 3 transcription complex. Mol Cell Biol 16:6957–6964
89. Li X, Leung S, Burns C, Stark GR (1998) Cooperative binding of Stat1-2 heterodimers and ISGF3 to tandem DNA elements. Biochimie 80:703–710
90. Darnell J, Kerr IM, Stark GR (1994) Jak-STAT pathways and transcriptional activation in response to IFNs and other extracellular signaling proteins. Science 264:1415–1421
91. Sykes SM et al (2006) Acetylation of the p53 DNA-binding domain regulates apoptosis induction. Mol Cell 24:841–851
92. Sakamoto S, Potla R, Larner AC (2004) Histone deacetylase activity is required to recruit RNA polymerase II to the promoters of selected interferon-stimulated early response genes. J Biol Chem 279:40362–40367
93. Hebenstreit D, Wirnsberger G, Horejs-Hoeck J, Duschl A (2006) Signaling mechanisms, interaction partners, and target genes of STAT6. Cytokine Growth Factor Rev 17:173–188
94. Yu CR et al (2004) Cell proliferation and STAT6 pathways are negatively regulated in T cells by STAT1 and suppressors of cytokine signaling. J Immunol 173:737–746
95. Gooch JL, Christy B, Yee D (2002) STAT6 mediates interleukin-4 growth inhibition in human breast cancer cells. Neoplasia 4:324–331
96. Shankaranarayanan P, Chaitidis P, Kühn H, Nigam S (2001) Acetylation by histone acetyltransferase CREB-binding protein/p300 of STAT6 is required for transcriptional activation of the 15-lipoxygenase-1 gene. J Biol Chem 276:42753–42760
97. Yamamoto K et al (1994) Stat4, a novel gamma interferon activation site-binding protein expressed in early myeloid differentiation. Mol Cell Biol 14:4342–4349
98. Hemmi H et al (2000) A Toll-like receptor recognizes bacterial DNA. Nature 408:740–745
99. Kaplan MH, Sun YL, Hoey T, Grusby MJ (1996) Impaired IL-12 responses and enhanced development of Th2 cells in Stat4-deficient mice. Nature 382:174–177
100. Brierley MM, Fish EN (2002) Review: IFN-alpha/beta receptor interactions to biologic outcomes: understanding the circuitry. J Interferon Cytokine Res 22:835–845
101. Schroder K, Hertzog PJ, Ravasi T, Hume DA (2004) Interferon-gamma: an overview of signals, mechanisms and functions. J Leukoc Biol 75:163–189

102. Jaruga B, Hong F, Kim W-H, Gao B (2004) IFN-γ/STAT1 acts as a proinflammatory signal in T cell-mediated hepatitis via induction of multiple chemokines and adhesion molecules: a critical role of IRF-1. Am J Physiol Gastrointest Liver Physiol 287:G1044–G1052
103. Dustin ML, Singer KH, Tuck DT, Springer TA (1988) Adhesion of T lymphoblasts to epidermal keratinocytes is regulated by interferon gamma and is mediated by intercellular adhesion molecule 1 (ICAM-1). J Exp Med 167:1323–1340
104. Strieter RM, Kunkel SL, Arenberg DA, Burdick MD, Polverini PJ (1995) Interferon gamma-inducible protein 10 (IP-10), a member of the C-X-C chemokine family, is an inhibitor of angiogenesis. Biochem Biophys Res Commun 210:51–57
105. Yoshimura A (2006) Signal transduction of inflammatory cytokines and tumor development. Cancer Sci 97:439–447
106. Hutchins AP, Poulain S, Miranda-Saavedra D (2012) Genome-wide analysis of STAT3 binding in vivo predicts effectors of the anti-inflammatory response in macrophages. Blood 119:e110–e119
107. Vallania F et al (2009) Genome-wide discovery of functional transcription factor binding sites by comparative genomics: the case of Stat3. Proc Natl Acad Sci USA 106:5117–5122
108. Ito S et al (1999) Interleukin-10 inhibits expression of both interferon alpha- and interferon gamma- induced genes by suppressing tyrosine phosphorylation of STAT1. Blood 93:1456–1463
109. Aoki Y, Feldman GM, Tosato G (2003) Inhibition of STAT3 signaling induces apoptosis and decreases survivin expression in primary effusion lymphoma. Blood 101:1535–1542
110. Epling-Burnette PK et al (2001) Inhibition of STAT3 signaling leads to apoptosis of leukemic large granular lymphocytes and decreased Mcl-1 expression. J Clin Invest 107:351–362
111. Sinibaldi D et al (2000) Induction of p21WAF1/CIP1 and cyclin D1 expression by the Src oncoprotein in mouse fibroblasts: role of activated STAT3 signaling. Oncogene 19: 5419–5427
112. Niu G et al (2002) Constitutive Stat3 activity up-regulates VEGF expression and tumor angiogenesis. Oncogene 21:2000–2008
113. Hung W, Elliott B (2001) Co-operative effect of c-Src tyrosine kinase and Stat3 in activation of hepatocyte growth factor expression in mammary carcinoma cells. J Biol Chem 276:12395–12403
114. Xie TX et al (2006) Activation of stat3 in human melanoma promotes brain metastasis. Cancer Res 66:3188–3196
115. Yoshimura A, Suzuki M, Sakaguchi R, Hanada T, Yasukawa H (2012) SOCS, inflammation, and autoimmunity. Front Immunol 3:20
116. Cooney RN (2002) Suppressors of cytokine signaling (SOCS): inhibitors of the JAK/STAT pathway. Shock 17:83–90
117. Wormald S, Hilton DJ (2004) Inhibitors of cytokine signal transduction. J Biol Chem 279:821–824
118. Xiong H et al (2012) Trichostatin A, a histone deacetylase inhibitor, suppresses JAK2/STAT3 signaling via inducing the promoter-associated histone acetylation of SOCS1 and SOCS3 in human colorectal cancer cells. Mol Carcinog 51:174–184
119. Brown R, Imran SA, Ur E, Wilkinson M (2008) Valproic acid and CEBPalpha-mediated regulation of adipokine gene expression in hypothalamic neurons and 3T3-L1 adipocytes. Neuroendocrinology 88:25–34
120. Hirose N, Maekawa T, Shinagawa T, Ishii S (2009) ATF-2 regulates lipopolysaccharide-induced transcription in macrophage cells. Biochem Biophys Res Commun 385:72–77
121. You M, Yu D-H, Feng G-S (1999) Shp-2 tyrosine phosphatase functions as a negative regulator of the interferon-stimulated Jak/STAT pathway. Mol Cell Biol 19:2416–2424
122. Shuai K (2006) Regulation of cytokine signaling pathways by PIAS proteins. Cell Res 16:196–202
123. Heinrich PC, Behrmann I, Muller-Newen G, Schaper F, Graeve L (1998) Interleukin-6-type cytokine signalling through the gp130/Jak/STAT pathway. Biochem J 334(Pt 2):297–314

124. Neel BG, Gu H, Pao L (2003) The 'Shp'ing news: SH2 domain-containing tyrosine phosphatases in cell signaling. Trends Biochem Sci 28:284–293
125. Wang H et al (2011) Histone deacetylase inhibitor LAQ824 augments inflammatory responses in macrophages through transcriptional regulation of IL-10. J Immunol 186:11
126. Leoni F et al (2002) The antitumor histone deacetylase inhibitor suberoylanilide hydroxamic acid exhibits antiinflammatory properties via suppression of cytokines. Proc Natl Acad Sci USA 99:2995–3000
127. Reddy P et al (2008) Histone deacetylase inhibition modulates indoleamine 2,3-dioxygenase-dependent DC functions and regulates experimental graft-versus-host disease in mice. J Clin Invest 118:2562–2573
128. Sun Y et al (2009) Cutting edge: negative regulation of dendritic cells through acetylation of the nonhistone protein STAT-3. J Immunol 182:5899–5903
129. Glauben R, Siegmund B (2011) Inhibition of histone deacetylases in inflammatory bowel diseases. Mol Med 17:426–433
130. Han SB, Lee JK (2009) Anti-inflammatory effect of Trichostatin-A on murine bone marrow-derived macrophages. Arch Pharm Res 32:613–624
131. Bode KA et al (2007) Histone deacetylase inhibitors decrease Toll-like receptor-mediated activation of proinflammatory gene expression by impairing transcription factor recruitment. Immunology 122:596–606
132. Wu C, Li A, Leng Y, Li Y, Kang J (2012) Histone deacetylase inhibition by sodium valproate regulates polarization of macrophage subsets. DNA Cell Biol 31:592–599
133. Jung ID et al (2009) Apicidin, the histone deacetylase inhibitor, suppresses Th1 polarization of murine bone marrow-derived dendritic cells. Int J Immunopathol Pharmacol 22:501–515
134. Wang B, Morinobu A, Horiuchi M, Liu J, Kumagai S (2008) Butyrate inhibits functional differentiation of human monocyte-derived dendritic cells. Cell Immunol 253:54–58
135. Mishra N, Reilly CM, Brown DR, Ruiz P, Gilkeson GS (2003) Histone deacetylase inhibitors modulate renal disease in the MRL-lpr/lpr mouse. J Clin Invest 111:539–552
136. Bosisio D et al (2008) Blocking TH17-polarizing cytokines by histone deacetylase inhibitors in vitro and in vivo. J Leukoc Biol 84:1540–1548
137. Salvi V et al (2010) Trichostatin A blocks type I interferon production by activated plasmacytoid dendritic cells. Immunobiology 215:756–761
138. Villagra A et al (2009) The histone deacetylase HDAC11 regulates the expression of interleukin 10 and immune tolerance. Nat Immunol 10:92–100
139. Wang H et al (2011) Histone deacetylase inhibitor LAQ824 augments inflammatory responses in macrophages through transcriptional regulation of IL-10. J Immunol 186:3986–3996
140. Kobayashi T et al (2012) IL-10 regulates Il12b expression via histone deacetylation: implications for intestinal macrophage homeostasis. J Immunol 189:1792–1799
141. Frikeche J et al (2012) Impact of valproic acid on dendritic cells function. Immunobiology 217:704–710
142. Zhou X, Hua X, Ding X, Bian Y, Wang X (2011) Trichostatin differentially regulates Th1 and Th2 responses and alleviates rheumatoid arthritis in mice. J Clin Immunol 31:395–405
143. Koenen HJ SR, Vink PM, van Rijssen E, Boots AM, Joosten I (2008) Human CD25highFoxp3pos regulatory T cells differentiate into IL-17-producing cells. Blood 112:2340–2352
144. Han S et al (2007) HDAC inhibitors TSA and sodium butyrate enhanced the human IL-5 expression by altering histone acetylation status at its promoter region. Immunol Lett 108:143–150
145. Su RC, Becker AB, Kozyrskyj AL, Hayglass KT (2008) Epigenetic regulation of established human type 1 versus type 2 cytokine responses. J Allergy Clin Immunol 121:57–63 e53
146. Edens RE, Dagtas S, Gilbert KM (2006) Histone deacetylase inhibitors induce antigen specific anergy in lymphocytes: a comparative study. Int Immunopharmacol 6:1673–1681
147. Rossi LE et al (2012) Histone deacetylase inhibitors impair NK cell viability and effector functions through inhibition of activation and receptor expression. J Leukoc Biol 91:321–331

148. Chang S, Collins PL, Aune TM (2008) T-bet dependent removal of Sin3A-histone deacetylase complexes at the Ifng locus drives Th1 differentiation. J Immunol 181:8372–8381
149. Lee CG et al (2012) Interaction of Ets-1 with HDAC1 represses IL-10 expression in Th1 cells. J Immunol 188:2244–2253
150. Gillespie J et al (2012) Histone deacetylases are dysregulated in rheumatoid arthritis and a novel histone deacetylase 3-selective inhibitor reduces interleukin-6 production by peripheral blood mononuclear cells from rheumatoid arthritis patients. Arthritis Rheum 64:418–422
151. Choo QY, Ho PC, Tanaka Y, Lin HS (2010) Histone deacetylase inhibitors MS-275 and SAHA induced growth arrest and suppressed lipopolysaccharide-stimulated NF-kappaB p65 nuclear accumulation in human rheumatoid arthritis synovial fibroblastic E11 cells. Rheumatology (Oxford) 49:1447–1460
152. Segain JP, Raingeard de la Blétière D, Bourreille A, Leray V, Gervois N, Rosales C, Ferrier L, Bonnet C, Blottière HM, Galmiche JP (2000) Butyrate inhibits inflammatory responses through NFkappaB inhibition: implications for Crohn's disease. Gut 47:397–403
153. Lewis EC et al (2011) The oral histone deacetylase inhibitor ITF2357 reduces cytokines and protects islet beta cells in vivo and in vitro. Mol Med 17:369–377
154. Mishra N, Brown DR, Olorenshaw IM, Kammer GM (2001) Trichostatin A reverses skewed expression of CD154, interleukin-10, and interferon-gamma gene and protein expression in lupus T cells. Proc Natl Acad Sci USA 98:2628–2633
155. Nusinzon I, Horvath CM (2006) Positive and negative regulation of the innate antiviral response and beta interferon gene expression by deacetylation. Mol Cell Biol 26:3106–3113
156. Song W et al (2011) HDAC inhibition by LBH589 affects the phenotype and function of human myeloid dendritic cells. Leukemia 25:161–168
157. Mishra N, Reilly CM, Brown DR, Ruiz P, Gilkeson GS (2003) Histone deacetylase inhibitors modulate renal disease in the MRL-lpr/lpr mouse. J Clin Invest 111:539–552
158. Lu J et al (2005) Interleukin-12 p40 promoter activity is regulated by the reversible acetylation mediated by HDAC1 and p300. Cytokine 31:46–51
159. Halili MA et al (2010) Differential effects of selective HDAC inhibitors on macrophage inflammatory responses to the Toll-like receptor 4 agonist LPS. J Leukoc Biol 87:1103–1114
160. Gillespie J et al (2012) Histone deacetylases are dysregulated in rheumatoid arthritis and a novel histone deacetylase 3-selective inhibitor reduces interleukin-6 production by peripheral blood mononuclear cells from rheumatoid arthritis patients. Arthritis Rheum 64:418–422
161. Choo QY, Ho PC, Tanaka Y, Lin HS (2010) Histone deacetylase inhibitors MS-275 and SAHA induced growth arrest and suppressed lipopolysaccharide-stimulated NF-kappaB p65 nuclear accumulation in human rheumatoid arthritis synovial fibroblastic E11 cells. Rheumatology (Oxford) 49:1447–1460
162. Chang S, Collins PL, Aune TM (2008) T-bet dependent removal of Sin3a-histone deacetylase complexes at the ifng locus drives Th1 differentiation. J Immunol 181:8372–8381
163. Nusinzon I, Horvath CM (2006) Positive and negative regulation of the innate antiviral response and beta interferon gene expression by deacetylation. Mol Cell Biol 26:3106–3113
164. Kim YH, Lee JK (2012) Suppression of T cell functions by hydroxamic acid-based histone deacetylase inhibitors. Arch Pharm Res 35:929–936

# Chapter 8
# Receptor Tyrosine Kinases in the Nucleus: Nuclear Functions and Therapeutic Implications in Cancers

**Longfei Huo, Jennifer L. Hsu, and Mien-Chie Hung**

**Abstract** Receptor tyrosine kinases (RTKs), originally considered as cell surface receptors responsible for transmitting external signals into the interior of cells through diverse signaling cascades, have now been demonstrated to also localize within cells to different compartments such as the nucleus and the mitochondria in addition to their cell surface localization. They also exhibit biological functions when located at the particular cellular compartments. There is increasing evidence indicating that nuclear RTKs, like their cell surface counterparts, are also involved directly in regulating gene expression and DNA repair and replication, leading to cell proliferation, survival, and insensitivity to therapeutic agents. Here, we discuss the advances of our knowledge of nuclear localization of RTKs and their novel functions in the nucleus and describe their implications in cancer therapy.

**Keywords** EGFR • FGFR • IGF1R • c-Met • VEGFR • Tyrosine kinase inhibitor • Drug- resistance • DNA repair • Gene transcription • rRNA biogenesis

---

L. Huo, Ph.D.
Department of Molecular & Cellular Oncology, The University of Texas
MD Anderson Cancer Center, Houston, USA

J.L. Hsu, Ph.D. • M.-C. Hung, Ph.D. (✉)
Department of Molecular and Cellular Oncology, The University of Texas
MD Anderson Cancer Center, Houston, USA

Center for Molecular Medicine and Graduate Institute of Cancer Biology,
China Medical University, Taichung, Taiwan

Department of Biotechnology, Asia University, Taichung, Taiwan
e-mail: mhung@mdanderson.org

R. Kumar (ed.), *Nuclear Signaling Pathways and Targeting Transcription in Cancer*,
Cancer Drug Discovery and Development, DOI 10.1007/978-1-4614-8039-6_8,

## Abbreviations

| | |
|---|---|
| ALL | Acute lymphocytic leukemia |
| BCRP | Breast cancer resistance protein |
| CBP | CREB binding protein |
| CCND1 | Cyclin D1 |
| ChIP | Chromatin immunoprecipitation |
| COX2 | Cyclooxygenase 2 |
| CTF | C-terminal fragment |
| DSB | Double-strand breaks |
| EGFR | Epidermal growth factor receptor |
| ER | Estrogen receptor |
| ERE | Estrogen response element |
| ERSE | Endoplasmic reticulum stress response element |
| FGFR | Fibroblast growth factor receptor |
| ICD | Intracellular domain |
| IGF1R | Insulin-like growth factor 1 receptor |
| IR | Insulin receptor |
| NRG1 | Neuregulin 1 |
| PCNA | Proliferating cell nuclear antigen |
| POMC | Proopiomelanocortin |
| RSK1 | Ribosomal S6 kinase 1 |
| STAT3 | Signal transducer and activator of transcription 3 |
| TACE | Tumor necrosis factor-alpha converting enzyme |
| TKI | Tyrosine kinase inhibitor |
| TS | Thymidylate synthase |
| PR | Progesterone receptor |
| VEGFR | Vascular endothelial growth factor receptor |
| BAMC | Bovine adrenal medullary chromaffin cells |
| CML | Chronic myeloid leukemia |
| HCMEC | Human cardiac microvascular endothelial cells |
| HDMEC | Human dermal microvascular endothelial cells |
| HNSCC | Head and neck squamous cell carcinoma |
| HPEC | Human pulmonary endothelial cells |
| HUVEC | Human umbilical vein endothelial cells |
| NSCLC | Non-small cell lung cancer |
| OSCC | Oral squamous cell carcinoma |
| SCLC | Small cell lung cancer |
| SRCC | Sarcomatoid renal cell carcinoma |

## 8.1 Introduction

Receptor tyrosine kinases (RTKs), the high-affinity cell surface receptors for polypeptide hormones, growth factors, and cytokines, are key regulators for important cellular processes, including cell growth, proliferation, differentiation, metabolism, migration, invasion, and survival [1–5]. The activation of RTKs is strictly regulated both temporally and spatially in normal cells, and their deregulation is usually associated with numerous human diseases, including cancer. During the last several decades, 58 RTKs have been identified and classified into 20 different subfamilies. All RTKs share similar molecular structures, including an extracellular region for ligand binding, a single hydrophobic transmembrane helix, and an intracellular region containing a tyrosine kinase domain [1].

RTKs are highly conserved during evolution, and many of them share similar activation mechanisms and common downstream key components. Generally, the binding of ligand to the extracellular domain of a given RTK induces receptor dimerization and activation through trans-autophosphorylation, which recruits cytoplasmic substrates to initiate such downstream signaling cascades as PI3K/Akt, Ras/MEK/ERK, PLCγ/PKC, and JAK/STATs [1, 6, 7]. Following ligand activation, cell membrane-associated RTKs are subjected to rapid endocytosis for degradation in the lysosomes, or recycling back to the cell membrane, or translocation to other organelles such as the nucleus. A number of studies from different laboratories have repeatedly demonstrated nuclear expression of RTKs in cells from either normal tissues or diseased specimens including cancer. In fact, 19 of these 58 RTKs have been detected in the nucleus (Table 8.1). Importantly, the nuclear expression levels of RTKs are associated with cancer progression and correlated with clinical outcome of patients with different cancer types (discussed in further detail below). Consistent with these observations, nuclear RTKs have been shown to play important roles in regulating gene transcription and DNA repair and replication, thereby contributing to tumor cell proliferation, survival, and therapeutic resistance (Fig. 8.1). These novel functions of RTKs in the nucleus are hereinafter referred to as "nuclear functions" to distinguish them from those canonical functions of membrane-associated RTKs. The mechanism of RTK trafficking from the cell surface to the nucleus has been recently reviewed [8–10] and therefore will not be discussed here. In this review, we focus on the nuclear functions of RTKs and their clinical implications.

## 8.2 Clinicopathological and Prognostic Significance of the Nuclear RTKs

Although the nuclear expression of RTKs in cancer cells has been frequently identified and reported in multiple human tumors (Table 8.1), the clinicopathological import and prognostic significance of nuclear RTKs have not been widely studied. Nevertheless,

**Table 8.1** RTKs in the nucleus of various tissues and/or cells

| RTK | Tissue/cell types observed with nuclear RTKs | References |
|---|---|---|
| **EGFR** | **Tumor tissues/cells** | [11–18, 41, 42, 45, 46, 49, 50, 55, 57, 60, 61, 63, 64, 123, 126, 165, 181, 184, 194–196, 249] |
| | Breast cancer, lung cancer, ovarian cancer, esophageal squamous cell carcinoma, prostate cancer, head and neck tumor, gastric cancer, T-cell lymphoblastoma, pancreatic cancer, colorectal carcinoma, bladder carcinoma, pituitary adenoma, nasopharyngeal carcinoma, glioma, choriocarcinoma | |
| | **Normal tissues/cells** | [46, 100, 128, 139, 194, 197] |
| | Keratinocytes, hepatocytes, thyroid tissue, placenta, mouse lung epithelial cells, umbilical endothelial cells | |
| **ErbB2** | **Tumor tissues/cells** | [19, 20, 48, 65, 66, 132, 133, 167, 169, 198–201] |
| | Breast cancer, prolactinoma, colon cancer, ovarian cancer | |
| | **Normal tissues/cells** | [100, 139, 202] |
| | Mouse lung epithelial cells, umbilical endothelial cells, rat Schwann cells | |
| **ErbB3** | **Tumor tissues/cells** | [17, 22, 23, 68, 203] |
| | Prostate cancer, non-small cell lung carcinoma | |
| | **Normal tissues/cells** | [67, 100, 139, 144, 202] |
| | Mouse lung epithelial cells, umbilical endothelial cells, rat Schwann cells, mammary epithelial cells | |
| **ErbB4** | **Tumor tissues/cells** | [17, 24, 25, 69, 72–74, 204, 205] |
| | Prostate cancer, breast cancer | |
| | **Normal tissues/cells** | [73, 79–81, 100, 134, 139] |
| | Mouse lung epithelial cells, umbilical endothelial cells, rat ventricular myocytes, embryonic neural precursor cells, fetal murine type II epithelial cells, hippocampal neurons, breast epithelium | |
| **FGFR1** | **Tumor tissues/cells** | [90, 206, 207] |
| | Medulloblastoma, neuroblastoma, breast cancer | |
| | **Normal tissues/cells** | [84–87] |
| | 3T3 cells, BAMCs, neuronal progenitor cells | |
| **FGFR2** | **Tumor tissues/cells** | [29, 30, 91, 208–211] |
| | NSCLC, breast cancer, endometrial carcinoma | |
| | **Normal tissues/cells** | [212–215] |
| | Sertoli cells, immortalized rat osteoblastic Py1a cells, stomach progenitor cells, mammary gland terminal end bud cells | |
| **FGFR3** | **Tumor tissues/cells** | [31, 32, 216–218] |
| | Breast cancer, prostate cancer, bladder cancer | |
| | **Normal tissues/cells** | [219] |
| | Choroid plexus | |
| **IR** | **Tumor tissues/cells** | [95, 220, 221] |
| | Liver cancer, colonic tumors, NSCLC | |
| | **Normal tissues/cells** | [94, 99, 137, 145, 222–224] |
| | Hepatocytes, osteoblast-like cells, corneal epithelial cells, adipocytes | |

(continued)

**Table 8.1** (continued)

| RTK | Tissue/cell types observed with nuclear RTKs | References |
|---|---|---|
| **IGF1R** | **Tumor tissues/cells** | [27, 28, 96, 171] |
| | Prostate cancer, breast cancer, soft tissue sarcomas, Ewing sarcoma, hepatocellular carcinoma | |
| | **Other diseases** | [225] |
| | Graves' disease | |
| | **Normal tissues/cells** | [99, 135] |
| | Corneal epithelial cells | |
| **VEGFR1/ VEGFR2/ VEGFR3** | **Tumor tissues/cells** | [33–35, 226–232] |
| | AML, ALL, CML, lymphoma, NSCLC, breast cancer, cervical carcinoma, colon cancer, glioma, ovarian granulosa cell tumor, myxoid liposarcoma, HNSCC | |
| | **Normal tissues/cells** | [232–238] |
| | HUVEC, HCMEC, HDMEC, HPEC, cardiac capillary endothelial cells, human thyroid follicular cells, vascular endothelial cells, rodent Brunner's gland epithelial cells | |
| **Met** | **Tumor tissues/cells** | [36, 37, 102–104, 239–242] |
| | Melanoma, uveal melanoma, NSCLC, SCLC, breast cancer, colon cancer, testicular cancer, lymphoma, OSCC | |
| | **Normal tissues/cells** | [36, 146, 243] |
| | Liver cells, melanocytes, colon, skin, testis, lymph node | |
| **TrkA** | **Tumor tissues/cells** | [38, 112, 113, 244, 245] |
| | Pheochromocytoma, glioma, melanoma, colorectal carcinoma | |
| | **Other diseases** | [246] |
| | Parkinson's disease | |
| | **Normal tissues/cells** | [114, 247, 248] |
| | Liver, mouse brain, blastocysts | |
| **Ryk** | **Tumor tissues/cells** | |
| | ? | |
| | **Normal tissues/cells** | [107, 109] |
| | Neural progenitor cells | |
| **Ron** | **Tumor tissues/cells** | [105] |
| | Bladder cancer cells | |
| | **Normal tissues/cells** | |
| | ? | |
| **Mer** | **Tumor tissues/cells** | [136] |
| | Leukemia | |
| | **Normal tissues/cells** | |
| | ? | |
| **Ror1** | Using overexpress system to demonstrate Ror1 in the cell nucleus | [110, 111] |
| c-**Kit** | **Tumor tissues/cells** | [250–252] |
| | Gastrointestinal stromal tumors, SRCC, extrahepatic bile duct carcinomas | |

? unknown

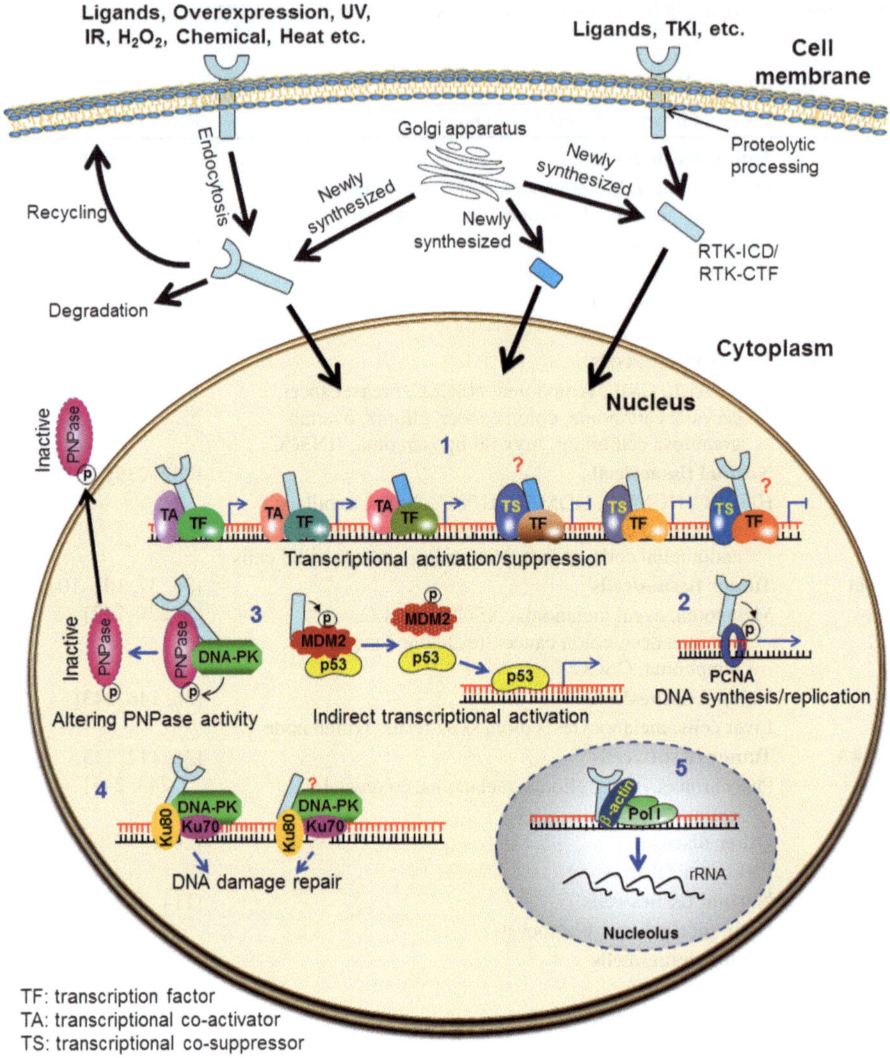


**Fig. 8.1** Diverse functions of nuclear RTKs. EGFR and its ErbB subfamily members are the most extensively studied receptors among 58 RTKs and therefore used to depict the functions of nuclear RTKs. Upon ligand stimulation or other stimulation, cell surface RTKs (including in-frame deletion variants such as EGFRvIII) can undergo endocytosis. RTKs, in addition to traditional degradation and recycling, can translocate into the nucleus either directly or after proteolytic cleavage first to release their intracellular domains (ICD) (e.g., ErbB4-ICD). In addition, newly synthesized RTKs (e.g., FGFR1) or RTK variants either from in-frame splicing (e.g., mLEEK, nucErbB3) or from alternative translation initiation (e.g., ErbB2-CTFs) can go to the nucleus directly without any stimulation. Once in the nucleus, nuclear RTKs or RTK variants can (*1*) associate with nuclear transcription-related proteins such as transcription factors and/or transcriptional co-activators or transcriptional co-suppressors to regulate gene transcription; (*2*) interact with and phosphorylate DNA replication-related protein(s) such as PCNA to modulate DNA synthesis; (*3*) interact with nuclear kinase/phosphorylase protein(s) such as MDM2 or PNPase to change the phosphorylation status of nuclear kinase/phosphorylase protein(s), leading to alteration of their stability/activity; (*4*) interact with DNA-PKs and other DNA repair complex proteins to regulate DNA damage repair; (*5*) associate with β-actin and Pol I in the nucleolus to stimulate rRNA biogenesis; The scale of the diagram does not reflect the relative sizes of different molecules or subcellular structures

existing studies have shed light on the importance of nuclear RTK expression as a prognostic indicator for clinical outcome of human cancers. Among these RTKs, EGFR subfamily receptors have been the most extensively studied to date.

The first report from an immunohistochemical analysis of EGFR in 130 breast carcinomas indicated that 37.7 % of tumor samples were positive for nuclear staining of EGFR, among which 6.9 % had high expression levels (by staining). Interestingly, the high expression level of nuclear EGFR rather than cytoplasmic EGFR in cancer cells was found to be significantly correlated with poor overall survival of cancer patients [11]. Consistently, another analysis of 113 breast carcinomas detected nuclear EGFR expression in 40 % of cases, and among these12 % had high nuclear EGFR staining [12]. In addition, nuclear EGFR correlated not only with tumor size, lymph node metastasis, and Nottingham prognostic index but also with shorter overall survival [12]. Similar to the findings in breast cancer, positive nuclear EGFR staining was also detected in 24.3 % of 37 cases of oral squamous cell carcinomas, and those with high nuclear EGFR expression showed a tendency toward poor survival [11]. Psyrri et al. analyzed 95 oropharyngeal cancer cases by quantitative immunohistochemistry and found 33 cases (35 %) with high nuclear EGFR staining and 34 with low nuclear EGFR expression. Patients with high nuclear EGFR expression also had a higher local recurrence rate and inferior 5-year disease-free survival than patients with low nuclear EGFR expression [13]. In patients with esophageal squamous cell carcinoma, nuclear expression of phosphorylated EGFR (i.e., the activated form) was identified in 19 (36.5 %) of 52 cases, which correlated significantly with TNM stage and lymph node metastasis as well as with poor prognosis of cancer patients [14]. The prognostic value of nuclear EGFR has also been corroborated in ovarian cancer. High nuclear expression of EGFR was found in 28.3 % of cases in a cohort of 221 ovarian cancer tissues, and a significant inverse correlation between high level of nuclear EGFR and overall survival was shown [15]. Recently, overexpression of nuclear EGFR was found to be significantly associated with advanced AJCC (American Joint Committee on Cancer) stages and vascular invasion in gallbladder carcinomas [16]. This study also showed that phosphorylated EGFR, which is predominantly overexpressed in those cases marked by both vascular and perineurial invasion, significantly correlated with adverse disease-specific survival [16]. Similar to wild-type EGFR, the study of EGFRvIII, a constitutively active mutant of EGFR, in a cohort of 74 cases from patients with matched hormone-sensitive and hormone-refractory prostate tumors revealed that nuclear EGFRvIII expression in hormone-refractory tumors was associated with quick death from biochemical relapse and therefore correlated with poor overall survival [17]. However, the prognostic value of nuclear EGFR reported in the cancers mentioned above was not found for colorectal cancers, in which nuclear EGFR expression did not associate with any clinicopathological characteristics tested, including tumor grade, tumor size, and tumor stage. However, a high percentage (57 %) of strong nuclear EGFR expression was identified and significantly related with CCND1, an important indicator for cell proliferation [18]. Taken together, these findings indicate that high expression of nuclear EGFR may be a useful clinical prognosticator for human cancers.

The clinical relevance of ErbB2 nuclear expression in human cancer was first noted in studies of breast cancer. Dillon et al. first reported that nuclear ErbB2 expression was associated with tumor size and tumor grade in a cohort study of 560 breast cancer patients. In addition, nuclear ErbB2 expression level was significantly correlated with poor 5-year disease-free survival [19]. This finding was confirmed by a recent study using a tissue microarray containing 273 primary invasive breast carcinomas in which nuclear expression of ErbB2 was detected in 33.6 % of tumor samples and identified not only as a significant independent predictor of worse overall survival in patients with ErbB2 membrane overexpression but also as a biomarker of lower overall survival in patients with tumors that overexpress membrane ErbB2 and lack steroid hormone receptors [20].

Although few reports are currently available on the clinical relevance of nuclear ErbB3 in human cancers, one study indicated that the expression of ErbB3 in prostate cancer cells depends on androgen status and bone microenvironment and is dynamically regulated during prostate cancer metastasis in in vivo xenograft models [21]. Moreover, a predominant nuclear staining of ErbB3 was identified in 50 % of 24 lymph node/bone metastatic specimens expressing ErbB3 in a total 45 human prostate cancers, implying a positive link between nuclear translocation of ErbB3 and prostate cancer metastasis. Another study also showed that nuclear ErbB3 expression is more frequently identified in prostate cancerous tissues than in normal or benign prostatic hyperplasia tissues, and the increased nuclear ErbB3 staining was associated with increasing Gleason grade of prostate cancers. Among prostate cancer tissues, nuclear expression of ErbB3 was more frequent in hormone-refractory (17/17, 100 %) than in hormone-sensitive (37/92, 40.2 %) tissues [22]. Interestingly, a 3-year biochemical-recurrence-free survival probability study from the same research group later showed that a low frequency of nuclear ErbB3 is associated with increasing biochemical recurrence in patients with prostate cancer and positive surgical margins after radical prostatectomy [23]. Therefore, the prognostic value of nuclear ErbB3 in human cancers seems to be much complex.

Similar to ErbB2, the clinicopathological significance of nuclear ErbB4 was investigated primarily in breast cancer. Both the full-length ErbB4 and proteolytically cleaved ErbB4 (ErbB4 intracellular domain; ErbB4-ICD) were found in the nucleus of cancer cells, and their expression in the nucleus was linked to poorer survival outcome [24, 25]. In contrast to nuclear ErbB4, the prognostic value of total ErbB4 in breast cancer is much more controversial [26], highlighting the significance of nuclear ErbB4 expression in clinical prognosis.

The prognostic value of nuclear IGF1R on disease outcome is still inconclusive among studies from different cancers and therefore needs further investigation. Nuclear expression of IGF1R was identified in 48 % (94/195) of clear cell renal cell carcinomas and associated with adverse prognosis [27]. However, a recent study in sarcomas indicated that the high percentage of nuclear IGF1R expression (75 % in 16 cases) was significantly correlated with better progression-free and overall survivals upon IGF1R antibody therapy [28].

In lung squamous cell carcinomas, nuclear expression of FGFR1 has a significant correlation with worse recurrence-free survival while nuclear expression of FGFR2is strongly related with worse recurrence-free and overall survival [29].

Similar to the findings in lung cancer, nuclear FGFR2 expression in breast cancer was also correlated with worse overall and disease-free survival [30]. In contrast to FGFR2, the prognostic value for nuclear FGFR3 expression seems to be controversial. During the development from normal urothelium to bladder cancer, FGFR3 translocates from the cytoplasm to the nucleus [31], suggesting its potential role in bladder carcinogenesis. However, a small cohort study of 55 primary bladder cancers via immunohistochemical staining identified nuclear expression of FGFR3 in 32.7 % of bladder tumors, and patients with tumor cells positive for nuclear FGFR3 expression had less tendency to have recurrence within 2 years after primary bladder cancer treatment [32].

High nuclear expression of phosphorylated VEGFR2 was reported in VEGFR2-positive ovarian granulosa cell tumors, and this high level nuclear expression was significantly correlated with the increased number of blood vessels in tumor tissues and inclined to a positive association with tumor recurrence ($p=0.006$) [33]. In addition, nuclear expression of phosphorylated VEGFR2 was also detected in 87.4 % of AML cases and 65.8 % of ALL cases and closely correlated with increased microvessel density [34]. However, the prognostic value of nuclear phosphorylated VEGFR2 in acute leukemia remains uninvestigated. In contrast to VEGFR2, the nuclear expression of VEGFR3 seems to be a predictor for better survival in patients with NSCLC [35].

The clinicopathological feature of nuclear c-Met is variable among different cancer types. For example, a high level of nuclear c-Met (a truncated form) expression was significantly correlated with poorer 5-year survival in a cohort study of 640 cases of invasive breast cancer [36]; however, it seemed to have a good prognosis in oral squamous cell carcinomas in a small cohort study of 55 cases [37].

Despite the reported discrepancies in the clinicopathological significance of nuclear RTKs, such as IGF1R, FGFR3, and c-Met, cancers with high nuclear expression of RTKs are inclined to have unfavorable prognosis. Given that multiple forms of RTKs might exist (e.g., full length vs. truncated form and wild type vs. constitutive activated mutant in the case of EGFR), it remains to be seen if they might contribute to these discrepancies. Therefore, to fully understand the clinical importance of nuclear RTKs in human cancers, specific antibody to distinguish different forms of RTKs in the nucleus and more studies with large cohorts of samples will be required. Elucidating the functional roles of RTKs in the nucleus will not only help us understand the clinicopathological significance but also explain the discrepancies in the prognostic value of nuclear RTK expression among different cancer types.

## 8.3 Functions of RTKs in the Nucleus

In the 1980s, several growth factors and their receptors were detected in the nucleus and found to associate with chromatin [38–43]. For example, nerve growth factor (NGF) receptor was shown to associate with *Eco*RI-digested chromatin [39]. Another study demonstrated that $^{125}$I-EGF was taken up by cells into the nucleus to bind to chromatin [43]. Interestingly, the uptake and subsequent binding of $^{125}$I-EGF to chromatin are receptor dependent and correlated with the expression level of EGFR, indicating the possibility of EGFR localization in the nucleus. This idea was supported by a later study that specific mAb targeting EGFR, like EGF ligand, is able to bind a chromatin-associated receptor when taken up by cells and localized in the nucleus [41], suggesting EGFR was associated with chromatin in the nucleus. Soon afterwards, EGFR was confirmed in the nucleus of hepatocytes by both immunoelectron microscopy and nuclear isolation followed by immunoprecipitation [44], and EGF/EGFR complex was detected in the nucleus through $^{125}$I-EGF crosslinking to EGFR [45]. Moreover, nuclear EGFR was demonstrated to functionally induce some nuclear proteins' phosphorylation in response to EGF [46]. In 1994, ErbB2, another EGFR family receptor, was unexpectedly found to have transcriptional activity and be able to translocate into the nucleus [47]. Together, these studies provide a rationale for the phenomena of growth factor receptors' association with chromatin in the nucleus and open a new direction for possible roles of RTK in the nucleus. The functions of nuclear RTKs, however, remained largely forgotten until 2001 when the first genomic target of nuclear EGFR was identified and the mechanism of how EGFR transactivates CCND1 gene expression elucidated [45]. Since then, more and more RTKs have been shown to have functions in the nucleus (Table 8.2). Indeed, investigation of the nuclear functions of RTKs has gradually received more attention than ever before and has become an increasingly attractive and important research area in human cancer.

## *8.3.1 Regulation of Gene Transcription*

### 8.3.1.1 Transcriptional Activation by ErbB Family Receptors

The neu-encoded p185 protein, a rat homologue of human HER2/ErbB2, was the first RTK shown to have nuclear function, namely transcriptional activity, in 1994 [47]. However, the first physiological relevant transcriptional target (COX2) of HER2/ErbB2/neu was not identified until a decade later [48]. EGFR was the first RTK found to have nuclear function in cancer cells [45], and its nuclear function is the most well-characterized among all nuclear RTKs. Similar to the rat neu-encoded p185, the C-terminus of EGFR contains transactivation function, and EGFR binds to an AT-rich consensus sequence (i.e., ATRS) of CCND1 gene promoter to stimulate its transcription [45, 49–51]. CCND1 is a key regulator for cell-cycle progression, and its overexpression is usually related to tumorigenesis. The pathological significance of nuclear EGFR-stimulated CCND1 gene expression was further validated by a positive correlation between nuclear EGFR expression and CCND1 protein level in several cancer types [11, 12, 15, 16, 18]. EGFR has been identified as a DNA-binding protein through an unbiased protein-DNA interactome analysis [52]. However, because EGFR does not contain any known DNA-binding domains, it is plausible that nuclear EGFR binds the specific dsDNA through other nuclear protein(s) that have DNA-binding domains. This

**Table 8.2** Diverse functions of RTKs and their associated proteins in the nucleus

| RTK | Detected form in the nucleus | Possible nuclear functions | Target genes/proteins | Nuclear-associated proteins | References |
|---|---|---|---|---|---|
| **EGFR/ErbB1** | Intact EGFR | Transcriptional regulation | CCND1↑ | RHA | [45, 53] |
| | | | iNOS↑, STAT1↑, COX2↑, c-Myc↑ | STAT3 | [54–57] |
| | | | b-Myb↑ | E2F1 | [58] |
| | | | Aurora A↑ | STAT5 | [59] |
| | | | BCRP↑ | | [62] |
| | | | TS↑ | | [63] |
| | | DNA repair | | DNA-PK, rH2AX | [119, 122, 125, 177] |
| | | DNA replication | | PCNA | [115] |
| | | Protein activity regulation | | DNA-PK, PNPase | [161] |
| | EGFR VIII | Transcriptional regulation | STAT1↑, COX2↑ | STAT3 | [55, 56] |
| | | ? | ? | STAT3 | [196] |
| | mLEEK | Transcriptional regulation | GRP78↑, UPR↑ | ? | [64] |
| | EGFR-ICD | ? | ? | ? | [253] |
| **ErbB2/HER2** | Intact ErbB2 | Transcriptional regulation | COX2 ↑, PRPK↑, FANCC↑, | ? | [48] |
| | | | CCND1↑ | PR, STAT3 | [65] |
| | | | TS↑ | | [63] |
| | | | rRNA↑ | β-actin, RNA Pol I | [140] |
| | | | ErbB2/HER2↑ | mccroH2A1.2 | [66] |
| | | DNA repair | | ? | [133] |
| | ErbB2-CTF/P95L | Therapeutic resistance to TKI | | | [167, 169] |
| **ErbB3/HER3** | Intact ErbB3 | ? | ? | ? | [21–23, 144, 202] |
| | nucErbB3 | Transcriptional regulation | Ezrin↑, HMGB1↓ | ? | [67] |
| | ErbB3-80 kDa | Transcriptional regulation | CCND1↑ | ? | [68] |

(continued)

**Table 8.2** (continued)

| RTK | Detected form in the nucleus | Possible nuclear functions | Target genes/proteins | Nuclear-associated proteins | References |
|---|---|---|---|---|---|
| **ErbB4/HER4** | ErbB4-ICD | Transcriptional regulation | | YAP | [71] |
| | | | β-casein↑, Sftpb↑ | STAT5A | [78, 79] |
| | | | ERE-element gene↑, PgR↑ | ER | [24, 73] |
| | | | P53 target gene↑ | Hdm2 | [76] |
| | | | Eto 2-mediated gene transcription↓ | Eto-2 | [75] |
| | | | GFAP↓, S100β↓ | Tab2 and N-CoR | [81] |
| | | | HIF-1α-mediated gene transcription↑ | HIF-1α | [74] |
| | | | Others | | [80] |
| | Intact ErbB4 | DNA damage response | | ? | [134] |
| **FGFR1** | Intact FGFR1 | Transcriptional regulation | C-Jun↑, CCND1↑ | ? | [85] |
| | | | FGF2↑ | ? | [84] |
| | | | NF-L↑, NSE↑ | | [86] |
| | | | TH↑ | RSK1 and CBP | [87, 206] |
| | | | TH↑, FGF2↑, FGFR1↑ | Nurr1, Nur77, RXR, RAR | [88, 89] |
| | Intracellular C-terminal FGFR1 | Transcriptional regulation | KRTAP5-6↑, PRSS27↑, SFN↑. EBI3↓, GRINA↓ | ? | [90] |
| **FGFR2** | Intact FGFR2 | Transcriptional regulation | Myc↑ | PR and STAT5 | [91] |
| **FGFR3** | Intact FGFR3 | ? | ? | ? | [31, 32, 217, 218] |
| | FGFR3 intracellular domain | ? | ? | ? | [254] |
| **IR** | Intact IR | Transcriptional regulation | GCK↑, EGR-1↑ | IR signaling complex | [95] |
| | | | Malic enzyme↑ | 30 kDa transcription factor | [94] |
| | | | IGF-1R↓ | ? | [96] |

| | | | | | |
|---|---|---|---|---|---|
| **IGF1R** | Intact IGF1R | Transcriptional regulation | IGF-1R↑ | ? | [96] |
| | | | CCND1↑, Axin2↑ | LEF1 | [98] |
| | | | Binding to enhancer regions of chromatin to activate transcription | ? | [97] |
| **IR/IGF1R hybrid** | Intact forms | Transcriptional regulation | Affecting a clustering genes by genome-wide ChIP sequence | ? | [99] |
| **Met** | Intact Met | ? | ? | PAX5 | [102] |
| | Met C-terminal fragment (60 kDa) | Transcriptional regulation | Transactivation of Gal4Luc | YAP | [103] |
| | | | | β-catenin | [104] |
| **Ron** | Intact Ron | Transcriptional regulation | c-Jun↑, FLJ46072↑, SSTR1↑ | nEGFR | [105] |
| **VEGFR1** | Intact VEGFR1 | ? | ? | ? | [231, 232, 255] |
| | | ICD acting as transcriptional activator in yeast two-hybrid system | ? | ? | [256] |
| **VEGFR2** | Intact VEGFR2 | Transcriptional regulation | VEGFR2↑ | SP1 | [106] |
| **VEGFR3** | ? | ? | ? | ? | [35, 228, 236] |
| **Ror1** | Intact Ror1 | Transcriptional regulation | RDX↑, EZR↑, SOS2↑, CALD1↑ | ? | [111] |
| **Ryk** | Intracellular domain | Transcriptional regulation | Dlx2↑, Olig2↓ | ? | [109] |
| **Trk A** | Intact TrkA | ? | ? | ERβ | [114] |
| **Mer** | Intact Mer | Chromatin bound with unknown function | ? | ? | [136] |

? unknown; *nEGFR* nulcear EGFR; *up arrow* up-regulation; *down arrow* down-regulation hypothesis is consistent with mounting evidence showing that most RTKs including EGFR mediate gene transcription in the nucleus through their associated nuclear partners such as transcription factors/cofactors (see Table 8.2). For example, transactivation of CCND1 gene expression by nuclear EGFR has been demonstrated through its association with RHA, a transcription cofactor with AT-rich dsDNA binding ability [53]. Moreover, nuclear EGFR also interacts with STAT3, E2F1, and STAT5 to activate the transcription of genes such as iNOS [54], COX2 [55], STAT1 [56], c-Myc [57], b-Myb [58], and Aurora A [59] in different cancer cells. Nuclear EGFR may be also involved in stimulating POMC transcription in pituitary adenomas [60]. Consistent with the roles of these genes in regulating tumor cell growth, survival, and invasiveness, nuclear EGFR is linked to cancer cell proliferation in vitro [49, 50] and tumor progression in vivo [60, 61]. Moreover, nuclear EGFR has been recently demonstrated to contribute to tumor cell drug resistance by activating the expression of genes such as BCRP [62] and TS [63], further substantiating the importance of EGFR nuclear translocation in cancer progression. However, it is not yet clear whether EGFR mediates BCRP and TS gene expression through its interaction with the nuclear factors mentioned above or with other unknown factors. In addition to wild-type EGFR, the constitutively active variant of EGFR, EGFRvIII, can transactivate STAT1 and COX2 genes in cancer cells by interacting with STAT3 in the nucleus [55, 56]. Interestingly, mLEEK, a newly identified in-frame splicing variant of EGFR that lacks extracytoplasmic, transmembrane, and tyrosine kinase domains, has been reported to also translocate into the nucleus and act as a transcriptional activator to bind ERSE DNA motif and regulate the expression of endoplasmic reticulum stress-related genes such as GRP78 in cancer cells [64].

As previously mentioned, COX2 was the first characterized genomic target of nuclear ErbB2. ErbB2 binds to the HER2/ErbB2-associated sequence (HAS) of COX2 gene promoter to activate its transcription [48]. Supporting this mechanism, a strong positive correlation between nuclear expression of ErbB2 and COX2 protein levels was observed in human primary tumors [19, 48]. Both nuclear ErbB2 and COX2 were later shown to be predictors for poor disease-free survival in breast cancer patients on endocrine treatment [19]. Upon progestin stimulation, PR induces nuclear translocation of ErbB2 to promote the assembly of nuclear STAT3/ErbB2/PR transcriptional complex in which ErbB2 acts as a coactivator of STAT3 [65]. This complex then binds to the CCND1 promoter through STAT3 to stimulate CCND1 expression and therefore increases breast cancer cell growth [65]. In addition, ErbB2 increases its own gene transcription in the nucleus by specifically interacting with atypical histone variant mH2A1.2 to enhance the binding of mH2A1.2 to ErbB2 promoter, thereby facilitating ErbB2 mRNA expression [66]. Moreover, this feedback stimulation of ErbB2 transcription by nuclear ErbB2 itself is dependent on its own kinase activity. This finding is clinically supported by a high correlation between the mRNA expression of mH2A1.2 and ErbB2 as observed in both breast cancers and ovarian cancers through human tumor microarray analysis [66]. However, the transcription factor(s) involved in this process remains unknown.

Although the role of full-length ErbB3 in regulating gene transcription in the nucleus has not yet been reported, two ErbB3 variants have recently been discovered to function in gene expression in the nucleus [67, 68]. An alternative transcript of ErbB3, corresponding to the intracellular domain of full-length ErbB3 with a molecular weight of 80 kDa (ErbB3-80 kDa), was identified in cancer cells and found to mainly localize in the nucleus where it binds to CCND1 promoter to activate CCND1 expression in a heregulin-independent manner [68]. Another nuclear ErbB3 variant with a molecular weight of 50 kDa (nucErbB3), which is generated by an alternative transcription initiation containing of exons 23–28 of full-length ErbB3 but with an independent 5'-UTR, was recently identified in rat Schwann cells as a transcriptional cofactor in the nucleus. Through ChIP-chip and microarray assays, nucErbB3 was shown to regulate the expression of 63 genes, including Ezrin and HMGB1 [67]. However, whether these nuclear variants (nucErbB3 or ErbB3-80 kDa) mediate gene transcription directly by binding nuclear ErbB3 to DNA or indirectly by interacting with other nuclear cofactor(s) is still unclear. In addition, the functional role of nucErbB3 in human cancers remains to be elucidated.

It is generally accepted that, after a two-step successive proteolytic processing by TACE and γ-secretase, an 80-kDa intracellular domain of ErbB4 (ErbB4-ICD) translocates into the nucleus and executes its nuclear functions to regulate cell proliferation, differentiation, and apoptosis [26, 69, 70]. The association of ErbB4-ICD with transcriptional coactivator YAP enhances the ability of nuclear ErbB4 in gene transactivation [71]. Upon E2 treatment, ErbB4-ICD increases the expression of genes containing ERE half-site such as PR, SDF-1, and ErbB4 itself through its interaction with ERα to stimulate a growth-promoting signal in breast cancer cells [72]. This finding was supported by the observation of a significant correlation between PR and nuclear ErbB4-ICD in human $ER^+$ invasive breast ductal carcinomas [73], highlighting the oncogenic effect of nuclear ErbB4-ICD in ER-mediated breast cancer development. Moreover, nuclear ErbB4-ICD might also promote tumor progression through its interaction with and stabilization of HIF1α, therefore leading to increasing HIF1α target gene expression in breast cancer cells [74]. ErbB4-ICD also associates with the transcriptional corepressor ETO2 (a tumor suppressor) in the nucleus and blocks its transcriptional suppression function independently of ErbB4-ICD kinase activity [75]. Therefore, ErbB4-ICD may contribute to tumor progression through its inhibiting tumor suppressor ETO2-mediated gene suppression. In contrast to its function in promoting tumors, nuclear ErbB4 has also been shown to activate the genes involved in cell differentiation and apoptosis, which may lead to tumor growth suppression. For example, heregulin induces the association of ErbB4-ICD with Hdm2 to stimulate Hdm2 tyrosine phoshporylation and ubiquitination and consequently stabilizes p53 and increases expression of its target gene p21 [76]. Nuclear ErbB4-ICD, through its interaction with and activation of STAT5A, was found to increase the expression of β-casein and surfactant protein B (Sftpb) in normal epithelial cells, leading to cell differentiation and maturation [77–79]. In addition, a recent study found that the expression of 47 genes, including PFN1, NEDD4, and DCTN4, in hippocampal neuronal cultures was upregulated by ErbB4-ICD upon NRG1 stimulation [80]. Given that most of these genes are involved in both neuronal function and schizophrenia, it is reasonable to believe that nuclear ErbB4-ICD may contribute to normal and schizophrenic neuronal function. On the basis of these results, it has become increasingly clear that nuclear ErbB4-ICD-mediated gene expression is complex and may be cell type- and cell context-dependent.

#### 8.3.1.2 Transcriptional Inhibition by ErbB Family Receptors

In addition to their transactivation function as mentioned above, ErbB family receptors also play an important role in suppressing gene expression through association with nuclear co-repressors. nucErbB3 has been shown to regulate the transcription of both Ezrin and HMGB1 genes in rat Schwann cells. Interestingly, nucErbB3 stimulates Ezrin promoter activity while suppressing HMGB1 transcription [67]. Although the mechanisms for the different regulation effects of nucErbB3 remain uninvestigated, it is conceivable that nucErbB3 may interact with different nuclear factors to execute its different functions in gene transcription. ErbB4-ICD, on the other hand, was shown to associate with N-CoR through TAB2 to form ErbB4-ICD/TAB2/N-CoR transcriptional suppressor complex, which moves into the nucleus to repress the expression of genes (e.g., GFAP and S100β) to prevent differentiation of neuronal progenitor cells [81].

#### 8.3.1.3 Transcriptional Functions of Other RTKs

FGFR Subfamily

Like EGFR, full-length FGFR1 can enter into the nucleus and bind to the nuclear matrix [82]. Nuclear FGFR1 also localizes to the nuclear speckles, which are associated with the sites of RNA synthesis and processing. These data indicate that nuclear FGFR1 may act as a direct and global transcriptional regulator [83]. The first target gene of nuclear FGFR1 is FGF2. In bovine adrenal medullary chromaffin cells (BAMCs), angiotensin II (AII), a peptide hormone, stimulates its membrane receptors to induce tyrosine phosphorylation of nuclear FGFR1, which then binds to and transactivates the FGF2 promoter through an AII-response element at the FGF2 promoter [84]. Nuclear FGFR1 has also been shown to increase the transcription of c-Jun and CCND1, suggesting a possible role for FGFR1 in regulating cell proliferation [85]. The expression of neurofilament-L (NF-L) and neuron specific enolase (NSE) genes can also be activated by nuclear FGFR1 during cAMP-induced neuronal differentiation, indicating that nuclear FGFR1 may also have a role in cell differentiation [86]. In 2005, nuclear FGFR1 was shown to upregulate tyrosine hydroxylase (TH) and FGF2 expression through its interaction with CPB and RSK1 to activate CPB by releasing CPB C-terminal domain from RSK1 inhibition. Interestingly, the function of nuclear FGFR1 in activating TH and FGF2 expression that results in cell differentiation is independent of its kinase activity [87]. Recently, nuclear FGFR1 was further shown to associate with nuclear receptor transcription factors (such as RXR, RAR, Nur77, and Nurr1) to stimulate TH, FGF2, and its own gene expression in embryonic stem cells (ESCs) to induce cell differentiation [88, 89].

A strong link between nuclear FGFR1 and human cancers came from a recent study that showed that granzyme B-cleaved FGFR1 (molecular weight 55 kDa, with tyrosine kinase domain and C-terminus of FGFR1) after activation by its ligand translocates to the nucleus to regulate the expression of several genes responsible for cell migration in breast cancer cells [90]. For example, genes (KRTAP5-6, PRSS27, SFN, EBI3, GRINA) regulated by the 55-kDa nuclear FGFR1 were shown to be involved in cell migration. Knockdown of KRTAP5-6, PRSS27, and SFN, which are upregulated by FGFR1, decreased tumor cell migration whereas knockdown of EBI3 and GRINA, which are downregulated by FGFR1, increased tumor cell migration [90]. Importantly, the presence of nuclear FGFR1 in the invading cells in both human tumor tissues and a three-dimensional culture model of breast cancer [90] highlights the clinical importance of this nuclear FGFR1 in tumor progression.

Similar to FGFR1, FGFR2 also localizes in the nucleus to transactivate gene expression in breast cancer cells [91]. Treatment of cells with medroxyprogesterone acetate (MPA) and FGF2 induces FGFR2 translocation into the nucleus to form a complex with PR and STAT5, and the PR/FGFR2/STAT5 complex then activates the progesterone response element (PRE)- and gamma activated site (GAS)-containing reporter genes. Furthermore, this complex was found to increase the expression of Myc gene through binding to the PRE sequence of the Myc promoter. Forced expression of constitutively active FGFR2 increased tumor cell proliferation in vitro and tumor growth in vivo (mouse model) [91]. Colocalization of FGFR2, PR, and STAT5 was confirmed in human tumor tissues [91]. Taken together, these findings suggest that nuclear translocation of FGFR2 contributes to breast cancer progression.

### IR Subfamily

As early as 1995, nuclear IR was proposed to function as a gene expression regulator based on its structure analysis, which showed that the insulin receptor α-subunit contains several zinc finger-like motifs and an RGG box [92]; therefore, a model of an insulin/IR/RB complex to regulate gene expression was hypothesized [93]. This hypothesis has not been yet validated, but a later study showed that nuclear IR induces malic enzyme gene expression through regulating the phosphorylation of insulin response element (IRE) transcription factors in hepatocytes from mice treated with glucose meal [94]. More recently, IR and its associated signaling complex, SHC/Grb2/SOS, were found to bind to the same insulin-inducible gene loci (e.g., EGR-1 gene locus) to increase transcription upon insulin stimulation [95], indicating that the gene regulation mediated by nuclear IR is much more complex than previously expected. In addition, IR was found to bind to the IGF1R promoter to downregulate IGF1R expression in ER-depleted C4.12.5 cells (derived from MCF7) due to a possible competition between IR and limited ER for Sp1 transcription factor [96]. In contrast, IGF1R binds to its own gene promoter to increase its own gene expression in such ER-depleted C4.12.5 cells, suggesting an autoregulation loop in ER-negative breast cancer cells [96]. Using electrophoretic mobility shift assays (EMSAs) with randomly synthesized dsDNA probes and ChIP-based cloning strategy, Larsson's group found that nuclear IGF1R binds mostly to the enhancer regions of chromatin to regulate gene transcription [97]. Moreover, a recent study demonstrated that nuclear IGF1R associates with β-catenin and transcription factor LEF1 to increase the promoter activity of LEF1 target genes, such as CCND1 and axin2, in several human cancer cells [98], further supporting the important roles of nuclear IGF1R in cancer biology. Interestingly, another study detected a hybrid receptor of IGF1R and IR (IGF1R/IR hybrid-R) in the nucleus of human telomerized corneal epithelial (hTCEpi) cells. Genome-wide ChIP-sequencing assay with antibodies against IGF1R and IR identified 52 target genes for IGF1R and 31 genes for IR; of these 52, 11 genes are involved in cell proliferation/cell cycle control and six are related to cell death/apoptosis [99]. This study suggested that IGF1R or IR translocates into the nucleus as an IGF1R/IR heterotetrameric complex that, once in the nucleus, could co-regulate corneal epithelial homeostatic pathways. One interesting question to be addressed is what the different functions are between their nuclear homodimer receptors and heterotetrameric complex. In addition, it is worth noting that heterodimers among ErbB family receptors are common phenomena and have been detected in the nucleus [100]. It would be of interest to further investigate whether these heterodimer receptors of ErbB family proteins have a similar transactivation function in the nucleus as that of IGF1R/IR.

### Met Subfamily

PAX5 is a nuclear transcription factor that is believed to promote cancer cell survival and metastasis and which is highly expressed in SCLC cells but not in NSCLC cells [101]. In addition, PAX5 was found to be the main regulator of c-Met (full-length) expression in SCLC cells [102]. Interestingly, phosphorylated full-length c-Met was shown to interact and colocalize with PAX5 in the nucleus of HGF-treated SCLC cells [102]. However, the functions of phosphorylated full-length c-Met and its association with PAX5 in the nucleus of SCLC cells has yet to be determined. In the highly aggressive MDA-MB-231 breast cancer cell line, a C-terminal fragment of c-Met (Met-CTF) rather than the intact c-Met is the major form of c-Met identified in the nucleus and is constitutively activated independently of HGF. In addition, Met-CTF was shown to have transcriptional activity when the C-terminal Met fragment containing the juxtamenbrane domain was fused with GAL4-DNA binding domain in a Gal4Luc assay, and its gene transactivity was further enhanced by YAP [103]. Furthermore, this Met-CTF was also found to interact with β-catenin in the nucleus of aggressive breast cancer cells [104], suggesting the possibility that Met-CTF promotes β-catenin/Wnt signaling-mediated gene transcription and tumor progression.

Ron, another member of the c-Met subfamily of receptors, translocates from the cell surface into the nucleus of bladder cancer cells upon physiological stress, e.g., serum starvation or oxidative stress [105]. Interestingly, nuclear Ron, in a complex with EGFR, functions as a transcriptional regulator in response to acute disturbances such as serum starvation. ChIP-chip assay indicated that at least 134 genes are potentially targeted by the Ron/EGFR nuclear complex, and motif scanning of the promoters from ChIP-sequence analysis identified a consensus sequence of GCA(G)GGGGCAGCG as the target binding sequence of nuclear Ron upon serum starvation in bladder cancer cells [105]. Moreover, several genes, including c-Jun, FLJ46072, and SSTR1, have been confirmed to contain the consensus binding sequence of Ron and validated as the real targets of nuclear Ron/EGFR complex by ChIP-qPCR [105].

#### VEGFR Subfamily

Among the members of VEGFR subfamily, only VEGFR2 has been shown to have transactivation function to date. Using immunoprecipitation combined with mass spectrometry, Domingues et al. identified Sp1 as one of the nuclear partners of VEGFR2. The VEGFR2/Sp1 complex binds to Sp1 binding sites in the VEGFR2 promoter to upregulate its expression [106], suggesting a role of VEGFR2 in positive feedback regulation of its own expression.

#### Ryk Subfamily

Cleavage at the intracellular domain of Ryk by γ-secretase releases its intracellular domain (Ryk-ICD). Ryk-ICD then translocates into the nucleus to function as a nuclear transcriptional cofactor to regulate Wnt target gene expression [107, 108]. In particular, nuclear Ryk-ICD has been shown to promote GABAergic neuron formation and inhibit oligodendrogenesis by transcriptionally regulating key cell-fate determinant genes such as Dlx2 and Oligo2 during ventral embryonic brain development [109].

#### Ror Subfamily

Ror1, an orphan receptor tyrosine kinase, was recently demonstrated to enter the nucleus of gastric cancer cells transfected with Ror1 plasmid [110]. Nuclear Ror1 could activate several genes involved in regulating actin cytoskeleton, e.g., radixin (RDX), ezrin (EZR), son of sevenless homologue 2 (SOS2), and caldesmon 1 (CALD1) [111]. These results suggest a potential role of nuclear Ror1 in regulating cell migration and cytoskeleton remodeling.

Trk Subfamily

Although the NGF receptor (TrkA) was detected in the nucleus by the early 1980s [38, 39, 112], its genomic functions remain largely unclear. In fact, among the members of the Trk subfamily of RTKs, only TrkA is frequently observed in the nucleus. Nuclear TrkA likely has a role in regulating cell mitosis, because phosphorylated TrkA was found to colocalize with α-tubulin at the mitotic spindle in the glioma cell line U251 upon NGF treatment [113]. In addition, TrkA was reported to interact with the ligand-binding domain of ERβ, and their interaction changed with age in a mouse model, indicating that TrkA may be involved in ERβ-mediated gene expression during aging [114].

### *8.3.2 Regulation of DNA Synthesis and Replication*

Nuclear EGFR has been detected in highly proliferative cells including regenerating liver, uterus of pregnant mouse, and tumor tissues [45, 115], suggesting a role of nuclear EGFR in DNA synthesis and replication. PCNA is an important cofactor of replicative DNA polymerases and an essential component of eukaryotic chromosomal DNA replisome and is involved in DNA replication, DNA repair, and cell cycle control [116, 117]. A study from Wang et al. indicated that nuclear EGFR interacts with and phosphorylates PCNA at the Tyr-211 residue, leading to its protein stability and activity enhancement, which is correlated with an increase of DNA synthesis and tumor cell proliferation [115]. A positive correlation among nuclear EGFR, PCNA, and phospho-Y211 PCNA was observed in a cohort of primary breast tumor samples. In addition, phospho-Y211 PCNA expression was also significantly correlated with poor survival of breast cancer patients [115]. These results are supported by a study in oropharyngeal squamous cell carcinoma in which the expression of PCNA was the only one that correlated positively with nuclear EGFR among the pathological factors tested in anoropharyngeal tumor tissue microarray [118].

### *8.3.3 Regulation of DNA Repair*

DNA-dependent protein kinase catalytic subunit (DNA-PK), a serine/threonine protein kinase, plays a crucial role in regulating the non-homologous end joining (NHEJ) DNA repair process that is activated by DSB. Several lines of evidence suggest an important role of nuclear EGFR in DSB repair. The first clue of the role of nuclear EGFR in DSB-induced NHEJ DNA repair came from the physical interaction between EGFR and DNA-PK in the nucleus, with a consequent increase of DNA-PK activity [119]. Nuclear translocation of EGFR has been linked to an increase of radiation-induced Thr2609 phosphorylation of DNA-PK [119–121], which is essential for the activity and function of DNA-PK in DSB repair and radioresistance. Furthermore, nuclear EGFR also co-localizes with γ-H2AX loci and associates with γ-H2AX complex upon radiation treatment [122]. In agreement with these findings, NSCLC cells with EGFR somatic mutations (including L858R and ΔE476-E750) which are defective in radiation-induced nuclear translocation failed to activate DNA-PK, and NSCLC cells with expression of these mutant EGFRs were more sensitive to radiation treatment than those with wild-type EGFR [123]. Consistently, EGFR with NLS mutation, which is unable to translocate to the nucleus, has reduced association with DNA-PK, which consequently decreases DNA-PK activity and DSB repair in response to radiation treatment [124]. On the other hand, EGFRvIII (a somatic EGFR mutant without the loss of nuclear translocation ability) can be detected in the nucleus and is able to activate DNA-PK to increase DSB repair [124]. Additionally, nuclear EGFR has been shown to promote DSB repair induced by DNA-alkylating agents (including cisplatin) through its interaction with DNA-PK [124–126].

In addition to irradiation, UV and $H_2O_2$ are also known to induce DNA damage, and both have been reported to induce EGFR nuclear translocation [127, 128], implying that nuclear EGFR may be also involved in the repair process of other types of DNA damage (e.g., single-strand damage) besides DSB and be a part of the cellular defense system to promote cell survival from lethal stress environments [129, 130]. Moreover, nuclear EGFR has been demonstrated to regulate DNA mismatch repair through activating PCNA [115] that is important for both DNA replication [117] and nucleotide excision repair [131].

ErbB2 is translocated into the nucleus of breast cancer cells upon irradiation and cisplatin treatment, and blockage of ErbB2 by trastuzumab (Herceptin) reduces nuclear translocation of ErbB2, which is accompanied by a decreased cell survival population [132]. In line with this finding, wild-type ErbB2, when translocated to the nucleus, increased DNA repair induced by cisplatin treatment in breast cancer cells, whereas ErbB2-ΔNLS mutant with loss of nuclear translocation ability had no effect on DNA repair. Furthermore, trastuzumab treatment blocked the nuclear ErbB2-stimulated cisplatin-induced DNA damage repair [133]. Taken together, results of these studies suggest that nuclear ErbB2 is important for promoting DSB repair in cancer cells even though its exact mechanism is not yet clear. It should be mentioned that nuclear ErbB2 did not affect the repair of DNA damage induced by etoposide [133], which requires homologous recombination, suggesting the functional specificity of ErbB2 in NHEJ-mediated DSB repair.

Intact ErbB4, rather than cleaved ErbB4-ICD, was recently identified in the nucleus of cardiac myocytes. In contrast to findings for EGFR and ErbB2, treatment with the chemotherapeutic agent doxorubicin reduced the nuclear expression of ErbB4 and increased expression and activation of p53; however, knockdown of ErbB4 in doxorubicin-treated cells surprisingly abolished the activation of p53, indicating that nuclear ErbB4 may be responsible for doxorubicin-induced p53 activation and DSB repair [134]. These findings need to be further validated before a clear conclusion can be drawn.

No information is currently available on the roles of other nuclear RTKs besides the ErbB subfamily receptors in DNA damage repair. Such studies are important and necessary for us to understand the entire role of nuclear RTKs in repairing DNA damage in cancer cells.

## 8.3.4 Other Functions of Nuclear RTKs

### 8.3.4.1 rRNA Biogenesis

Within the nucleus, RTKs associate with either chromatin [39, 41, 135, 136] or the nuclear matrix [83, 137, 138]. In addition, RTKs are localized at specific sub-nuclear compartments such as nucleoli [100, 139–141] and nuclear speckles [83, 84] in different type cells. The nucleolus is the site of rRNA biogenesis, including rRNA transcription, rRNA processing and modification, and rRNA assembly and maturation [142, 143]. The nucleolar localization of RTKs indicates that RTKs may have a role in rRNA biogenesis. In supporting this idea, Li et al. recently demonstrated that nuclear ErbB2 interacts with RNA Pol I through β-actin and activates rRNA transcription, with resultant increasing protein synthesis in breast cancer cells [140]. Interestingly, this activation of rRNA expression by ErbB2 seems to be specific for nuclear ErbB2 rather than for cell surface receptor, as blocking traditional downstream pathways of cell surface ErbB2 did not change the rRNA expression and protein synthesis induced by the nuclear receptor. EGFR-associated proteins in the nucleus have been investigated by a proteomic strategy, and some of the proteins determined to associate with EGFR are nucleolar proteins, such as NOL6 (NP_075068), NOC2L (NP_056473), and NOC4L (NP_076983) [53]. Further validation and study is required for the interaction between nuclear EGFR and these nucleolar proteins and their potential roles in rRNA biogenesis. Interestingly, ErbB3 (full length) was found to localize to the nucleolus when cells become polarized and differentiated by growing on permeable filters. Moreover, ErbB3 translocates from the nucleolus to the nucleoplasm and subsequently to the cytoplasm in response to stimulation by its ligand heregulin in human mammary epithelial cells [144], indicating that the nucleolus may be a temporary storage site for cellular ErbB3 to inhibit its cellular function or that ErbB3 in the nucleolus may play a role in regulating nucleolar structure or rRNA biogenesis under specific conditions, e.g., polarization and differentiation. Indeed, a recent report indicated that $p14^{ARF}$, a tumor suppressor, interacts and sequesters ErbB3-80 kDa in the nucleolus to inhibit its target gene transcription by prevent ErbB3-80 kDa from its binding to the gene promoter [68].

### 8.3.4.2 Nuclear Calcium Signaling

Both IR and c-Met have been found to stimulate nuclear calcium signaling through the inositol 1,4,5-triphosphate (InsP3)/$Ca^{2+}$ pathway when they are translocated into the nucleus [145, 146]. Given the important roles of nuclear calcium in the regulation of gene expression [147] and cell proliferation [148], it is plausible that nuclear IR and c-Met also regulate cell growth and proliferation by stimulating the nuclear InsP3/$Ca^{2+}$ pathway in tumor cells. Further studies are needed to validate the roles of IR and c-Met in nuclear calcium signaling in cancer cells.

## 8.4 Roles of Nuclear RTKs in Therapeutic Resistance of Cancer Cells

Aberrant activation of RTK signaling has been implicated in most human cancers [3, 7], which provides a rationale for anti-cancer drug development. Indeed, RTK has become one of most attractive therapeutic targets for human cancers. Many anti-RTK therapeutic agents have been designed, developed, and used in clinical trials, and some of them have been approved by the FDA in the last 20 years, including anti-RTK antibodies, antagonists of RTK ligands, small molecule inhibitors of RTK kinase activity, and many others [3, 149–152]. Although promising therapeutic effects have been demonstrated in some human tumors, resistance to anti-RTK therapy occurs frequently in other cancers (intrinsic resistance) or in a set of cancers after long-term treatment (acquired resistance) [153–157], which is reminiscent of other cancer treatments such as chemotherapy [158].

It is clear that multiple mechanisms are responsible for cancer cell resistance to these drugs [153, 159]. DNA damage repair and drug efflux are two of the major mechanisms contributing to cancer cell resistance to chemotherapeutic agents. Several nuclear RTKs have been implicated in therapeutic resistance. As mentioned above, nuclear EGFR associates with DNA repair complex and activates DNA-PK to promote the repair of irradiation- or chemo-induced DSB with a subsequent increase in tumor cell survival [119, 122, 129, 160], and nuclear EGFR also directly phosphorylates PCNA to regulate mismatch repair and increase DNA synthesis, leading to cell proliferation [115]. In addition, nuclear EGFR was recently found to induce irradiation resistance through its association with DNA-PK and PNPase to enhance DNA-PK-mediated phosphorylation and inactivation of PNPase [161]. PNPase is a 3'→5' exoribonuclease responsible for c-Myc mRNA degradation. After phosphorylation by DNA-PK upon irradiation, PNPase losses its exoribonuclease activity and is unable to degrade Myc mRNA when translocated to the cytoplasm, leading to an increase of Myc expression and resistance to irradiation-induced apoptosis [161]. It is worth noting that these aforementioned mechanisms are not mutually exclusive. It is possible that they cooperate with each other when cancer cells receive irradiation to confer cell survival and resistance. In addition to irradiation, cisplatin treatment also stimulates nuclear translocation of EGFR, and once in the nucleus, EGFR contributes to tumor cell resistance to cisplatin treatment [124, 125]. By the use of in vitro established drug-resistant stable clones, nuclear EGFR was shown to be responsible for cancer cell resistance to TKI treatment. Nuclear EGFR transcriptionally activates the expression of the chemo-resistance gene BCRP/ABCG2, which in turn targets gefitinib to pump it out of cells, leading to cancer cell resistance to gefitinib treatment [62, 162]. Similarly, treatment of cells with cetuximab induces EGFR nuclear translocation [163], which has been shown to render cancer cells resistant to cetuximab treatment in both in vitro and in vivo models [164]. However, the cetuximab-induced EGFR nuclear translocation requires further validation due to controversial reports in the literature [127, 165], although different cell systems and experimental conditions may be responsible for the discrepancies between these studies. In support of the role of nuclear EGFR in drug resistance, the expression level of nuclear EGFR has been shown to correlate significantly with poor survival in several human tumors including breast cancer [11, 12], oral squamous carcinomas [11, 13, 14], ovarian cancer [15], and prostate cancer [17].

Repairing the cisplatin-induced DNA damage requires nuclear translocation of intact ErbB2, which leads to cancer cell survival with cisplatin treatment [133]. Phosphorylation of $p34^{Cdc2}$ at Y15 by ErbB2, probably by nuclear ErbB2, inactivates $p34^{Cdc2}$, leading to M phase entry delay and therefore resistance to paclitaxel-induced apoptosis [166]. In addition to intact ErbB2, ErbB2 variant forms (from either alternative initiation of translation, e.g., ErbB2-CTFs/p95HER2 or from proteolytic cleavage, e.g., p95L) have been shown to induce cancer cells to become resistant to ErbB2 target therapy. Scaltriti et al. first identified ErbB2 C-terminal fragments (ErbB2-CTFs/p95HER2; molecular weight: 95 kDa), which are generated by alternative translation initiation from methionines located near the transmembrane domain of full-length molecule [167]. In breast cancer cells, ErbB2-CTFs localize to the nucleus, the cytoplasm, and the plasma membrane, and those with ErbB2-CTF expression are resistant to trastuzumab (a monoclonal antibody of ErbB2) but not to lapatinib (a dual TKI of EGFR and ErbB2) [167]. Consistently, a xenograft tumor model indicated that tumors expressing ErbB2-CTFs are sensitive to lapatinib but resistant to trastuzumab [168]. This finding was further supported by a clinical study in which 46 patients with metastatic breast cancer and treated with trastuzumab were investigated; patients with tumors expressing ErbB2-CTFs were found to have little response to trastuzumab [167]. However, it is not clear whether the observed trastuzumab resistance was caused by nuclear ErbB2-CTFs or cytosol/plasma membrane-ErbB2-CTFs. In contrast to ErbB2-CTFs, p95L, a cleaved form of intact ErbB2 with molecular weight similar to that of ErbB2-CTFs, was recently found to confer resistance to ErbB2-targeted therapy when it is found in the nucleus. Lapatinib induces p95L expression that is likely through proteasome cleavage to stimulate p95L nuclear translocation. Nuclear p95L is activated by phosphorylation, and cells with this lapatinib-induced p95L nuclear expression have been shown to be resistant to both trastuzumab and lapatinib [169]. On the basis of these studies, the expression of ErbB2 variants could be a useful predicator for resistance to ErbB2-targeted therapy in cancer cells.

Activation of IGF1R signaling and altered chromatin status were found to be important for cancer cells to acquire reversible drug tolerance [170]. In line with this idea, a recent study showed that nuclear IGF1R expression is increased in the gefitinib-resistant hepatocellular carcinoma (HCC) cell line Mahlavu, and gefitinib treatment enhances not only the activity of IGF1R signaling but also the nuclear translocation of IGF1R in a dose-dependent manner that is accompanied by elevated expression of CD133 [171], an important liver cancer stem cell marker [172, 173]. These findings indicate that nuclear translocation of IGF1R may contribute to gefitinib resistance in HCC tumor cells by increasing the cancer stem cell population. However, the exact function of nuclear IGF1R in mediating gefitinib resistance in HCC warrants further investigation. Nevertheless, whether nuclear IGF1R regulates CD133 gene expression and/or other genes involved in increasing the cancer stem cell population and whether nuclear IGF1R, like nuclear EGFR as mentioned above, activates drug-resistance genes such as ATP binding cassette (ABC) transporters to induce drug excretion, are all interesting questions waiting to be addressed to further our understanding of the role of nuclear IGF1R in driving tumor cells to become resistant to therapeutic agents.

## 8.5 Strategies Targeting Nuclear Functions of RTKs for Cancer Therapy

As mentioned earlier, nuclear translocation of RTKs is involved in tumor cell growth and therapeutic resistance, and their nuclear expression levels are correlated with poor survival in multiple human cancer types. Thus, blocking the nuclear translocation of RTKs or their nuclear functions may provide a new direction to improve therapeutic efficacy or to overcome therapeutic resistance of cancers. Although this has not yet been extensively explored, several in vitro and in vivo animal studies have shed light on the promising prospects and feasibility of this strategy.

### *8.5.1 Blockage of Nuclear Translocation of RTKs*

While the mechanisms of RTKs nuclear translocation have become much clearer with the efforts by several research groups [8–10, 174–176], the trafficking is a complicated process, and more work is needed to completely understand how RTKs travel to the nucleus under certain cellular conditions. On the basis of the identified trafficking mechanism, several studies have demonstrated that blocking the nuclear translocation of EGFR could be a strategy for cancer therapy and overcoming the acquired resistance of cancer cells to therapeutic agents. For example, inhibition of nuclear translocation of EGFR by tyrosine kinase inhibitor (TKI) has been shown to sensitize tumor cells to irradiation [127]. An artificial system utilizing the $EGFR_{\Delta NLS}$ mutant, which lacks nuclear translocation, significantly suppressed tumor cell growth in softer agar [55] and sensitized cancer cells to cisplatin [125], suggesting the feasibility of inhibiting tumor growth or reversing the drug resistance by blocking nuclear EGFR translocation. Indeed, inhibition of nuclear translocation of EGFR and ErbB2 by lapatinib, a dual TKI of EGFR and ErbB2, sensitized xenografted tumors of gastric cancer cells (N87) to chemotherapy [63], at least, in part, by the inhibition of nuclear translocation of EGFR and ErbB2, which led to a reduction of EGFR- and ErbB2-mediated TS expression. Several studies have indicated that the activity of Src family kinases is crucial for nuclear transport of EGFR in response to irradiation [177, 178] and cetuximab [164, 179]. Therefore, inhibition of Src family kinase activity may also block nuclear translocation of EGFR. Indeed, dasatinib, a potent inhibitor of multiple tyrosine kinases (including the Src family kinases) was shown to inhibit both cetuximab- and radiation-induced EGFR nuclear translocation in HNSCC [180]. Consistently, treatment of cetuximab-resistant cancer cells with dasatinib resensitized the cells to cetuximab [164]. In addition, irradiation combined with the poly(ADP-ribose) polymerase (PARP) inhibitor ABT-888 significantly increased the susceptibility of HNSCC cells to irradiation, which correlated with the reduction of EGFR nuclear translocation [181], suggesting that the outcomes of patients with HNSCC can be improved when patients are treated by a combination of radiotherapy and PARP inhibitor.

### 8.5.2 Blockage of Nuclear Functions of RTKs

Blocking the nuclear functions of RTKs is another approach for cancer therapy and overcoming the therapeutic resistance of cancer cells. A few studies have shown the feasibility of this promising strategy to reverse the acquired resistance of cancer cells. For example, a second methionine mutation of the EGFR L858R mutant at T790 (T790M) is known to be responsible for the resistance of cancer cells to TKI treatment in NSCLC [182, 183]. In contrast to the single L858R EGFR mutant, this double EGFR mutant (L858R/T790M) has been detected in the nucleus and interacts with myosin II (MYH9) and β-actin in H1975 NSCLC cells, suggesting that its nuclear translocation may contribute to TKI resistance. Indeed, disruption of cytoskeleton integrity caused by the myosin inhibitor blebbistatin reduced the nuclear interaction of EGFR with MYH9 and β-actin, and cells treated with both blebbistatin and the T790M-specific TKI CL-387,785 showed decreased COX2 gene transcription and increased the sensitivity to CL-387,785, leading to enhanced apoptosis [184]. Similarly, multidrug-resistant cells Kbvin10, Kbtax50, and CEM/VBL, which overexpress P-glycoprotein and are resistant to vincristine and doxorubicin, were found to be more sensitive to the bifunctional alkylating derivatives of 3a-aza-cyclopenta[a]indene (an DNA interstrand crosslinking agent) treatment through a possible mechanism that blocks the Src/nuclear EGFR cascade and subsequently inactivates DNA-PK and reduces DSB repair [126].

### 8.5.3 Utilization of Nuclear Translocation to Enhance Cellular Radiotoxicity

In contrast to the therapeutic strategies of blocking nuclear translocation and nuclear functions of RTKs, targeted radiotherapy with Auger electrons is another approach that uses RTK-specific ligands conjugated with radiopharmaceuticals to deliver the radiotoxic effects directly into the nucleus, inducing DNA damage through nuclear translocation of the ligand/receptor complex.

Twenty years ago, $^{125}$I-labeled growth factor peptide was used to investigate the nuclear translocation of receptor ligands after binding to their specific cell surface receptors [42, 43], and this mechanism has now been extended to target radiotherapy for tumors with EGFR overexpression. Reilly et al. first tested the hypothesis that EGF (after binding to its cell surface receptor) is subjected to internalization followed by nuclear translocation, which would deliver the Auger electron emitter $^{111}$In into the nucleus of EGFR-positive breast cancer cells for targeted radiotherapy. They found that after internalization and nuclear translocation, $^{111}$In-DTPA-hEGF associates with chromatin to induce DNA damage and subsequently reduce cell survival [185]. Moreover, this radiotoxic effects highly selective in human breast cancer cells and shows a good correlation between EGFR expression and $^{111}$In-DTPA-hEGF uptake,$^{111}$In-DTPA-hEGF nuclear localization, DNA damage, and cell-killing activity, which was consistently demonstrated in a panel of breast cancer cell lines, including MDA-MB-468 cells ($1.3\times10^{6}$ EGFR per cell), MDA-MB-231 cells ($2\times10^{5}$ EGFR per cell), and MCF7 cells ($1.5\times10^{4}$ EGFR per cell) [185–187]. In addition, $^{111}$In-DTPA-hEGF was shown to effectively inhibit tumor growth of MDA-MB-468 xenografts in athymic mice, with little toxicity to normal tissues [187, 188]. Furthermore, the combination of EGFR TKI (e.g., gefitinib) with $^{111}$In-DTPA-hEGF significantly increased the cytotoxic effects of $^{111}$In [189], implying the feasibility of combination therapy. Indeed, the combination of $^{111}$In-DTPA-hEGF with either anti-ErbB2 antibody (trastuzumab) or with farnesyltransferase inhibitor (L-778,123) increased the nuclear uptake of $^{111}$In-DTPA-hEGF in breast cancer cells. Co-treatment of mice bearing 231-H2N (MDA-MB-231 ErbB2-stable transfectants) xenografts with $^{111}$In-DTPA-hEGF and L-788,123 significantly prolonged animal survival [190]. Moreover, systemic administration of $^{111}$In-DTPA-hEGF in preclinical studies demonstrated that it has no major toxicity to normal tissues and is safe to animals [191], suggesting good potential for further clinical trials.

It is worth noting that EGFR antibody nimotuzumab labeled with Auger electron emitter $^{111}$In has also been shown to have a potential for killing EGFR positive and transtuzumab-resistant breast cancer cells recently [192]. Given that EGFR antibody cetuximab is known to induce EGFR nuclear translocation, which leads to the therapeutic resistance of cancer cells [163, 164], it is possible that a combinational treatment consisting of unlabeled normal antibody and $^{111}$In-labeled antibody may be a good strategy to overcome tumor cell resistance to EGFR family antibody treatment.

## 8.6 Conclusion

Given the importance of RTK signaling in regulating crucial cellular processes and the fact that RTKs are considered to be temporally and spatially regulated, the spatial deregulation of RTKs is likely to contribute totumorigenesis and tumor progression and may further affect the sensitivity and resistance of cancer cells to therapeutic agents [7, 193]. In contrast to the canonical signaling cascades of cell surface RTKs, which have been extensively investigated and well documented in the last few decades, the nuclear translocation and nuclear functions of RTKs are still far from clear. Despite much initial skepticism, the concept of nuclear localization and nuclear functions of RTKs, especially intact RTKs, has now been demonstrated and supported by mounting evidence from numerous investigators and disciplines, and it is being gradually recognized due to the substantial number of research articles published since 2001.

Although the presence of nuclear RTKs might not be absolutely essential for cell growth, the fact that RTKs can translocate into the nucleus under stimulation by their ligands [10] or under certain stress conditions (e.g., liver regeneration, UV, heat, $H_2O_2$ in the case of EGFR [119, 128]) indicates that nuclear translocation of RTKs may provide cells with some growth and survival advantages. Indeed, one of the major functions of nuclear RTKs identified to date is to regulate the expression of genes involved in cell growth and proliferation, as demonstrated by different research groups using multiple cell types. However, the full consequences of nuclear translocation of RTKs for cancer development remain unclear. For example: (1) Do the nuclear RTKs directly contribute to tumorigenesis? (2) How do cells regulate the nuclear functions and canonical functions of RTKs during tumor development if any differences exist between these two signaling cascades? (3) Is there any cross-talk among nuclear RTKs similar to what is well documented in the literature for cell surface RTKs? (4) Multiple types of nuclear RTKs such as full-length receptors, receptor variants, and receptor fragments have been found in the nucleus (see Table 8.2). Do they share common nuclear targets and/or nuclear functions? (5) How do we develop therapeutic strategies to efficiently target the different locations and diverse functions of RTK in human cancers?

To answer such questions, a good transgenic animal model with RTKs lacking their nuclear translocation ability but without affecting their other functional activities will be helpful in clarifying the exact roles of nuclear RTKs in tumorigenesis. Another approach to understand the nuclear functions of RTKs is to use non-biased approaches to identify their nuclear interacting proteins and transcriptional targets. Currently, most of the functions and/or targets of nuclear RTK identified to date could be explained by the functions of its canonical signaling cascades, which is not sufficient to address the importance of nuclear RTK in cancer development. More in-depth investigation of nuclear RTKs will help us understand the need for RTKs to translocate into the nucleus if they cannot execute their functions at the cell surface. Finally, such newly gained information will provide substantial rationale for us to develop new effective targeted therapies for RTKs in human cancers and other diseases caused by RTK deregulation.

**Acknowledgements** We thank Michael Worley at Department of Scientific Publications for editing this manuscript. The study was supported by the following grants: National Institutes of Health (RO1 CA 109311; PO1 CA 099031; CCSG CA16672); National Breast Cancer Foundation, Inc.; The University of Texas MD Anderson-China Medical University and Hospital Sister Institution Fund; Cancer Research Center of Excellence (DOH102-TD-C-111-005, Taiwan); Program for Stem Cell and Regenerative Medicine Frontier Research (NSC101-2321-B-039-001, Taiwan); International Research-Intensive Centers of Excellence in Taiwan (NSC102-2911-I-002-303, Taiwan); Private University grant (NSC99-2632-B-039-001-MY3, Taiwan).

## References


1. Lemmon MA, Schlessinger J (2010) Cell signaling by receptor tyrosine kinases. Cell 141(7):1117–1134
2. Hubbard SR, Miller WT (2007) Receptor tyrosine kinases: mechanisms of activation and signaling. Curr Opin Cell Biol 19(2):117–123
3. Gschwind A, Fischer OM, Ullrich A (2004) The discovery of receptor tyrosine kinases: targets for cancer therapy. Nat Rev Cancer 4(5):361–370
4. Avraham R, Yarden Y (2011) Feedback regulation of EGFR signalling: decision making by early and delayed loops. Nat Rev Mol Cell Biol 12(2):104–117
5. Hynes NE, MacDonald G (2009) ErbB receptors and signaling pathways in cancer. Curr Opin Cell Biol 21(2):177–184
6. Gordus A, Krall JA, Beyer EM, Kaushansky A, Wolf-Yadlin A, Sevecka M et al (2009) Linear combinations of docking affinities explain quantitative differences in RTK signaling. Mol Syst Biol 5:235
7. Casaletto JB, McClatchey AI (2012) Spatial regulation of receptor tyrosine kinases in development and cancer. Nat Rev Cancer 12(6):387–400
8. Wang YN, Hung MC (2012) Nuclear functions and subcellular trafficking mechanisms of the epidermal growth factor receptor family. Cell Biosci 2(1):13
9. Wang YN, Yamaguchi H, Hsu JM, Hung MC (2010) Nuclear trafficking of the epidermal growth factor receptor family membrane proteins. Oncogene 29(28):3997–4006
10. Carpenter G, Liao HJ (2009) Trafficking of receptor tyrosine kinases to the nucleus. Exp Cell Res 315(9):1556–1566
11. Lo HW, Xia W, Wei Y, Ali-Seyed M, Huang SF, Hung MC (2005) Novel prognostic value of nuclear epidermal growth factor receptor in breast cancer. Cancer Res 65(1):338–348
12. Hadzisejdic I, Mustac E, Jonjic N, Petkovic M, Grahovac B (2010) Nuclear EGFR in ductal invasive breast cancer: correlation with cyclin-D1 and prognosis. Mod Pathol 23(3):392–403
13. Psyrri A, Yu Z, Weinberger PM, Sasaki C, Haffty B, Camp R et al (2005) Quantitative determination of nuclear and cytoplasmic epidermal growth factor receptor expression in oropharyngeal squamous cell cancer by using automated quantitative analysis. Clin Cancer Res 11(16):5856–5862
14. Hoshino M, Fukui H, Ono Y, Sekikawa A, Ichikawa K, Tomita S et al (2007) Nuclear expression of phosphorylated EGFR is associated with poor prognosis of patients with esophageal squamous cell carcinoma. Pathobiology 74(1):15–21
15. Xia W, Wei Y, Du Y, Liu J, Chang B, Yu YL et al (2009) Nuclear expression of epidermal growth factor receptor is a novel prognostic value in patients with ovarian cancer. Mol Carcinog 48(7):610–617
16. Li CF, Fang FM, Wang JM, Tzeng CC, Tai HC, Wei YC et al (2012) EGFR nuclear import in gallbladder carcinoma: nuclear phosphorylated EGFR upregulates iNOS expression and confers independent prognostic impact. Ann Surg Oncol 19(2):443–454
17. Edwards J, Traynor P, Munro AF, Pirret CF, Dunne B, Bartlett JM (2006) The role of HER1-HER4 and EGFRvIII in hormone-refractory prostate cancer. Clin Cancer Res 12(1):123–130
18. Dekanic A, Dintinjan RD, Budisavljevic I, Pecanic S, Butorac MZ, Jonjic N (2011) Strong nuclear EGFR expression in colorectal carcinomas is associated with cyclin-D1 but not with gene EGFR amplification. Diagn Pathol 6:108
19. Dillon MF, Stafford AT, Kelly G, Redmond AM, McIlroy M, Crotty TB et al (2008) Cyclooxygenase-2 predicts adverse effects of tamoxifen: a possible mechanism of role for nuclear HER2 in breast cancer patients. Endocr Relat Cancer 15(3):745–753
20. Schillaci R, Guzman P, Cayrol F, Beguelin W, Diaz Flaque MC, Proietti CJ et al (2012) Clinical relevance of ErbB-2/HER2 nuclear expression in breast cancer. BMC Cancer 12:74
21. Cheng CJ, Ye XC, Vakar-Lopez F, Kim J, Tu SM, Chen DT et al (2007) Bone microenvironment and androgen status modulate subcellular localization of ErbB3 in prostate cancer cells. Mol Cancer Res 5(7):675–684

22. Koumakpayi IH, Diallo JS, Le Page C, Lessard L, Gleave M, Begin LR et al (2006) Expression and nuclear localization of ErbB3 in prostate cancer. Clin Cancer Res 12(9): 2730–2737
23. Koumakpayi IH, Diallo JS, Le Page C, Lessard L, Filali-Mouhim A, Begin LR et al (2007) Low nuclear ErbB3 predicts biochemical recurrence in patients with prostate cancer. BJU Int 100(2):303–309
24. Junttila TT, Sundvall M, Lundin M, Lundin J, Tanner M, Harkonen P et al (2005) Cleavable ErbB4 isoform in estrogen receptor-regulated growth of breast cancer cells. Cancer Res 65(4):1384–1393
25. Tovey SM, Dunne B, Witton CJ, Cooke TG, Bartlett JM (2006) HER4 In breast cancer: comparison of antibodies against intra- and extra-cellular domains of HER4. Breast Cancer Res 8(2):R19
26. Sundvall M, Iljin K, Kilpinen S, Sara H, Kallioniemi OP, Elenius K (2008) Role of ErbB4 in breast cancer. J Mammary Gland Biol Neoplasia 13(2):259–268
27. Aleksic T, Chitnis MM, Perestenko OV, Gao S, Thomas PH, Turner GD et al (2010) Type 1 insulin-like growth factor receptor translocates to the nucleus of human tumor cells. Cancer Res 70(16):6412–6419
28. Asmane I, Watkin E, Alberti L, Duc A, Marec-Berard P, Ray-Coquard I et al (2012) Insulin-like growth factor type 1 receptor (IGF-1R) exclusive nuclear staining: a predictive biomarker for IGF-1R monoclonal antibody (Ab) therapy in sarcomas. Eur J Cancer 48:3027–3035
29. Behrens C, Lin HY, Lee JJ, Raso MG, Hong WK, Wistuba II et al (2008) Immunohistochemical expression of basic fibroblast growth factor and fibroblast growth factor receptors 1 and 2 in the pathogenesis of lung cancer. Clin Cancer Res 14(19):6014–6022
30. Sun S, Jiang Y, Zhang G, Song H, Zhang X, Zhang Y et al (2012) Increased expression of fibroblastic growth factor receptor 2 is correlated with poor prognosis in patients with breast cancer. J Surg Oncol 105(8):773–779
31. Rotterud R, Fossa SD, Nesland JM (2007) Protein networking in bladder cancer: immunoreactivity for FGFR3, EGFR, ERBB2, KAI1, PTEN, and RAS in normal and malignant urothelium. Histol Histopathol 22(4):349–363
32. Maeng YH, Eun SY, Huh JS (2010) Expression of fibroblast growth factor receptor 3 in the recurrence of non-muscle-invasive urothelial carcinoma of the bladder. Korean J Urol 51(2):94–100
33. Farkkila A, Pihlajoki M, Tauriala H, Butzow R, Leminen A, Unkila-Kallio L et al (2011) Serum vascular endothelial growth factor a (VEGF) is elevated in patients with ovarian granulosa cell tumor (GCT), and VEGF inhibition by bevacizumab induces apoptosis in GCT in vitro. J Clin Endocrinol Metab 96(12):E1973–E1981
34. Zhang Y, Pillai G, Gatter K, Blazquez C, Turley H, Pezzella F et al (2005) Expression and cellular localization of vascular endothelial growth factor a and its receptors in acute and chronic leukemias: an immunohistochemical study. Hum Pathol 36(7):797–805
35. Carrillo de Santa Pau E, Arias FC, Caso Pelaez E, Munoz Molina GM, Sanchez Hernandez I, Muguruza Trueba I et al (2009) Prognostic significance of the expression of vascular endothelial growth factors a, B, C, and D and their receptors R1, R2, and R3 in patients with nonsmall cell lung cancer. Cancer 115(8):1701–1712
36. Pozner-Moulis S, Cregger M, Camp RL, Rimm DL (2007) Antibody validation by quantitative analysis of protein expression using expression of met in breast cancer as a model. Lab Invest 87(3):251–260
37. Brusevold IJ, Soland TM, Khuu C, Christoffersen T, Bryne M (2010) Nuclear and cytoplasmic expression of met in oral squamous cell carcinoma and in an organotypic oral cancer model. Eur J Oral Sci 118(4):342–349
38. Rakowicz-Szulczynska EM, Herlyn M, Koprowski H (1988) Nerve growth factor receptors in chromatin of melanoma cells, proliferating melanocytes, and colorectal carcinoma cells in vitro. Cancer Res 48(24 Pt 1):7200–7206
39. Rakowicz-Szulczynska EM, Koprowski H (1986) Identification of NGF receptor in chromatin of melanoma cells using monoclonal antibody to cell surface NGF receptor. Biochem Biophys Res Commun 140(1):174–180

40. Rakowicz-Szulczynska EM, Linnenbach AJ, Koprowski H (1989) Intracellular receptor binding and nuclear transport of nerve growth factor in intact cells and a cell-free system. Mol Carcinog 2(1):47–58
41. Rakowicz-Szulczynska EM, Otwiaska D, Rodeck U, Koprowski H (1989) Epidermal growth factor (EGF) and monoclonal antibody to cell surface EGF receptor bind to the same chromatin receptor. Arch Biochem Biophys 268(2):456–464
42. Rakowicz-Szulczynska EM, Rodeck U, Herlyn M, Koprowski H (1986) Chromatin binding of epidermal growth factor, nerve growth factor, and platelet-derived growth factor in cells bearing the appropriate surface receptors. Proc Natl Acad Sci USA 83(11):3728–3732
43. Savion N, Vlodavsky I, Gospodarowicz D (1981) Nuclear accumulation of epidermal growth factor in cultured bovine corneal endothelial and granulosa cells. J Biol Chem 256(3):1149–1154
44. Marti U, Burwen SJ, Wells A, Barker ME, Huling S, Feren AM et al (1991) Localization of epidermal growth factor receptor in hepatocyte nuclei. Hepatology 13(1):15–20
45. Lin SY, Makino K, Xia W, Matin A, Wen Y, Kwong KY et al (2001) Nuclear localization of EGF receptor and its potential new role as a transcription factor. Nat Cell Biol 3(9):802–808
46. Cao H, Lei ZM, Bian L, Rao CV (1995) Functional nuclear epidermal growth factor receptors in human choriocarcinoma JEG-3 cells and normal human placenta. Endocrinology 136(7):3163–3172
47. Xie Y, Hung MC (1994) Nuclear localization of p185neu tyrosine kinase and its association with transcriptional transactivation. Biochem Biophys Res Commun 203(3):1589–1598
48. Wang SC, Lien HC, Xia W, Chen IF, Lo HW, Wang Z et al (2004) Binding at and transactivation of the COX-2 promoter by nuclear tyrosine kinase receptor ErbB-2. Cancer Cell 6(3):251–261
49. Kim J, Jahng WJ, Di Vizio D, Lee JS, Jhaveri R, Rubin MA et al (2007) The phosphoinositide kinase PIKfyve mediates epidermal growth factor receptor trafficking to the nucleus. Cancer Res 67(19):9229–9237
50. Tao Y, Song X, Deng X, Xie D, Lee LM, Liu Y et al (2005) Nuclear accumulation of epidermal growth factor receptor and acceleration of G1/S stage by epstein-barr-encoded oncoprotein latent membrane protein 1. Exp Cell Res 303(2):240–251
51. Bitler BG, Goverdhan A, Schroeder JA (2010) MUC1 regulates nuclear localization and function of the epidermal growth factor receptor. J Cell Sci 123(Pt 10):1716–1723
52. Hu S, Xie Z, Onishi A, Yu X, Jiang L, Lin J et al (2009) Profiling the human protein-DNA interactome reveals ERK2 as a transcriptional repressor of interferon signaling. Cell 139(3):610–622
53. Huo L, Wang YN, Xia W, Hsu SC, Lai CC, Li LY et al (2010) RNA helicase a is a DNA-binding partner for EGFR-mediated transcriptional activation in the nucleus. Proc Natl Acad Sci USA 107(37):16125–16130
54. Lo HW, Hsu SC, Ali-Seyed M, Gunduz M, Xia W, Wei Y et al (2005) Nuclear interaction of EGFR and STAT3 in the activation of the iNOS/NO pathway. Cancer Cell 7(6):575–589
55. Lo HW, Cao X, Zhu H, Ali-Osman F (2010) Cyclooxygenase-2 is a novel transcriptional target of the nuclear EGFR-STAT3 and EGFRvIII-STAT3 signaling axes. Mol Cancer Res 8(2):232–245
56. Han W, Carpenter RL, Cao X, Lo HW (2012) STAT1 gene expression is enhanced by nuclear EGFR and HER2 via cooperation with STAT3. Mol Carcinog 1–11; doi:10.1002/mc.21936
57. Jaganathan S, Yue P, Paladino DC, Bogdanovic J, Huo Q, Turkson J (2011) A functional nuclear epidermal growth factor receptor, SRC and Stat3 heteromeric complex in pancreatic cancer cells. PLoS One 6(5):e19605
58. Hanada N, Lo HW, Day CP, Pan Y, Nakajima Y, Hung MC (2006) Co-regulation of B-Myb expression by E2F1 and EGF receptor. Mol Carcinog 45(1):10–17
59. Hung LY, Tseng JT, Lee YC, Xia W, Wang YN, Wu ML et al (2008) Nuclear epidermal growth factor receptor (EGFR) interacts with signal transducer and activator of transcription 5 (STAT5) in activating aurora-a gene expression. Nucleic Acids Res 36(13):4337–4351
60. Fukuoka H, Cooper O, Ben-Shlomo A, Mamelak A, Ren SG, Bruyette D et al (2011) EGFR as a therapeutic target for human, canine, and mouse ACTH-secreting pituitary adenomas. J Clin Invest 121(12):4712–4721

61. Gu Y, Zhang S, Wu Q, Xu S, Cui Y, Yang Z et al (2010) Differential expression of decorin, EGFR and cyclin D1 during mammary gland carcinogenesis in TA2 mice with spontaneous breast cancer. J Exp Clin Cancer Res 29:6
62. Huang WC, Chen YJ, Li LY, Wei YL, Hsu SC, Tsai SL et al (2011) Nuclear translocation of epidermal growth factor receptor by Akt-dependent phosphorylation enhances breast cancer-resistant protein expression in gefitinib-resistant cells. J Biol Chem 286(23):20558–20568
63. Kim HP, Yoon YK, Kim JW, Han SW, Hur HS, Park J et al (2009) Lapatinib, a dual EGFR and HER2 tyrosine kinase inhibitor, downregulates thymidylate synthase by inhibiting the nuclear translocation of EGFR and HER2. PLoS One 4(6):e5933
64. Piccione EC, Lieu TJ, Gentile CF, Williams TR, Connolly AJ, Godwin AK et al (2012) A novel epidermal growth factor receptor variant lacking multiple domains directly activates transcription and is overexpressed in tumors. Oncogene 31(24):2953–2967
65. Beguelin W, Diaz Flaque MC, Proietti CJ, Cayrol F, Rivas MA, Tkach M et al (2010) Progesterone receptor induces ErbB-2 nuclear translocation to promote breast cancer growth via a novel transcriptional effect: ErbB-2 function as a coactivator of Stat3. Mol Cell Biol 30(23):5456–5472
66. Li X, Kuang J, Shen Y, Majer MM, Nelson CC, Parsawar K et al (2012) The atypical histone MacroH2A1.2 Interacts with HER-2 protein in cancer cells. J Biol Chem 287(27): 23171–23183
67. Adilakshmi T, Ness-Myers J, Madrid-Aliste C, Fiser A, Tapinos N (2011) A nuclear variant of ErbB3 receptor tyrosine kinase regulates ezrin distribution and Schwann cell myelination. J Neurosci 31(13):5106–5119
68. Andrique L, Fauvin D, El Maassarani M, Colasson H, Vannier B, Seite P (2012) ErbB3(80 kDa), a nuclear variant of the ErbB3 receptor, binds to the cyclin D1 promoter to activate cell proliferation but is negatively controlled by p14ARF. Cell Signal 24(5):1074–1085
69. Jones FE (2008) HER4 Intracellular domain (4ICD) activity in the developing mammary gland and breast cancer. J Mammary Gland Biol Neoplasia 13(2):247–258
70. Carpenter G (2003) Nuclear localization and possible functions of receptor tyrosine kinases. Curr Opin Cell Biol 15(2):143–148
71. Komuro A, Nagai M, Navin NE, Sudol M (2003) WW domain-containing protein YAP associates with ErbB-4 and acts as a co-transcriptional activator for the carboxyl-terminal fragment of ErbB-4 that translocates to the nucleus. J Biol Chem 278(35):33334–33341
72. Zhu Y, Sullivan LL, Nair SS, Williams CC, Pandey AK, Marrero L et al (2006) Coregulation of estrogen receptor by ERBB4/HER4 establishes a growth-promoting autocrine signal in breast tumor cells. Cancer Res 66(16):7991–7998
73. Rokicki J, Das PM, Giltnane JM, Wansbury O, Rimm DL, Howard BA et al (2010) The ERalpha coactivator, HER4/4ICD, regulates progesterone receptor expression in normal and malignant breast epithelium. Mol Cancer 9:150
74. Paatero I, Jokilammi A, Heikkinen PT, Iljin K, Kallioniemi OP, Jones FE et al (2012) Interaction with ErbB4 promotes hypoxia-inducible factor-1alpha signaling. J Biol Chem 287(13):9659–9671
75. Linggi B, Carpenter G (2006) ErbB-4 s80 intracellular domain abrogates ETO2-dependent transcriptional repression. J Biol Chem 281(35):25373–25380
76. Arasada RR, Carpenter G (2005) Secretase-dependent tyrosine phosphorylation of Mdm2 by the ErbB-4 intracellular domain fragment. J Biol Chem 280(35):30783–30787
77. Long W, Wagner KU, Lloyd KC, Binart N, Shillingford JM, Hennighausen L et al (2003) Impaired differentiation and lactational failure of Erbb4-deficient mammary glands identify ERBB4 as an obligate mediator of STAT5. Development 130(21):5257–5268
78. Williams CC, Allison JG, Vidal GA, Burow ME, Beckman BS, Marrero L et al (2004) The ERBB4/HER4 receptor tyrosine kinase regulates gene expression by functioning as a STAT5A nuclear chaperone. J Cell Biol 167(3):469–478
79. Zscheppang K, Dork T, Schmiedl A, Jones FE, Dammann CE (2011) Neuregulin receptor ErbB4 functions as a transcriptional cofactor for the expression of surfactant protein B in the fetal lung. Am J Respir Cell Mol Biol 45(4):761–767

80. Allison JG, Das PM, Ma J, Inglis FM, Jones FE (2011) The ERBB4 intracellular domain (4ICD) regulates NRG1-induced gene expression in hippocampal neurons. Neurosci Res 70(2):155–163
81. Sardi SP, Murtie J, Koirala S, Patten BA, Corfas G (2006) Presenilin-dependent ErbB4 nuclear signaling regulates the timing of astrogenesis in the developing brain. Cell 127(1):185–197
82. Stachowiak MK, Maher PA, Joy A, Mordechai E, Stachowiak EK (1996) Nuclear accumulation of fibroblast growth factor receptors is regulated by multiple signals in adrenal medullary cells. Mol Biol Cell 7(8):1299–1317
83. Somanathan S, Stachowiak EK, Siegel AJ, Stachowiak MK, Berezney R (2003) Nuclear matrix bound fibroblast growth factor receptor is associated with splicing factor rich and transcriptionally active nuclear speckles. J Cell Biochem 90(4):856–869
84. Peng H, Moffett J, Myers J, Fang X, Stachowiak EK, Maher P et al (2001) Novel nuclear signaling pathway mediates activation of fibroblast growth factor-2 gene by type 1 and type 2 angiotensin II receptors. Mol Biol Cell 12(2):449–462
85. Reilly JF, Maher PA (2001) Importin beta-mediated nuclear import of fibroblast growth factor receptor: role in cell proliferation. J Cell Biol 152(6):1307–1312
86. Stachowiak EK, Fang X, Myers J, Dunham S, Stachowiak MK (2003) CAMP-induced differentiation of human neuronal progenitor cells is mediated by nuclear fibroblast growth factor receptor-1 (FGFR1). J Neurochem 84(6):1296–1312
87. Fang X, Stachowiak EK, Dunham-Ems SM, Klejbor I, Stachowiak MK (2005) Control of CREB-binding protein signaling by nuclear fibroblast growth factor receptor-1: a novel mechanism of gene regulation. J Biol Chem 280(31):28451–28462
88. Baron O, Forthmann B, Lee YW, Terranova C, Ratzka A, Stachowiak EK et al (2012) Cooperation of nuclear fibroblast growth factor receptor 1 and Nurr1 offers new interactive mechanism in postmitotic development of mesencephalic dopaminergic neurons. J Biol Chem 287(24):19827–19840
89. Lee YW, Terranova C, Birkaya B, Narla S, Kehoe D, Parikh A et al (2012) A novel nuclear FGF receptor-1 partnership with retinoid and Nur receptors during developmental gene programming of embryonic stem cells. J Cell Biochem 113(9):2920–2936
90. Chioni AM, Grose R (2012) FGFR1 cleavage and nuclear translocation regulates breast cancer cell behavior. J Cell Biol 197(6):801–817
91. Cerliani JP, Guillardoy T, Giulianelli S, Vaque JP, Gutkind JS, Vanzulli SI et al (2011) Interaction between FGFR-2, STAT5, and progesterone receptors in breast cancer. Cancer Res 71(10):3720–3731
92. Radulescu RT (1995) Insulin receptor alpha-subunit: a putative gene regulatory molecule. Med Hypotheses 45(2):107–111
93. Radulescu RT (1995) From insulin, retinoblastoma protein and the insulin receptor to a new model on growth factor specificity: the nucleocrine pathway. J Endocrinol 146(3):365–368
94. Gletsu N, Dixon W, Clandinin MT (1999) Insulin receptor at the mouse hepatocyte nucleus after a glucose meal induces dephosphorylation of a 30-kDa transcription factor and a concomitant increase in malic enzyme gene expression. J Nutr 129(12):2154–2161
95. Nelson JD, LeBoeuf RC, Bomsztyk K (2011) Direct recruitment of insulin receptor and ERK signaling cascade to insulin-inducible gene loci. Diabetes 60(1):127–137
96. Sarfstein R, Pasmanik-Chor M, Yeheskel A, Edry L, Shomron N, Warman N et al (2012) Insulin-like growth factor-I receptor (IGF-IR) translocates to nucleus and autoregulates IGF-IR gene expression in breast cancer cells. J Biol Chem 287(4):2766–2776
97. Sehat B, Tofigh A, Lin Y, Trocme E, Liljedahl U, Lagergren J et al (2010) SUMOylation mediates the nuclear translocation and signaling of the IGF-1 receptor. Sci Signal 3(108):ra10
98. Warsito D, Sjostrom S, Andersson S, Larsson O, Sehat B (2012) Nuclear IGF1R is a transcriptional co-activator of LEF1/TCF. EMBO Rep 13(3):244–250
99. Wu YC, Zhu M, Robertson DM (2012) Novel nuclear localization and potential function of insulin-like growth factor-1 receptor/insulin receptor hybrid in corneal epithelial cells. PLoS One 7(8):e42483

100. Zscheppang K, Korenbaum E, Bueter W, Ramadurai SM, Nielsen HC, Dammann CE (2006) ErbB receptor dimerization, localization, and co-localization in mouse lung type II epithelial cells. Pediatr Pulmonol 41(12):1205–1212
101. Baumann Kubetzko FB, Di Paolo C, Maag C, Meier R, Schafer BW, Betts DR et al (2004) The PAX5 oncogene is expressed in N-type neuroblastoma cells and increases tumorigenicity of a S-type cell line. Carcinogenesis 25(10):1839–1846
102. Kanteti R, Nallasura V, Loganathan S, Tretiakova M, Kroll T, Krishnaswamy S et al (2009) PAX5 Is expressed in small-cell lung cancer and positively regulates c-Met transcription. Lab Invest 89(3):301–314
103. Matteucci E, Bendinelli P, Desiderio MA (2009) Nuclear localization of active HGF receptor Met in aggressive MDA-MB231 breast carcinoma cells. Carcinogenesis 30(6):937–945
104. Previdi S, Maroni P, Matteucci E, Broggini M, Bendinelli P, Desiderio MA (2010) Interaction between human-breast cancer metastasis and bone microenvironment through activated hepatocyte growth factor/Met and beta-catenin/Wnt pathways. Eur J Cancer 46(9): 1679–1691
105. Liu HS, Hsu PY, Lai MD, Chang HY, Ho CL, Cheng HL et al (2010) An unusual function of RON receptor tyrosine kinase as a transcriptional regulator in cooperation with EGFR in human cancer cells. Carcinogenesis 31(8):1456–1464
106. Domingues I, Rino J, Demmers JA, de Lanerolle P, Santos SC (2011) VEGFR2 translocates to the nucleus to regulate its own transcription. PLoS One 6(9):e25668
107. Lyu J, Yamamoto V, Lu W (2008) Cleavage of the Wnt receptor Ryk regulates neuronal differentiation during cortical neurogenesis. Dev Cell 15(5):773–780
108. Zhong WM (2008) Going nuclear is again a winning (Wnt) strategy. Dev Cell 15(5): 635–636
109. Zhong J, Kim HT, Lyu J, Yoshikawa K, Nakafuku M, Lu W (2011) The Wnt receptor Ryk controls specification of GABAergic neurons versus oligodendrocytes during telencephalon development. Development 138(3):409–419
110. Tseng HC, Lyu PC, Lin WC (2010) Nuclear localization of orphan receptor protein kinase (Ror1) is mediated through the juxtamembrane domain. BMC Cell Biol 11:48
111. Tseng HC, Kao HW, Ho MR, Chen YR, Lin TW, Lyu PC et al (2011) Cytoskeleton network and cellular migration modulated by nuclear-localized receptor tyrosine kinase ROR1. Anticancer Res 31(12):4239–4249
112. Yankner BA, Shooter EM (1979) Nerve growth factor in the nucleus: interaction with receptors on the nuclear membrane. Proc Natl Acad Sci USA 76(3):1269–1273
113. Zhang Z, Yang Y, Gong A, Wang C, Liang Y, Chen Y (2005) Localization of NGF and TrkA at mitotic apparatus in human glioma cell line U251. Biochem Biophys Res Commun 337(1):68–74
114. Thakur MK, Paramanik V (2012) Expression of Trk a and Src and their interaction with ERbeta ligand binding domain show age and sex dependent alteration in mouse brain. Neurochem Res 37(2):448–453
115. Wang SC, Nakajima Y, Yu YL, Xia W, Chen CT, Yang CC et al (2006) Tyrosine phosphorylation controls PCNA function through protein stability. Nat Cell Biol 8(12):1359–1368
116. Maga G, Hubscher U (2003) Proliferating cell nuclear antigen (PCNA): a dancer with many partners. J Cell Sci 116(Pt 15):3051–3060
117. Moldovan GL, Pfander B, Jentsch S (2007) PCNA, the maestro of the replication fork. Cell 129(4):665–679
118. Psyrri A, Egleston B, Weinberger P, Yu Z, Kowalski D, Sasaki C et al (2008) Correlates and determinants of nuclear epidermal growth factor receptor content in an oropharyngeal cancer tissue microarray. Cancer Epidemiol Biomarkers Prev 17(6):1486–1492
119. Dittmann K, Mayer C, Fehrenbacher B, Schaller M, Raju U, Milas L et al (2005) Radiation-induced epidermal growth factor receptor nuclear import is linked to activation of DNA-dependent protein kinase. J Biol Chem 280(35):31182–31189
120. Wanner G, Mayer C, Kehlbach R, Rodemann HP, Dittmann K (2008) Activation of protein kinase cepsilon stimulates DNA-repair via epidermal growth factor receptor nuclear accumulation. Radiother Oncol 86(3):383–390

121. Javvadi P, Makino H, Das AK, Lin YF, Chen DJ, Chen BP et al (2012) Threonine 2609 phosphorylation of the DNA-dependent protein Kinase is a critical prerequisite for epidermal growth factor receptor mediated radiation resistance. Mol Cancer Res 10(10):1359–1368
122. Dittmann K, Mayer C, Fehrenbacher B, Schaller M, Kehlbach R, Rodemann HP (2011) Nuclear epidermal growth factor receptor modulates cellular radio-sensitivity by regulation of chromatin access. Radiother Oncol 99(3):317–322
123. Das AK, Chen BP, Story MD, Sato M, Minna JD, Chen DJ et al (2007) Somatic mutations in the tyrosine kinase domain of epidermal growth factor receptor (EGFR) abrogate EGFR-mediated radioprotection in non-small cell lung carcinoma. Cancer Res 67(11):5267–5274
124. Liccardi G, Hartley JA, Hochhauser D (2011) EGFR nuclear translocation modulates DNA repair following cisplatin and ionizing radiation treatment. Cancer Res 71(3):1103–1114
125. Hsu SC, Miller SA, Wang Y, Hung MC (2009) Nuclear EGFR is required for cisplatin resistance and DNA repair. Am J Transl Res 1(3):249–258
126. Lee PC, Lee HJ, Kakadiya R, Sanjiv K, Su TL, Lee TC (2012) Multidrug-resistant cells overexpressing P-glycoprotein are susceptible to DNA crosslinking agents due to attenuated Src/nuclear EGFR cascade-activated DNA repair activity. Oncogene 32(9):1144–1154
127. Dittmann K, Mayer C, Rodemann HP (2005) Inhibition of radiation-induced EGFR nuclear import by C225 (cetuximab) suppresses DNA-PK activity. Radiother Oncol 76(2):157–161
128. Xu Y, Shao Y, Zhou J, Voorhees JJ, Fisher GJ (2009) Ultraviolet irradiation-induces epidermal growth factor receptor (EGFR) nuclear translocation in human keratinocytes. J Cell Biochem 107(5):873–880
129. Chen DJ, Nirodi CS (2007) The epidermal growth factor receptor: a role in repair of radiation-induced DNA damage. Clin Cancer Res 13(22 Pt 1):6555–6560
130. Szumiel I (2006) Epidermal growth factor receptor and DNA double strand break repair: the cell's self-defence. Cell Signal 18(10):1537–1548
131. Umar A, Buermeyer AB, Simon JA, Thomas DC, Clark AB, Liskay RM et al (1996) Requirement for PCNA in DNA mismatch repair at a step preceding DNA resynthesis. Cell 87(1):65–73
132. Luo B, Yu S, Zhuang L, Xia S, Zhao Z, Rong L (2009) Induction of ERBB2 nuclear transport after radiation in breast cancer cells. J Huazhong Univ Sci Technol Med Sci 29(3):350–353
133. Boone JJ, Bhosle J, Tilby MJ, Hartley JA, Hochhauser D (2009) Involvement of the HER2 pathway in repair of DNA damage produced by chemotherapeutic agents. Mol Cancer Ther 8(11):3015–3023
134. Icli B, Bharti A, Pentassuglia L, Peng X, Sawyer DB (2012) ErbB4 Localization to cardiac myocyte nuclei, and its role in myocyte DNA damage response. Biochem Biophys Res Commun 418(1):116–121
135. Robertson DM, Zhu M, Wu YC (2012) Cellular distribution of the IGF-1R in corneal epithelial cells. Exp Eye Res 94(1):179–186
136. Migdall-Wilson J, Bates C, Schlegel J, Brandao L, Linger RM, DeRyckere D et al (2012) Prolonged exposure to a Mer ligand in leukemia: Gas6 favors expression of a partial Mer glycoform and reveals a novel role for Mer in the nucleus. PLoS One 7(2):e31635
137. Seol KC, Kim SJ (2003) Nuclear matrix association of insulin receptor and IRS-1 by insulin in osteoblast-like UMR-106 cells. Biochem Biophys Res Commun 306(4):898–904
138. Stachowiak MK, Maher PA, Joy A, Mordechai E, Stachowiak EK (1996) Nuclear localization of functional FGF receptor 1 in human astrocytes suggests a novel mechanism for growth factor action. Brain Res Mol Brain Res 38(1):161–165
139. Bueter W, Dammann O, Zscheppang K, Korenbaum E, Dammann CE (2006) ErbB receptors in fetal endothelium – a potential linkage point for inflammation-associated neonatal disorders. Cytokine 36(5–6):267–275
140. Li LY, Chen H, Hsieh YH, Wang YN, Chu HJ, Chen YH et al (2011) Nuclear ErbB2 enhances translation and cell growth by activating transcription of ribosomal RNA genes. Cancer Res 71(12):4269–4279
141. Thompson M, Lauderdale S, Webster MJ, Chong VZ, McClintock B, Saunders R et al (2007) Widespread expression of ErbB2, ErbB3 and ErbB4 in non-human primate brain. Brain Res 1139:95–109

142. Hernandez-Verdun D, Roussel P, Thiry M, Sirri V, Lafontaine DL (2010) The nucleolus: structure/function relationship in RNA metabolism. Wiley Interdiscip Rev RNA 1(3): 415–431
143. Mao YS, Zhang B, Spector DL (2011) Biogenesis and function of nuclear bodies. Trends Genet 27(8):295–306
144. Offterdinger M, Schofer C, Weipoltshammer K, Grunt TW (2002) c-erbB-3: a nuclear protein in mammary epithelial cells. J Cell Biol 157(6):929–939
145. Rodrigues MA, Gomes DA, Andrade VA, Leite MF, Nathanson MH (2008) Insulin induces calcium signals in the nucleus of rat hepatocytes. Hepatology 48(5):1621–1631
146. Gomes DA, Rodrigues MA, Leite MF, Gomez MV, Varnai P, Balla T et al (2008) c-Met must translocate to the nucleus to initiate calcium signals. J Biol Chem 283(7):4344–4351
147. Hardingham GE, Chawla S, Johnson CM, Bading H (1997) Distinct functions of nuclear and cytoplasmic calcium in the control of gene expression. Nature 385(6613):260–265
148. Rodrigues MA, Gomes DA, Leite MF, Grant W, Zhang L, Lam W et al (2007) Nucleoplasmic calcium is required for cell proliferation. J Biol Chem 282(23):17061–17068
149. Bennasroune A, Gardin A, Aunis D, Cremel G, Hubert P (2004) Tyrosine kinase receptors as attractive targets of cancer therapy. Crit Rev Oncol Hematol 50(1):23–38
150. Fischer OM, Gschwind A, Ullrich AA (2004) Cell surface growth factor receptor molecules as targets for cancer therapy. Discov Med 4(22):166–171
151. Takeuchi K, Ito F (2011) Receptor tyrosine kinases and targeted cancer therapeutics. Biol Pharm Bull 34(12):1774–1780
152. Seshacharyulu P, Ponnusamy MP, Haridas D, Jain M, Ganti AK, Batra SK (2012) Targeting the EGFR signaling pathway in cancer therapy. Expert Opin Ther Targets 16(1):15–31
153. Rosenzweig SA (2012) Acquired resistance to drugs targeting receptor tyrosine kinases. Biochem Pharmacol 83(8):1041–1048
154. Patel R, Leung HY (2012) Targeting the EGFR-family for therapy: biological challenges and clinical perspective. Curr Pharm Des 18(19):2672–2679
155. Lurje G, Lenz HJ (2009) EGFR signaling and drug discovery. Oncology 77(6):400–410
156. Di Maio M, Gridelli C, Normanno N, Perrone F, Ciardiello F (2005) Trying to compose the puzzle with all the pieces: epidermal growth factor receptor tyrosine kinase inhibitors in non-small cell lung cancer. J Cell Physiol 205(3):355–363
157. Normanno N, De Luca A, Maiello MR, Mancino M, D'Antonio A, Macaluso M et al (2005) Epidermal growth factor receptor (EGFR) tyrosine kinase inhibitors in breast cancer: current status and future development. Front Biosci 10:2611–2617
158. Redmond KM, Wilson TR, Johnston PG, Longley DB (2008) Resistance mechanisms to cancer chemotherapy. Front Biosci 13:5138–5154
159. Giaccone G, Wang Y (2011) Strategies for overcoming resistance to EGFR family tyrosine kinase inhibitors. Cancer Treat Rev 37(6):456–464
160. Dittmann K, Mayer C, Kehlbach R, Rodemann HP (2008) The radioprotector Bowman-Birk proteinase inhibitor stimulates DNA repair via epidermal growth factor receptor phosphorylation and nuclear transport. Radiother Oncol 86(3):375–382
161. Yu YL, Chou RH, Wu CH, Wang YN, Chang WJ, Tseng YJ et al (2012) Nuclear EGFR suppresses ribonuclease activity of polynucleotide phosphorylase through DNAPK-mediated phosphorylation at serine 776. J Biol Chem 287(37):31015–31026
162. Chen YJ, Huang WC, Wei YL, Hsu SC, Yuan P, Lin HY et al (2011) Elevated BCRP/ABCG2 expression confers acquired resistance to gefitinib in wild-type EGFR-expressing cells. PLoS One 6(6):e21428
163. Liao HJ, Carpenter G (2009) Cetuximab/C225-induced intracellular trafficking of epidermal growth factor receptor. Cancer Res 69(15):6179–6183
164. Li C, Iida M, Dunn EF, Ghia AJ, Wheeler DL (2009) Nuclear EGFR contributes to acquired resistance to cetuximab. Oncogene 28(43):3801–3813
165. Ryu SH, Lee SW, Yang YJ, Song SY, Kim JH, Choi EK et al (2012) Intracytoplasmic epidermal growth factor receptor shows poor response to the cetuximab antitumor effect in irradiated non-small cell lung cancer cell lines. Lung Cancer 77(3):482–487

166. Tan M, Jing T, Lan KH, Neal CL, Li P, Lee S et al (2002) Phosphorylation on tyrosine-15 of p34(Cdc2) by ErbB2 inhibits p34(Cdc2) activation and is involved in resistance to taxol-induced apoptosis. Mol Cell 9(5):993–1004
167. Scaltriti M, Rojo F, Ocana A, Anido J, Guzman M, Cortes J et al (2007) Expression of p95HER2, a truncated form of the HER2 receptor, and response to anti-HER2 therapies in breast cancer. J Natl Cancer Inst 99(8):628–638
168. Anido J, Scaltriti M, Bech Serra JJ, Santiago Josefat B, Todo FR, Baselga J et al (2006) Biosynthesis of tumorigenic HER2 C-terminal fragments by alternative initiation of translation. EMBO J 25(13):3234–3244
169. Xia W, Liu Z, Zong R, Liu L, Zhao S, Bacus SS et al (2011) Truncated ErbB2 expressed in tumor cell nuclei contributes to acquired therapeutic resistance to ErbB2 kinase inhibitors. Mol Cancer Ther 10(8):1367–1374
170. Sharma SV, Lee DY, Li B, Quinlan MP, Takahashi F, Maheswaran S et al (2010) A chromatin-mediated reversible drug-tolerant state in cancer cell subpopulations. Cell 141(1):69–80
171. Bodzin AS, Wei Z, Hurtt R, Gu T, Doria C (2012) Gefitinib resistance in HCC mahlavu cells: upregulation of CD133 expression, activation of IGF-1R signaling pathway, and enhancement of IGF-1R nuclear translocation. J Cell Physiol 227(7):2947–2952
172. Clevers H (2011) The cancer stem cell: premises, promises and challenges. Nat Med 17(3):313–319
173. Yu Y, Ramena G, Elble RC (2012) The role of cancer stem cells in relapse of solid tumors. Front Biosci 4:1528–1541 (Elite Ed)
174. Stachowiak MK, Maher PA, Stachowiak EK (2007) Integrative nuclear signaling in cell development–a role for FGF receptor-1. DNA Cell Biol 26(12):811–826
175. Wang YN, Lee HH, Lee HJ, Du Y, Yamaguchi H, Hung MC (2012) Membrane-bound trafficking regulates nuclear transport of integral epidermal growth factor receptor (EGFR) and ErbB-2. J Biol Chem 287(20):16869–16879
176. Du Y, Shen J, Hsu J, Han Z, Hsu M, Yang C et al (2013) Syntaxin 6-mediated Golgi translocation plays an important role in nuclear functions of EGFR through microtubule-dependent trafficking. Oncogene. In press
177. Dittmann K, Mayer C, Kehlbach R, Rodemann HP (2008) Radiation-induced caveolin-1 associated EGFR internalization is linked with nuclear EGFR transport and activation of DNA-PK. Mol Cancer 7:69
178. Dittmann K, Mayer C, Kehlbach R, Rothmund MC, Rodemann Peter H (2009) Radiation-induced lipid peroxidation activates src kinase and triggers nuclear EGFR transport. Radiother Oncol 92(3):379–382
179. Iida M, Brand TM, Campbell DA, Li C, Wheeler DL (2013) Yes and Lyn play a role in nuclear translocation of the epidermal growth factor receptor. Oncogene 32(6):759–767
180. Li C, Iida M, Dunn EF, Wheeler DL (2010) Dasatinib blocks cetuximab- and radiation-induced nuclear translocation of the epidermal growth factor receptor in head and neck squamous cell carcinoma. Radiother Oncol 97(2):330–337
181. Nowsheen S, Bonner JA, Yang ES (2011) The poly(ADP-ribose) polymerase inhibitor ABT-888 reduces radiation-induced nuclear EGFR and augments head and neck tumor response to radiotherapy. Radiother Oncol 99(3):331–338
182. Godin-Heymann N, Bryant I, Rivera MN, Ulkus L, Bell DW, Riese DJ 2nd et al (2007) Oncogenic activity of epidermal growth factor receptor kinase mutant alleles is enhanced by the T790M drug resistance mutation. Cancer Res 67(15):7319–7326
183. Conway-Campbell BL, Brooks AJ, Robinson PJ, Perani M, Waters MJ (2008) The extracellular domain of the growth hormone receptor interacts with coactivator activator to promote cell proliferation. Mol Endocrinol 22(9):2190–2202
184. Chiu HC, Chang TY, Huang CT, Chao YS, Hsu JT (2012) EGFR and myosin II inhibitors cooperate to suppress EGFR-T790M-mutant NSCLC cells. Mol Oncol 6(3):299–310
185. Reilly RM, Kiarash R, Cameron RG, Porlier N, Sandhu J, Hill RP et al (2000) 111In-labeled EGF is selectively radiotoxic to human breast cancer cells overexpressing EGFR. J Nucl Med 41(3):429–438

186. Cai Z, Chen Z, Bailey KE, Scollard DA, Reilly RM, Vallis KA (2008) Relationship between induction of phosphorylated H2AX and survival in breast cancer cells exposed to 111In-DTPA-hEGF. J Nucl Med 49(8):1353–1361
187. Hu M, Scollard D, Chan C, Chen P, Vallis K, Reilly RM (2007) Effect of the EGFR density of breast cancer cells on nuclear importation, in vitro cytotoxicity, and tumor and normal-tissue uptake of [111In]DTPA-hEGF. Nucl Med Biol 34(8):887–896
188. Chen P, Cameron R, Wang J, Vallis KA, Reilly RM (2003) Antitumor effects and normal tissue toxicity of 111In-labeled epidermal growth factor administered to athymic mice bearing epidermal growth factor receptor-positive human breast cancer xenografts. J Nucl Med 44(9):1469–1478
189. Bailey KE, Costantini DL, Cai Z, Scollard DA, Chen Z, Reilly RM et al (2007) Epidermal growth factor receptor inhibition modulates the nuclear localization and cytotoxicity of the auger electron emitting radiopharmaceutical 111In-DTPA human epidermal growth factor. J Nucl Med 48(9):1562–1570
190. Cornelissen B, Darbar S, Hernandez R, Kersemans V, Tullis I, Barber PR et al (2011) ErbB-2 blockade and prenyltransferase inhibition alter epidermal growth factor and epidermal growth factor receptor trafficking and enhance (111)in-DTPA-hEGF auger electron radiation therapy. J Nucl Med 52(5):776–783
191. Reilly RM, Chen P, Wang J, Scollard D, Cameron R, Vallis KA (2006) Preclinical pharmacokinetic, biodistribution, toxicology, and dosimetry studies of 111In-DTPA-human epidermal growth factor: an auger electron-emitting radiotherapeutic agent for epidermal growth factor receptor-positive breast cancer. J Nucl Med 47(6):1023–1031
192. Fasih A, Fonge H, Cai Z, Leyton JV, Tikhomirov I, Done SJ et al (2012) In-Bn-DTPA-nimotuzumab with/without modification with nuclear translocation sequence (NLS) peptides: an auger electron-emitting radioimmunotherapeutic agent for EGFR-positive and trastuzumab (herceptin)-resistant breast cancer. Breast Cancer Res Treat 135(1):189–200 (1)(1)(1)
193. Hung MC, Link W (2011) Protein localization in disease and therapy. J Cell Sci 124 (Pt 20):3381–3392
194. De Angelis Campos AC, Rodrigues MA, de Andrade C, de Goes AM, Nathanson MH, Gomes DA (2011) Epidermal growth factor receptors destined for the nucleus are internalized via a clathrin-dependent pathway. Biochem Biophys Res Commun 412(2):341–346
195. Li A, Zhang C, Gao S, Chen F, Yang C, Luo R et al (2013) TIP30 loss enhances cytoplasmic and nuclear EGFR signaling and promotes lung adenocarcinogenesis in mice. Oncogene 32(18):2273–2281
196. de la Iglesia N, Konopka G, Puram SV, Chan JA, Bachoo RM, You MJ et al (2008) Identification of a PTEN-regulated STAT3 brain tumor suppressor pathway. Genes Dev 22(4):449–462
197. Marti U, Ruchti C, Kampf J, Thomas GA, Williams ED, Peter HJ et al (2001) Nuclear localization of epidermal growth factor and epidermal growth factor receptors in human thyroid tissues. Thyroid 11(2):137–145
198. Redmond AM, Bane FT, Stafford AT, McIlroy M, Dillon MF, Crotty TB et al (2009) Coassociation of estrogen receptor and p160 proteins predicts resistance to endocrine treatment; SRC-1 is an independent predictor of breast cancer recurrence. Clin Cancer Res 15(6):2098–2106
199. Valdez KE, Fan F, Smith W, Allred DC, Medina D, Behbod F (2011) Human primary ductal carcinoma in situ (DCIS) subtype-specific pathology is preserved in a mouse intraductal (MIND) xenograft model. J Pathol 225(4):565–573
200. Menendez JA, Vellon L, Mehmi I, Oza BP, Ropero S, Colomer R et al (2004) Inhibition of fatty acid synthase (FAS) suppresses HER2/neu (erbB-2) oncogene overexpression in cancer cells. Proc Natl Acad Sci USA 101(29):10715–10720
201. Fukuoka H, Cooper O, Mizutani J, Tong Y, Ren SG, Bannykh S et al (2011) HER2/ErbB2 Receptor signaling in rat and human prolactinoma cells: strategy for targeted prolactinoma therapy. Mol Endocrinol 25(1):92–103
202. Raabe TD, Deadwyler G, Varga JW, Devries GH (2004) Localization of neuregulin isoforms and erbB receptors in myelinating glial cells. Glia 45(2):197–207

203. Koumakpayi IH, Le Page C, Delvoye N, Saad F, Mes-Masson AM (2011) Macropinocytosis inhibitors and Arf6 regulate ErbB3 nuclear localization in prostate cancer cells. Mol Carcinog 50(11):901–912
204. Naresh A, Thor AD, Edgerton SM, Torkko KC, Kumar R, Jones FE (2008) The HER4/4ICD estrogen receptor coactivator and BH3-only protein is an effector of tamoxifen-induced apoptosis. Cancer Res 68(15):6387–6395
205. Thor AD, Edgerton SM, Jones FE (2009) Subcellular localization of the HER4 intracellular domain, 4ICD, identifies distinct prognostic outcomes for breast cancer patients. Am J Pathol 175(5):1802–1809
206. Hu Y, Fang X, Dunham SM, Prada C, Stachowiak EK, Stachowiak MK (2004) 90-kDa ribosomal S6 kinase is a direct target for the nuclear fibroblast growth factor receptor 1 (FGFR1): role in FGFR1 signaling. J Biol Chem 279(28):29325–29335
207. Dunham-Ems SM, Lee YW, Stachowiak EK, Pudavar H, Claus P, Prasad PN et al (2009) Fibroblast growth factor receptor-1 (FGFR1) nuclear dynamics reveal a novel mechanism in transcription control. Mol Biol Cell 20(9):2401–2412
208. Giulianelli S, Cerliani JP, Lamb CA, Fabris VT, Bottino MC, Gorostiaga MA et al (2008) Carcinoma-associated fibroblasts activate progesterone receptors and induce hormone independent mammary tumor growth: a role for the FGF-2/FGFR-2 axis. Int J Cancer 123(11):2518–2531
209. Cerliani JP, Vanzulli SI, Pinero CP, Bottino MC, Sahores A, Nunez M et al (2012) Associated expressions of FGFR-2 and FGFR-3: from mouse mammary gland physiology to human breast cancer. Breast Cancer Res Treat 133(3):997–1008
210. Gatius S, Velasco A, Azueta A, Santacana M, Pallares J, Valls J et al (2011) FGFR2 Alterations in endometrial carcinoma. Mod Pathol 24(11):1500–1510
211. Martin AJ, Grant A, Ashfield AM, Palmer CN, Baker L, Quinlan PR et al (2011) FGFR2 Protein expression in breast cancer: nuclear localisation and correlation with patient genotype. BMC Res Notes 4:72
212. Lu P, Ewald AJ, Martin GR, Werb Z (2008) Genetic mosaic analysis reveals FGF receptor 2 function in terminal end buds during mammary gland branching morphogenesis. Dev Biol 321(1):77–87
213. Nyeng P, Norgaard GA, Kobberup S, Jensen J (2007) FGF10 Signaling controls stomach morphogenesis. Dev Biol 303(1):295–310
214. Sabbieti MG, Marchetti L, Gabrielli MG, Menghi M, Materazzi S, Menghi G et al (2005) Prostaglandins differently regulate FGF-2 and FGF receptor expression and induce nuclear translocation in osteoblasts via MAPK kinase. Cell Tissue Res 319(2):267–278
215. Schmahl J, Kim Y, Colvin JS, Ornitz DM, Capel B (2004) Fgf9 Induces proliferation and nuclear localization of FGFR2 in sertoli precursors during male sex determination. Development 131(15):3627–3636
216. Gowardhan B, Douglas DA, Mathers ME, McKie AB, McCracken SR, Robson CN et al (2005) Evaluation of the fibroblast growth factor system as a potential target for therapy in human prostate cancer. Br J Cancer 92(2):320–327
217. Kuroso K, Imai Y, Kobayashi M, Yanagimoto K, Suzuki T, Kojima M et al (2010) Immunohistochemical detection of fibroblast growth factor receptor 3 in human breast cancer: correlation with clinicopathological/molecular parameters and prognosis. Pathobiology 77(5):231–240
218. Zammit C, Barnard R, Gomm J, Coope R, Shousha S, Coombes C et al (2001) Altered intracellular localization of fibroblast growth factor receptor 3 in human breast cancer. J Pathol 194(1):27–34
219. Reid S, Ferretti P (2003) Differential expression of fibroblast growth factor receptors in the developing murine choroid plexus. Brain Res Dev Brain Res 141(1–2):15–24
220. Kiunga GA, Raju J, Sabljic N, Bajaj G, Good CK, Bird RP (2004) Elevated insulin receptor protein expression in experimentally induced colonic tumors. Cancer Lett 211(2):145–153
221. Mattarocci S, Abbruzzese C, Mileo AM, Visca P, Antoniani B, Alessandrini G et al (2009) Intracellular presence of insulin and its phosphorylated receptor in non-small cell lung cancer. J Cell Physiol 221(3):766–770

222. Gletsu NA, Field CJ, Clandinin MT (1999) Obese mice have higher insulin receptor levels in the hepatocyte cell nucleus following insulin stimulation in vivo with an oral glucose meal. Biochim Biophys Acta 1454(3):251–260
223. Podlecki DA, Smith RM, Kao M, Tsai P, Huecksteadt T, Brandenburg D et al (1987) Nuclear translocation of the insulin receptor. A possible mediator of insulin's long term effects. J Biol Chem 262(7):3362–3368
224. Kim SJ, Kahn CR (1993) Insulin induces rapid accumulation of insulin receptors and increases tyrosine kinase activity in the nucleus of cultured adipocytes. J Cell Physiol 157(2):217–228
225. Hoa N, Tsui S, Afifiyan NF, Sinha Hikim A, Li B, Douglas RS et al (2012) Nuclear targeting of IGF-1 receptor in orbital fibroblasts from Graves' disease: apparent role of ADAM17. PLoS One 7(4):e34173
226. Blazquez C, Cook N, Micklem K, Harris AL, Gatter KC, Pezzella F (2006) Phosphorylated KDR can be located in the nucleus of neoplastic cells. Cell Res 16(1):93–98
227. Blazquez C, Gonzalez-Feria L, Alvarez L, Haro A, Casanova ML, Guzman M (2004) Cannabinoids inhibit the vascular endothelial growth factor pathway in gliomas. Cancer Res 64(16):5617–5623
228. Mylona E, Alexandrou P, Mpakali A, Giannopoulou I, Liapis G, Markaki S et al (2007) Clinicopathological and prognostic significance of vascular endothelial growth factors (VEGF)-C and -D and VEGF receptor 3 in invasive breast carcinoma. Eur J Surg Oncol 33(3):294–300
229. Santos SC, Dias S (2004) Internal and external autocrine VEGF/KDR loops regulate survival of subsets of acute leukemia through distinct signaling pathways. Blood 103(10):3883–3889
230. Stewart M, Turley H, Cook N, Pezzella F, Pillai G, Ogilvie D et al (2003) The angiogenic receptor KDR is widely distributed in human tissues and tumours and relocates intracellularly on phosphorylation. An immunohistochemical study. Histopathology 43(1):33–39
231. Andersson MK, Goransson M, Olofsson A, Andersson C, Aman P (2010) Nuclear expression of FLT1 and its ligand PGF in FUS-DDIT3 carrying myxoid liposarcomas suggests the existence of an intracrine signaling loop. BMC Cancer 10:249
232. Zhang Z, Neiva KG, Lingen MW, Ellis LM, Nor JE (2010) VEGF-dependent tumor angiogenesis requires inverse and reciprocal regulation of VEGFR1 and VEGFR2. Cell Death Differ 17(3):499–512
233. Dardik R, Inbal A (2006) Complex formation between tissue transglutaminase II (tTG) and vascular endothelial growth factor receptor 2 (VEGFR-2): proposed mechanism for modulation of endothelial cell response to VEGF. Exp Cell Res 312(16):2973–2982
234. Feng Y, Venema VJ, Venema RC, Tsai N, Caldwell RB (1999) VEGF induces nuclear translocation of Flk-1/KDR, endothelial nitric oxide synthase, and caveolin-1 in vascular endothelial cells. Biochem Biophys Res Commun 256(1):192–197
235. Santos SC, Miguel C, Domingues I, Calado A, Zhu Z, Wu Y et al (2007) VEGF and VEGFR-2 (KDR) internalization is required for endothelial recovery during wound healing. Exp Cell Res 313(8):1561–1574
236. Susarla R, Gonzalez AM, Watkinson JC, Eggo MC (2012) Expression of receptors for VEGFs on normal human thyroid follicular cells and their role in follicle formation. J Cell Physiol 227(5):1992–2002
237. Thuringer D, Maulon L, Frelin C (2002) Rapid transactivation of the vascular endothelial growth factor receptor KDR/Flk-1 by the bradykinin B2 receptor contributes to endothelial nitric-oxide synthase activation in cardiac capillary endothelial cells. J Biol Chem 277(3):2028–2032
238. Hagedorn A, Germann PG, Junker-Walker U, Tomovic A, Seewald W, Polkinghorne A et al (2005) Immunohistochemical study about the Flt-1/VEGFR1 expression in the gastrointestinal tract of mouse, rat, dog, swine and monkey. Exp Toxicol Pathol 57(2):149–159
239. Ma PC, Jagadeeswaran R, Jagadeesh S, Tretiakova MS, Nallasura V, Fox EA et al (2005) Functional expression and mutations of c-Met and its therapeutic inhibition with SU11274 and small interfering RNA in non-small cell lung cancer. Cancer Res 65(4):1479–1488

240. Pozner-Moulis S, Pappas DJ, Rimm DL (2006) Met, the hepatocyte growth factor receptor, localizes to the nucleus in cells at low density. Cancer Res 66(16):7976–7982
241. Abdel-Rahman MH, Boru G, Massengill J, Salem MM, Davidorf FH (2010) MET oncogene inhibition as a potential target of therapy for uveal melanomas. Invest Ophthalmol Vis Sci 51(7):3333–3339
242. Ye M, Hu D, Tu L, Zhou X, Lu F, Wen B et al (2008) Involvement of PI3K/Akt signaling pathway in hepatocyte growth factor-induced migration of uveal melanoma cells. Invest Ophthalmol Vis Sci 49(2):497–504
243. Saitoh K, Takahashi H, Sawada N, Parsons PG (1994) Detection of the c-met proto-oncogene product in normal skin and tumours of melanocytic origin. J Pathol 174(3):191–199
244. Moughal NA, Waters C, Sambi B, Pyne S, Pyne NJ (2004) Nerve growth factor signaling involves interaction between the Trk a receptor and lysophosphatidate receptor 1 systems: nuclear translocation of the lysophosphatidate receptor 1 and Trk a receptors in pheochromocytoma 12 cells. Cell Signal 16(1):127–136
245. Gong A, Zhang Z, Xiao D, Yang Y, Wang Y, Chen Y (2007) Localization of phosphorylated TrkA in carrier vesicles involved in its nuclear translocation in U251 cell line. Sci China C Life Sci 50(2):141–146
246. Mufson EJ, Presley LN, Kordower JH (1991) Nerve growth factor receptor immunoreactivity within the nucleus basalis (Ch4) in Parkinson's disease: reduced cell numbers and co-localization with cholinergic neurons. Brain Res 539(1):19–30
247. Bonacchi A, Taddei ML, Petrai I, Efsen E, Defranco R, Nosi D et al (2008) Nuclear localization of TRK-a in liver cells. Histol Histopathol 23(3):327–340
248. Munoz M, Rodriguez A, Diez C, Caamano JN, Fernandez-Sanchez MT, Perez-Gomez A et al (2009) Tyrosine kinase a, C and fibroblast growth factor-2 receptors in bovine embryos cultured in vitro. Theriogenology 71(6):1005–1010
249. Marti U, Wells A (2000) The nuclear accumulation of a variant epidermal growth factor receptor (EGFR) lacking the transmembrane domain requires coexpression of a full-length EGFR. Mol Cell Biol Res Commun 3(1):8–14
250. Castillo M, Petit A, Mellado B, Palacin A, Alcover JB, Mallofre C (2004) C-kit expression in sarcomatoid renal cell carcinoma: potential therapy with Imatinib. J Urol 171(6 Pt 1):2176–2180
251. Hong SM, Hwang I, Song DE, Choi J, Yu E (2007) Clinical and prognostic significances of nuclear and cytoplasmic KIT expressions in extrahepatic bile duct carcinomas. Mod Pathol 20(5):562–569
252. Di Matteo G, Pescarmona E, Peparini N, Di Matteo FM, Zeri KP, Mascagni D et al (2002) Histopathological features and clinical course of the gastrointestinal stromal tumors. Hepatogastroenterology 49(46):1013–1016
253. Liao HJ, Carpenter G (2012) Regulated intramembrane cleavage of the EGF receptor. Traffic 13(8):1106–1112
254. Degnin CR, Laederich MB, Horton WA (2011) Ligand activation leads to regulated intramembrane proteolysis of fibroblast growth factor receptor 3. Mol Biol Cell 22(20):3861–3873
255. Cai J, Jiang WG, Grant MB, Boulton M (2006) Pigment epithelium-derived factor inhibits angiogenesis via regulated intracellular proteolysis of vascular endothelial growth factor receptor 1. J Biol Chem 281(6):3604–3613
256. Wu LW (ed) (2010) Evidence and actions for nuclear localization of vascular endothelial growth factor 1 (VEGFR1) in endothelial and cancer cells. Proceedings of the 101st annual meeting of the American Association for Cancer Research, April 2010. AACR, Washington, DC, Philadelphia, pp 17–21

# Chapter 9
# Nucleolar Signaling Determines Cell Fate: The RP-Mdm2-p53 Axis Fine-Tunes Cellular Homeostasis

**Yong Liu and Yanping Zhang**

**Abstract** One of the main functions of the nucleolus is to conduct ribosome biogenesis, which is the most energy-consuming process in growing cells. Not surprisingly, this process is highly regulated during cellular proliferation when energy consumption levels are relatively high. Various stresses such as genotoxic, osmotic and oncogenic stress as well as metabolic fluctuations converge at the nucleolus and impinge on ribosome biogenesis. Depending on the severity of the insult, nucleolar stress signaling can induce cell cycle arrest, apoptosis or metabolic adaptation. Although mechanisms associated with the nucleolar stress response are complex and remain to be fully elucidated, many of the pathways that convert stress signals into a cellular response link the nucleolus to the tumor suppressor and guardian of the genome known as p53. It is now known that the activation of p53 upon nucleolar signaling occurs predominantly through the ribosomal protein (RP)-Mdm2-p53 axis.

**Keywords** Ribosome protein • Ribosome biosynthesis • Mdm2 • p53 • c-Myc • Cancer • Metabolism

Y. Liu, Ph.D.
Department of Radiation Oncology, Lineberger Comprehensive Cancer Center, School of Medicine, University of North Carolina at Chapel Hill, Chapel Hill, NC 27514, USA
e-mail: ypzhang@med.unc.edu; yongl@email.unc.edu

Y. Zhang, Ph.D. (✉)
Department of Radiation Oncology, Lineberger Comprehensive Cancer Center, School of Medicine, University of North Carolina at Chapel Hill, Chapel Hill, NC 27514, USA

Laboratory of Biological Cancer Therapy, Xuzhou Medical College, Xuzhou, Jiangsu, China
e-mail: ypzhang@med.unc.edu

R. Kumar (ed.), *Nuclear Signaling Pathways and Targeting Transcription in Cancer*, Cancer Drug Discovery and Development, DOI 10.1007/978-1-4614-8039-6_9,

## 9.1 Ribosome Biogenesis

The nucleolus is a non-membrane-bound subnuclear structure found in eukaryotic cells. The nucleolus serves as the cellular location for the assembly of small and large ribosomal subunits, which is crucial to ensure adequate protein synthesis and to maintain cellular homeostasis. The entire process of ribosome biogenesis has been estimated to occupy nearly 60 % of the cellular resources [1]. Ribosome biogenesis consists of three events, which occur in different subregions of the nucleolus: rRNA transcription, rRNA processing and ribosome assembly. Based on the morphology and composition as observed by electron microscopy (EM), the nucleolus can be divided into three functional compartments: the fibrillar center (FC), the dense fibrillar component (DFC) and the granular component (GC) (Fig. 9.1) [2].

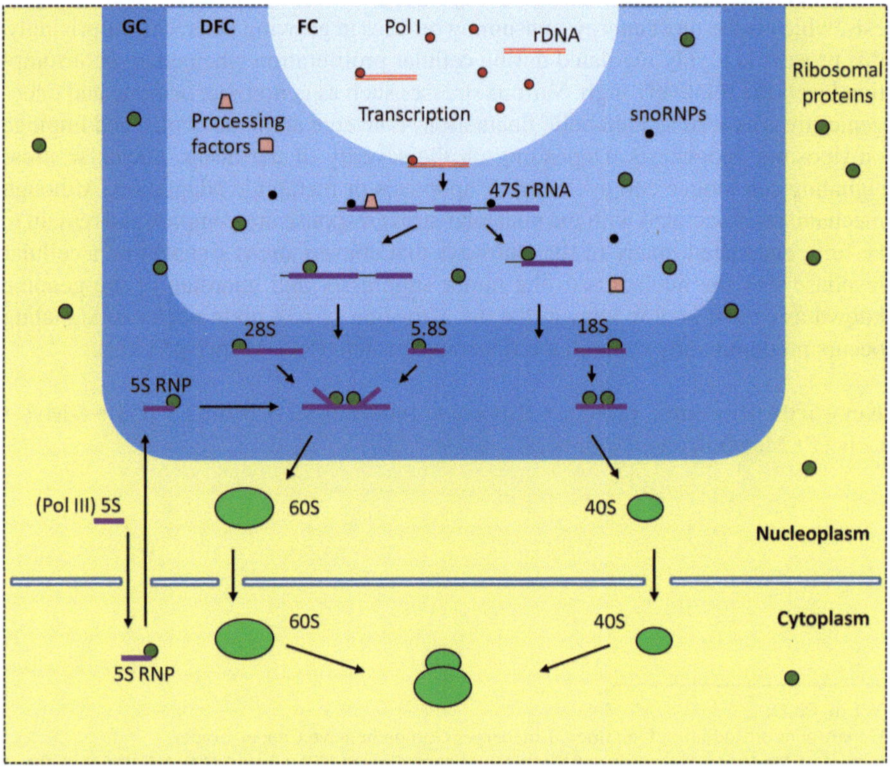


**Fig. 9.1 Ribosome biogenesis in the nucleolus.** RNA Pol I catalyzes the initial transcription of clusters of rDNA tandem repeat genes into the 47S precursor rRNA either in the FC or at the border between the FC and the DFC. After transcription, the nascent precursors are post-transcriptionally modified through their interaction with snoRNPs and additional processing factors, and the nascent precursors are subsequently processed to form the 5.8S, 18S and 28S mature rRNAs in the DFC. 5S rRNA is transcribed separately by RNA polymerase III (Pol III) in the nucleus, exported to the cytoplasm where it binds to the ribosomal protein L5 to form a complex (5S RNP) and then imported into the nucleolus for incorporation into the large subunit. Once processed or modified, the rRNAs enter the GC region for final maturation and then assemble with many additional ribosomal proteins before becoming transported to the cytoplasm to facilitate protein synthesis

The FC is enriched in components of the RNA polymerase I (PolI) machinery, whereas the DFC harbors rRNA processing factors such as the small nucleolar ribonucleic acids (snoRNAs) and the small nucleolar ribonucleoproteins (snRNPs). The FC and the DFC are enclosed by the GC region, where the final steps of ribosome subunit assembly are conducted.

Based on current evidence, ribosome biogenesis is thought to require the production of roughly equimolar amounts of rRNAs and ribosomal proteins (RPs) to generate 60S and 40S subunits, which mature into 80S polysomes. This process requires the coordinated regulation of all three nuclear RNA polymerases. In humans, RNA PolI catalyzes the initial transcription of clusters of rDNA tandem repeat genes into the 47S precursor rRNA either in the FC or at the border between the FC and the DFC. After transcription, the nascent precursors are post-transcriptionally modified through interactions with snoRNPs and additional processing factors, and the rRNA transcripts are subsequently processed to form the 5.8S, 18S and 28S mature rRNAs in the DFC. In contrast, the smallest component of the large ribosomal subunit, 5S rRNA is transcribed separately by RNA polymerase III (PolIII) in the nucleus. After transcription, 5S rRNA is exported to the cytoplasm where it binds to the ribosomal protein L5 to form a complex (5S RNP), after which the 5S RNP is imported into the nucleolus to be incorporated into the large ribosomal subunit [3]. After processing and modification, the rRNAs enter the GC region where they mature and become assembled along with several ribosomal proteins before being transported to the cytoplasm to conduct protein synthesis. In eukaryotes, 79 ribosomal proteins have been identified, 33 of which are incorporated into the 40S subunit and 46 of which are incorporated into the 60S subunit. All of the genes that encode the ribosomal proteins are exclusively transcribed by RNA PolII [2, 4].

Aside from the ribosomal proteins that are incorporated into the ribosomal subunits, our knowledge on the protein content of the nucleolus remains quite limited. In the past 10 years, proteomic analyses have identified more than 6,000 human proteins that stably co-purify with isolated nucleoli; however, less than 10 % of these proteins have been functionally characterized [5–8]. Among the nucleolus-associated proteins that have been characterized, approximately 30 % directly assist with the production of ribosomal subunits; however, the functions of the remaining 70 % of nucleolar proteins fall into several groups including RNA processing, DNA replication and repair, cell cycle control and apoptosis [2]. Interestingly, additional studies have indicated that the protein content of the nucleolus is dynamic and can be altered in response to various stress stimuli suggesting a complex reorganization of the nucleolus during stress responses [9–11].

## 9.2 Ribosome Biogenesis and Cancer

### *9.2.1 Oncogene- and Tumor Suppressor-Dependent Regulation of Ribosome Biogenesis*

Based on the fact that ribosome biogenesis is a highly complex and energy-demanding process, one may predict that this process is under constant surveillance

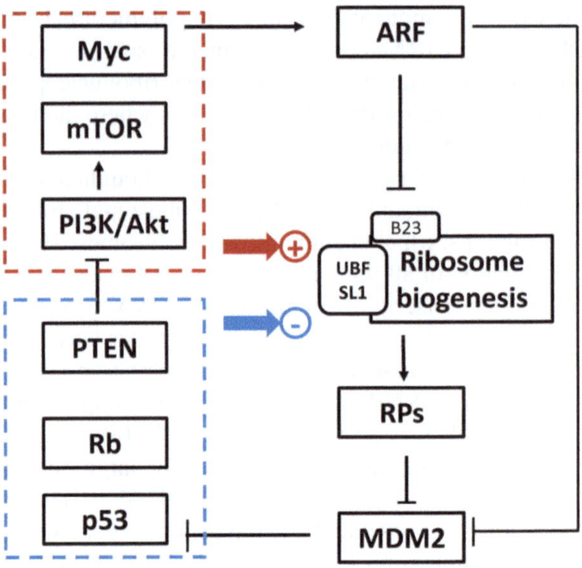


**Fig. 9.2 Crosstalk between oncogenes and tumor suppressor genes on ribosome biogenesis.** The oncogene Myc promotes Pol I-dependent rDNA transcription by directly binding to SL1 or UBF. At the same time, Myc activates Pol II and promotes RP translation. The PI3K/Akt pathway regulates ribosome biogenesis either through the phosphorylation of S6K to promote 5′-TOP gene expression (RPs) or by promoting rRNA processing and maturation. mTOR promotes rRNA synthesis by activating UBF. Akt also contributes to myc-mediated ribosome biogenesis, and PI3K/Akt signaling is antagonized by PTEN. Oncogenic expression of myc causes the induction of both ARF and RPs. ARF not only directly binds to B23 and inhibits ribosome biogenesis but also interacts with Mdm2 and inhibits Mdm2 E3 ligase activity resulting in the stabilization of p53.p53 interferes with the assembly of the UBF-SL1-Pol I complex and directly inhibits rRNA synthesis. Rb can accumulate in the nucleolus and inhibit ribosome biogenesis by binding to UBF. In a negative feedback loop, myc-induced RP expression results in the inhibition of Mdm2 and further stabilizes p53

at multiple levels [2]. Indeed, the cell must maintain a delicate balance between cell cycle progression and ribosome production, and several key checkpoints exist to assist the cell in keeping this balance. Several tumor suppressors and proto-oncogenes such as p53, Mdm2, ARF, Rb and Myc are sequestered in the nucleoli and regulate aberrant cell division by altering the protein synthesis machinery and coordinating other non-nucleolar signaling pathways such as the PI3K-Akt-mTORC1 signaling axis (Fig. 9.2).

#### 9.2.1.1 Oncogenes and Ribosome Biogenesis

Myc, PI3K-Akt-mTORC1 and Ras-ERK are the three major signaling networks known to promote cell growth when activated, and all of these pathways can become oncogenic if aberrantly activated suggesting a tight correlation between the deregulation of protein biosynthesis and malignant transformation.

Several studies have suggested that Myc regulates ribosome biogenesis, total protein synthesis and cell growth. Overexpression of Myc in normal cells increases the size of the nucleus, the nucleolus and the cell overall, which is associated with cellular proliferation and self-renewal and is consistent with the observation that Myc over expression leads to tumorigenesis [12–14]. Conversely, reduced Myc levels in cancer cells decreases the overall cellular transcription level and may lead to differentiation [15, 16], senescence [17] or apoptosis [18] and a more benign phenotype. In vertebrates, Myc increases the expression of rRNA and ribosomal proteins, and some compelling evidence has suggested that Myc also affects ribosome biogenesis by regulating the expression of the auxiliary factors that are required for rRNA processing, ribosome assembly and the maturation of the ribosomal subunits.

Myc regulates transcription through all three RNA polymerases by binding to DNA as a heterodimer with Max and enabling the recognition of the E-box sequence CACGTG [19]. Grandori et al. have found that when associated with the Pol I-specific factor SL1 (TIF-IB), Myc localizes to the nucleoli at sites of active rDNA transcription and enhances Pol I-dependent gene transcription [20]. Myc also facilitates rDNA transcription by enhancing the expression and the recruitment of the Pol I cofactor known as the upstream binding transcription factor (UBF). SL1 and UBF are indispensible for the augmentation of Pol I-dependent transcription. Myc also directly binds to and activates TFIIIB and augments Pol III-mediated 5S and tRNA transcription [21]. Additionally, Myc activates Pol II-dependent transcription of a large number of genes that encode ribosomal proteins, ribosome assembly proteins and proteins involved in translation initiation and elongation [22–27]. These auxiliary factors include nucleolar protein 56 (NOP56), fibrillarin (FBL), dyskerin (DKC1), block of proliferation 1(BOP1), nucleolin (NCL) and B23 (NPM1 or nucleophosmin) [28]. Notably, B23 is involved in Myc-induced proliferation and transformation by directly interacting with Myc at target promoters of downstream genes. However, B23 has been shown to play an antagonistic role in proliferation by enhancing the stability and transactivation activity of p53, which is important for the maintenance of genomic integrity [29].

The ability of Myc to initiate and promote tumorigenesis is associated with its role in enhancing ribosome biogenesis, and this relationship is particularly evident with the role of Myc in the expression of the ribosomal genes [30, 31]. Several ribosomal proteins such as L19 [32], L7a, L37 [33], L15 [34] and L36a [35] are highly expressed in many different cancers. In addition, the deletion of one allele of L24 or L38 can significantly decrease the incidence of Eμ-Myc-induced lymphoma, which results in a significant delay in tumor onset [31], indicating that even a modest reduction of a single ribosomal protein may attenuate Myc-induced tumorigenesis. However, whether ribosomal proteins exclusively contribute to the transformation of cancer will require further investigation. Interestingly, ribosomal protein L11 has been shown to be a negative feedback regulator of Myc through multiple mechanisms. L11 can directly interact with Myc at the promoters of target genes reducing histone acetylation and consequently inhibiting the translation of Myc target genes [36]. L11 may also inhibit the transcriptional activity of Myc by sequestering Myc to the nucleolus while concomitantly reducing Myc transcript levels. Similarly, L5,

L23 and S7 also bind to Myc, although the functional consequences of these interactions have not been fully elucidated. Interestingly, the ribosomal proteins that bind to Myc have also been shown to interact with Mdm2 to promote p53 stability and activity. Taken together, ribosomal proteins appear to play a dual role in regulating cell proliferation and genomic instability [37, 38].

The PI3K-Akt-mTORC1 pathway regulates the multistep process of mRNA translation at every stage from ribosome biogenesis to translation initiation and elongation. The direct activation of mTOR (mammalian target of rapamycin) by Akt results in the activation of S6 kinase (S6K), which leads to the phosphorylation of ribosomal protein S6 [39]. S6 phosphorylation has been associated with the translation of 5′-TOP genes (terminal oligopyrimidine or tract of oligopyrimidine genes), which have been shown to predominantly encode RP genes, translation initiation factors and translation elongation factors [40].In addition, mTORC1 components RAPTOR (regulatory associated protein of mTOR)and mTOR are both present in the nucleolus. Through the function of mTOR, mTORC1 not only positively regulates the transcription of rRNA and the synthesis of ribosomal proteins, but it is also involved in the processing and maturation of rRNA molecules [41]. mTOR promotes rRNA synthesis through the activation of either transcription initiation factor IA (TIF-IA) or UBF [42]. A recent study also showed that continuous Akt-dependent, mTORC1-independent signaling is required for rRNA transcription elongation. Furthermore, Akt activity is required for the maximal activation of Myc-induced rRNA synthesis in lymphomas, and decreased Akt-dependent rRNA transcription is associated with enhanced apoptosis in lymphoma cells. Collectively, these studies define a network involving Akt, mTORC1 and Myc as master controllers of cell growth and transformation [43]. Additionally, Ras-Erk signaling intersects with both Myc and Akt pathways to regulate each other to promote cell survival [44, 45]. The Ras-Erk-RSKs (p90 ribosomal S6 kinase) axis has also been shown to phosphorylate S6 and to regulate the assembly of the translation preinitiation complex independent of mTOR [46].

#### 9.2.1.2 Tumor Suppressors Involved in Ribosome Biogenesis

Presumably due to the central role of ribosome biogenesis in cell growth, proliferation and tumorigenic transformation, several tumor suppressors such as Rb, p53 and PTEN closely monitor the fidelity of ribosome biogenesis. The tumor suppressor PTEN is mutated in many different types of tumors. Mice that are heterozygous for the PTEN gene are prone to developing tumors of different histological origins [47, 48]. Loss of PTEN function correlates with increases in Akt and S6K activity in tumors. PTEN acts as a phosphatase and is responsible for the dephosphorylation of phosphatidylinositol-3,4,5-triphosphate, which consequently results in the down regulation of the PI3K-Akt pathway. Through the inhibition of the PI3K-Akt-mTORC1 axis, PTEN inactivates S6K and inhibits ribosome biogenesis. Another tumor suppressor known to regulate ribosome biogenesis is the retinoblastoma tumor suppressor gene (Rb). Rb regulates cell cycle progression by directly inhibiting the transcriptional activity of E2F, which is necessary for S phase progression [49]. Rb has also been shown to modulate ribosome biogenesis. In differentiated cells, Rb accumulates in the nucleolus, and nonphosphorylated active Rb inhibits rRNA synthesis by directly binding to UBF and down regulating Pol I activity [50, 51]. Similarly to Rb, p53 has been shown to repress Pol I transcription by directly interfering with the assembly of the UBF-SL1-Pol I complex [52, 53].In addition to Pol I inhibition, p53 has also been shown to inhibit Pol III activity by binding to TFIIIB thereby compromising the function of TFIIIB in 5S rRNA and tRNA synthesis [54].In addition to modulating ribosome biogenesis, p53 plays a pivotal role in safeguarding deregulated ribosome biogenesis, and the mechanisms through which this occurs will be further discussed below.

## *9.2.2 Ribosomopathies and Their Associations with Cancer*

Almost all ribosome biogenesis factors including the ribosomal proteins themselves are essential for cell survival. Deficiencies in ribosome biogenesis interfere with developmental processes and cause genetic abnormalities, which eventually lead to a collection of human pathological conditions known as ribosomopathies. Over the past few decades, a growing variety of ribosomopathies have been reported, and nearly all of them appear to be associated with an increased incidence of cancer, although the type and frequency vary considerably.

### 9.2.2.1 Ribosomopathies of Ribosomal Proteins

The first characterized ribosomopathy is known as Diamond Blackfan anemia (DBA) and was originally described and categorized as a congenital hypoplastic anemia in the 1930s [55, 56]. DBA is associated with anemia, macrocytosis, reticulocytopenia and a selective decrease or absence of erythroid precursors in an otherwise normocellular bone marrow [57, 58]. In 1999, recurrent mutations were discovered in the ribosomal protein gene *RPS19* in patients with DBA suggesting the existence an association between mutations in genes that encode ribosomal proteins and DBA [59]. Subsequently, mutations in several other ribosomal proteins including RPS24, RPS17, RPL35A, RPS7, RPS15, RPS27A, RPL36, RPL5 and RPL11 have been identified in approximately 50 % of DBA patients [60–63]. Additional studies revealed that mutations in different ribosomal genes lead to distinct clinical phenotypes by impairing pre-rRNA processing and inhibiting the production of the corresponding ribosomal subunit [64, 65]. The predisposition of DBA patients to cancer remains unclear. Among the 568 DBA patients registered in Diamond-Blackfan Anemia Registry of North America, 15 cancers have been identified in 13 of these patients, which includes four patients with myelodysplastic syndrome (MDS)/acute myeloid leukemia (AML), three patients with osteosarcomas, two patients with colon cancer, two patients with squamous cell carcinomas, two patients with breast cancers, one patient with a soft tissue sarcoma and one patient with uterine cancer (A. Clachos, Schneider Children's Hospital, Division of Hematology/Oncology, personal written communication, January 2010).

The 5q- syndrome is a subtype of MDS that is preponderant in adult females and is characterized by severe macrocytic anemia, normal/elevated platelet levels with hypolobulated micromegakaryocytes and progression to AML [66]. As implied by the name, 5q- syndrome is characterized by a deletion in the long arm of chromosome 5, which is a critical region containing approximately 40 genes including *RPS14*. RPS14 is a component of the small subunit of the ribosome and RPS14 is required for the processing of 18S rRNA [67]. A recent study that used an RNA interference-based functional screen showed that a partial loss of function of RPS14 phenocopies 5q- syndrome in normal hematopoietic progenitor cells. Furthermore, the forced expression of RPS14 rescues the 5q- syndrome phenotype in patient-derived bone marrow cells harvested from patients with the disease [68]. Notably, both DBA and 5q- syndrome result in a predilection for the development of AML, although the age of onset is quite different between the two diseases.

#### 9.2.2.2 Ribosomopathies of Non-ribosomal Proteins

The small subunit (SSU) processome, composed of the U3 snoRNA and over 40 proteins, is a large ribonucleoprotein (RNP) that is required for the maturation of the 18S rRNA of the SSU. In addition to this complex, many other proteins are required for the maturation of the large and small ribosomal subunits, although most of them are not present in the mature ribosome. Mutations in these ribosome biogenesis factors can act either as a causative agent or as a modifying agent of a particular ribosomopathy such as Shwachman-Diamond syndrome (SDS) and Treacher Collins syndrome (TCS) [57]. SDS is an autosomal recessive disease characterized by exocrine pancreatic insufficiency, ineffective hematopoiesis and an increased risk of leukemia. The *SBDS* gene has been found to be biallelically mutated in approximately 90 % of SDS patients [55]. Although the structure and function of *SBDS*is not fully understood, increasing evidence suggests that this gene plays an important role in ribosome biogenesis and RNA processing [56]. In addition, some ribosomal protein genes such as *RPS9, RPS20, RPL6, RPL15, RPL22, RPL23* and *RPL29* have also shown decreased expression in cells derived from SDS patients suggesting a multiple gene deficiency in SDS. For patients with SDS, a report from the French Severe Chronic Neutropenia Registry estimated the risk of MDS at 19 % at 20 years of age, whereas the risk of AML was estimated at up to 36 % at 30 years of age [58]. TCS is an autosomal dominant craniofacial disorder that includes abnormalities of the eyes, ears and facial bones. The gene *TCOF1* encodes a protein known as treacle and was previously identified as the gene responsible for TCS [69]. Treacle is a nucleolar phosphoprotein that plays a role in rDNA transcription and 18S rRNA gene methylation [70].

Small nucleolar ribonucleoproteins (snoRNPs) are complexes composed of RNA and proteins that localize to the nucleolus. The snoRNPs cause the endonucleolytic cleavage and chemical modification of pre-rRNA and facilitate the proper folding of pre-rRNA. Mutations in genes encoding either the non-coding RNA components or the protein components of the RNPs tend to lead to ribosomopathies which include cartilage-hair dysplasia (CHH) and dyskeratosiscongenita (DKC) [57]. CHH is an autosomal recessive syndrome characterized by hypoplastic hair, immune dysfunction and a predisposition to various malignancies [71]. The *RMRP* gene encodes a nucleolars no RNA composed of the mitochondrial RNA processing complex (RNase MRP). Mutations within the *RMRP* gene have been suggested to be causative of CHH [72]. In a study of CHH patients with a mean follow-up of 19.2 years, 14 out of 123 patients with CHH were diagnosed with cancer. Of the patients who developed cancer, non-Hodgkin lymphoma was the most common type (nine cases) followed by squamous cell carcinoma, leukemia and Hodgkin lymphoma. In addition, ten patients in CHH cohort also developed basal cell carcinoma of the skin. In general, there was a sevenfold increase in the overall cancer rate in patients with CHH when compared with the normal population [73]. DKC is an X-linked genetic disease characterized by bone marrow failure, mucocutaneous abnormalities and a predisposition to a variety of cancers. In some cases, DKC has also been associated with immune-deficiency, growth retardation and neurological symptoms. DKC can be caused by mutations in different components of the telomerase complex such as telomerase reverse transcriptase (TERT), telomerase RNA (TERC) and dyskerin [74]. A report from the National Cancer Institute DKC cohort revealed an 11-fold increase in the ratio of observed to expected cancers relative to the general population [59].

In addition to these observations in human pathology, in vivo experimental data from animal models also support a model whereby alterations in ribosome biogenesis may promote neoplastic transformation. In *Drosophila*, reduced expression of RPS6 can cause aberrant cell growth and neoplastic transformation in the hematopoietic system [75]. In a zebrafish screen of hundreds of lines of zebrafish with heterozygous embryonic lethal mutations, Amsterdam et al. have shown that 11 of 12 lines with an elevated rate of cancer development harbored a mutation in various ribosomal protein genes [76]. Of note, in these zebrafish lines, the distribution and penetrance of cancers phenocopies the malignancies in p53-null zebrafish. Further analysis revealed that the tumors in zebrafish with ribosomal haploinsufficiency also lost expression of p53 [77]. Moreover, increasing evidence from mouse models further tie together the deficiencies of ribosome biogenesis and tumorigenesis [78–80].

Although each ribosomopathy displays unique characteristics, they share certain features. The most notable similarity between the ribosomopathies is the presence of hypoplastic behavior characterized by decreased cell proliferation and increased apoptosis. In recent years, increasing evidence has demonstrated that the tumor suppressor p53 becomes stabilized and activated in response to impaired ribosome biogenesis. p53 in turn exerts control on cell proliferation and viability providing a link between the activation of p53 and the induction of ribosomopathies [38]. Indeed, evidence is accumulating from studies in animal models that the untimely activation of p53 is responsible for perturbations in tissue homeostasis that cause the development of ribosomopathies such as TCS and 5q-syndrome [81].

## 9.3 p53 Surveillance of Ribosome Biogenesis

p53 is the principal guardian of genomic stability preventing the initiation and progression of cancer. The tumor suppressor p53 is activated by a broad range of cellular stressors including oncogenic activation, DNA damage, metabolic stress, hypoxia and nucleolar stress [82]. In response to different stresses, p53 transactivates a set of target genes that induces cell cycle arrest, senescence, autophagy and/or apoptosis. Among the growing complexity of roles in cell fate, p53 is known to play a fundamental role in the surveillance of ribosome biogenesis and protein translation. The murine double minute 2 protein (Mdm2, or HDM2 in humans) is the central regulator of p53 that functions as a link between ribosome biogenesis and the p53 pathway. Mdm2 has been shown to bind specifically to several free ribosomal proteins including RPL5, RPL23, RPL11, RPL26, RPS3, RPS7, RPS14, RPS20, RPS25 and RPS27. In an elegant series of experiments, nucleolar disruption, ribosome protein depletion or chemical treatment by agents such as actinomycin D has been shown to cause the release of RPL11 and other ribosomal proteins into the nucleoplasm, which are able to bind to Mdm2 effectively inhibiting Mdm2 activity and eliciting the accumulation of p53 [38, 83].

### *9.3.1 Function and Regulation of p53*

Dubbed the "guardian of the genome" and the "cellular gatekeeper," p53 is the most frequently mutated gene in human cancers, and the mutational spectrum of p53 has been thought to represent a molecular link explaining the etiological causes of cancer. Somatic mutations occur in almost every type of cancer, and even in cancers harboring wild-type p53, the p53 signaling pathway becomes altered [84–86]. p53 functions largely as a transcription factor triggering several different anti-proliferative programs by transcriptionally activating or repressing key effector genes. During the past several years, the p53 transcriptome has emerged as a complex and intriguing field with the discovery of a variety of new and bewildering p53-dependent responses. p53 is responsible for the transcriptional activation or suppression of thousands of genes in stimulus-specific, promoter-dependent and/or cell type-dependent manners. These genes can be functionally categorized into several subsets, which are determined based on the involvement of the particular type of p53 response, including cell cycle arrest, senescence, apoptosis, autophagy and metabolic regulation.

In response to cellular stress, p53 activation consists of three integrated events including p53 stabilization, DNA binding and transcriptional activation of target genes. These processes are regulated by multiple post-translational modifications including phosphorylation, acetylation, methylation, glycosylation, neddylation, sumolyation and poly-ribosylation [87]. p53 stabilization is mainly achieved by disrupting the ability of p53 to interact with Mdm2. As the primary negative regulator of p53, Mdm2-mediated ubiquitination of p53 plays a pivotal role in regulating both p53 turnover and cellular localization. Monoubiquitination of p53 results in the export of p53 to the cytoplasm, whereas the polyubiquitination of p53 by Mdm2 stimulates the proteasome-mediated degradation of p53 [88]. On one hand, stress signals such as DNA damage cause the activation of various kinases including ATM, ATR, DNA-PK, Chk1 and Chk2, which phosphorylate p53 at specific amino acids. The phosphorylation of p53 at amino-terminal residues such as Ser15 inhibits the interaction between p53 and Mdm2, which results in the stabilization of p53 [89]. Additionally, Mdm2 phosphorylation by ATR has been shown to reduce the Mdm2-dependent exportation of p53 from the nucleus to the cytoplasm and to facilitate p53-induced cell cycle arrest [90]. On the other hand, p53 stabilization can also occur in response to oncogenic challenges to the cell, which is primarily mediated through the antagonism of the p53-Mdm2 interaction by the tumor suppressor ARF [91]. In addition to the phosphorylation of p53, some recent studies have revealed the critical involvement of acetylation events in the selective stimulation of p53-dependent transactivation. Several p53 carboxy-terminal lysine residues are acetylated by CBP/p300, which results in the stabilization and activation of p53 [92], and this mechanism of activation appears to be partially due to the fact that the acetylated lysine residues cannot be ubiquitinated by Mdm2 [93]. Interestingly, lysine 120, which resides within the DNA binding domain of p53, is acetylated by Tip60/hMOF in response to DNA damage resulting in the preferential induction of pro-apoptotic genes instead of those involved in cell cycle arrest, indicating that different modifications of p53 may allow cells to tailor their response to different stress signals [94, 95].

## *9.3.2 p53 as a Nucleolar Stress Effector Through the RP-Mdm2-p53Axis*

### 9.3.2.1 RP-Mdm2-p53 Axis

As mentioned above, Mdm2 negatively regulates p53 by either conjugating ubiquitin to p53 or by directly concealing the transactivation domain of p53 from the cellular transcriptional machinery [96]. Extensive modifications of Mdm2 through ubiquitination, sumoylation and phosphorylation have also been shown to determine the subcellular localization of Mdm2 and consequently regulate p53 activity [97]. Recently, a subset of ribosomal proteins was shown to inhibit Mdm2 and activate p53 through the extra ribosomal functions of RPs.

The first evidence of RP interaction with Mdm2 involved RPL5 binding to Mdm2 in a 5S rRNA-RPL5-Mdm2-p53 ribonucleoprotein complex [98]. However, the functional consequences of the RP-Mdm2 interaction has not been fully appreciated until recently when several additional studies revealed that other RPs including RPL11, RPL23 and RPL5 could activate p53 through their interaction with Mdm2 [99–103]. Mdm2 is shuttled between the nuclear and cytoplasmic compartments, whereas the RPs are translated in the cytosol and shuttled to the nucleolus to be assembled into the ribosome. Thus, the initiation and maintenance of the interaction between Mdm2 and the RPs requires further investigation. Studies on RPL11 suggested two possible models that may explain the RP-Mdm2 interaction. One model holds that the disruption of the nucleolus promotes the release of RPL11 from the nucleolus to the nucleoplasm, and then RPL11 binds to Mdm2 and stabilizes p53. Another model suggested by Fumagalli et al. proposes that in response to nucleolar stress, cells increase the translation of 5′-TOP mRNAs, which consists of most of the ribosomal proteins including RPL11 [104]. In this model, nucleolar stress causes an increase in RPL11 expression, which increases the concentration of free RPL11 available to bind Mdm2.

Additional evidence for the roles of RPL26 [105], RPS3 [106], RPS7 [107, 108], RPS14 [109], RPS25 [110], RPS27 [111], RPS27L [111–113] and possibly RPS20 [106] as Mdm2 binding partners has also been presented further validating the fact that the Mdm2-p53 interaction is regulated by RPs. Unlike the current understanding of RPL5 and RPL11, some RPs are involved with the Mdm2-p53 interaction in various other mechanisms. RPL26 binds to p53 mRNA and augments the translation of p53, whereas Mdm2 binds to RPL26 and drives the polyubiquitination and proteasomal degradation of RPL26. Moreover, the binding of Mdm2 to RPL26 attenuates the association between RPL26 and p53 mRNA and represses the RPL26-mediated augmentation of p53 gene translation. This balance helps dictate the cellular p53 level and activity in unstressed cells. Genotoxic stress interrupts this balance by enabling a rapid increase in p53 synthesis [105]. Besides RPL26, RPS7 and RPS27L, an RPS27-like protein, are both substrates for Mdm2 E3 ligase activity. Furthermore, RPS25 is a transcriptional target of p53, as p53 directly binds to the RPS25 promoter region and suppresses RPS25 expression. Similarly, RPS27L is a direct p53-inducible target, whereas RPS27 is a p53-repressible target. However, the in vivo significance of the contradictory inter-regulation among RPs, Mdm2 and p53 remains to be determined.

Notably, some recent controversial findings have shown that MdmX (also known as Mdm4) is involved in the regulation of the RP-Mdm2 axis. MdmX, a homologue of Mdm2, is a p53-binding protein. MdmX forms a heterodimer with Mdm2 through their C-terminal RING domains and it greatly increases the capacity of Mdm2 to ubiquitinate and degrade p53 [114]. The activation of p53 by ribosomal stress requires the down regulation of MdmX, and the overexpression of MdmX abrogates p53 activation and prevents growth arrest [115]. In contrast, MdmX facilitates the ability of RPS7 and RPS25 to inhibit Mdm2 E3 ligase activity [108, 110] suggesting a positive feedback loop mediated by MdmX. In addition to MdmX, several novel regulators of the RP-Mdm2 complex have been recently discovered. Promyelocytic leukemia (PML) gene expression was recently shown to enhance p53 stability by sequestering Mdm2 to the nucleolus. Interestingly, the loss of RPL11 expression impairs the ability of PML to localize to the nucleolus and regulate Mdm2. Another RP-Mdm2 binding regulator that has been recently discovered is PICT1 (protein interacting with the C terminus 1, also known as GLTSCR2). PICT1 is a nucleolar protein and has emerged as a key regulator of the nucleolar stress response. PICT1 directly binds to RPL11 and traps L11 in the nucleolus. As a result, PICT1 loss leads to RPL11 release from the nucleolus, inhibition of Mdm2 E3 activity and the accumulation of p53. In human cancers, a lower level of PICT1 expression correlates with a better prognosis, suggesting that PICT1 is a potent regulator of the Mdm2-p53 pathway and promotes tumor progression by retaining RPL11 in the nucleolus [116]. Furthermore, another nucleolar protein MYBBP1A (Myb-binding protein 1a) functions as a link between the cellular energy status and the p53-mediated cell cycle machinery. Upon glucose starvation, eNoSC (energy-dependent nucleolar silencing complex) inhibits rRNA transcription, which results in a reduction in nucleolar RNA content. As a consequence, MYBBP1A, which is anchored to the nucleolus via RNA interactions, translocates from the nucleolus to the nucleoplasm. The translocated MYBBP1A induces the acetylation and accumulation of p53 by enhancing the interaction between p300 and p53 [117]. In addition to RPs, different nucleolar proteins such as B23 [118, 119], PAK1IP1 [120], NEDD8 [121, 122]and several others mentioned above participate in the regulation of p53 stability and activity with or without the involvement of Mdm2 highlighting the importance of the nucleolus in the surveillance of cell proliferation and the determination of cell fate.

Mdm2 contains three conserved regions: an N-terminal p53-binding domain, a central acidic region encompassing a C4 zinc finger and a C-terminal RING domain possessing E3 ligase activity. Previous studies have shown that many cancer-associated alternative and aberrant splicing events in Mdm2 mRNA retain the N-terminal and C-terminal domains but splice out the central acidic domain [123]. In a study involving 23 primary tumors of four types (osteosarcoma, non-Hodgkin's lymphoma, hepatocellular carcinoma and leukemia), eight samples were found to contain mutations in the coding region of MDM2. Notably, most of these mutations target the central zinc finger of MDM2 [124]. Not surprisingly, many RPs including RPL5 and RPL11 form a stable complex with Mdm2 through the direct binding to Mdm2's zinc finger domain [102, 125, 126]. Further analysis demonstrated that the MDM2 C4 zinc finger is critical for the binding of RPL5 and RPL11 to MDM2. A single point mutation converting a cysteine residue to a phenylalanine residue at codon 305 in the zinc finger domain of Mdm2 blocks the binding interaction between Mdm2 and RPL5 and RPL11 but not RPL23[125, 127]. The physiological significance of the RP-Mdm2 interaction was further investigated through the generation of $Mdm2^{C305F}$ knock-in mice [128]. Interestingly, $Mdm2^{C305F}$ mice retain an intact p53 response to DNA damage but fail to stabilize p53 in response to nucleolar stress, which is most likely due to the disruption of the binding between Mdm2 and RPL5 and RPL11. Importantly, the loss of RP-Mdm2 interaction significantly accelerates Eμ-Myc-induced lymphomagenesis, indicating that the RPs-Mdm2-p53 pathway plays a critical role in safeguarding against tumorigenesis. Remarkably, when full-length Mdm2 possessing an acidic domain mutation was used as bait, binding to RPS3 was not affected, suggesting that the interaction between RPS3 and Mdm2 involves more than just the Mdm2 acidic domain [106].

#### 9.3.2.2 Nucleolar Function of p53: Guarding Ribosome Biogenesis

Nucleolar stress specifically refers to perturbations in ribosome biogenesis resulting in the subsequent breakdown of the nucleolar structure. Ribosome biogenesis can be inhibited by serum starvation, contact inhibition, depletion of nucleotides, treatment with chemotherapeutic compounds such as 5-FU (5-fluorouracil), and the dysfunction of nucleolar proteins involved in ribosome biogenesis. All of these stressors have been shown to generate nucleolar stress that signals to p53 and allows cells to cease proliferation under conditions that sustain low quality or insufficient ribosome biogenesis. According to the disruption that occurs at the different stages of ribosome biogenesis, phenomena that result in nucleolar stress can be placed into three categories: disruption of rRNA synthesis, disruption of rRNA processing and maturation and ribosomal protein imbalance.

The disruption of rRNA synthesis activates p53. As mentioned above, RNA Pol I and Pol III are responsible for the transcription of the rDNA genes. Modifications of RNA Pol I and RNA Pol III have been reported to induce the breakdown of the nucleolar structure and activate a p53 stress response. Several well-established reagent-based experimental systems have been used to mimic the inhibition of precursor rRNA synthesis. Low concentration of actinomycin D (<10nM) specifically disrupts ribosome biogenesis by intercalating into the GC-rich regions of rDNA to inhibit PolI-mediated transcription of nascent 47S rRNA [103, 129].In addition,5-FU, a uracil analog antimetabolite that becomes mis incorporated into nascent RNA and blocks complete RNA synthesis [130],and mycophenolic acid(MPA), an agent that selectively inhibits inosine monophosphate dehydrogenase and depletes the guanine nucleotide pool, which disrupts pre-ribosomal RNA synthesis [131], have also been used to induce nucleolar stress responses to demonstrate p53 stabilization throughRPL5- and RPL11-directed inhibition of Mdm2 [132, 133]. Genetic models such as the deletion of the RNA Pol I transcription cofactor TIF-1A [134], ablation of BAP28 [135] and the inhibition of SL1 recruitment to rDNA promoters by activating PTEN [136]also reduce rRNA synthesis and trigger p53-dependent phenotypes such as apoptosis.

Additionally, the disruption of rRNA processing activates p53. Infidelity in rRNA processing can lead to the accumulation of unprocessed intermediate transcripts, which can retard subunit assembly thereby triggering a nucleolar stress event. One example of this comes from the study of the nucleolar protein Bop1, which is involved in rRNA processing and ribosome assembly. A dominant-negative form of Bop1 (DN-Bop1), which inhibits the transcription of 28S and 5.8S rRNA and causes a deficiency of newly synthesized 60S ribosomal subunits, induces p53-dependent cell cycle arrest in 3T3 fibroblasts. Inactivation of functional p53 reversed the DN-Bop1-induced cell cycle arrest but did not restore normal rRNA processing [137]. Additional study of Wrd36, which is required for 18S rRNA processing, in zebrafish provided in vivo evidence that the disruption of pre-rRNA processing and the inappropriate accumulation of mature rRNA is sufficient to signal a stress response to p53.

Finally, ribosomal protein imbalances can activate p53. A continuous supply of ribosomal proteins is essential for maintaining ribosome biosynthesis and translational machinery. Thus, insufficient RP production can trigger a ribosomal stress response that leads to p53 accumulation and stabilization. Based on in vitro knockdown or in vivo deletion models, the deficiency of several RPs results in the induction of p53-dependent stress responses [138], which indicates that the activation of p53 seems to be a general response of cells to RP insufficiency. However, interestingly, the phenotypes associated with an RP deficiency are quite heterogeneous. In vertebrates, depending on the specific RP deleted, the resulting effect can either be lethal or pleiotropic manifesting as a specific defect in a particular tissue alongside additional phenotypes, or the deletion of an RP may have no significanteffect. p53 appears to have contrasting roles in determining cell survival or cell death in response to RP insufficiency. Activation of p53 causes embryonic lethality in embryos deficient for RPS6 [139], RPS19 [140] or RPL11 [141]; however, p53 promotes the survival of RPL24-deficient mice during the embryonic stage by delaying the cell cycle and induces a *Bst* phenotype (belly spot and tail) in adults due to apoptosis [142]. Taken together, the RP-Mdm2-p53 axis acts as a checkpoint for ribosome biogenesis and consequently determines cell fate. The stress response could vary between different stages of development and between different tissues in vivo, which is consistent with the differential effects of p53 in different cell types, as p53 is capable of differentiating between cofactors and between different promoters of target genes.

### *9.3.3 p53 Regulation of Metabolism*

As a cellular gatekeeper and a tumor suppressor, p53 plays a major role in sensing and responding to a variety of internal and external stressors to maintain cellular homeostasis. In addition to its classical role in regulating cell cycle arrest, senescence and apoptosis, p53 has recently been shown to regulate metabolism through the transcriptional activation of genes involved in glucose transport, glycolysis, oxidative phosphorylation, glutamine hydrolysis and genes upstream of the mTOR and autophagy pathways.

The primary carbon source for ATP production is glucose. Glycolysis, the enzymatic breakdown of glucose to pyruvate in the cytosol, is an important energy-generating process in cells and is the only alternative to oxidative phosphorylation (OXPHOS) for ATP production. Oxidative phosphorylation is a mitochondrial process in which ADP is phosphorylated to ATP as a direct consequence of oxidizing NADH and FADH2 by the electron transport chain (ETC). Interestingly, in the presence of wild type p53, ATP production is skewed in favor of mitochondrial respiration. However, when p53 is absent, ATP is primarily produced by anaerobic glycolysis [143]. In cancers, tumor cells predominantly metabolize glucose through glycolysis regardless of the surrounding concentration of oxygen (Warburg effect), and this increase in glycolysis correlates with the malignancy of the tumor. Recent findings revealed that p53, which is extensively mutated in cancers, plays a role in governing the switch from aerobic to anaerobic metabolism. The first metabolic gene identified as a p53 transcriptional target is phosphoglycerate mutase (PGM). PGM is a glycolytic enzyme that catalyzes the conversion of 3-phosphoglycerate to 2-phosphoglycerate and drives glycolytic flux [144]. p53 suppress the transcription of PGM and inhibits glycolysis albeit not completely, as PGM is not the rate-limiting enzyme for glycolysis. Later, Vousden et al. identified TIGAR (Tp53-induced glycolysis and apoptosis regulator) as a novel p53-inducible target gene.

TIGAR functions as a potent inhibitor of glycolysis by decreasing the concentration of fructose-2, 6-bisphosphate, which abrogates the activity of phosphofructokinase 1 (PFK1), which is the rate-limiting enzyme for glycolysis [145]. Furthermore, the induction of TIGAR favors the accumulation of fructose-1, 6-bisphosphate, which is effectively shunted into the pentose phosphate pathway to produce NADPH, which acts as a reactive oxygen species (ROS) scavenger and facilitates anabolism. On the other hand, p53 directly regulates OXPHOS by inducing SCO2 (cytochrome c oxidase subunit 2), which regulates the cytochrome *c* oxidase complex of the ETC. Additional studies have shown that the single-allelic deletion of *sco2* is sufficient to recapitulate the glycolytic phenotype observed in p53-null mice, and the expression of SCO2 in p53-null HCT116 cells successfully shifts HCT116 dependence on glycolysis towards the use of OXPHOS substantiating the significance of p53 in the regulation of glucose metabolism [143].

In animals, fatty acids become the main source of energy as blood glucose levels decrease such as when the organism is fasted. Recent studies have begun to unveil the function of p53 in the regulation of lipid homeostasis, which is an important metabolic transition when glucose is not readily available. p53has been shown to inhibit lipogenesis by transcriptionally suppressing SREBP-1c (sterol regulatory element binding protein 1) [146, 147], which is a transcription factor involved in the expression of genes responsible for triglyceride synthesis and lipid accumulation. Furthermore, several genes involved in fatty acid oxidation (FAO) have been characterized as p53 target genes. Upon nutritional stress, p53 becomes phosphorylated at serine residue 18 (equivalent S15 in human), which activates p53 and results in the induction of Lpin1 and the promotion of FAO in mice [148]. p53 has also been demonstrated to promote FAO in response to nutritional stress by inducing GAMT (guanidinoacetate methyltransferase), a critical enzyme involved in creatine biosynthesis and fatty acid oxidation [149]. GAMT catalyzes the conversion of guanidinoacetate to creatine using glycine, arginine or methionine as the substrate. Creatine is produced from these amino acids primarily in the kidneys and the liver, where it is secreted into the peripheral blood circulation for use by myocytes for energy production. Once it is taken up by myocytes, creatine enhances ATP recycling by using a phosphocreatine intermediate to convert ADP to the energy-rich ATP. Under glucose-deprived conditions, GAMT enhances FAO, thereby enhancing this alternative fuel source to maintain constant energy production. The involvement of GAMT in creatine-based energy production indicates that p53 may be involved in energy homeostasis and metabolic shifts between glucose, lipid and amino acid metabolism. In addition to GAMT, other lipid metabolism-associated genes such as CROT (carnitine O-octanoyltransferase), CPT1a (carnitine palmitoyltransferase 1A) [150, 151] and MCD (malonyl CoA decarboxylase) (our unpublished data) have been shown by microarray to be potential p53 transcriptional targets, suggesting that p53 controls lipid metabolism through the mitochondrial and peroxisome oxidation systems.

p53 regulates protein synthesis and degradation mainly through crosstalk with the AMPK and mTOR signaling pathways. As mentioned above, p53 can directly bind to rRNA transcriptional factors and inhibit ribosome biogenesis and protein biosynthesis. Alternatively, p53 transcriptionally up regulates sestrin1 and sestrin2 [152], which are known to protect cells from ROS damage through the regeneration of overoxidized peroxiredoxins. Sestrin1/2 can also activate AMPK and consequently inhibit mTOR signaling thus promoting the transition from anabolism to catabolism and reserve energy utilization in the interest of cell survival. Interestingly, although the precise signaling pathway remains far from defined, p53 apparently regulates autophagy by two different mechanisms. Nuclear-localized p53 promotes autophagy through the transcriptional activation of the autophagy-inducing protein DRAM (damage-regulated autophagy modulator) and through the inactivation of the mTOR pathway [153–155], whereas cytoplasmic p53 appears to play an opposite role by inhibiting autophagy [152]. Notably, p53-dependent induction of autophagy is considered a stress response. However, at basal levels, p53 regulates autophagy directly at the endoplasmic reticulum (ER). Loss of p53 in the ER causes an ER-specific unfolded protein response (UPR) and results in an increase in autophagy. Interestingly, p53-dependent inhibition of autophagy does not depend on wild-type p53, as several tumor-associated p53 mutants including several gain-of-function mutants can also inhibit autophagy [156]. These observations raise the question of what is the real function of autophagy in the context of tumorigenesis. Clearly, more studies will be required to address this issue.

On a whole cell scale, the cellular redox system is the master regulator responsible for controlling the global cellular metabolism. Reactive oxygen species (ROS) represent an important signaling messenger and stressor. ROS can cause DNA damage and genomic instability leading to the activation of p53 and various stress responses such as apoptosis, cell cycle arrest and senescence. On the other hand, the promotion of OXPHOS is a major metabolic alteration that is brought about by p53 signaling, and interestingly, OXPHOS is the primary intracellular source of ROS. Remarkably, p53 also produces antioxidant products and protects the genome from the oxidation by ROS [157]. Several antioxidants have been reported to be downstream transcriptional targets of p53, including catalase (CAT), manganese superoxide dismutase 2 (MnSOD2), glutathione peroxidase (GPX1), sestrin1/2 and glutaminase (GLS2; responsible for the generation of the major endogenous antioxidant reduced glutathione) [158, 159].As mentioned before, NADPH is also regulated by p53 making p53-dependent antioxidant targets pervasive throughout all cellular redox systems. The involvement of p53 in the production of antioxidants is important in the context of the classical genomic guardian role of p53, as these antioxidants provide potent reducing power to safeguard genomic stability and to prevent senescence and tumorigenesis.

## 9.4 Conclusions and Perspectives

### *9.4.1 RP-Mdm2-p53 Pathway Senses Metabolic Alterations and Regulates Metabolic Adaptations*

In general, ribosome biogenesis consumes a major portion of the cellular energy supply and resources and plays a key role in the life cycle of the cell. Hence, the cell has developed ways to alter ribosomal biogenesis in the presence of subtle metabolic fluctuations such as the daily feeding-fasting cycle between meals. Previous studies performed on in vitro cultured cells have shown that metabolic stress such as glucose deprivation can transactivate p53 [117] and induce downstream target genes such as Lpin1 [148] and GAMT [149]. In previous work on the Mdm2$^{C305F}$mouse model, in which the Mdm2 interaction with RPL5 and RPL11 is specifically disrupted, we have shown that this mouse tends to accumulate lipid relative to its wild-type littermate when fed a normal diet. However, when fasted, which indeed inhibits rRNA synthesis, Mdm2$^{C305F}$ mice harbor a phenotype reminiscent of diet-induced obesity, which includes impaired lipid oxidation (our unpublished data). These findings suggest the following: (1) Metabolic fluctuation is a physiological trigger of the RP-Mdm2-p53 signaling pathway; (2) The RP-Mdm2-p53 axis fine-tunes cellular energy homeostasis in low stress or even so-called unstressed conditions.

Other studies have shown that p53 not only performs as a stress-induced player that activates acute responses such as apoptosis but also acts as a housekeeping gene to maintain cellular metabolism within a range that promotes cell survival under low or unstressed conditions. First, p53 functions through stress – and promoter-specific recruitment of transcription initiation. P53 can bind to the promoters of several p53 target genes even in unstressed conditions [160, 161], which suggests that p53 is structurally capable of binding target genes in the absence of stress-induced modifications. Second, previous studies have shown that certain metabolic outcomes of p53 such as the inhibition of autophagy and the induction of antioxidants specifically responds to low or unstressed conditions. Our unpublished data also show that acute stress such as treatment with a low concentration of actinomycin D and other DNA damage inducers only causes the induction of acute response genes such as p21 or Bax, whereas relatively low stress like dietary restriction exclusively induces metabolic target genes. In light of these observations, the physiological function of the RP-Mdm2-p53 pathway appears to involve the detection of metabolic alterations to promote DNA repair (by antioxidants) or to regulate energy homeostasis (by metabolic adaptation) in the interest of cell survival. If this pathway fails to prevent cellular damage, p53-dependent acute responses become initiated by the accumulated stress signals such as DNA damage and ROS resulting in the induction of apoptosis.

### *9.4.2 RP-Mdm2-p53 Pathway Contributes to the Tumor Suppression Function of p53*

Ever since p53 was first categorized as a tumor suppressor, researchers have sought to understand the mechanism of the tumor suppression function of p53. A large amount of findings have shown that acute p53 responses such as cell cycle arrest, senescence and apoptosis could be the key to unlocking several mysteries of cancer treatment, as these processes have been well accepted to be responsible for the success of chemotherapy and radiotherapy. However, several recent studies have raised different issues. By working on an inducible p53$^{ER}$ mouse model combined with

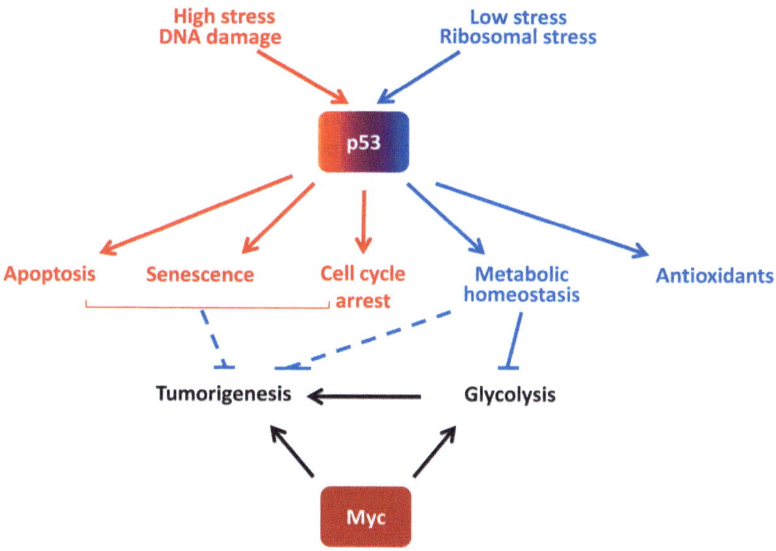


**Fig. 9.3 Potential mechanism for p53 to suppress tumorigenesis.** *Red color* indicates acute responses induced by intense or accumulated stress; *Blue color* indicates signaling under unstressed or low stress conditions such as metabolic fluctuations. *Dashed lines* indicate signaling that is not well elucidated

irradiation-induced lymphoma, Christophorou et al. have demonstrated that acute radiation responses do not contribute to p53-mediated tumor suppression. Instead, they show that ARF, a tumor suppressor induced by oncogenic expression, plays a key role in p53-mediated tumor suppression [162]. Very recently, another group generated p53 knock-in mice harboring three amino acid substitutions at known acetylation sites ($p53^{3KR}$), which is unable to undergo cell cycle arrest, senescence and apoptosis. Surprisingly, unlike p53-null mice, which succumb to spontaneous thymic lymphomas, early onset tumor formation does not occur in $p53^{3KR}$ mice. Notably, $p53^{3KR}$mice retain the ability to regulate energy metabolism and ROS production, indicating that unconventional activities of p53 such as metabolic regulation and antioxidant function are critical for p53-mediated tumor suppression [163] (Fig. 9.3).

Emerging evidence suggests that metabolic reprogramming is a prerequisite for the rapid cell proliferation of cancer cells. In contrast to differentiated cells, which rely primarily on mitochondrial oxidative phosphorylation to generate the ATP needed for cellular processes, rapidly proliferating cells and cancer cells tend to convert most glucose to lactate through anaerobic glycolysis, which rapidly provides ATP while simultaneously creating the macromolecules necessary for unrestrained biosynthesis typical of cancer cells. As a consequence, RP-Mdm2-p53 axis-mediated metabolic alterations could contribute significantly to tumor suppression. A recent study has provided evidence that nucleolar stress could inhibit tumorigenesis by specifically activating p53 [164]. Verified by both genetic and pharmacological models, this study demonstrated that accelerated rRNA synthesis and nucleolar integrity are necessary for oncogenic activity. Furthermore, this study showed that the selective inhibition of Pol I is effective in vivo for the treatment of lymphoma and leukemia through the nongenotoxic activation of p53-dependent apoptosis, while sparing normal cells. Combined with our previous findings that inactivation of the RP-Mdm2-p53 pathway increases the incidence of oncogene-induced lymphomas in an ARF-independent fashion [128], the RP-Mdm2-p53 axis conceivably contributes significantly to p53-mediated tumor suppression. However, additional studies are required to address whether metabolic alterations could be the major contribution of the RP-Mdm2-p53 axis to tumor suppression.

## References

1. Warner JR (1999) The economics of ribosome biosynthesis in yeast. Trends Biochem Sci 24(11):437–440
2. Boisvert FM, van Koningsbruggen S, Navascues J, Lamond AI (2007) The multifunctional nucleolus. Nat Rev Mol Cell Biol 8(7):574–585
3. Szymanski M, Barciszewska MZ, Erdmann VA, Barciszewski J (2003) 5S RRNA: structure and interactions. Biochem J 371(Pt 3):641–651
4. Kressler D, Hurt E, Bassler J (2010) Driving ribosome assembly. Biochim Biophys Acta 1803(6):673–683
5. Ahmad Y, Boisvert FM, Gregor P, Cobley A, Lamond AI (2009) NOPdb: nucleolar proteome database – 2008 update. Nucleic Acids Res 37(Database issue):D181–D184
6. Andersen JS, Lam YW, Leung AK, Ong SE, Lyon CE, Lamond AI et al (2005) Nucleolar proteome dynamics. Nature 433(7021):77–83
7. Andersen JS, Lyon CE, Fox AH, Leung AK, Lam YW, Steen H et al (2002) Directed proteomic analysis of the human nucleolus. Curr Biol 12(1):1–11
8. François-Michel B, Yasmeen A, Lamond AI (2011) The dynamic proteome of the nucleolus. In: Olson MOJ (ed) The nucleolus. Springer, New York, p 29
9. Boisvert FM, Lamond AI (2010) p53-Dependent subcellular proteome localization following DNA damage. Proteomics 10(22):4087–4097
10. Lam YW, Evans VC, Heesom KJ, Lamond AI, Matthews DA (2010) Proteomics analysis of the nucleolus in adenovirus-infected cells. Mol Cell Proteomics 9(1):117–130
11. Moore HM, Bai B, Boisvert FM, Latonen L, Rantanen V, Simpson JC et al (2011) Quantitative proteomics and dynamic imaging of the nucleolus reveal distinct responses to UV and ionizing radiation. Mol Cell Proteomics 10(10):M111 009241
12. Schuhmacher M, Staege MS, Pajic A, Polack A, Weidle UH, Bornkamm GW et al (1999) Control of cell growth by c-Myc in the absence of cell division. Curr Biol 9(21):1255–1258
13. Iritani BM, Eisenman RN (1999) c-Myc enhances protein synthesis and cell size during B lymphocyte development. Proc Natl Acad Sci USA 96(23):13180–13185
14. Schuhmacher M, Kohlhuber F, Holzel M, Kaiser C, Burtscher H, Jarsch M et al (2001) The transcriptional program of a human B cell line in response to Myc. Nucleic Acids Res 29(2):397–406
15. Jain M, Arvanitis C, Chu K, Dewey W, Leonhardt E, Trinh M et al (2002) Sustained loss of a neoplastic phenotype by brief inactivation of MYC. Science 297(5578):102–104
16. Shachaf CM, Kopelman AM, Arvanitis C, Karlsson A, Beer S, Mandl S et al (2004) MYC inactivation uncovers pluripotent differentiation and tumour dormancy in hepatocellular cancer. Nature 431(7012):1112–1117
17. Wu CH, van Riggelen J, Yetil A, Fan AC, Bachireddy P, Felsher DW (2007) Cellular senescence is an important mechanism of tumor regression upon c-Myc inactivation. Proc Natl Acad Sci USA 104(32):13028–13033

18. Felsher DW, Bishop JM (1999) Reversible tumorigenesis by MYC in hematopoietic lineages. Mol Cell 4(2):199–207
19. Adhikary S, Eilers M (2005) Transcriptional regulation and transformation by Myc proteins. Nat Rev Mol Cell Biol 6(8):635–645
20. Grandori C, Gomez-Roman N, Felton-Edkins ZA, Ngouenet C, Galloway DA, Eisenman RN et al (2005) c-Myc binds to human ribosomal DNA and stimulates transcription of rRNA genes by RNA polymerase I. Nat Cell Biol 7(3):311–318
21. Gomez-Roman N, Grandori C, Eisenman RN, White RJ (2003) Direct activation of RNA polymerase III transcription by c-Myc. Nature 421(6920):290–294
22. Coller HA, Grandori C, Tamayo P, Colbert T, Lander ES, Eisenman RN et al (2000) Expression analysis with oligonucleotide microarrays reveals that MYC regulates genes involved in growth, cell cycle, signaling, and adhesion. Proc Natl Acad Sci USA 97(7):3260–3265
23. Guo QM, Malek RL, Kim S, Chiao C, He M, Ruffy M et al (2000) Identification of c-myc responsive genes using rat cDNA microarray. Cancer Res 60(21):5922–5928
24. Menssen A, Hermeking H (2002) Characterization of the c-MYC-regulated transcriptome by SAGE: identification and analysis of c-MYC target genes. Proc Natl Acad Sci USA 99(9):6274–6279
25. Boon K, Caron HN, van Asperen R, Valentijn L, Hermus MC, van Sluis P et al (2001) N-myc enhances the expression of a large set of genes functioning in ribosome biogenesis and protein synthesis. EMBO J 20(6):1383–1393
26. Orian A, van Steensel B, Delrow J, Bussemaker HJ, Li L, Sawado T et al (2003) Genomic binding by the drosophila Myc, Max, Mad/Mnt transcription factor network. Genes Dev 17(9):1101–1114
27. Fernandez PC, Frank SR, Wang L, Schroeder M, Liu S, Greene J et al (2003) Genomic targets of the human c-Myc protein. Genes Dev 17(9):1115–1129
28. Watson JD, Oster SK, Shago M, Khosravi F, Penn LZ (2002) Identifying genes regulated in a Myc-dependent manner. J Biol Chem 277(40):36921–36930
29. Li J, Sejas DP, Burma S, Chen DJ, Pang Q (2007) Nucleophosmin suppresses oncogene-induced apoptosis and senescence and enhances oncogenic cooperation in cells with genomic instability. Carcinogenesis 28(6):1163–1170
30. Wu CH, Sahoo D, Arvanitis C, Bradon N, Dill DL, Felsher DW (2008) Combined analysis of murine and human microarrays and ChIP analysis reveals genes associated with the ability of MYC to maintain tumorigenesis. PLoS Genet 4(6):e1000090
31. Barna M, Pusic A, Zollo O, Costa M, Kondrashov N, Rego E et al (2008) Suppression of Myc oncogenic activity by ribosomal protein haploinsufficiency. Nature 456(7224):971–975
32. Henry JL, Coggin DL, King CR (1993) High-level expression of the ribosomal protein L19 in human breast tumors that overexpress erbB-2. Cancer Res 53(6):1403–1408
33. Vaarala MH, Porvari KS, Kyllonen AP, Mustonen MV, Lukkarinen O, Vihko PT (1998) Several genes encoding ribosomal proteins are over-expressed in prostate-cancer cell lines: confirmation of L7a and L37 over-expression in prostate-cancer tissue samples. Int J Cancer 78(1):27–32
34. Wang Q, Yang C, Zhou J, Wang X, Wu M, Liu Z (2001) Cloning and characterization of full-length human ribosomal protein L15 cDNA which was overexpressed in esophageal cancer. Gene 263(1–2):205–209
35. Kim JH, You KR, Kim IH, Cho BH, Kim CY, Kim DG (2004) Over-expression of the ribosomal protein L36a gene is associated with cellular proliferation in hepatocellular carcinoma. Hepatology 39(1):129–138
36. Dai MS, Arnold H, Sun XX, Sears R, Lu H (2007) Inhibition of c-Myc activity by ribosomal protein L11. EMBO J 26(14):3332–3345
37. Dai MS, Sears R, Lu H (2007) Feedback regulation of c-Myc by ribosomal protein L11. Cell Cycle 6(22):2735–2741
38. Zhang Y, Lu H (2009) Signaling to p53: ribosomal proteins find their way. Cancer Cell 16(5):369–377
39. Hsieh AC, Truitt ML, Ruggero D (2011) Oncogenic AKTivation of translation as a therapeutic target. Br J Cancer 105(3):329–336

40. Levy S, Avni D, Hariharan N, Perry RP, Meyuhas O (1991) Oligopyrimidine tract at the 5′ end of mammalian ribosomal protein mRNAs is required for their translational control. Proc Natl Acad Sci USA 88(8):3319–3323
41. Iadevaia V, Zhang Z, Jan E, Proud CG (2012) MTOR signaling regulates the processing of pre-rRNA in human cells. Nucleic Acids Res 40(6):2527–2539
42. Mayer C, Zhao J, Yuan X, Grummt I (2004) MTOR-dependent activation of the transcription factor TIF-IA links rRNA synthesis to nutrient availability. Genes Dev 18(4):423–434
43. Chan JC, Hannan KM, Riddell K, Ng PY, Peck A, Lee RS et al (2011) AKT promotes rRNA synthesis and cooperates with c-MYC to stimulate ribosome biogenesis in cancer. Sci Signal 4(188):ra56
44. Kapeli K, Hurlin PJ (2011) Differential regulation of N-Myc and c-Myc synthesis, degradation, and transcriptional activity by the Ras/mitogen-activated protein kinase pathway. J Biol Chem 286(44):38498–38508
45. Mendoza MC, Er EE, Blenis J (2011) The Ras-ERK and PI3K-mTOR pathways: cross-talk and compensation. Trends Biochem Sci 36(6):320–328
46. Roux PP, Shahbazian D, Vu H, Holz MK, Cohen MS, Taunton J et al (2007) RAS/ERK signaling promotes site-specific ribosomal protein S6 phosphorylation via RSK and stimulates cap-dependent translation. J Biol Chem 282(19):14056–14064
47. Di Cristofano A, Pesce B, Cordon-Cardo C, Pandolfi PP (1998) Pten is essential for embryonic development and tumour suppression. Nat Genet 19(4):348–355
48. Suzuki A, de la Pompa JL, Stambolic V, Elia AJ, Sasaki T, del Barco BI et al (1998) High cancer susceptibility and embryonic lethality associated with mutation of the PTEN tumor suppressor gene in mice. Curr Biol 8(21):1169–1178
49. Harbour JW, Dean DC (2000) The Rb/E2F pathway: expanding roles and emerging paradigms. Genes Dev 14(19):2393–2409
50. Cavanaugh AH, Hempel WM, Taylor LJ, Rogalsky V, Todorov G, Rothblum LI (1995) Activity of RNA polymerase I transcription factor UBF blocked by Rb gene product. Nature 374(6518):177–180
51. Hannan KM, Kennedy BK, Cavanaugh AH, Hannan RD, Hirschler-Laszkiewicz I, Jefferson LS et al (2000) RNA polymerase I transcription in confluent cells: Rb downregulates rDNA transcription during confluence-induced cell cycle arrest. Oncogene 19(31):3487–3497
52. Zhai W, Comai L (2000) Repression of RNA polymerase I transcription by the tumor suppressor p53. Mol Cell Biol 20(16):5930–5938
53. Ruggero D, Pandolfi PP (2003) Does the ribosome translate cancer? Nat Rev Cancer 3(3):179–192
54. Cairns CA, White RJ (1998) p53 is a general repressor of RNA polymerase III transcription. EMBO J 17(11):3112–3123
55. Boocock GR, Morrison JA, Popovic M, Richards N, Ellis L, Durie PR et al (2003) Mutations in SBDS are associated with Shwachman-Diamond syndrome. Nat Genet 33(1):97–101
56. Ganapathi KA, Austin KM, Lee CS, Dias A, Malsch MM, Reed R et al (2007) The human Shwachman-Diamond syndrome protein, SBDS, associates with ribosomal RNA. Blood 110(5):1458–1465
57. Freed EF, Bleichert F, Dutca LM, Baserga SJ (2010) When ribosomes go bad: diseases of ribosome biogenesis. Mol Biosyst 6(3):481–493
58. Donadieu J, Leblanc T, Bader Meunier B, Barkaoui M, Fenneteau O, Bertrand Y et al (2005) Analysis of risk factors for myelodysplasias, leukemias and death from infection among patients with congenital neutropenia. Experience of the French severe chronic neutropenia study group. Haematologica 90(1):45–53
59. Alter BP, Giri N, Savage SA, Rosenberg PS (2009) Cancer in dyskeratosis congenita. Blood 113(26):6549–6557
60. Gazda HT, Grabowska A, Merida-Long LB, Latawiec E, Schneider HE, Lipton JM et al (2006) Ribosomal protein S24 gene is mutated in Diamond-Blackfan anemia. Am J Hum Genet 79(6):1110–1118
61. Cmejla R, Cmejlova J, Handrkova H, Petrak J, Pospisilova D (2007) Ribosomal protein S17 gene (RPS17) is mutated in Diamond-Blackfan anemia. Hum Mutat 28(12):1178–1182

62. Farrar JE, Nater M, Caywood E, McDevitt MA, Kowalski J, Takemoto CM et al (2008) Abnormalities of the large ribosomal subunit protein, Rpl35a, in Diamond-Blackfan anemia. Blood 112(5):1582–1592
63. Gazda HT, Sheen MR, Vlachos A, Choesmel V, O'Donohue MF, Schneider H et al (2008) Ribosomal protein L5 and L11 mutations are associated with cleft palate and abnormal thumbs in Diamond-Blackfan anemia patients. Am J Hum Genet 83(6):769–780
64. Choesmel V, Bacqueville D, Rouquette J, Noaillac-Depeyre J, Fribourg S, Cretien A et al (2007) Impaired ribosome biogenesis in Diamond-Blackfan anemia. Blood 109(3): 1275–1283
65. Robledo S, Idol RA, Crimmins DL, Ladenson JH, Mason PJ, Bessler M (2008) The role of human ribosomal proteins in the maturation of rRNA and ribosome production. RNA 14(9):1918–1929
66. Vardiman JW, Harris NL, Brunning RD (2002) The World Health Organization (WHO) classification of the myeloid neoplasms. Blood 100(7):2292–2302
67. Bernstein KA, Gallagher JE, Mitchell BM, Granneman S, Baserga SJ (2004) The small-subunit processome is a ribosome assembly intermediate. Eukaryot Cell 3(6):1619–1626
68. Ebert BL, Pretz J, Bosco J, Chang CY, Tamayo P, Galili N et al (2008) Identification of RPS14 as a 5q- syndrome gene by RNA interference screen. Nature 451(7176):335–339
69. Jill D, Sara JE, Amanda JG, Michael JD, Stacie KL, Cynthia AB et al (1996) Positional cloning of a gene involved in the pathogenesis of Treacher Collins Syndrome. The Treacher Collins Syndrome Collaborative Group. Nat Genet, 12(2):130–136
70. Valdez BC, Henning D, So RB, Dixon J, Dixon MJ (2004) The treacher Collins syndrome (TCOF1) gene product is involved in ribosomal DNA gene transcription by interacting with upstream binding factor. Proc Natl Acad Sci USA 101(29):10709–10714
71. Ganapathi KA, Shimamura A (2008) Ribosomal dysfunction and inherited marrow failure. Br J Haematol 141(3):376–387
72. Ridanpaa M, van Eenennaam H, Pelin K, Chadwick R, Johnson C, Yuan B et al (2001) Mutations in the RNA component of RNase MRP cause a pleiotropic human disease, cartilage-hair hypoplasia. Cell 104(2):195–203
73. Taskinen M, Ranki A, Pukkala E, Jeskanen L, Kaitila I, Makitie O (2008) Extended follow-up of the Finnish cartilage-hair hypoplasia cohort confirms high incidence of non-Hodgkin lymphoma and basal cell carcinoma. Am J Med Genet Part A 146A(18):2370–2375
74. Kirwan M, Dokal I (2008) Dyskeratosis congenita: a genetic disorder of many faces. Clin Genet 73(2):103–112
75. Watson KL, Konrad KD, Woods DF, Bryant PJ (1992) *Drosophila* homolog of the human S6 ribosomal protein is required for tumor suppression in the hematopoietic system. Proc Natl Acad Sci USA 89(23):11302–11306
76. Amsterdam A, Sadler KC, Lai K, Farrington S, Bronson RT, Lees JA et al (2004) Many ribosomal protein genes are cancer genes in zebrafish. PLoS Biol 2(5):E139
77. MacInnes AW, Amsterdam A, Whittaker CA, Hopkins N, Lees JA (2008) Loss of p53 synthesis in zebrafish tumors with ribosomal protein gene mutations. Proc Natl Acad Sci USA 105(30):10408–10413
78. Ruggero D, Grisendi S, Piazza F, Rego E, Mari F, Rao PH et al (2003) Dyskeratosis congenita and cancer in mice deficient in ribosomal RNA modification. Science 299(5604):259–262
79. Grisendi S, Bernardi R, Rossi M, Cheng K, Khandker L, Manova K et al (2005) Role of nucleophosmin in embryonic development and tumorigenesis. Nature 437(7055):147–153
80. Montanaro L, Trere D, Derenzini M (2008) Nucleolus, ribosomes, and cancer. Am J Pathol 173(2):301–310
81. Fumagalli S, Thomas G (2011) The role of p53 in ribosomopathies. Semin Hematol 48(2):97–105
82. Lane D, Levine A (2010) p53 Research: the past thirty years and the next thirty years. Cold Spring Harb Perspect Biol 2(12):a000893
83. Deisenroth C, Zhang Y (2010) Ribosome biogenesis surveillance: probing the ribosomal protein-Mdm2-p53 pathway. Oncogene 29(30):4253–4260

84. Hollstein M, Sidransky D, Vogelstein B, Harris CC (1991) p53 Mutations in human cancers. Science 253(5015):49–53
85. Robles AI, Harris CC (2010) Clinical outcomes and correlates of TP53 mutations and cancer. Cold Spring Harb Perspect Biol 2(3):a001016
86. Zilfou JT, Lowe SW (2009) Tumor suppressive functions of p53. Cold Spring Harb Perspect Biol 1(5):a001883
87. Meek DW, Anderson CW (2009) Posttranslational modification of p53: cooperative integrators of function. Cold Spring Harb Perspect Biol 1(6):a000950
88. Kruse JP, Gu W (2009) Modes of p53 regulation. Cell 137(4):609–622
89. Shieh SY, Ikeda M, Taya Y, Prives C (1997) DNA damage-induced phosphorylation of p53 alleviates inhibition by MDM2. Cell 91(3):325–334
90. Shinozaki T, Nota A, Taya Y, Okamoto K (2003) Functional role of Mdm2 phosphorylation by ATR in attenuation of p53 nuclear export. Oncogene 22(55):8870–8880
91. Sherr CJ (2006) Divorcing ARF and p53: an unsettled case. Nat Rev Cancer 6(9):663–673
92. Gu W, Roeder RG (1997) Activation of p53 sequence-specific DNA binding by acetylation of the p53 C-terminal domain. Cell 90(4):595–606
93. Ito A, Lai CH, Zhao X, Saito S, Hamilton MH, Appella E et al (2001) p300/CBP-mediated p53 acetylation is commonly induced by p53-activating agents and inhibited by MDM2. EMBO J 20(6):1331–1340
94. Sykes SM, Mellert HS, Holbert MA, Li K, Marmorstein R, Lane WS et al (2006) Acetylation of the p53 DNA-binding domain regulates apoptosis induction. Mol Cell 24(6): 841–851
95. Tang Y, Luo J, Zhang W, Gu W (2006) Tip60-dependent acetylation of p53 modulates the decision between cell-cycle arrest and apoptosis. Mol Cell 24(6):827–839
96. Oliner JD, Pietenpol JA, Thiagalingam S, Gyuris J, Kinzler KW, Vogelstein B (1993) Oncoprotein MDM2 conceals the activation domain of tumour suppressor p53. Nature 362(6423):857–860
97. Meek DW, Knippschild U (2003) Posttranslational modification of MDM2. Mol Cancer Res 1(14):1017–1026
98. Marechal V, Elenbaas B, Piette J, Nicolas JC, Levine AJ (1994) The ribosomal L5 protein is associated with mdm-2 and mdm-2-p53 complexes. Mol Cell Biol 14(11):7414–7420
99. Lohrum MA, Ludwig RL, Kubbutat MH, Hanlon M, Vousden KH (2003) Regulation of HDM2 activity by the ribosomal protein L11. Cancer Cell 3(6):577–587
100. Zhang Y, Wolf GW, Bhat K, Jin A, Allio T, Burkhart WA et al (2003) Ribosomal protein L11 negatively regulates oncoprotein MDM2 and mediates a p53-dependent ribosomal-stress checkpoint pathway. Mol Cell Biol 23(23):8902–8912
101. Dai MS, Lu H (2004) Inhibition of MDM2-mediated p53 ubiquitination and degradation by ribosomal protein L5. J Biol Chem 279(43):44475–44482
102. Dai MS, Zeng SX, Jin Y, Sun XX, David L, Lu H (2004) Ribosomal protein L23 activates p53 by inhibiting MDM2 function in response to ribosomal perturbation but not to translation inhibition. Mol Cell Biol 24(17):7654–7668
103. Jin A, Itahana K, O'Keefe K, Zhang Y (2004) Inhibition of HDM2 and activation of p53 by ribosomal protein L23. Mol Cell Biol 24(17):7669–7680
104. Fumagalli S, Di Cara A, Neb-Gulati A, Natt F, Schwemberger S, Hall J et al (2009) Absence of nucleolar disruption after impairment of 40S ribosome biogenesis reveals an rpL11-translation-dependent mechanism of p53 induction. Nat Cell Biol 11(4):501–508
105. Ofir-Rosenfeld Y, Boggs K, Michael D, Kastan MB, Oren M (2008) Mdm2 Regulates p53 mRNA translation through inhibitory interactions with ribosomal protein L26. Mol Cell 32(2):180–189
106. Yadavilli S, Mayo LD, Higgins M, Lain S, Hegde V, Deutsch WA (2009) Ribosomal protein S3: a multi-functional protein that interacts with both p53 and MDM2 through its KH domain. DNA Repair (Amst) 8(10):1215–1224
107. Chen D, Zhang Z, Li M, Wang W, Li Y, Rayburn ER et al (2007) Ribosomal protein S7 as a novel modulator of p53-MDM2 interaction: binding to MDM2, stabilization of p53 protein, and activation of p53 function. Oncogene 26(35):5029–5037

108. Zhu Y, Poyurovsky MV, Li Y, Biderman L, Stahl J, Jacq X et al (2009) Ribosomal protein S7 is both a regulator and a substrate of MDM2. Mol Cell 35(3):316–326
109. Zhou X, Hao Q, Liao J, Zhang Q, Lu H (2013) Ribosomal protein S14 unties the MDM2-p53 loop upon ribosomal stress. Oncogene 32(3):388–396
110. Zhang X, Wang W, Wang H, Wang MH, Xu W, Zhang R (2013) Identification of ribosomal protein S25 (RPS25)-MDM2-p53 regulatory feedback loop. Oncogene 32(22):2782–2791
111. Xiong X, Zhao Y, He H, Sun Y (2011) Ribosomal protein S27-like and S27 interplay with p53-MDM2 axis as a target, a substrate and a regulator. Oncogene 30(15):1798–1811
112. He H, Sun Y (2007) Ribosomal protein S27L is a direct p53 target that regulates apoptosis. Oncogene 26(19):2707–2716
113. Li J, Tan J, Zhuang L, Banerjee B, Yang X, Chau JF et al (2007) Ribosomal protein S27-like, a p53-inducible modulator of cell fate in response to genotoxic stress. Cancer Res 67(23):11317–11326
114. Gu J, Kawai H, Nie L, Kitao H, Wiederschain D, Jochemsen AG et al (2002) Mutual dependence of MDM2 and MDMX in their functional inactivation of p53. J Biol Chem 277(22):19251–19254
115. Gilkes DM, Chen L, Chen J (2006) MDMX regulation of p53 response to ribosomal stress. EMBO J 25(23):5614–5625
116. Sasaki M, Kawahara K, Nishio M, Mimori K, Kogo R, Hamada K et al (2011) Regulation of the MDM2-P53 pathway and tumor growth by PICT1 via nucleolar RPL11. Nat Med 17(8):944–951
117. Kumazawa T, Nishimura K, Kuroda T, Ono W, Yamaguchi C, Katagiri N et al (2011) Novel nucleolar pathway connecting intracellular energy status with p53 activation. J Biol Chem 286(23):20861–20869
118. Itahana K, Bhat KP, Jin A, Itahana Y, Hawke D, Kobayashi R et al (2003) Tumor suppressor ARF degrades B23, a nucleolar protein involved in ribosome biogenesis and cell proliferation. Mol Cell 12(5):1151–1164
119. Kurki S, Peltonen K, Latonen L, Kiviharju TM, Ojala PM, Meek D et al (2004) Nucleolar protein NPM interacts with HDM2 and protects tumor suppressor protein p53 from HDM2-mediated degradation. Cancer Cell 5(5):465–475
120. Yu W, Qiu Z, Gao N, Wang L, Cui H, Qian Y et al (2011) PAK1IP1, a ribosomal stress-induced nucleolar protein, regulates cell proliferation via the p53-MDM2 loop. Nucleic Acids Res 39(6):2234–2248
121. Sundqvist A, Liu G, Mirsaliotis A, Xirodimas DP (2009) Regulation of nucleolar signalling to p53 through NEDDylation of L11. EMBO Rep 10(10):1132–1139
122. Mahata B, Sundqvist A, Xirodimas DP (2011) Recruitment of RPL11 at promoter sites of p53-regulated genes upon nucleolar stress through NEDD8 and in an Mdm2-dependent manner. Oncogene 31(25):3060–3071
123. Bartel F, Taubert H, Harris LC (2002) Alternative and aberrant splicing of MDM2 mRNA in human cancer. Cancer Cell 2(1):9–15
124. Schlott T, Reimer S, Jahns A, Ohlenbusch A, Ruschenburg I, Nagel H et al (1997) Point mutations and nucleotide insertions in the MDM2 zinc finger structure of human tumours. J Pathol 182(1):54–61
125. Lindstrom MS, Jin A, Deisenroth C, White Wolf G, Zhang Y (2007) Cancer-associated mutations in the MDM2 zinc finger domain disrupt ribosomal protein interaction and attenuate MDM2-induced p53 degradation. Mol Cell Biol 27(3):1056–1068
126. Zhang Q, Xiao H, Chai SC, Hoang QQ, Lu H (2011) Hydrophilic residues are crucial for ribosomal protein L11 (RPL11) interaction with zinc finger domain of MDM2 and p53 protein activation. J Biol Chem 286(44):38264–38274
127. Dai MS, Shi D, Jin Y, Sun XX, Zhang Y, Grossman SR et al (2006) Regulation of the MDM2-p53 pathway by ribosomal protein L11 involves a post-ubiquitination mechanism. J Biol Chem 281(34):24304–24313
128. Macias E, Jin A, Deisenroth C, Bhat K, Mao H, Lindstrom MS et al (2010) An ARF-independent c-MYC-activated tumor suppression pathway mediated by ribosomal protein-Mdm2 interaction. Cancer Cell 18(3):231–243

129. Bhat KP, Itahana K, Jin A, Zhang Y (2004) Essential role of ribosomal protein L11 in mediating growth inhibition-induced p53 activation. EMBO J 23(12):2402–2412
130. Longley DB, Harkin DP, Johnston PG (2003) 5-Fluorouracil: mechanisms of action and clinical strategies. Nat Rev Cancer 3(5):330–338
131. Huang M, Ji Y, Itahana K, Zhang Y, Mitchell B (2008) Guanine nucleotide depletion inhibits pre-ribosomal RNA synthesis and causes nucleolar disruption. Leuk Res 32(1):131–141
132. Sun XX, Dai MS, Lu H (2007) 5-Fluorouracil activation of p53 involves an MDM2-ribosomal protein interaction. J Biol Chem 282(11):8052–8059
133. Sun XX, Dai MS, Lu H (2008) Mycophenolic acid activation of p53 requires ribosomal proteins L5 and L11. J Biol Chem 283(18):12387–12392
134. Yuan X, Zhou Y, Casanova E, Chai M, Kiss E, Grone HJ et al (2005) Genetic inactivation of the transcription factor TIF-IA leads to nucleolar disruption, cell cycle arrest, and p53-mediated apoptosis. Mol Cell 19(1):77–87
135. Azuma M, Toyama R, Laver E, Dawid IB (2006) Perturbation of rRNA synthesis in the bap28 mutation leads to apoptosis mediated by p53 in the zebrafish central nervous system. J Biol Chem 281(19):13309–13316
136. Zhang C, Comai L, Johnson DL (2005) PTEN represses RNA polymerase I transcription by disrupting the SL1 complex. Mol Cell Biol 25(16):6899–6911
137. Pestov DG, Strezoska Z, Lau LF (2001) Evidence of p53-dependent cross-talk between ribosome biogenesis and the cell cycle: effects of nucleolar protein Bop1 on G(1)/S transition. Mol Cell Biol 21(13):4246–4255
138. Chakraborty A, Uechi T, Kenmochi N (2011) Guarding the 'translation apparatus': defective ribosome biogenesis and the p53 signaling pathway. Wiley Interdiscip Rev RNA 2(4):507–522
139. Panic L, Tamarut S, Sticker-Jantscheff M, Barkic M, Solter D, Uzelac M et al (2006) Ribosomal protein S6 gene haploinsufficiency is associated with activation of a p53-dependent checkpoint during gastrulation. Mol Cell Biol 26(23):8880–8891
140. Danilova N, Sakamoto KM, Lin S (2008) Ribosomal protein S19 deficiency in zebrafish leads to developmental abnormalities and defective erythropoiesis through activation of p53 protein family. Blood 112(13):5228–5237
141. Chakraborty A, Uechi T, Higa S, Torihara H, Kenmochi N (2009) Loss of ribosomal protein L11 affects zebrafish embryonic development through a p53-dependent apoptotic response. PLoS One 4(1):e4152
142. Barkic M, Crnomarkovic S, Grabusic K, Bogetic I, Panic L, Tamarut S et al (2009) The p53 tumor suppressor causes congenital malformations in Rpl24-deficient mice and promotes their survival. Mol Cell Biol 29(10):2489–2504
143. Matoba S, Kang JG, Patino WD, Wragg A, Boehm M, Gavrilova O et al (2006) p53 Regulates mitochondrial respiration. Science 312(5780):1650–1653
144. Kondoh H, Lleonart ME, Gil J, Wang J, Degan P, Peters G et al (2005) Glycolytic enzymes can modulate cellular life span. Cancer Res 65(1):177–185
145. Bensaad K, Tsuruta A, Selak MA, Vidal MN, Nakano K, Bartrons R et al (2006) TIGAR, a p53-inducible regulator of glycolysis and apoptosis. Cell 126(1):107–120
146. Yahagi N, Shimano H, Matsuzaka T, Najima Y, Sekiya M, Nakagawa Y et al (2003) p53 Activation in adipocytes of obese mice. J Biol Chem 278(28):25395–25400
147. Yahagi N, Shimano H, Matsuzaka T, Sekiya M, Najima Y, Okazaki S et al (2004) p53 Involvement in the pathogenesis of fatty liver disease. J Biol Chem 279(20):20571–20575
148. Assaily W, Rubinger DA, Wheaton K, Lin Y, Ma W, Xuan W et al (2011) ROS-mediated p53 induction of Lpin1 regulates fatty acid oxidation in response to nutritional stress. Mol Cell 44(3):491–501
149. Ide T, Brown-Endres L, Chu K, Ongusaha PP, Ohtsuka T, El-Deiry WS et al (2009) GAMT, a p53-inducible modulator of apoptosis, is critical for the adaptive response to nutrient stress. Mol Cell 36(3):379–392
150. Goldstein I, Ezra O, Rivlin N, Molchadsky A, Madar S, Goldfinger N et al (2012) p53, a novel regulator of lipid metabolism pathways. J Hepatol 56(3):656–662

151. Goldstein I, Rotter V (2012) Regulation of lipid metabolism by p53 – fighting two villains with one sword. Trends Endocrinol Metab 23(11):567–575
152. Budanov AV, Karin M (2008) p53 target genes sestrin1 and sestrin2 connect genotoxic stress and mTOR signaling. Cell 134(3):451–460
153. Feng Z, Zhang H, Levine AJ, Jin S (2005) The coordinate regulation of the p53 and mTOR pathways in cells. Proc Natl Acad Sci USA 102(23):8204–8209
154. Crighton D, Wilkinson S, O'Prey J, Syed N, Smith P, Harrison PR et al (2006) DRAM, a p53-induced modulator of autophagy, is critical for apoptosis. Cell 126(1):121–134
155. Abida WM, Gu W (2008) p53-dependent and p53-independent activation of autophagy by ARF. Cancer Res 68(2):352–357
156. Morselli E, Tasdemir E, Maiuri MC, Galluzzi L, Kepp O, Criollo A et al (2008) Mutant p53 protein localized in the cytoplasm inhibits autophagy. Cell Cycle 7(19):3056–3061
157. Sablina AA, Budanov AV, Ilyinskaya GV, Agapova LS, Kravchenko JE, Chumakov PM (2005) The antioxidant function of the p53 tumor suppressor. Nat Med 11(12):1306–1313
158. Tolstonog GV, Deppert W (2010) Metabolic sensing by p53: keeping the balance between life and death. Proc Natl Acad Sci USA 107(30):13193–13194
159. Deisenroth C, Zhang Y (2011) The ribosomal protein-Mdm2-p53 pathway and energy metabolism: bridging the Gap between feast and famine. Genes & Cancer 2(4):392–403
160. Espinosa JM, Verdun RE, Emerson BM (2003) p53 functions through stress- and promoter-specific recruitment of transcription initiation components before and after DNA damage. Mol Cell 12(4):1015–1027
161. Liu Y, Lagowski JP, Vanderbeek GE, Kulesz-Martin MF (2004) Facilitated search for specific genomic targets by p53 C-terminal basic DNA binding domain. Cancer Biol Ther 3(11):1102–1108
162. Christophorou MA, Ringshausen I, Finch AJ, Swigart LB, Evan GI (2006) The pathological response to DNA damage does not contribute to p53-mediated tumour suppression. Nature 443(7108):214–217
163. Li T, Kon N, Jiang L, Tan M, Ludwig T, Zhao Y et al (2012) Tumor suppression in the absence of p53-mediated cell-cycle arrest, apoptosis, and senescence. Cell 149(6): 1269–1283
164. Bywater MJ, Poortinga G, Sanij E, Hein N, Peck A, Cullinane C et al (2012) Inhibition of RNA polymerase I as a therapeutic strategy to promote cancer-specific activation of p53. Cancer Cell 22(1):51–65

# Chapter 10
# Transcriptional Regulation of Lipogenesis as a Therapeutic Target for Cancer Treatment

**Chenguang Wang, Jun-Yuan Ji, Lifeng Tian, and Richard G. Pestell**

**Abstract** A significant increase in lipogenesis is a metabolic hallmark of proliferating tumor cells and is required for oncogenic transformation of epithelial cells. Although most normal cells acquire the bulk of their fatty acids from the circulation, tumor cells synthesize more than 90 % of required lipids de novo. Consistent with an increased demand for lipid synthesis, diverse human cancer cells express high levels of lipogenic enzymes, such as fatty acid synthase (FASN) and stearoyl-CoA desaturase 1 (SCD1). The sterol regulatory element-binding protein 1 (SREBP1) and peroxisome proliferator-activated receptor γ (PPARγ) are master regulators of lipogenesis in diverse organisms. Previous studies have established that FASN and SCD1, the major transcriptional targets of SREBP1 and PPARγ, promote synthesis of fatty acids, which then serve as ligands for PPARγ activation. This review focuses on the potential therapeutic value of these lipogenic transcription factors as targets in cancer treatment.

**Keywords** Lipogenesis • Transcription factor • Sterol regulatory element-binding protein 1 (SREBP1) • Peroxisome proliferator-activated receptor gamma (PPARγ) • Cancer • Lipid metabolism • Cell proliferation

---

C. Wang, Ph.D. • L. Tian, Ph.D. • R.G. Pestell, M.D., Ph.D. (✉)
Cancer Biology, Thomas Jefferson University, Philadelphia, PA, USA
e-mail: chenguang.wang@jefferson.edu; lifeng.tian@jefferson.edu; richard.pestell@jefferson.edu

J.-Y. Ji, Ph.D.
Molecular and Cellular Medicine, College of Medicine, Texas A&M Health Science Center, College Station, TX, USA
e-mail: Ji@medicine.tamhsc.edu

R. Kumar (ed.), *Nuclear Signaling Pathways and Targeting Transcription in Cancer*, Cancer Drug Discovery and Development, DOI 10.1007/978-1-4614-8039-6_10,

## 10.1 Introduction

Lipogenesis is almost universally upregulated in human cancers [1]. Consistent with an essential role of the sterol regulatory element-binding protein 1 (SREBP1) in sensing and regulating intracellular lipid homeostasis, increased expression of SREBP1 has been detected in colorectal carcinoma, breast and prostate cancer, and hepatocarcinoma [2–5]. Moreover, elevated expression of SREBP1 is closely correlated with malignant transformation, cancer progression, and metastasis for several cancer types, particularly hormone responsive tissue-derived cancers, such as breast and prostate cancers [2, 4, 6, 7]. SREBP1 expression correlates with the expression of *FASN* (encodes fatty acid synthase or FAS) and Ki-67(a nuclear marker for cellular proliferation) in colorectal cancer, suggesting a role for SREBP1 in supporting rapid cellular proliferation [7]. SREBP1 is elevated in clinical prostate cancer samples compared to benign prostatic hypertrophy [3]. Gene expression profiling of hepatocellular carcinoma (HCC) tissue and non-cancerous liver tissue showed increased lipogenic signaling in HCC. ElevatedSREBP1 expression in hepatocellular carcinoma is a known predictor of increased mortality [4, 6]. Overexpression of SREBP1 in human hepatoma HuH7 and Hep3B cells enhanced cellular proliferation and foci formation, while knockdown of SREBP1 in these cells reduced cell replication and anchorage-independent cell growth [6]. A dramatic increase of SREBP1 has been correlated with the progression of prostate cancer towards androgen-independence [3]. Oncogenic transformation of normal breast epithelial cells was accompanied by increased *SREBP1* and *FASN* expression, consistent with the observation of increased SREBP1 levels in human breast cancers [8–10]. Previous studies have established that SREBP1, through induction of *FASN* and subsequent fatty acids production, regulates PPARγ transactivation [11, 12].

Dietary carbohydrates are digested into glucose, the major source of energy for many tissues. Once transported into cells, glucose is converted into pyruvate through glycolysis and subsequently acetyl Co-enzyme A (acetyl-coA), which is then re-engineered into palmitate, the major fatty acid, by Acetyl-coA carboxylase (ACC), the rate-limiting enzyme, and FAS, the major enzyme, both of which are required for fatty acid biosynthesis. Palmitate is further converted into triglycerides for energy storage and phospholipids, the major components of cell membrane. The key steps in lipogenesis in mammalian hepatocytes are summarized in Fig. 10.1. The enzymatic reactions that govern carbohydrate and lipid metabolism, as well as the allosteric regulation of the activities of these enzymes, also known as the "short-term regulation", have been elucidated by many pioneering biochemists during the first half of the twentieth century.

Compared to the short-term quick regulation of the enzymes, however, the transcriptional regulation of the metabolic enzymes in vivo, known as the "long-term regulation" of metabolism, is less well-understood [13]. Since defects in short-term regulation of enzymes are likely detrimental to survival at the cellular or organismal levels, the aberrant regulation of the long-term regulation contributes to a number of major diseases in adults, collectively known as the metabolic syndrome [14–17].

Because of the fundamental importance of FAS, ACC, ACS and SCD1 in regulating lipid metabolism, it is essential to understand the transcriptional regulation of

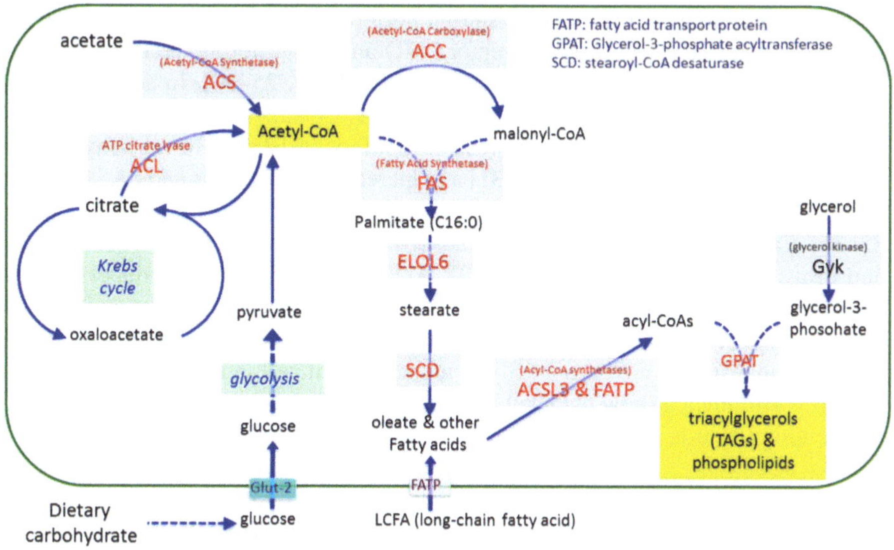


**Fig. 10.1 The key biochemical reactions and enzymes involved in de novo lipogenesis in mammalian hepatocytes.** This process is highly conserved in evolution. Transcription of many metabolic enzymes in this process is directly regulated by several transcription factors, such as PPARγ, SREBP, ChREBP, and LXR etc.

these enzymes in response to physiological stimuli by key transcription factors, including SREBP, PPARγ, liver X receptor (LXR), and carbohydrate-responsive element-binding protein (ChREBP) [13, 18–20]. In this chapter, we summarize the recent advances in studies linking deregulated lipogenesis in cancers, and then focus on our understanding of SREBPs and PPARγ in regulating lipid homeostasis. Finally, we will discuss potential therapeutic approaches to target lipid metabolism in treating cancer.

## 10.2 Deregulation of Lipogenic Signaling in Cancer

### *10.2.1 Elevated FASN Expression and Enhanced De Novo Fatty Acid Synthesis in Cancer*

Most normal human tissues preferentially use circulating lipids for synthesis of new structural lipids, and de novo fatty-acid synthesis is normally suppressed due to the low levels of *FASN* expression. In cancer cells, however, fatty-acid supply becomes highly dependent on de novo lipogenesis. Deregulation of de novo fatty-acid synthesis leads to cellular fatty-acid accumulation and affects cellular processes, including signal transduction and gene expression.

*FASN* over-expression occurs in a variety of human cancers [21–25]. In cancer cells, *FASN* gene expression is upregulated in response to multiple signaling pathways, including growth factors, steroid hormone receptors such as the estrogen receptor alpha (ERα), androgen receptor (AR) and progesterone receptor (PR), as well as oncoproteins including ErbB2, Ras and Akt [9, 25–30]. In addition to the essential role in cancer cell growth and survival, FAS is involved in other phases of cancer development. FAS over-expression confers resistance to adriamycin and mitoxantrone in breast cancer cells [31] and increased lipogenesis and FAS has been reported to be associated with invasive phenotype and cancer metastasis [24, 27, 32–36]. Elevated expression of *FASN* leads to increased cell proliferation, migration and invasion of prostate cancer cells [27, 32] and FAS inhibition reduces cellular migration and invasiveness [25, 26, 29, 35]. For example, Orlistat, an anti-obesity drug, inhibits FAS function and suppresses endothelial cell proliferation and angiogenesis, suggesting a novel role of FAS in endothelial cell in tumor growth in vivo [37]. It is still unclear how the level and activity of FAS are regulated during tumor progression towards metastasis.

### *10.2.2 Stearoyl-CoA Desaturase (SCD) and Cancer*

SCD is a regulatory enzyme in lipogenesis, catalyzing the rate-limiting step in the de novo synthesis of monounsaturated fatty acids (MUFAs), mainly palmitic and stearic acids. Increased content of the MUFA products, palmitoleic and oleic acids, occurs in a variety of transformed cells and cancers [38–41], suggesting that the high rate of fatty acid synthesis in cancer is coupled to the conversion of saturated fatty acids (SFAs) into MUFAs. Elevated expression and activity of SCD1, the endoplasmic reticulum-resident Δ9 desaturase that converts SFA into MUFA, has been reported in several types of cancers, including colonic and oesophageal carcinoma, liver cancer, and mammary gland tumor [42–45]. SV40-transformed human lung fibroblasts show significantly increased protein and activity levels of SCD1 compared to their parental normal cell line [46]. This is consistent with a model in which a high rate of MUFA synthesis is required for producing membrane lipids in order to sustain the proliferation of transformed cells. Deficiency or inhibition of SCD1 reduces cell proliferation and anchorage-independent growth, and enhances apoptosis in several different cancer cell types [40, 47]. We and others have shown that SCD1 is a transcriptional target of SREBP1 and PPARγ [48–54].

## 10.3 Cellular Regulation of SREBP1 Function

### *10.3.1 SREBP1 Signaling in Lipogenesis and Tumorigenesis*

SREBPs are a family of transcription factors that control lipid homeostasis by regulating the expression of enzymes required for cholesterol and fatty acids (FAs) synthesis. The three SREBP isoforms, SREBP-1a, SREBP-1c and SREBP-2, have distinct roles in lipid synthesis [55, 56]. In vivo studies using transgenic and knockout mice suggest that SREBP-1c is involved in FA synthesis and insulin-induced glucose metabolism (particularly in lipogenesis), whereas SREBP-2 is relatively specific in controlling cholesterol synthesis. The SREBP-1a isoform is implicated in regulating both cholesterol and FA pathways [57, 58].

Extensive studies in the past two decades have revealed an elegant paradigm to understand how SREBPs maintain the intracellular lipid and cholesterol homeostasis. SREBP transcription factors are synthesized as inactive precursors bound to the endoplasmic reticulum (ER) membranes and their processing is mainly controlled by cellular sterol content: when sterol level decreases, the precursor undergoes a sequential two-step cleavage process to release the NH2-terminal active domain in the nucleus (designated as the nSREBPs), which then activates SREBP target genes whose products are required for the de novo biosynthesis of cholesterol and FAs [59–66]. The major SREBP targets include FASN [12, 67] and stearoyl-CoA desaturase (SCD) [49–51, 68]. This sterol-sensitive process appears to be a major point of regulation for the SREBP-1a and SREBP-2 isoforms, but not for SREBP-1c. Moreover, the SREBP-1c isoform is mainly regulated at the transcriptional level by insulin. The unique regulation and activation properties of each SREBP isoform facilitate the coordinated regulation of lipid and energy metabolism.

## *10.3.2 Regulation of the Transcriptional Activity of SREBP1*

As summarized above, SREBPs are family of transcription factors that play critical roles in regulating intracellular lipid and cholesterol homeostasis. Using SREBP-1a/-1c as an example, here we focus on the recent advances in our understanding of how SREBP-1 activates lipogenic gene expression and how the transcriptional activity of SREBP is regulated.

### 10.3.2.1 Transcription Activation by SREBP

In response to cholesterol depletion, the N-terminus of SREBP that contains the transactivation domain and the basic helix-loop-helix leucine zipper (bHLH-Zip) DNA binding domain, is cleaved from its precursor, which is localized in ER and Golgi apparatus, and then translocates to the nucleus and activates the expression of SREBP-target genes [69, 70]. Through the bHLH-Zip DNA-binding domain, the nuclear SREBP fragments bind to the SREBP-target gene promoters that contain either palindromic E-boxes (CAXXTG) or nonpalindromic sterol regulatory elements (SREs) [71].

The transactivation domain of SREBPs can directly interact with transcription coactivators including CBP/p300, PGC-1β, MED14/DRIP150, and MED15/ARC105 [72]. Recruitment of CBP/p300 via the KIX domain of SREBP may alter chromatin structure through the intrinsic histone acetyltransferase activity of CBP/p300, thereby facilitating gene activation [73]. The interaction between PGC-1β and SREBP is required for SREBP-dependent lipogenic gene expression and contributes to the effect of saturated fat in stimulating hyperlipidemia and atherogenesis [74]. In addition, SREBPs directly interact with the MED14/DRIP150 and the MED15/ARC105 subunits of the Mediator complex in mammals and *C. elegans*, which provides an elegant model to explain how transcription activator SREBPs interact with the general transcription machinery [73, 75–77]. The interactions between SREBP-1c and MED14 or MED15 are weaker than the interactions between SREBP-1a and MED14 or MED15, which may explain why SREBP-1a is more potent than SREBP-1c in activating gene expression [73, 75]. Since the nuclear SREBPs bind to DNA as homo-dimers, it is unclear whether the two trans-activation domains of the SREBP homo-dimer can bind to MED14 and MED15 simultaneously.

#### 10.3.2.2 Inactivation of SREBP-Mediated Transcription

Because of the fundamental roles of SREBPs in regulating the expression of lipogenic and cholesterogenic genes, the mechanisms that restrain SREBP transactivation are also important. CDK8, the enzymatic subunit of the Mediator complex, directly phosphorylates a conserved Threonine residue in SREBP (Thr402 in SREBP-1c), thereby promoting nuclear SREBP degradation [78]. Consistent with this model, the mutants of CDK8 and its regulatory partner Cyclin C (CycC) in *Drosophila* larvae, as well as depletion of CDK8 in cultured mammalian cells and mouse liver, display significantly increased expression of SREBP-target genes and dramatic increase of triglyceride accumulation [78]. Feeding and activation of the insulin-signaling pathway can down-regulate CDK8-CycC thus allow the activation of nuclear SRBEP, providing a mechanism for the lipogenic effect of insulin [78]. Together with the previous works on MED14 and MED15 in activating SREBP-dependent gene expression, this recent work on the inhibitory effect of CDK8-CycC on SREBP-regulated de novo lipogenes further highlights the importance of the Mediator complexes in modulating the activation and subsequent degradation of nuclear SREBPs.

Interestingly, GSK3β also negatively regulates SREBP by phosphorylating SREBP-1a at Thr 426 and Ser430 (corresponding to Thr402 and Ser406 in SREBP-1c), thereby providing a docking site for the ubiquitin ligase FBW7 [79–81]. It is still not known whether CDK8 and GSK3β play redundant roles in phosphorylating and thereby promoting SREBP destruction, however, these studies suggest a model to explain how activation of SREBP-dependent transcription is coupled to its degradation. This mechanism is consistent with a general theme by coupling transactivation with their destruction for many transactivators in yeasts and multicellular organisms [82].

Both CDK8 and CycC are amplified, mutated or deleted in a variety of cancers, and CDK8 is identified as an oncoprote in melanoma and colorectal cancers [83]. In addition, the MED12 subunit of the CDK8 module, which is composed of CDK8, CycC, MED12 and MED13, is mutated in prostate cancer, colorectal cancer, and

~70 % of uterine leiomyomas [84–87]. Importantly, MED12, but not MED13, is required for human CDK8 kinase activity [88]. Therefore, although it is still unclear how dysregulation of the CDK8 module contributes to tumorigenesis, it is conceivable that dysregulation of CDK8 submodules may compromise CDK8 activity, thereby potentiating SREBP activity, increasing SREBP target gene expression and promoting lipogenesis in cancer cells. This model may explain the mechanisms underlying aberrantly increased lipogenesis in human cancer cells and provide the rationale for developing pharmaceutical approaches to block de novo lipogenesis in tumor cells.

## 10.4 Modulation of PPARγ Activation for Cancer Therapeutics

### *10.4.1 The Function of PPARγ in Lipogenesis*

Besides SREBPs, the peroxisome proliferator-activator receptor gamma (PPARγ) also plays a critical role in both lipid metabolism and tumorigenesis. The PPARs are ligand-activated nuclear receptors, which include PPARα, PPARγ and PPARδ [89]. Their modular structure resembles other nuclear hormone receptors with an N-terminal activation function 1 (AF-1), a DNA binding domain, and a C-terminal ligand-binding domain that harbors AF2. PPARγ was initially cloned as a transcription factor involved in adipocyte differentiation. Subsequent studies suggested a broad spectrum of PPARγ functions in lipid metabolism, inflammation, atherogenesis, cell differentiation, as well as tumorigenesis. The endogenous PPARγ ligands include derivatives of fatty acids produced through lipogenesis (Fig. 10.1).

PPARγ regulates lipogenesis and adipocyte differentiation, and ectopic PPARγ expression promotes cell adipogenesis in an NIH 3 T3 cell model [90, 91]. Synthetic PPARγ ligands enhance de novo lipogenesis [92, 93], which was further supported by the genetic evidence that $PPAR\gamma^{-/-}$ ES cells and embryonic fibroblasts are resistant to induction of adipogenesis [94, 95]. Aberrant hepatic expression of PPARγ2 stimulates murine hepatic lipogenesis [96, 97]. The screening for PPARγ-regulated genes in mammary epithelial cells, identified that Scd1 (*SCD1*) as a transcriptional target of PPARγ [98]. SCD1 production of unsaturated fatty acids may thereby serve as PPARγ agonist ligands, providing a feedback loop to PPARγ. Reciprocal up-regulation of PPARγ and SREBP-1 has been reported. Ectopic expression of SREBP-1 in pre-adipocyte 3 T3-L1 cells and hepatic cancer HepG2 cells induced endogenous PPARγ mRNA expression [99]. SREBP-1 activation increased the production of lipids as endogenous ligands for PPARγ, which binds to PPARγ and augments the transcriptional activity of PPARγ [11, 12]. PPARγ, upon ligand binding, up-regulates the expression of *INSIG1*, the key regulator in the processing of SREBPs [100].

### 10.4.2 Contradictory Role of PPARγ in Tumorigenesis

PPARγ has been shown to function either as an oncoprote in, or as a tumor suppressor. PPARγ is expressed in breast, prostate and colonic epithelium and ligand-dependent activation of PPARγ in cell lines derived from these tumors inhibits cellular proliferation [101–105]. Consistent with the role of PPARγ as a tumor suppressor, PPARγ ligand reduced tumorigenesis in the $Apc^{Min}$ model of familial adenomatous polyposis. Carcinogen, N-nitroso-N-methylurea (NMU)-induced mammary tumorigenesis is prevented by PPARγ agonists [106], and 7, 12-dimethylbenz(a)anthracene (DMBA)-induced mammary tumorigenesis was inhibited by troglitazone [107–109]. A chromosomal translocation between PAX8 and PPARγ in follicular thyroid cancer served as a dominant inhibitor of endogenous PPARγ expression [110]. PPARγ expression is reduced in human breast cancers compared with normal breast tissue [111, 112] and PPARγ over expression in tumor cells inhibits cell proliferation in tissue culture [112]. PPARγ levels are reduced in mouse transgenic mammary tumors induced by distinct oncogenes, compared with normal adjacent non tumorous mammary epithelium [111, 112].

In contrast, evidence that PPARγ is an oncogene includes observations that PPARγ ligands promote colonic tumor growth in $Apc^{Min}$ mice when fed a high fat diet [113]. Heterozygous mutations of PPARγ have been detected in 4/55 patients with colon cancer [114]. Although genetic analyses failed to show that PPARγ-deficient mice develop enhanced tumor phenotypes in prostate epithelium induced by the SV40 large-T antigen oncogene [113]. A constitutively active mutant of PPARγ (PγCA) enhanced ErbB2-induced tumor in vivo in immune-competent animals (Fig. 10.2) and in transgenic mice [115]. PγCA promoted ErbB2-induced tumor growth in immune-competent animals. Increased angiogenesis is associated with enhanced tumor growth in vivo [116]. Collectively, these studies suggest cell-type specific functions of PPARγ in the tumor induction versus inhibition.

### 10.4.3 Can PPARγ Be Targeted to Block the Tumor Growth?

In cell culture, PPARγ expression and/or activation repressed tumor cell growth by inhibiting cell proliferation, promoting apoptotic and autophagic cell death, and inducing terminal differentiation of cancer cells [112, 117, 118]. In whole animal studies, the picture appears more complex, which is consistent with the importance of heterotypic signals in cancer progression, and the presence of PPARγ in a variety of cell types including the inflammatory system. Clinical trials have been undertaken in a variety of tumor types including liposarcomas, prostate, pancreatic, colorectal, breast, thyroid, head and neck cancers, as well as melanoma and leukemia [119]. Overall, PPARγ agonists failed to yield positive clinical outcome in most cancer types. PPARγ is increased in ERα-negative breast cancer, but reduced expression in ERα-positive breast cancers [120]. Breast cancer genetic subtypes

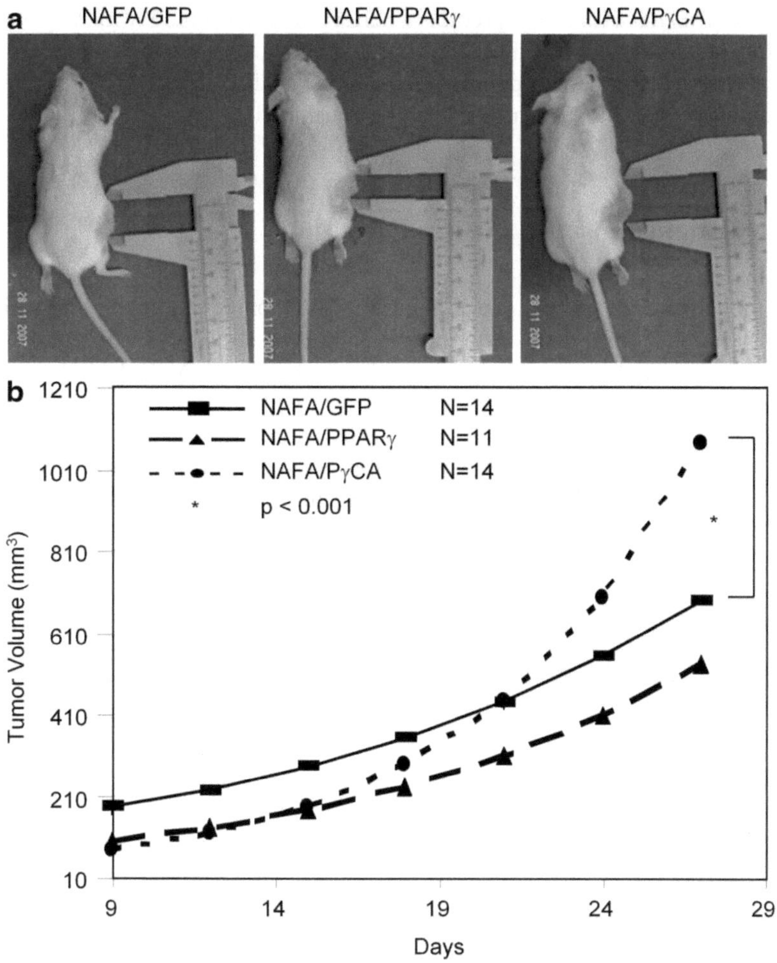


**Fig. 10.2 PγCA promotes tumor growth in vivo.** (**a**) NAFA cells transduced with MSCV-IRES-GFP vector encoding either PPARγ, PγCA, or empty vector were implanted into FVB by injecting $2 \times 10^6$ cells subcutaneously. Tumor growth was measured every 3 days by digital caliper and tumor volume was calculated. (**b**) Tumor volumes were logarithm-transformed and analyzed using a linear mixed model. Separate slope and intercepts were computed for each group (GFP, PPARγ, and PγCA), then compared across groups using a global test followed by pair-wise comparisons via linear contrasts (This figure was reproduced from our previous publication [116])

(Luminal A, Luminal B, Triple negative/basal-like, HER2 subtypes [121–123]) may have different response to PPARγ ligands. Our analysis of *PPARG* gene expression in a combined dataset comprising of over 2,000 breast cancers [124] showed a strong heterogeneous distribution of *PPARG* expression among the subtypes (unpublished data). Consistent with our previous IHC result showing reduced PPARγ expression in breast cancer comparing to normal breast tissue, this analysis demonstrated that the gene expression of *PPARG* was also reduced (data not shown).

The higher *PPARG* expression predicts a better clinical outcome, which again holds the promise that PPARγ could serve as a therapeutic target. Given the variability in PPARγ expression in patient populations, clinical trials using PPARγ expression or function as a companion diagnostic may be warranted.

## 10.5 Conclusions and Future Directions

Given the importance of lipogenesis in cancer development, targeting lipogenic signaling, particularly lipogenic enzymes, is an attractive strategy. The inhibitors of the rate-limiting or key lipogenic enzymes, including HMGCR, ACC, FASN, and SCD, are summarized in Table 10.1. In addition to inhibiting these lipogenic enzymes, pharmaceutical inhibition of SREBPs and PPARγ, may also be effective.

Taken together, these studies suggest that dysregulated lipogenic signaling in cancer is required for oncogenic transformation, thus targeting the dysregulated lipogenesis in câncer cells may represent an attractive therapeutic approach. Current

**Table 10.1** Chemical modulators of lipogenic signaling

| Modulator | Targeting molecule/pathway | Mechanism of function | References |
|---|---|---|---|
| Statins | HMG-CoA reductase (HMGCR)/mevalonate pathway | Structural analogs of HMG-CoA reductase, lipid-lowing agent | Review in [125] |
| Soraphen A | Acetyl CoA carboxylase (ACC) | Interferes with fatty acid elongation | [126, 127] |
| benzofuranyl alpha-pyrone (TEI-B00422) | Acetyl CoA carboxylase (ACC) | Competitive inhibition of ACC | [128] |
| 5-(tetradecyloxy)-2-furoic acid (TOFA) | Acetyl CoA carboxylase (ACC) | Long chain fatty acid analogues | [129] |
| CP-640186 | Acetyl CoA carboxylase (ACC) | Interacts with ACC | [130] |
| Cerulenin | Fatty acid synthase (FASN) | | [131, 132] |
| C75 | Fatty acid synthase (FASN) | Interacts and inhibits FASN | [133, 134] |
| C93 | Fatty acid synthase (FASN) | | [135] |
| Orlistat | Fatty acid synthase (FASN) | | [136] |
| EGCG | Fatty acid synthase (FASN) | | [137] |
| G28UCM | Fatty acid synthase (FASN) | | [138] |
| GSK837149A | Fatty acid synthase (FASN) | Target the beta-ketoacyl reductase reaction | [139] |
| MK-8245 | Stearoyl-CoA desaturase (SCD) | | [140] |
| Compound 9 | Stearoyl-CoA desaturase (SCD) | | [141] |
| Fatostatin (and derivatives) | SREBP-1 | Inhibits SREBP-1 processing | [142, 143] |
| BF175 (and derivatives) | SREBP-1 | N.D. | [144] |
| TZDs | PPARγ | Interacts with and activate PPARγ | [145] |

research efforts have been focused on repressing the activity of lipogenic enzymes (such as FASN, HMG-CoA reductase, ACC, ACLY, and SCD). Future studies are required to provide a deeper understanding of the following three major aspects. First, it would be important to understand how alterations in molecular mechanisms of lipogenic signaling occur in cancer. Second, a compendium of metabolic profiling in different cancer types and subtypes may allow for more accurate patient selection for specific lipogenic pathway targeted therapies. Third, it may be important to simultaneously target multiple lipogenic factors rather than a single molecule, to ensure therapy effectiveness.

**Acknowledgements** This work was supported in part by R01CA70896, R01CA75503, and R01CA86072 (R.G.P.). The Kimmel Cancer Center was supported by the NIH Cancer Center Core Grant P30CA56036 (R.G.P). This work is also funded in part from the Dr. Ralph and Marian C. Falk Medical Research Trust (R.G.P.), the Breast Cancer Research Foundation (R.G.P) and a grant from the Pennsylvania Department of Health (R.G.P., C.W.), the startup funds from Texas A&M Health Science Center and a grant from the American Heart Association (J.Y.J.).

# References

1. Swinnen JV, Brusselmans K, Verhoeven G (2006) Increased lipogenesis in cancer cells: new players, novel targets. Curr Opin Clin Nutr Metab Care 9:358
2. Yang YA (2003) Regulation of fatty acid synthase expression in breast cancer by sterol regulatory element binding protein-1c. Exp Cell Res 282:132
3. Ettinger SL (2004) Dysregulation of sterol response element-binding proteins and downstream effectors in prostate cancer during progression to androgen independence. Cancer Res 64:2212
4. Yahagi N (2005) Co-ordinate activation of lipogenic enzymes in hepatocellular carcinoma. Eur J Cancer 41:1316
5. Calvisi DF (2011) Increased lipogenesis, induced by AKT-mTORC1-RPS6 signaling, promotes development of human hepatocellular carcinoma. Gastroenterology 140:1071
6. Yamashita T (2009) Activation of lipogenic pathway correlates with cell proliferation and poor prognosis in hepatocellular carcinoma. J Hepatol 50:100
7. Li JN, Mahmoud MA, Han WF, Ripple M, Pizer ES (2000) Sterol regulatory element-binding protein-1 participates in the regulation of fatty acid synthase expression in colorectal neoplasia. Exp Cell Res 261:159
8. Furuta E (2008) Fatty acid synthase gene is up-regulated by hypoxia via activation of Akt and sterol regulatory element binding protein-1. Cancer Res 68:1003
9. Yoon S (2007) Up-regulation of acetyl-CoA carboxylase alpha and fatty acid synthase by human epidermal growth factor receptor 2 at the translational level in breast cancer cells. J Biol Chem 282:26122
10. Yang YA, Han WF, Morin PJ, Chrest FJ, Pizer ES (2002) Activation of fatty acid synthesis during neoplastic transformation: role of mitogen-activated protein kinase and phosphatidylinositol 3-kinase. Exp Cell Res 279:80
11. Kim JB, Wright HM, Wright M, Spiegelman BM (1998) ADD1/SREBP1 activates PPAR gamma through the production of endogenous ligand. Proc Natl Acad Sci USA 95:4333
12. Kim JB, Spiegelman BM (1996) ADD1/SREBP1 promotes adipocyte differentiation and gene expression linked to fatty acid metabolism. Genes Dev 10:1096
13. Desvergne B, Michalik L, Wahli W (2006) Transcriptional regulation of metabolism. Physiol Rev 86:465

14. Lusis AJ, Attie AD, Reue K (2008) Metabolic syndrome: from epidemiology to systems biology. Nat Rev Genet 9:819
15. Eckel RH, Grundy SM, Zimmet PZ (2005) The metabolic syndrome. Lancet 365:1415
16. Cornier MA et al (2008) The metabolic syndrome. Endocr Rev 29:777
17. Braun S, Bitton-Worms K, LeRoith D (2011) The link between the metabolic syndrome and cancer. Int J Biol Sci 7:1003
18. Postic C, Dentin R, Denechaud PD, Girard J (2007) ChREBP, a transcriptional regulator of glucose and lipid metabolism. Annu Rev Nutr 27:179
19. Laurencikiene J, Ryden M, Liver X (2012) Receptors and fat cell metabolism. Int J Obes (Lond). Dec; 36(12):1494–502
20. Horton JD, Goldstein JL, Brown MS (2002) SREBPs: activators of the complete program of cholesterol and fatty acid synthesis in the liver. J Clin Invest 109:1125
21. Mashima T, Seimiya H, Tsuruo T (2009) De novo fatty-acid synthesis and related pathways as molecular targets for cancer therapy. Br J Cancer 100:1369
22. Little JL, Kridel SJ (2008) Fatty acid synthase activity in tumor cells. Subcell Biochem 49:169
23. Kridel SJ, Lowther WT, Pemble CW (2007) Fatty acid synthase inhibitors: new directions for oncology. Expert Opin Investig Drugs 16:1817
24. Menendez JA, Lupu R (2007) Fatty acid synthase and the lipogenic phenotype in cancer pathogenesis. Nat Rev Cancer 7:763
25. Lupu R, Menendez JA (2006) Targeting fatty acid synthase in breast and endometrial cancer: an alternative to selective estrogen receptor modulators? Endocrinology 147:4056
26. Grunt TW et al (2009) Interaction between fatty acid synthase- and ErbB-systems in ovarian cancer cells. Biochem Biophys Res Commun 385:454
27. Migita T et al (2009) Fatty acid synthase: a metabolic enzyme and candidate oncogene in prostate cancer. J Natl Cancer Inst 101:519
28. Vazquez-Martin A, Ortega-Delgado FJ, Fernandez-Real JM, Menendez JA (2008) The tyrosine kinase receptor HER2 (erbB-2): from oncogenesis to adipogenesis. J Cell Biochem 105:1147
29. Uddin S et al (2008) Fatty acid synthase and AKT pathway signaling in a subset of papillary thyroid cancers. J Clin Endocrinol Metab 93:4088
30. Swinnen JV et al (2000) Stimulation of tumor-associated fatty acid synthase expression by growth factor activation of the sterol regulatory element-binding protein pathway. Oncogene 19:5173
31. Liu H, Liu Y, Zhang JT (2008) A new mechanism of drug resistance in breast cancer cells: fatty acid synthase overexpression-mediated palmitate overproduction. Mol Cancer Ther 7:263
32. Fiorentino M et al (2008) Overexpression of fatty acid synthase is associated with palmitoylation of Wnt1 and cytoplasmic stabilization of beta-catenin in prostate cancer. Lab Invest 88:1340
33. Ashida S et al (2004) Molecular features of the transition from prostatic intraepithelial neoplasia (PIN) to prostate cancer: genome-wide gene-expression profiles of prostate cancers and PINs. Cancer Res 64:5963
34. Vazquez-Martin A, Colomer R, Brunet J, Lupu R, Menendez JA (2008) Overexpression of fatty acid synthase gene activates HER1/HER2 tyrosine kinase receptors in human breast epithelial cells. Cell Prolif 41:59
35. Carvalho MA et al (2008) Fatty acid synthase inhibition with orlistat promotes apoptosis and reduces cell growth and lymph node metastasis in a mouse melanoma model. Int J Cancer 123:2557
36. Montgomery RB et al (2008) Maintenance of intratumoral androgens in metastatic prostate cancer: a mechanism for castration-resistant tumor growth. Cancer Res 68:4447
37. Browne CD, Hindmarsh EJ, Smith JW (2006) Inhibition of endothelial cell proliferation and angiogenesis by orlistat, a fatty acid synthase inhibitor. FASEB J J20:2027

38. Bagnato C, Igal RA (2003) Overexpression of diacylglycerol acyltransferase-1 reduces phospholipid synthesis, proliferation, and invasiveness in simian virus 40-transformed human lung fibroblasts. J Biol Chem 278:52203
39. Scaglia N, Chisholm JW, Igal RA (2009) Inhibition of stearoylCoA desaturase-1 inactivates acetyl-CoA carboxylase and impairs proliferation in cancer cells: role of AMPK. PLoS One 4:e6812
40. Scaglia N, Igal RA (2008) Inhibition of stearoyl-CoA desaturase 1 expression in human lung adenocarcinoma cells impairs tumorigenesis. Int J Oncol 33:839
41. Tsubura A, Yuri T, Yoshizawa K, Uehara N, Takada H (2009) Role of fatty acids in malignancy and visual impairment: epidemiological evidence and experimental studies. Histol Histopathol 24:223
42. Li J et al (1994) Partial characterization of a cDNA for human stearoyl-CoA desaturase and changes in its mRNA expression in some normal and malignant tissues. Int J Cancer 57:348
43. Paton CM, Ntambi JM (2008) Role of stearoyl-CoA desaturase-1 expression in cancer proliferation. FASEB J 22:794.1
44. Lu J, Pei H, Kaeck M, Thompson HJ (1997) Gene expression changes associated with chemically induced rat mammary carcinogenesis. Mol Carcinog 20:204
45. Falvella FS et al (2002) Stearoyl-CoA desaturase 1 (Scd1) gene overexpression is associated with genetic predisposition to hepatocarcinogenes is in mice and rats. Carcinogenesis 23:1933
46. Scaglia N, Caviglia JM, Igal RA (2005) High stearoyl-CoA desaturase protein and activity levels in simian virus 40 transformed-human lung fibroblasts. Biochim Biophys Acta 1687:141
47. Morgan-Lappe SE et al (2007) Identification of Ras-related nuclear protein, targeting protein for xenopus kinesin-like protein 2, and stearoyl-CoA desaturase 1 as promising cancer targets from an RNAi-based screen. Cancer Res 67:4390
48. Shimano H et al (1999) Sterol regulatory element-binding protein-1 as a key transcription factor for nutritional induction of lipogenic enzyme genes. J Biol Chem 274:35832
49. Ntambi JM (1999) Regulation of stearoyl-CoA desaturase by polyunsaturated fatty acids and cholesterol. J Lipid Res 40:1549
50. Tabor DE, Kim JB, Spiegelman BM, Edwards PA (1999) Identification of conserved cis-elements and transcription factors required for sterol-regulated transcription of stearoyl-CoA desaturase 1 and 2. J Biol Chem 274:20603
51. Shimomura I, Shimano H, Korn BS, Bashmakov Y, Horton JD (1998) Nuclear sterol regulatory element-binding proteins activate genes responsible for the entire program of unsaturated fatty acid biosynthesis in transgenic mouse liver. J Biol Chem 273:35299
52. Tabor DE, Kim JB, Spiegelman BM, Edwards PA (1998) Transcriptional activation of the stearoyl-CoA desaturase 2 gene by sterol regulatory element-binding protein/adipocyte determination and differentiation factor 1. J Biol Chem 273:22052
53. Kim YC, Gomez FE, Fox BG, Ntambi JM (2000) Differential regulation of the stearoyl-CoA desaturase genes by thiazolidinediones in 3T3-L1 adipocytes. J Lipid Res 41:1310
54. Yao-Borengasser A et al (2008) Stearoyl-coenzyme a desaturase 1 gene expression increases after pioglitazone treatment and is associated with peroxisomal proliferator-activated receptor-gamma responsiveness. J Clin Endocrinol Metab 93:4431
55. Hua X, Wu J, Goldstein JL, Brown MS, Hobbs HH (1995) Structure of the human gene encoding sterol regulatory element binding protein-1 (SREBF1) and localization of SREBF1 and SREBF2 to chromosomes 17p11.2 And 22q13. Genomics 25:667
56. Edwards PA, Tabor D, Kast HR, Venkateswaran A (2000) Regulation of gene expression by SREBP and SCAP. Biochim Biophys Acta 1529:103
57. Shimano H et al (1996) Overproduction of cholesterol and fatty acids causes massive liver enlargement in transgenic mice expressing truncated SREBP-1a. J Clin Invest 98:1575
58. Horton JD et al (1998) Activation of cholesterol synthesis in preference to fatty acid synthesis in liver and adipose tissue of transgenic mice overproducing sterol regulatory element-binding protein-2. J Clin Invest 101:2331

59. Wang X et al (1996) Cleavage of sterol regulatory element binding proteins (SREBPs) by CPP32 during apoptosis. EMBO J J15:1012
60. Wang X, Sato R, Brown MS, Hua X, Goldstein JL (1994) SREBP-1, a membrane-bound transcription factor released by sterol-regulated proteolysis. Cell 77:53
61. Hua X et al (1993) SREBP-2, a second basic-helix-loop-helix-leucine zipper protein that stimulates transcription by binding to a sterol regulatory element. Proc Natl Acad Sci USA A90:11603
62. Yokoyama C et al (1993) SREBP-1, a basic-helix-loop-helix-leucine zipper protein that controls transcription of the low density lipoprotein receptor gene. Cell 75:187
63. Briggs MR, Yokoyama C, Wang X, Brown MS, Goldstein JL (1993) Nuclear protein that binds sterol regulatory element of low density lipoprotein receptor promoter. I. Identification of the protein and delineation of its target nucleotide sequence. J Biol Chem 268:14490
64. Wang X et al (1993) Nuclear protein that binds sterol regulatory element of low density lipoprotein receptor promoter. II. Purification and characterization. J Biol Chem 268:14497
65. Shimano H (2001) Sterol regulatory element-binding proteins (SREBPs): transcriptional regulators of lipid synthetic genes. Prog Lipid Res 40:439
66. Horton JD, Shimomura I (1999) Sterol regulatory element-binding proteins: activators of cholesterol and fatty acid biosynthesis. Curr Opin Lipidol 10:143
67. Liang G et al (2002) Diminished hepatic response to fasting/refeeding and liver X receptor agonists in mice with selective deficiency of sterol regulatory element-binding protein-1c. J Biol Chem 277:9520
68. Bene H, Lasky D, Ntambi JM (2001) Cloning and characterization of the human stearoyl-CoA desaturase gene promoter: transcriptional activation by sterol regulatory element binding protein and repression by polyunsaturated fatty acids and cholesterol. Biochem Biophys Res Commun 284:1194
69. Goldstein JL, DeBose-Boyd RA, Brown MS (2006) Protein sensors for membrane sterols. Cell 124:35
70. Jeon TI, Osborne TF (2012) SREBPs: metabolic integrators in physiology and metabolism. Trends Endocrinol Metab 23:65
71. Kim JB et al (1995) Dual DNA binding specificity of ADD1/SREBP1 controlled by a single amino acid in the basic helix-loop-helix domain. Mol Cell Biol 15:2582
72. Bengoechea-Alonso MT, Ericsson J (2007) SREBP in signal transduction: cholesterol metabolism and beyond. Curr Opin Cell Biol 19:215
73. Toth JI, Datta S, Athanikar JN, Freedman LP, Osborne TF (2004) Selective coactivator interactions in gene activation by SREBP-1a and -1c. Mol Cell Biol 24:8288
74. Lin J et al (2005) Hyperlipidemic effects of dietary saturated fats mediated through PGC-1beta coactivation of SREBP. Cell 120:261
75. Yang F et al (2006) An ARC/mediator subunit required for SREBP control of cholesterol and lipid homeostasis. Nature 442:700
76. Osborne TF, Espenshade PJ (2009) Evolutionary conservation and adaptation in the mechanism that regulates SREBP action: what a long, strange tRIP it's been. Genes Dev 23:2578
77. Taubert S, Van Gilst MR, Hansen M, Yamamoto KR (2006) A mediator subunit, MDT-15, integrates regulation of fatty acid metabolism by NHR-49-dependent and -independent pathways in C. Elegans. Genes Dev 20:1137
78. Zhao X et al (2012) Regulation of lipogenesis by cyclin-dependent kinase 8-mediated control of SREBP-1. J Clin Inves 122:2417
79. Punga T, Bengoechea-Alonso MT, Ericsson J (2006) Phosphorylation and ubiquitination of the transcription factor sterol regulatory element-binding protein-1 in response to DNA binding. J Biol Chem 281:25278
80. Sundqvist A et al (2005) Control of lipid metabolism by phosphorylation-dependent degradation of the SREBP family of transcription factors by SCF(Fbw7). Cell Metab 1:379
81. Kim KH et al (2004) Regulatory role of glycogen synthase kinase 3 for transcriptional activity of ADD1/SREBP1c. J Biol Chem 279:51999

82. Tansey WB (2001) Transcriptional activation: risky business. Genes Dev 15:1045
83. Xu W, Ji JY (2011) Dysregulation of CDK8 and cyclin C in tumorigenesis. J Genet Genomics 38:439
84. Kampjarvi K et al (2012) Somatic MED12 mutations in uterine leiomyosarcoma and colorectal cancer. Br J Cancer 107:1761
85. Barbieri CE et al (2012) Exome sequencing identifies recurrent SPOP, FOXA1 and MED12 mutations in prostate cancer. Nat Genet 44:685
86. McGuire MM, et al (2012) Whole exome sequencing in a random sample of North American women with leiomyomas identifies MED12 mutations in majority of uterine leiomyomas. PLoS One 7, e33251
87. Makinen N et al (2011) MED12, the mediator complex subunit 12 gene, is mutated at high frequency in uterine leiomyomas. Science 334:252
88. Knuesel MT, Meyer KD, Donner AJ, Espinosa JM, Taatjes DJ (2009) The human CDK8 subcomplex is a histone kinase that requires Med12 for activity and can function independently of mediator. Mol Cell Biol 29:650
89. Rosen ED, Spiegelman BM (2001) PPARg: a nuclear regulator of metabolism, differentiation, and cell growth. J Biol Chem 276:37731
90. Tontonoz P, Hu E, Spiegelman BM (1994) Stimulation of adipogenesis in fibroblasts by PPARg2, a lipid-activated transcription factor. Cell 79:1147
91. Yu S et al (2003) Adipocyte-specific gene expression and adipogenic steatosis in the mouse liver due to peroxisome proliferator-activated receptor 1 (PPAR1) overexpression. J Biol Chem 278:498
92. Teruel T, Hernandez R, Rial E, Martin-Hidalgo A, Lorenzo M (2005) Rosiglitazone up-regulates lipoprotein lipase, hormone-sensitive lipase and uncoupling protein-1, and down-regulates insulin-induced fatty acid synthase gene expression in brown adipocytes of Wistar rats. Diabetologia 48:1180
93. Watkins SM, Reifsnyder PR, Pan HJ, German JB, Leiter EH (2002) Lipid metabolome-wide effects of the PPAR gamma agonist rosiglitazone. J Lipid Res 43:1809
94. Kubota N et al (1999) PPAR gamma mediates high-fat diet-induced adipocyte hypertrophy and insulin resistance. Mol Cell 4:597
95. Rosen ED et al (1999) PPAR gamma is required for the differentiation of adipose tissue in vivo and in vitro. Mol Cell 4:611
96. Poritsanos NJ, Wong D, Vrontakis ME, Mizuno TM (2008) Regulation of hepatic PPARgamma2 and lipogenic gene expression by melanocortin. Biochem Biophys Res Commun 376:384
97. Schroeder-Gloeckler JM et al (2007) CCAAT/enhancer-binding protein beta deletion reduces adiposity, hepatic steatosis, and diabetes in lepr(db/db) mice. J Biol Chem 282:15717
98. J. Zhou et al (2009) PPARgamma activation induces autophagy in breast cancer cells. Int J Biochem Cell Biol Nov; 41(11):2334–42
99. Fajas L et al (1999) Regulation of peroxisome proliferator-activated receptor gamma expression by adipocyte differentiation and determination factor 1/sterol regulatory element binding protein 1: implications for adipocyte differentiation and metabolism. Mol Cell Biol 19:5495
100. Kast-Woelbern HR et al (2004) Rosiglitazone induction of insig-1 in white adipose tissue reveals a novel interplay of peroxisome proliferator-activated receptor gamma and sterol regulatory element-binding protein in the regulation of adipogenesis. J Biol Chem 279:23908
101. Brockman JA, Gupta RA, Dubois RN (1998) Activation of PPARgamma leads to inhibition of anchorage-independent growth of human colorectal cancer cells. Gastroenterology 115:1049
102. Coopman PJ, Do MT, Thompson EW, Mueller SC (1998) Phagocytosis of cross-linked gelatin matrix by human breast carcinoma cells correlates with their invasive capacity. Clin Cancer Res 4:507
103. Elstner E et al (1998) Ligands for peroxisome proliferator-activated receptorg and retinoic acid receptor inhibit growth and induce apoptosis of human breast cancer cells in vitro and in BNX mice. Proc Natl Acad Sci USA 95:8806

104. Ricote M et al (1998) Expression of the peroxisome proliferator-activated receptor gamma (PPARgamma) in human atherosclerosis and regulation in macrophages by colony stimulating factors and oxidized low density lipoprotein. Proc Natl Acad Sci USA 95:7614
105. Sarraf P et al (1998) Differentiation and reversal of malignant changes in colon cancer through PPARgamma. Nat Med 4:1046
106. Suh N et al (1999) A new ligand for the peroxisome proliferator-activated receptor-gamma (PPAR-gamma), GW7845, inhibits rat mammary carcinogenesis. Cancer Res 59:5671
107. Mehta RG, Williamson E, Patel MK, Koeffler HP (2000) A ligand of peroxisome proliferator-activated receptor gamma, retinoids, and prevention of preneoplastic mammary lesions. J Natl Cancer Inst 92:418
108. Pighetti GM et al (2001) Therapeutic treatment of DMBA-induced mammary tumors with PPAR ligands. Anticancer Res 21:825
109. Nicol CJ et al (2004) PPARgamma influences susceptibility to DMBA-induced mammary, ovarian and skin carcinogenesis. Carcinogenesis 25:1747
110. Kroll TG et al (2000) PAX8-PPARgamma1 Fusion oncogene in human thyroid carcinoma. Science 289:1357
111. Wang C et al (2003) Cyclin D1 repression of peroxisome proliferator-activated receptor gamma expression and transactivation. Mol Cell Biol 23:6159
112. Wang C et al (2001) Inhibition of cellular proliferation through IkappaB kinase-independent and peroxisome proliferator-activated receptor gamma-dependent repression of cyclin D1. Mol Cell Biol 21:3057
113. Saez E et al (1998) Activators of the nuclear receptor PPARgamma enhance colon polyp formation. Nat Med 4:1058
114. Sarraf P et al (1999) Loss-of-function mutations in PPAR gamma associated with human colon cancer. Mol Cell 3:799
115. Saez E et al (2004) PPAR gamma signaling exacerbates mammary gland tumor development. Genes Dev 18:528
116. Tian L et al (2009) Activating peroxisome proliferator-activated receptor gamma mutant promotes tumor growth in vivo by enhancing angiogenesis. Cancer Res 69:9236
117. Elstner E et al (1998) Ligands for peroxisome proliferator-activated receptor gamma and retinoic acid receptor inhibit growth and induce apoptosis of human breast cancer cells in vitro and in BNX mice. Proc Natl Acad Sci USA A95:8806
118. Zhou J et al (2009) PPAR gamma activation induces autophagy in breast cancer cells. Int J Biochem Cell Biol 41:2334
119. Vamecq J et al (2012) PPARs: interference with Warburg' effect and clinical anticancer trials. PPAR Res 2012:304760
120. Mueller E et al (1998) Terminal differentiation of human breast cancer through PPAR gamma. Mol Cell 1:465
121. Perou CM et al (2000) Molecular portraits of human breast tumours. Nature 406:747
122. Sorlie T (2004) Molecular portraits of breast cancer: tumour subtypes as distinct disease entities. Eur J Cancer 40:2667
123. Sorlie T et al (2001) Gene expression patterns of breast carcinomas distinguish tumor subclasses with clinical implications. Proc Natl Acad Sci USA A98:10869
124. Casimiro MC et al (2012) ChIP sequencing of cyclin D1 reveals a transcriptional role in chromosomal instability in mice. J Clin Invest 122:833
125. Osmak M (2012) Statins and cancer: current and future prospects. Cancer Lett 324:1
126. Beckers A et al (2007) Chemical inhibition of acetyl-CoA carboxylase induces growth arrest and cytotoxicity selectively in cancer cells. Cancer Res 67:8180
127. Jump B, Torres-Gonzalez M, Olson LK, Soraphen A (2011) An inhibitor of acetyl CoA carboxylase activity, interferes with fatty acid elongation. Biochem Pharmacol 81:649
128. Sugimoto Y et al (2007) A novel acetyl-CoA carboxylase inhibitor reduces de novo fatty acid synthesis in HepG2 cells and rat primary hepatocytes. Arch Biochem Biophys 468:44
129. McCune SA, Harris RA (1979) Mechanism responsible for 5-(tetradecyloxy)-2-furoic acid inhibition of hepatic lipogenesis. J Biol Chem 254:10095

130. Harwood HJ Jr et al (2003) Isozyme-nonselective N-substituted bipiperidylcarboxamide acetyl-CoA carboxylase inhibitors reduce tissue malonyl-CoA concentrations, inhibit fatty acid synthesis, and increase fatty acid oxidation in cultured cells and in experimental animals. J Biol Chem 278:37099
131. Kuhajda FP et al (1994) Fatty acid synthesis: a potential selective target for antineoplastic therapy. Proc Natl Acad Sci USA A91:6379
132. Pizer ES et al (1996) Inhibition of fatty acid synthesis delays disease progression in a xenograft model of ovarian cancer. Cancer Res 56:1189
133. Pizer ES, Chrest FJ, DiGiuseppe JA, Han WF (1998) Pharmacological inhibitors of mammalian fatty acid synthase suppress DNA replication and induce apoptosis in tumor cell lines. Cancer Res 58:4611
134. Pizer ES et al (2000) Malonyl-coenzyme-a is a potential mediator of cytotoxicity induced by fatty-acid synthase inhibition in human breast cancer cells and xenografts. Cancer Res 60:213
135. Zhou W et al (2007) Fatty acid synthase inhibition activates AMP-activated protein kinase in SKOV3 human ovarian cancer cells. Cancer Res 67:2964
136. Kridel SJ, Axelrod F, Rozenkrantz N, Smith JW (2004) Orlistat is a novel inhibitor of fatty acid synthase with antitumor activity. Cancer Res 64:2070
137. Brusselmans K, De Schrijver E, Heyns W, Verhoeven G, Swinnen JV (2003) Epigallocatechin-3-gallate is a potent natural inhibitor of fatty acid synthase in intact cells and selectively induces apoptosis in prostate cancer cells. Int J Cancer 106:856
138. Oliveras G et al (2010) Novel anti-fatty acid synthase compounds with anti-cancer activity in HER2+ breast cancer. Ann N Y Acad Sci 1210:86
139. Vazquez MJ et al (2008) Discovery of GSK837149A, an inhibitor of human fatty acid synthase targeting the beta-ketoacyl reductase reaction. FEBS J J275:1556
140. Oballa RM et al (2011) Development of a liver-targeted stearoyl-CoA desaturase (SCD) inhibitor (MK-8245) to establish a therapeutic window for the treatment of diabetes and dyslipidemia. J Med Chem 54:5082
141. Uto Y et al (2011) Discovery of novel SCD1 inhibitors: 5-alkyl-4,5-dihydro-3H-spiro[1,5-benzoxazepine-2,4′-piperidine] analogs. Eur J Med Chem 46:1892
142. Kamisuki S et al (2009) A small molecule that blocks fat synthesis by inhibiting the activation of SREBP. Chem Biol 16:882
143. Kamisuki S et al (2011) Synthesis and evaluation of diarylthiazole derivatives that inhibit activation of sterol regulatory element-binding proteins. J Med Chem 54:4923
144. Das BC, Zhao X, Tang XY, Yang F (2011) Design, synthesis and biological study of pinacolyl boronate-substituted stilbenes as novel lipogenic inhibitors. Bioorg Med Chem Lett 21:5638
145. Wahli W, Michalik L (2012) PPARs at the crossroads of lipid signaling and inflammation. Trends Endocrinol Metab 23:351

# Part III
# Protein Domains and Cancer Therapeutics

# Chapter 11
# Selective Inhibition of Acetyl-Lysine Effector Domains of the Bromodomain Family in Oncology

**Susanne Müller, Hannah Lingard, and Stefan Knapp**

**Abstract** Acetylation of lysine residues is a posttranslational modification that plays a key role in the regulation of chromatin structure and transcription. In cancer, aberrant lysine acetylation often leads to changes in gene expression resulting in inactivation of tumour suppressor functions and the activation of pro-survival and proliferation promoting pathways. Enzymes that "write" (acetyltransferases, HATs) and "erase" (histone deacetylases, HDACs) ε-N-acetyl-lysine ($K_{ac}$) marks have therefore emerged as interesting targets for the development of novel drugs for cancer treatment. Recently also acetyl-lysine reader domains have gained interest as novel targets for pharmacological intervention. The acetyl-lysine mark is specifically recognized by the bromodomain family of protein interaction modules. Bromodomains are present in diverse nuclear proteins regulating the recruitment of transcriptional regulators and chromatin modifying enzymes and proteins to acetylated chromatin as well as proteins mediating the assembly of other nuclear protein complexes. Dysfunction of bromodomain containing proteins such as chromosomal rearrangements and aberrant expression of these proteins in cancer has been tightly linked to tumourigenesis. Recently identified inhibitors that selectively target bromodomains demonstrated potent anti-tumour activity, suggesting new avenues for the development of antineoplastic drugs. In this chapter we will review the current knowledge of the role of bromodomains in tumour development and identified selective inhibitors developed to disrupt acetyl-lysine dependent protein interactions mediated by this family of transcriptional regulators.

**Keywords** Bromodomain • Protein acetylation • Epigenetic reader modules • Transcription • Chromatin • BET • ATAD2 • BRG1 • TIF1alpha • BRD-NUT

---

S. Müller, Ph.D. (✉) • H. Lingard, M.Sci., M.A.(Cantab.), D.Phil.(Oxon.) • S. Knapp, Ph.D. (✉)
Nuffield Department of Clinical Medicine, Structural Genomics Consortium,
Medical Sciences Division, University of Oxford, Oxford, Oxon OX3 7DQ, UK
e-mail: susanne.muller-knapp@sgc.ox.ac.uk; hannah.lingard@sgc.ox.ac.uk;
Stefan.knapp@sgc.ox.ac.uk

R. Kumar (ed.), *Nuclear Signaling Pathways and Targeting Transcription in Cancer*,
Cancer Drug Discovery and Development, DOI 10.1007/978-1-4614-8039-6_11,

## Abbreviations

| | |
|---|---|
| AML | Acute myeloid leukemia |
| ASH1L | Absent, small or homeotic-like |
| ATAD2A/B | AAA domain-containing protein 2 |
| ATP | Adenosine triphosphate |
| BAZ | Bromodomain adjacent to zinc finger domain protein |
| BCL2 | B-cell lymphoma 2 |
| BET | Bromodomain and extra-terminal |
| BRD | Bromodomain-containing |
| BRPF | Bromodomain and PHD finger-containing protein |
| BRWD | Bromodomain and WD repeat-containing protein |
| CECR | Cat eye syndrome critical region |
| CML | Chronic myeloid leukemia |
| c-MYC | V-myc myelocytomatosis viral oncogene homolog |
| CREBBP | CREB-binding protein |
| EP300 | Histone acetyltransferase p300 |
| FALZ | Fetal Alzheimer antigene |
| FOSL1 | FOS-like antigen 1 |
| GCN5L2 | General control of amino acid synthesis protein 5-like 2 |
| HAT | Histone acetyltransferase |
| HBO1 | HAT bound to Orc1 |
| HCF-1 | Host cell factor 1 |
| HDAC | Histone deacetylase |
| HIV | Human immunodeficiency virus |
| hMOF | Human ortholog of fly mof, MOF, for males absent on the first |
| MLL | Mixed Linage Leukemia |
| MORF | MOZ-related factor |
| MOZ | Monocytic leukemic zinc finger protein |
| NF-κB | Nuclear factor kappa-light-chain-enhancer of activated B cells |
| NMC | NUT midline carcinoma |
| NSCLC | Non-small cell lung cancer |
| NUT | Nuclear protein in testis |
| PB1 | Polybromo 1, PBAF, Polybromo and Brg1-associated factor |
| PCAF | P300/CBP associated factor |
| PHIP | PH-interacting protein |
| PRKCBP1 | Protein kinase C-binding protein 1 |
| P-TEFb | Positive transcription elongation factor complex (cdk9/cyclinT) |
| Rara | Retinoic acid receptor alpha |
| RNA | Ribonucleic acid |
| siRNA | Small interfering RNA/Short interfering RNA |
| SIRT | Sirtuin |
| SMARCA | SWI/SNF-related, matrix-associated actin-dependent regulator of chromatin |
| SP100/110/140 | Nuclear body protein |

| | |
|---|---|
| SWI/SNF | SWItch/Sucrose NonFermentable |
| TAF1/TAF1L | Transcription initiation TFIID associated factor |
| Tat | Trans-activator of transcription |
| TIP60 | HIV Tat-interactive protein of 60 kDa |
| TRIM24/28/33/66 | Transcription intermediary factor 1-alpha (TIF1α) |
| WDR9 | Bromodomain and WD repeat-containing protein 1 |
| ZMYND11 | Zinc finger MYND domain-containing protein 11 |

Epigenetics has been defined as heritable regulatory mechanisms regulating gene transcription that are not encoded by the DNA sequence itself. These regulatory mechanisms comprise dynamic changes in chromatin structure mediated by post-translational modifications (PTMs) in proteins that organize chromatin such as histones.

Dense packing of chromatin requires neutralization of the high negative charge density of DNA by basic histones. Acetylation has a profound effect on the physico-chemical properties of the lysine side chain by neutralizing the charge of ε-*N* amine. This property is thought to favour an open, more loosely packed state of chromatin, leading to increased accessibility of regulatory regions on DNA and transcriptional activation. In cancer, inappropriate acetylation levels give rise to aberrant expression of genes that promote tumourigenesis. Both hyper- as well as hypoacetylation in promoter regions have been observed, leading to overexpression of growth and survival promoting genes as well as repression of tumour suppressor genes.

## 11.1 Role of Acetylation Homeostasis Regulating Transcription

ε-*N* lysine acetylation is one of the most frequently detected PTMs [1]. Acetylation homeostasis is principally controlled by enzymes that "write" acetylation marks (histone acetyltransferases (HATs)) or erase them (histone deacetylases, (HDACs)). Based on their homology with the corresponding yeast proteins the 18 human HDACs have been grouped into four different classes. Class I, II and IV are referred to as classical HDACs. They comprise HDAC1, HDAC2, HDAC3 and HDAC8 (class I), HDAC4, HDAC5, HDAC6, HDAC7, HDAC9 and HDAC10 (class II) and HDAC11 (class IV). These enzymes share a common mechanism of action and harbour a $Zn^{2+}$ ion in their active site. In contrast, class III enzymes require $NAD^+$ as cofactor and comprise the sirtuins SIRT1, SIRT2, SIRT3, SIRT4, SIRT5, SIRT6, and SIRT7 [2]. The HDAC subfamilies differ in their tissue specific and subcellular localisation, which has an influence on their substrates and biological function. While class I HDACs are uniformly nuclear, ubiquitously expressed proteins that deacetylate histones, other classes of HDACs like class IIa HDACs are expressed in a tissue specific manner and shuttle between nucleus and cytoplasm [3]. Accordingly, HDACs influence gene transcription on different levels and they have been reported to regulate chromatin as well as non-chromatin proteins. HDACs can drive carcinogenesis through a variety of different mechanisms: truncating mutations of HDAC2 have been identified in sporadic carcinomas with microsatellite instability and in tumours arising in individuals with hereditary nonpolyposis colorectal cancer syndrome [4], however most commonly dysregulated expression of HDACs has been observed in a wide variety of tumour types [2, 5]. HDAC overexpression can lead to upregulation of genes involved in development and proliferation [6] or downregulation of tumour suppressors [7]. HDACs may also regulate microRNAs (miRs) [8].

A variety of inhibitors against HDACs have been developed and currently two of these, suberoylanilidehydroxamic acid (SAHA, Vorinostat, Zolinza™) and depsipeptide (Romidepsin, Istodax™) have been approved by the US Food and Drug Administration. Most HDAC inhibitors are not target specific and inhibit most frequently class I enzymes. The precise mechanisms of action of HDAC inhibitors are not well understood and inhibition of the enzymatic activity as well as disruption of interactions with other proteins may play a role in the observed phenotypes [5, 9, 10].

Addition of acetyl groups to lysine residues is carried out by HATs which are classified into two different groups, class A HATs, which are nuclear and acetylate histone and non-histone proteins and the less characterised, cytoplasmic class B HATs, which are thought to acetylate newly synthesised histones and play a role in DNA repair and chromatin assembly [11, 12]. Nuclear HATs include 3 families, the GCN5L2/PCAF, the CREBBP/EP300 and the TAF families. In addition, the MYST family consisting of the five human HATs hMOF, TIP60, HBO1 (HAT bound to Orc1, also named MYST2), MOZ, and MORF belong to the class A HAT family [13]. HATs acetylate specific lysine residues on histone tails as well as non-histone proteins. At the chromatin level acetylation is generally thought to be associated with the formation of euchromatin and transcriptional activation. Acetylation of non-histone proteins can result in alteration of DNA binding, protein-protein interactions, protein stability or subcellular localisation [14].

TAF1, TAF1L, PCAF, GCN5L2 and EP300/CREBBP contain bromodomains in addition to the HAT domain, which will be discussed below and which offer alternative sites for pharmacological modulation of HAT function. Contrary to HDAC inhibitors, the development of inhibitors of HATs has not been as successful so far and only few natural products acting with low potency have been reported [15].

Until recently, little attention has been dedicated to inhibitor development and in understanding the function of protein interacting domains that "read" the epigenetic code but recent discoveries of highly potent and specific inhibitors for bromodomains strongly stimulated efforts targeting these domains. The recognition of the acetyl-lysine mark is principally mediated by bromodomains, a family of 61 diverse interaction modules in human that are present in 41 usually nuclear proteins [16]. Bromodomains have been named after the *Drosophila* gene "brahma" for which the core bromodomain sequence motif was first identified [17]. The proteins that contain one or more bromodomains comprise histone acetyl transferases such as PCAF, GCN5L2, CREBBP and EP300, ATP-dependent chromatin remodelling factors (SMARCA2/4 and ATAD2A/B), the methyl transferase ASH1L, transcriptional modulators such as BRPF, TAF1/TAF1L, TRIMs, BETs and the nuclear body proteins (SP100, SP110 and SP140) [18] (Table 11.1, Fig. 11.1).

**Table 11.1** Bromodomain containing proteins and their roles in cancer

| Protein | Role in cancer | Reference |
|---|---|---|
| ASH1L | Mutations or copy number changes detected in lung cancer | [19] |
| ATAD2A/B | Overexpressed in prostate, breast and lung cancer and high expression levels correlate with poor prognosis in some cancers | [20–25] |
| | Coactivator of E2F transcription factors | |
| | Controls the expression of B-MYB, the histone methyltransferase EZH2 as well as other growth promoting proteins | |
| | Involved in estrogen-induced cell proliferation and cell cycle progression of breast cancer cells | |
| BAZ1A/B | Aberrant expression in colon cancer | [26] |
| BAZ2A/B | Chromosomal rearrangements in paediatric pre-B acute lymphoblastic leukaemia | [27] |
| BRD1 | Chromosomal rearrangements detected in leukaemia with PAX5 transcription factors | [28] |
| BRD2 | Interacts with chromatin binding domain in Kaposi's Sarcoma-associated herpesvirus (KSHV) latency-associated nuclear antigen 1 (LANA-1). BRD2 expression is downregulated in nasopharyngeal carcinoma cells | [29, 30] |
| BRD3 | A chromosomal aberration involving BRD3 and NUT gives rise to a malignant subtype of midline carcinoma | [31] |
| BRD4 | Translocation t(15;9)(q14;q34) with NUT gives rise to a BRD4-NUT fusion protein and development of carcinomas | [32–36] |
| | BRD4 controls expression of growth promoting genes such as myc, aurora kinase and Bcl2 | |
| | Required for expression of target genes of the super elongation complex in MLL | |
| | Drives expression of FOSL1 in lung carcinoma | |
| | BRD4 is required for the replication of tumour viruses | |
| BRDT | Over-expression detected in non-small cell lung cancer and squamous cell carcinomas of the head and neck as well as of the oesophagus | [37] |
| BRD7 | Tumour suppressor required for efficient transcription of a subset of p53 target genes | [38] |
| | Deletion or low expression levels detected in lung cancer | |
| BRD8A/B | BRD8 regulates cellular survival and sensitivity to spindle poisons in colon cancer | [39] |
| BRD9 | DNA copy number changes in lung cancer | [40] |
| BRPF1 | Regulates HOX gene expression | [41, 42] |
| | Component of the monocytic leukemia zinc finger protein MOZ/MORF complex | |
| BRPF3 | Component of the monocytic leukemia zinc finger protein MOZ/MORF complex | [42] |
| BRWD3 | Chromosomal aberration leading to the disruption of the BRWD3 locus has been detected in B-cell chronic lymphocytic leukemia (B-CLL) | [43] |
| CECR2 | DNA damage response protein required for DSB repair | [44] |

(continued)

**Table 11.1** (continued)

| Protein | Role in cancer | Reference |
|---|---|---|
| CREBBP | Chromosomal aberrations and mutations detected in acute leukaemia | [45, 46] |
| | Inhibitor of RUNX1-mediated transcription | |
| | Required for self-renewal of leukaemic stem cells | |
| | Cofactor of androgen receptor in prostate cancer | |
| EP300 | Mutation identified in epithelial cancers | [47, 48] |
| | Chromosomal translocation detected in leukaemia | |
| FALZ | Overexpression is highly predictive of brain metastasis in early and advanced lung cancer | [49] |
| GCN5L2 | None identified as yet | |
| MLL | Mutation chromosomal translocations give rise to acute lymphoid leukemias and acute myeloid leukemias | [50] |
| PB1 | Tumour suppressor and regulator of p21 expression. Truncations and mutants in PB1 are a cause of renal cell carcinoma (RCC) and other cancer types | [51, 52] |
| PCAF | Inhibitor of the mitogenic activity of the adenoviral oncoprotein E1A | [53–55] |
| | Acetylates β-catenin enhancing its stability | |
| | Overexpressed in some paediatric tumours | |
| PHIP | Overexpression in an adverse predictor of survival in melanoma patients | [56, 57] |
| | Regulates PI3K/AKT signalling | |
| PRKCBP1 | T-cell lymphoma antigene | [58] |
| SMARCA2A/B | Loss of function mutations and down-regulation of this tumour suppressor have been detected in multiple tumour types | [59–61] |
| SMARCA4 | Germ line loss of function gives rise rhabdoid tumor predisposition syndrome type 2 (RTPS2) | [62] |
| | Mutation detected in many diverse cancers | |
| SP100/110/140 | SP bromodomain proteins are part of promyelocytic leukemia (PML) nuclear bodies | [63–66] |
| | SP100 has tumour suppressor function that is lost in some cancer types | |
| | SP100 inhibits ETS1 DNA binding and regulates ETS1 transcriptional activity | |
| | SP140 has been identified as a susceptibility gene for CML | |
| TAF1/TAF1L | None identified as yet | |
| TRIM24/28/33/66 | Overexpression of TRIM24 correlates with tumour progression in lung cancer and survival of breast cancer patients but promotes development of hepatocellular carcinoma by formation of a complex with TRIM33 and TRIM28 | [67–71] |
| | TRIM33 is overexpressed in colon cancer | |
| WDR9 | None identified as yet | |
| ZMYND11 | Copy number alteration in hematologic cancers | [72, 73] |
| | Linked to BRCA2 pathway | |

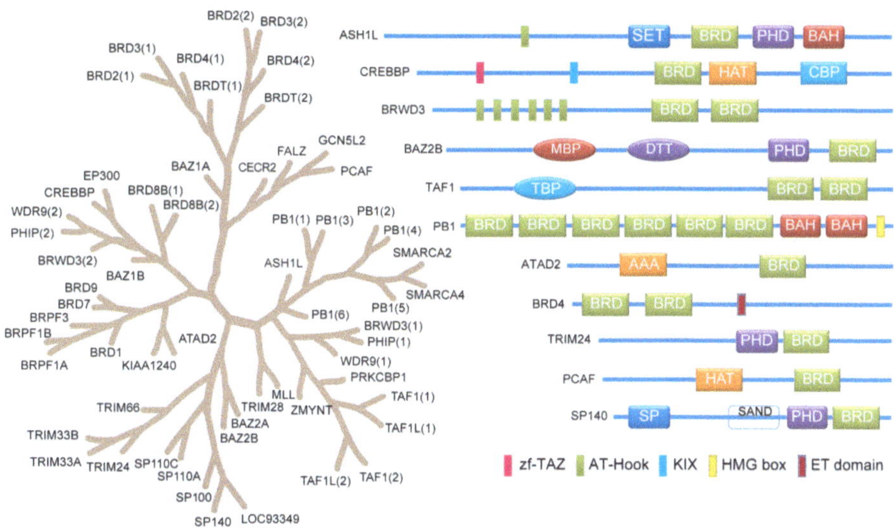


**Fig. 11.1** Phylogenetic tree of the bromodomain family and domain organization of representative bromodomain containing proteins

## 11.2 Molecular Architecture of Bromodomains

Despite the low level of sequence conservation, particularly in the terminal helices, all bromodomains share a conserved fold that comprises a left-handed bundle of four alpha helices (αZ, αA, αB, and αC). The four bromodomain core helices are linked by highly diverse loop regions (ZA and BC loops) that flank the acetyl-lysine binding site and determine substrate specificity (Fig. 11.2). Acetyl-lysine containing sequence motifs bind to a large central cavity created by the four canonical bromodomain helices. The bromodomain acetyl-lysine binding site contains largely hydrophobic and aromatic residues in addition to a conserved asparagine that anchors the carbonyl group of acetyl-lysine by a hydrogen bond. A number of alternative residues exist that can potentially act as hydrogen bond donors, but no complexes with acetyl-lysine containing peptides that would confirm the role of these alternative residues have been reported so far.

## 11.3 Bromodomain Recognition Sequences

The multi-domain architecture suggests that the reading process of epigenetic marks is a modular process which may involve interactions with diverse modifications that are required to precisely recruit a certain regulator to a specific site on chromatin. However, epigenetic marks that are simultaneously recognized by these

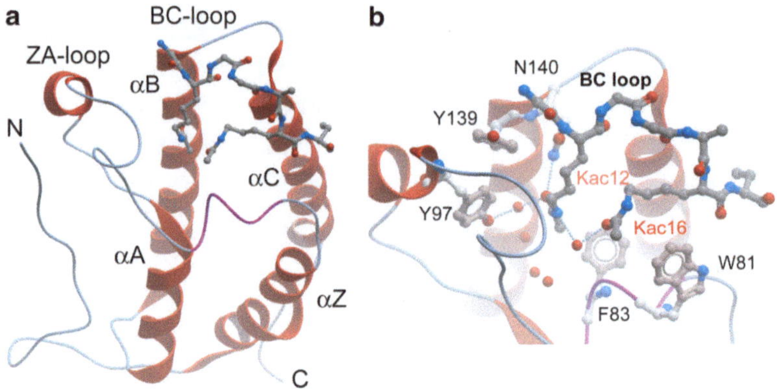


**Fig. 11.2 Molecular architecture of bromodomains.** (**a**) Shown is a peptide complex of the N-terminal bromodomain of human BRD4 with a diacetylated histone peptide. The main secondary structure elements are labelled and the N- and C- termini are indicated by N and C, respectively. (**b**) Details of the di-acetyl peptide interaction with the peptide binding site of BRD4. Peptide carbon atoms are highlighted in *grey* and acetyl-lysine residues on the H4 histone peptide are labelled in *red*. Water molecules are shown as *red spheres* and the main interacting residues are labelled in *black*

multi-domain proteins may be located in different histones, nuclear proteins or even on different nucleosomes. The complexity of the reading process poses a formidable challenge for our understanding of the epigenetic code.

Progress in peptide array technology has enabled systematic studies that have unravelled sequence specificity of individual reader domains. A recent comprehensive study revealed preferred histone interaction sequences for many members of the bromodomain family [16]. However, these data showed also that many bromodomains do not target histones and are probably recruited to other nuclear or non-nuclear protein complexes. For bromodomains that did interact with histones, another two interesting hallmarks of bromodomain peptide recognition were revealed: Firstly, many bromodomains seem to recognize poly-acetylated rather than singly acetylated lysine sites; and secondly post-translational modifications flanking the recognized acetylation site have a dominant effect on the bromodomain peptide recognition process.

Binding of a bromodomain to two acetylation sites was first reported for the bromodomain protein of murine BRDT [74]. Subsequent co-crystal structures and peptide array data showed that this is a shared feature of all N-terminal BET bromodomains [16] and significantly stronger interactions of poly-acetylated peptides with many bromodomains has suggested a widely distributed recognition mode for diacetyl-lysine containing sequences. A number of co-crystal structures of diacetyl-lysine containing peptides with BRD4 revealed a conserved interaction of both acetyl-lysines with the bromodomain peptide binding pocket. In all co-crystal structures the conserved bromodomain asparagine (N140 in BRD4(1)) forms a hydrogen bond with the carbonyl of the N-terminal peptide acetyl-lysine as it has been described for mono acetylated peptide complexes. In contrast the second acetyl-lysine carbonyl forms no direct polar contacts with the bromodomain but interacts via a water-mediated hydrogen bond with a conserved tyrosine residue (Y97 in BRD4(1)). Together both acetyl-lysine side chains show remarkable shape complementarity with the peptide binding pocket. No structural information of peptide complexes with additional post translational modifications have been reported so far. It is likely however that differences in surface potential (e.g. additional negative charges contributed by phosphorylated residues) significantly influence the recognition process of bromodomains for their recognition sequences. This feature suggests that bromodomains interpret post translational modifications generated by diverse pathways and function therefore as nuclear integrators of cellular signalling.

## 11.4 Development of Bromodomain Inhibitors

The acetyl-lysine binding pocket in bromodomains represents an attractive site for the development of inhibitors. In contrast to unmodified lysine side chains the acetylated residue is not charged. Thus, binding pockets that specifically recognize this PTM contain mainly neutral, hydrophobic and aromatic residues allowing the development of cell permeable acetyl-lysine competitive drug-like molecules. The size of the binding pocket, its high level of enclosure and its residue composition result in good predicted druggability values for most bromodomains [75]. However, inhibitor development efforts have focussed on few bromodomains so far.

The first bromodomain inhibitor, NP1, was described in 2005 by the Zhou laboratory as an inhibitor of the PCAF HIV-1 TAT interaction (PCAF $IC_{50}$ 1.6 μM) [76]. Subsequently a number of weakly binding fragments (MS2126 and MS7972) binding to the CREBBP bromodomains were discovered by the same laboratory using NMR screening methods and a library of putative acetyl-lysine mimetic compounds [77]. The most potent inhibitor of this series (MS7972) effectively repressed CREBBP recruitment to p53 suppressing p53 transcriptional activity demonstrating that bromodomains can be efficiently targeted in the cellular context. Interestingly NP1 and MS7972 do not seem to bind in an acetyl-lysine mimetic binding mode as they form no hydrogen bond with the conserved asparagine that typically ligates the acetyl-lysine carbonyl in peptidic complexes of bromodomains and their target sequences. Further development of the MS series led to ischemin which showed an $IC_{50}$ of 19 μM for CREBBP. Ischemin however mimics the acetylated lysine binding mode by forming a hydrogen bond to N1168 of CREBBP [78]. The tetrahydroquinoline Example 6 in WO2011/54848 can be seen as a larger analogue of MS2126 [79]. The acetyl-lysine mimetic binding mode has also been demonstrated in complexes of the fragment 3-methyl-3,4-dihydroquinazolin-2-(1*H*)-one with CREBBP and the BET inhibitor BIC [80].

Highly potent inhibitors were initially reported in the patent literature as potent BET inhibitors, a discovery that led to the development of the pan-BET thienodiazepine bromodomain inhibitor JQ1 [81] and benzodiazepines such as I-BET [82, 83].

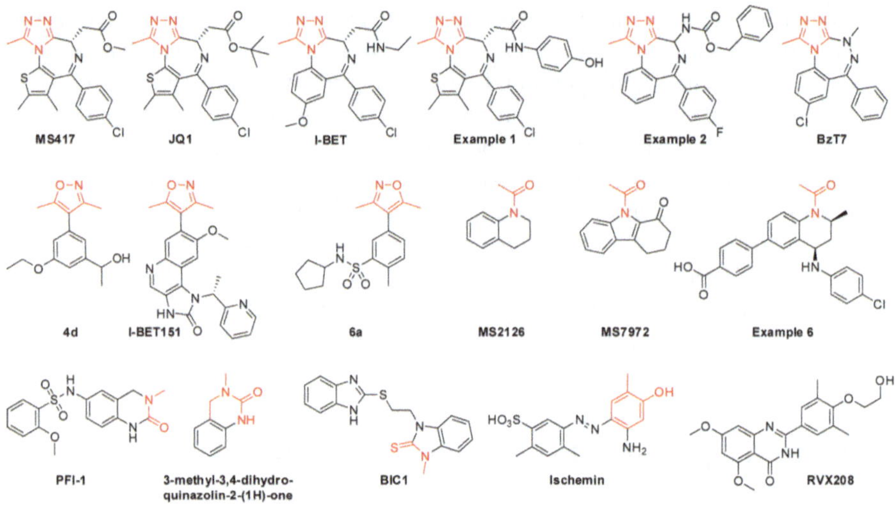


**Fig. 11.3 Structures of known bromodomain inhibitors.** In cases where experimental co-crystal structures have been determined we highlighted the acetyl-lysine mimetic moiety in *red*. Example 1 and 2 are from the patent WO2009/84693, WO/2011/054844, respectively and Example 6 (tetrahydroquinoline) has been described in the GSK patent application WO2011/054848. Compound 4d and 6a have been published in the paper by Hewings *et al.* All other inhibitors are described in the main text

Inhibitors of this class utilise the methyltriazolo-diazepine ring system as the acetyl-lysine mimetic. I-BET and JQ1 contain a stereo centre which makes synthesis or separation of the stereoisomers a challenging task. This problem was overcome by the development of benzotriazepines (for example Bzt7) where the asymmetric carbon atom is replaced by nitrogen [84]. Substitution of the methyltriazolo with the isosteric methylisoxazoles led to the development of a number of isoxazole based inhibitors with good ligand efficiency and potency [85, 86]. The most optimized inhibitor of this series, I-BET, showed good antiproliferative potency in cells as well as in mouse models of leukaemia [32]. A compilation of currently known bromodomain inhibitors is shown in Fig. 11.3.

## 11.5 Chromosomal Aberrations Lead to Oncogenic BRD Fusion Proteins

A number of bromodomains have been identified in highly oncogenic fusion proteins. One of the best characterized cancers with chromosomal translocations is NUT midline carcinoma (NMC), a highly aggressive rare, poorly differentiated squamous cell carcinoma, which predominantly occurs in children and mainly arises in the midline of the body, commonly in the head, neck or mediastinum. More recently BRD-NUT oncoproteins have also been detected in cancers with different tissue origin [33]. NMC is unusual because the single chromosomal translocation is often the only genetic aberration found in this cancer. In the majority of cases it is characterized by the rearrangement of the nuclear protein in testis (*NUT*) gene, located on chromosome 15q14 and the *BRD4* gene on chromosome 19p13.1 to create a *BRD4-NUT* fusion gene [87] whose product is driven by the *BRD4* promoter. The *BRD4-NUT* fusion gene contains the N-terminus of BRD4 or in rare cases BRD3 including both bromodomains and essentially the entire coding region for NUT [33, 88]. The NUT portion of the gene product is thought to be largely unstructured and contains an acidic binding domain for the HAT EP300. Little is known about the function of the NUT protein, though under normal conditions it is known to shuttle between the nucleus and the cytoplasm. When NUT is fused to BRD4 or BRD3 however it remains in the nucleus [31]. This implies that the BRD portion of the fusion protein tethers NUT to chromatin, affecting the function of one or both proteins and resulting in a knock-on effect on transcription. Genetic knock-down of BRD-NUT by siRNA in NMC cells results in dramatic terminal squamous differentiation and in G1 growth arrest consistent with the notion that BRD-NUT maintains an undifferentiated state of NMC cells [88, 89].

Another chromosomal aberration results in the disruption of the *EP300* and *CREBBP* locus in leukaemia [90]. The first of these translocations characterized was the t(8,16)(p11,p13) translocation which fuses the gene encoding the monocytic leukemia zinc-finger protein (MOZ) with the amino terminus of CREBBP, giving rise to the M4/M5 subtype of acute myeloid leukaemia (AML) [91, 92]. This translocation may result in activation of the HAT domain in the fusion protein, however the t(8,16) translocation is rare, accounting for only 0.4 % of patients with AML. A number of MOZ fusions with the CREBBP related HAT EP300 have been also described t(8,22)(p11,q13) [93, 94]. Fusion of the mixed lineage leukaemia gene products (MLL) with CREBBP and EP300 have been reported in relapsed MLL after treatment with topoisomerase inhibitors [95, 96]. *CREBBP* is more commonly involved in chromosomal rearrangements than *EP300*. Interestingly most of the *CREBBP* rearrangements target the same 13 kb genomic interval suggesting that this region contains an unstable genomic element accounting for the higher frequency of *CREBBP* rearrangements [97].

### *11.5.1 Aberrant Expression of Bromodomain Proteins in Cancer*

A number of genes coding for proteins containing bromodomains are overexpressed in cancers and expression levels have been linked to tumour progression and in some cases patient survival. An interesting example is TRIM24/TIF1α which has been reported as a tumour suppressor in liver cancer but as a tumour promoting factor in breast and lung cancer suggesting highly context dependent roles of reader domains.

Overexpression of TRIM24 has been shown to correlate with poor survival of breast cancer patients [67]. The expression of TRIM24 has also been examined in non-small-cell lung (NSCLC) cancer tissues where high expression levels were found in tumour cells compared to normal lung epithelium [68]. Cells from patients with high levels of TRIM24 expression were largely undifferentiated and showed high proliferation activity. Interestingly, TRIM24 levels did increase in patients with advanced NSCLC. Knockdown of TRIM24 by siRNA inhibited cell cycle progression, slowing proliferation and inducing apoptosis. TRIM24 has also been linked to chronic myeloid leukemia (CML) where expression was found to be low in the chronic phase but increasing gradually through the accelerated phase to the blastic phase [98]. This pattern of expression suggests that it is intrinsic to the progression of the disease. In the same study it was also observed that TRIM24 was overexpressed in the CD34-positive compartment, suggesting that the signalling pathways governing CML progression and CML stem cell homeostasis may be related.

In contrast, TRIM24 has also been reported as a potent liver-specific tumour suppressor. TRIM24 knockout mice are predisposed to the development of both spontaneous and chemically induced hepatocellular carcinoma [69]. Studies in TRIM24$^{-/-}$ mice showed upregulated retinoic acid receptor signalling, an effect that is compensated by simultaneous mono allelic deletion of the *Rara* gene [70].

BRDT has been associated with a number of different cancers. This usually testis specific BET family member has been found to be overexpressed in digestive tract tumours [99] and in non-small cell lung cancer where BRDT was found to be expressed in nearly half of the biopsies examined [37].

Overexpression of ATAD2 has been reported in a large variety of tumours, particularly in gastrointestinal tumours, large B-cell lymphoma, hepatocellular carcinoma and breast and lung cancers [100–103]. ATAD2 overexpression correlates with poor prognosis for lung and breast cancer patients and high expression levels have also been linked to distant recurrence [104]. Significantly, ATAD2 is highly expressed in >70 % of breast tumours and its expression correlates with highly aggressive, triple negative tumours, tumour metastasis and poor prognosis for the patient [20].

ATAD2 has been shown to bind to histone H4K5ac through its bromodomain, and to the transcription factor c-MYC (v-myc myelocytomatosis viral oncogene homolog) stimulating transcription of c-MYC target genes [105]. It is also a mediator of E2F transcription factors and is required for recruitment of the host cell factor 1 (HCF-1)-MLL histone methyltransferase complex to chromatin. In agreement with its key function regulating growth promoting transcription factors, genetic knockout studies showed that Atad2 is essential for proliferation and survival of tumour cells and for promoting cell proliferation, survival and cell migration.

## 11.6 Tumour Suppressor Roles of Bromodomain Proteins

The gene encoding for the central component of SWI/SNF (Switch/Sucrose nonfermenting) chromatin remodelling complexes SMARCA4 (/BRG1, Brahma related gene) was identified as a tumour suppressor gene [106, 107]. *SMARCA4* expression is silenced in several human tumour cell lines and tumour tissues and inactivating mutations have been detected in rhabdoid tumours. However, the frequency of SMARCA4 loss varies significantly across different tumour types and it is particularly frequent in lung cancer [106, 108]. Interestingly, *SMARCA4* expression never occurs concurrently with *cMYC* amplification, suggesting that these factors may play a common role in lung cancer. SMARCA4 was subsequently shown to inhibit c-Myc-mediated transcription in HeLa cells [109]. Loss of SMARCA4 has been observed to correlate with lung cancer tumour aggressiveness [110] and SMARCA4 inactivation is often accompanied by loss of the related gene product SMARCA2 (/BRM, Brahma gene) [111]. The combined loss of the ATPase activity of both proteins may lead to the observed aggressive phenotype of these tumours. Furthermore, cells that express a mutant SMARCA4 gene lacking ATPase activity increase in both overall volume and nuclear size and also in area of attachment [112]. However SMARCA4 and SMARCA2 are not functionally redundant, despite having a high degree of homology and several overlapping functions [108]. Recently SMARCA4 has been implicated as a tumour suppressor in melanoma, binding to $p16^{INK4a}$, an important melanoma susceptibility gene [113]. Other components of the SWI/SNF complexes are also frequently truncated, mutated or silenced in tumours. These components include the polybromo protein PB1, a component of the PBAF (Polybromo and Brg1-associated factor) subclass of SWI/SNF complexes which has also been found to be mutated in a high percentage of renal cell carcinomas. In support of its role as a tumour suppressor, genetic knockdown of more than 60 % of PB1 resulted in significant increase in proliferation in renal cell carcinoma cell lines.

### *11.6.1 Bromodomain Proteins as Driver of Genetic Programs of Tumourigenesis*

A number of bromodomain-containing proteins are key regulators that drive transcriptional programs that lead to cell proliferation. An interesting example of such a central regulatory function is the BET family member BRD4, which is required for efficient transcriptional elongation of diverse growth-promoting and anti-apoptotic genes. BRD4 and other BET family members recruit the positive transcription elongation factor complex (P-TEFb, cdk9/cyclinT) to transcriptional start sites during the M/G1 transition allowing CDK9 to phosphorylate the C-terminal domain of RNA polymerase. This leads to a unique marking of genes that remain to be expressed at the end of mitosis, functioning as a component of transcriptional memory during cell division. Intriguingly, genes that are controlled during cell division in a P-TEFb/BRD4 dependent manner include the mitotic kinase Aurora B, NF-κB and c-MYC as well as anti-apoptotic genes such as BCL2. Recently Zuber and colleagues showed that BRD4 is required for AML cell survival by upregulating a number of growth stimulating genes such as c-MYC providing a compelling strategy for targeting BRD4 in leukaemia. Chemical inhibition of the BRD4 bromodomain by the BET inhibitor JQ1 also resulted in significant antiproliferative effects, down-regulation of c-MYC transcription, cell-cycle arrest and cellular senescence in multiple myeloma. Similar to other bromodomain proteins the effect of BET inhibition on transcription is highly context dependent. In lung cancer for instance c-MYC levels are not affected. Instead BET inhibition has been linked to another set of affected genes that include the oncogenic transcription factor FOS-like antigen 1 (FOSL1). In breast cancer however, *BRD4* has been identified as a susceptibility gene for disease progression. In this cancer high expression levels have been associated with poor patient survival presumably due to transcriptional up-regulation of genes that promote cell migration and as a consequence metastasis. Similar to hormone receptors bromodomain-containing transcriptional regulators need to be comprehensively studied in different tissues to evaluate their role in normal physiology and disease. The excellent druggability of these domains however, together with their central role in tumourigenesis suggests that many bromodomain containing proteins will develop into attractive drug targets in the future.

**Acknowledgements** The authors are grateful for financial support from the SGC, a registered charity (number 1097737) that receives funds from the Canadian Institutes for Health Research, the Canada Foundation for Innovation, Genome Canada, GlaxoSmithKline, Pfizer, Eli Lilly, Takeda, AbbVie, the Novartis Research Foundation, Boehringer Ingelheim, the Ontario Ministry of Research and Innovation and the Wellcome Trust. We apologise to the researchers that we were not able to cite as a result of space constraints.

## References

1. Choudhary C, Kumar C, Gnad F, Nielsen ML, Rehman M, Walther TC et al (2009) Lysine acetylation targets protein complexes and co-regulates major cellular functions. Science 325(5942):834–840, Epub 2009/07/18
2. Barneda-Zahonero B, Parra M (2012) Histone deacetylases and cancer. Mol Oncol 6(6):579–589, Epub 2012/09/12
3. Parra M, Verdin E (2010) Regulatory signal transduction pathways for class IIa histone deacetylases. Curr Opin Pharmacol 10(4):454–460, Epub 2010/05/08
4. Ropero S, Fraga MF, Ballestar E, Hamelin R, Yamamoto H, Boix-Chornet M et al (2006) A truncating mutation of HDAC2 in human cancers confers resistance to histone deacetylase inhibition. Nat Genet 38(5):566–569, Epub 2006/04/28
5. Lee JH, Choy ML, Marks PA (2012) Mechanisms of resistance to histone deacetylase inhibitors. Adv Cancer Res 116:39–86, Epub 2012/10/24
6. Hagelkruys A, Sawicka A, Rennmayr M, Seiser C (2011) The biology of HDAC in cancer: the nuclear and epigenetic components. Handb Exp Pharmacol 206:13–37, Epub 2011/09/01
7. Magnaghi-Jaulin L, Groisman R, Naguibneva I, Robin P, Lorain S, Le Villain JP et al (1998) Retinoblastoma protein represses transcription by recruiting a histone deacetylase. Nature 391(6667):601–605, Epub 1998/02/19
8. Buurman R, Gurlevik E, Schaffer V, Eilers M, Sandbothe M, Kreipe H et al (2012) Histone deacetylases activate hepatocyte growth factor signaling by repressing microRNA-449 in hepatocellular carcinoma cells. Gastroenterology 143(3):811–820, e1–15. Epub 2012/05/30
9. Khan O, La Thangue NB (2012) HDAC inhibitors in cancer biology: emerging mechanisms and clinical applications. Immunol Cell Biol 90(1):85–94, Epub 2011/11/30
10. Licciardi PV, Ververis K, Tang ML, El-Osta A, Karagiannis TC (2012) Immunomodulatory effects of histone deacetylase inhibitors. Curr Mol Med 13(4):640–647, Epub 2012/10/16

11. Marmorstein R, Trievel RC (2009) Histone modifying enzymes: structures, mechanisms, and specificities. Biochim Biophys Acta 1789(1):58–68, Epub 2008/08/30
12. Yang X, Yu W, Shi L, Sun L, Liang J, Yi X et al (2011) HAT4, a Golgi apparatus-anchored B-type histone acetyltransferase, acetylates free histone H4 and facilitates chromatin assembly. Mol Cell 44(1):39–50, Epub 2011/10/11
13. Avvakumov N, Cote J (2007) The MYST family of histone acetyltransferases and their intimate links to cancer. Oncogene 26(37):5395–5407, Epub 2007/08/19
14. Kouzarides T (2000) Acetylation: a regulatory modification to rival phosphorylation? EMBO J 19(6):1176–1179, Epub 2000/03/16
15. Tohyama S, Tomura A, Ikeda N, Hatano M, Odanaka J, Kubota Y et al (2012) Discovery and characterization of NK13650s, naturally occurring p300-selective histone acetyltransferase inhibitors. J Org Chem 77(20):9044–9052, Epub 2012/09/19
16. Filippakopoulos P, Picaud S, Mangos M, Keates T, Lambert JP, Barsyte-Lovejoy D et al (2012) Histone recognition and large-scale structural analysis of the human bromodomain family. Cell 149(1):214–231, Epub 2012/04/03
17. Haynes SR, Dollard C, Winston F, Beck S, Trowsdale J, Dawid IB (1992) The bromodomain: a conserved sequence found in human, Drosophila and yeast proteins. Nucleic Acids Res 20(10):2603, Epub 1992/05/25
18. Muller S, Filippakopoulos P, Knapp S (2011) Bromodomains as therapeutic targets. Expert Rev Mol Med 13:e29, Epub 2011/09/22
19. Liu J, Lee W, Jiang Z, Chen Z, Jhunjhunwala S, Haverty PM et al (2012) Genome and transcriptome sequencing of lung cancers reveal diverse mutational and splicing events. Genome Res 22(12):2315–2327, Epub 2012/10/04
20. Kalashnikova EV, Revenko AS, Gemo AT, Andrews NP, Tepper CG, Zou JX et al (2010) ANCCA/ATAD2 overexpression identifies breast cancer patients with poor prognosis, acting to drive proliferation and survival of triple-negative cells through control of B-Myb and EZH2. Cancer Res 70(22):9402–9412, Epub 2010/09/25
21. Zou JX, Revenko AS, Li LB, Gemo AT, Chen HW (2007) ANCCA, an estrogen-regulated AAA+ ATPase coactivator for ERalpha, is required for coregulator occupancy and chromatin modification. Proc Natl Acad Sci USA 104(46):18067–18072, Epub 2007/11/14
22. Duan Z, Zou JX, Yang P, Wang Y, Borowsky AD, Gao AC et al (2012) Developmental and androgenic regulation of chromatin regulators EZH2 and ANCCA/ATAD2 in the prostate Via MLL histone methylase complex. Prostate 73(5):455–466, Epub 2012/10/06
23. Fouret R, Laffaire J, Hofman P, Beau-Faller M, Mazieres J, Validire P et al (2012) A comparative and integrative approach identifies ATPase family, AAA domain containing 2 as a likely driver of cell proliferation in lung adenocarcinoma. Clin Cancer Res 18(20):5606–5616, Epub 2012/08/24
24. Leachman NT, Brellier F, Ferralli J, Chiquet-Ehrismann R, Tucker RP (2010) ATAD2B is a phylogenetically conserved nuclear protein expressed during neuronal differentiation and tumorigenesis. Dev Growth Differ 52(9):747–755, Epub 2010/12/17
25. Revenko AS, Kalashnikova EV, Gemo AT, Zou JX, Chen HW (2010) Chromatin loading of E2F-MLL complex by cancer-associated coregulator ANCCA via reading a specific histone mark. Mol Cell Biol 30(22):5260–5272, Epub 2010/09/22
26. Hennig EE, Mikula M, Rubel T, Dadlez M, Ostrowski J (2012) Comparative kinome analysis to identify putative colon tumor biomarkers. J Mol Med (Berl) 90(4):447–456, Epub 2011/11/19
27. Panagopoulos I, Strombeck B, Isaksson M, Heldrup J, Olofsson T, Johansson B (2006) Fusion of ETV6 with an intronic sequence of the BAZ2A gene in a paediatric pre-B acute lymphoblastic leukaemia with a cryptic chromosome 12 rearrangement. Br J Haematol 133(3):270–275, Epub 2006/04/29
28. Nebral K, Denk D, Attarbaschi A, Konig M, Mann G, Haas OA et al (2009) Incidence and diversity of PAX5 fusion genes in childhood acute lymphoblastic leukemia. Leukemia 23(1):134–143, Epub 2008/11/21

29. Viejo-Borbolla A, Ottinger M, Bruning E, Burger A, Konig R, Kati E et al (2005) Brd2/RING3 interacts with a chromatin-binding domain in the Kaposi's Sarcoma-associated herpesvirus latency-associated nuclear antigen 1 (LANA-1) that is required for multiple functions of LANA-1. J Virol 79(21):13618–13629, Epub 2005/10/18
30. Zhou M, Peng C, Nie XM, Zhang BC, Zhu SG, Yu Y et al (2003) Expression of BRD7-interacting proteins, BRD2 and BRD3, in nasopharyngeal carcinoma tissues. Chin J Cancer 22(2):123–127
31. French CA, Ramirez CL, Kolmakova J, Hickman TT, Cameron MJ, Thyne ME et al (2008) BRD-NUT oncoproteins: a family of closely related nuclear proteins that block epithelial differentiation and maintain the growth of carcinoma cells. Oncogene 27(15):2237–2242, Epub 2007/10/16
32. Dawson MA, Prinjha RK, Dittmann A, Giotopoulos G, Bantscheff M, Chan WI et al (2011) Inhibition of BET recruitment to chromatin as an effective treatment for MLL-fusion leukaemia. Nature 478(7370):529–533, Epub 2011/10/04
33. French CA (2012) Pathogenesis of NUT midline carcinoma. Annu Rev Pathol 7:247–265, Epub 2011/10/25
34. Delmore JE, Issa GC, Lemieux ME, Rahl PB, Shi J, Jacobs HM et al (2011) BET bromodomain inhibition as a therapeutic strategy to target c-Myc. Cell 146(6):904–917, Epub 2011/09/06
35. Lockwood WW, Zejnullahu K, Bradner JE, Varmus H (2012) Sensitivity of human lung adenocarcinoma cell lines to targeted inhibition of BET epigenetic signaling proteins. Proc Natl Acad Sci USA 109(47):19408–19413, Epub 2012/11/07
36. You J, Srinivasan V, Denis GV, Harrington WJ Jr, Ballestas ME, Kaye KM et al (2006) Kaposi's sarcoma-associated herpesvirus latency-associated nuclear antigen interacts with bromodomain protein Brd4 on host mitotic chromosomes. J Virol 80(18):8909–8919, Epub 2006/08/31
37. Grunwald C, Koslowski M, Arsiray T, Dhaene K, Praet M, Victor A et al (2006) Expression of multiple epigenetically regulated cancer/germline genes in nonsmall cell lung cancer. Int J Cancer 118(10):2522–2528, Epub 2005/12/15
38. Drost J, Mantovani F, Tocco F, Elkon R, Comel A, Holstege H et al (2010) BRD7 is a candidate tumour suppressor gene required for p53 function. Nat Cell Biol 12(4):380–389, Epub 2010/03/17
39. Yamada HY, Rao CV (2009) BRD8 is a potential chemosensitizing target for spindle poisons in colorectal cancer therapy. Int J Oncol 35(5):1101–1109, Epub 2009/09/30
40. Kang JU, Koo SH, Kwon KC, Park JW, Kim JM (2008) Gain at chromosomal region 5p15.33, containing TERT, is the most frequent genetic event in early stages of non-small cell lung cancer. Cancer Genet Cytogenet 182(1):1–11, Epub 2008/03/11
41. Laue K, Daujat S, Crump JG, Plaster N, Roehl HH, Kimmel CB et al (2008) The multidomain protein Brpf1 binds histones and is required for Hox gene expression and segmental identity. Development 135(11):1935–1946, Epub 2008/05/13
42. Ullah M, Pelletier N, Xiao L, Zhao SP, Wang K, Degerny C et al (2008) Molecular architecture of quartet MOZ/MORF histone acetyltransferase complexes. Mol Cell Biol 28(22):6828–6843, Epub 2008/09/17
43. Kalla C, Nentwich H, Schlotter M, Mertens D, Wildenberger K, Dohner H et al (2005) Translocation t(X;11)(q13;q23) in B-cell chronic lymphocytic leukemia disrupts two novel genes. Genes Chromosomes Cancer 42(2):128–143, Epub 2004/11/16
44. Lee SK, Park EJ, Lee HS, Lee YS, Kwon J (2012) Genome-wide screen of human bromodomain-containing proteins identifies Cecr2 as a novel DNA damage response protein. Mol Cells 34(1):85–91, Epub 2012/06/16
45. Mullighan CG, Zhang J, Kasper LH, Lerach S, Payne-Turner D, Phillips LA et al (2011) CREBBP mutations in relapsed acute lymphoblastic leukaemia. Nature 471(7337):235–239, Epub 2011/03/11
46. Gong AY, Eischeid AN, Xiao J, Zhao J, Chen D, Wang ZY et al (2012) miR-17-5p targets the p300/CBP-associated factor and modulates androgen receptor transcriptional activity in cultured prostate cancer cells. BMC Cancer 12(1):492, Epub 2012/10/26

47. Gayther SA, Batley SJ, Linger L, Bannister A, Thorpe K, Chin SF et al (2000) Mutations truncating the EP300 acetylase in human cancers. Nat Genet 24(3):300–303, Epub 2000/03/04
48. Chaffanet M, Gressin L, Preudhomme C, Soenen-Cornu V, Birnbaum D, Pebusque MJ (2000) MOZ is fused to p300 in an acute monocytic leukemia with t(8;22). Genes Chromosomes Cancer 28(2):138–144, Epub 2000/05/29
49. Grinberg-Rashi H, Ofek E, Perelman M, Skarda J, Yaron P, Hajduch M et al (2009) The expression of three genes in primary non-small cell lung cancer is associated with metastatic spread to the brain. Clin Cancer Res 15(5):1755–1761, Epub 2009/02/05
50. Krivtsov AV, Armstrong SA (2007) MLL translocations, histone modifications and leukaemia stem-cell development. Nat Rev Cancer 7(11):823–833, Epub 2007/10/25
51. Varela I, Tarpey P, Raine K, Huang D, Ong CK, Stephens P et al (2011) Exome sequencing identifies frequent mutation of the SWI/SNF complex gene PBRM1 in renal carcinoma. Nature 469(7331):539–542, Epub 2011/01/21
52. Xia W, Nagase S, Montia AG, Kalachikov SM, Keniry M, Su T et al (2008) BAF180 is a critical regulator of p21 induction and a tumor suppressor mutated in breast cancer. Cancer Res 68(6):1667–1674, Epub 2008/03/15
53. Yang XJ, Ogryzko VV, Nishikawa J, Howard BH, Nakatani Y (1996) A p300/CBP-associated factor that competes with the adenoviral oncoprotein E1A. Nature 382(6589):319–324, Epub 1996/07/25
54. Ge X, Jin Q, Zhang F, Yan T, Zhai Q (2009) PCAF acetylates {beta}-catenin and improves its stability. Mol Biol Cell 20(1):419–427, Epub 2008/11/07
55. Armas-Pineda C, Arenas-Huertero F, Perezpenia-Diazconti M, Chico-Ponce de Leon F, Sosa-Sainz G, Lezama P et al (2007) Expression of PCAF, p300 and Gcn5 and more highly acetylated histone H4 in pediatric tumors. J Exp Clin Cancer Res 26(2):269–276, Epub 2007/08/30
56. De Semir D, Nosrati M, Bezrookove V, Dar AA, Federman S, Bienvenu G et al (2012) Pleckstrin homology domain-interacting protein (PHIP) as a marker and mediator of melanoma metastasis. Proc Natl Acad Sci USA 109(18):7067–7072, Epub 2012/04/19
57. Tokuda E, Fujita N, Oh-hara T, Sato S, Kurata A, Katayama R et al (2007) Casein kinase 2-interacting protein-1, a novel Akt pleckstrin homology domain-interacting protein, down-regulates PI3K/Akt signaling and suppresses tumor growth in vivo. Cancer Res 67(20):9666–9676, Epub 2007/10/19
58. Eichmuller S, Usener D, Dummer R, Stein A, Thiel D, Schadendorf D (2001) Serological detection of cutaneous T-cell lymphoma-associated antigens. Proc Natl Acad Sci USA 98(2):629–634, Epub 2001/01/10
59. Shain AH, Giacomini CP, Matsukuma K, Karikari CA, Bashyam MD, Hidalgo M et al (2012) Convergent structural alterations define SWItch/Sucrose NonFermentable (SWI/SNF) chromatin remodeler as a central tumor suppressive complex in pancreatic cancer. Proc Natl Acad Sci USA 109(5):E252–E259, Epub 2012/01/12
60. Liu G, Gramling S, Munoz D, Cheng D, Azad AK, Mirshams M et al (2011) Two novel BRM insertion promoter sequence variants are associated with loss of BRM expression and lung cancer risk. Oncogene 30(29):3295–3304, Epub 2011/04/12
61. Reisman D, Glaros S, Thompson EA (2009) The SWI/SNF complex and cancer. Oncogene 28(14):1653–1668, Epub 2009/02/24
62. Schneppenheim R, Fruhwald MC, Gesk S, Hasselblatt M, Jeibmann A, Kordes U et al (2010) Germline nonsense mutation and somatic inactivation of SMARCA4/BRG1 in a family with rhabdoid tumor predisposition syndrome. Am J Hum Genet 86(2):279–284, Epub 2010/02/09
63. Negorev DG, Vladimirova OV, Kossenkov AV, Nikonova EV, Demarest RM, Capobianco AJ et al (2010) Sp100 as a potent tumor suppressor: accelerated senescence and rapid malignant transformation of human fibroblasts through modulation of an embryonic stem cell program. Cancer Res 70(23):9991–10001, Epub 2010/12/02
64. Yordy JS, Moussa O, Pei H, Chaussabel D, Li R, Watson DK (2005) SP100 inhibits ETS1 activity in primary endothelial cells. Oncogene 24(5):916–931, Epub 2004/12/14

65. Nicewonger J, Suck G, Bloch D, Swaminathan S (2004) Epstein-Barr virus (EBV) SM protein induces and recruits cellular Sp110b to stabilize mRNAs and enhance EBV lytic gene expression. J Virol 78(17):9412–9422, Epub 2004/08/17
66. Di Bernardo MC, Crowther-Swanepoel D, Broderick P, Webb E, Sellick G, Wild R et al (2008) A genome-wide association study identifies six susceptibility loci for chronic lymphocytic leukemia. Nat Genet 40(10):1204–1210, Epub 2008/09/02
67. Chambon M, Orsetti B, Berthe ML, Bascoul-Mollevi C, Rodriguez C, Duong V et al (2011) Prognostic significance of TRIM24/TIF-1alpha gene expression in breast cancer. Am J Pathol 178(4):1461–1469, Epub 2011/03/26
68. Li H, Sun L, Tang Z, Fu L, Xu Y, Li Z et al (2012) Overexpression of TRIM24 correlates with tumor progression in non-small cell lung cancer. PLoS One 7(5):e37657, Epub 2012/06/06
69. Herquel B, Ouararhni K, Khetchoumian K, Ignat M, Teletin M, Mark M et al (2011) Transcription cofactors TRIM24, TRIM28, and TRIM33 associate to form regulatory complexes that suppress murine hepatocellular carcinoma. Proc Natl Acad Sci USA 108(20):8212–8217, Epub 2011/05/03
70. Khetchoumian K, Teletin M, Tisserand J, Mark M, Herquel B, Ignat M et al (2007) Loss of Trim24 (Tif1alpha) gene function confers oncogenic activity to retinoic acid receptor alpha. Nat Genet 39(12):1500–1506, Epub 2007/11/21
71. Tsai WW, Wang Z, Yiu TT, Akdemir KC, Xia W, Winter S et al (2010) TRIM24 links a non-canonical histone signature to breast cancer. Nature 468(7326):927–932, Epub 2010/12/18
72. Yang H, Zhang C, Zhao X, Wu Q, Fu X, Yu B et al (2010) Analysis of copy number variations of BS69 in multiple types of hematological malignancies. Ann Hematol 89(10):959–964, Epub 2010/04/29
73. Hughes-Davies L, Huntsman D, Ruas M, Fuks F, Bye J, Chin SF et al (2003) EMSY links the BRCA2 pathway to sporadic breast and ovarian cancer. Cell 115(5):523–535, Epub 2003/12/04
74. Moriniere J, Rousseaux S, Steuerwald U, Soler-Lopez M, Curtet S, Vitte AL et al (2009) Cooperative binding of two acetylation marks on a histone tail by a single bromodomain. Nature 461(7264):664–668
75. Vidler LR, Brown N, Knapp S, Hoelder S (2012) Druggability analysis and structural classification of bromodomain acetyl-lysine binding sites. J Med Chem 55(17):7346–7359, Epub 2012/07/14
76. Zeng L, Li J, Muller M, Yan S, Mujtaba S, Pan C et al (2005) Selective small molecules blocking HIV-1 Tat and coactivator PCAF association. J Am Chem Soc 127(8):2376–2377, Epub 2005/02/24
77. Sachchidanand, Resnick-Silverman L, Yan S, Mutjaba S, Liu WJ, Zeng L et al (2006) Target structure-based discovery of small molecules that block human p53 and CREB binding protein association. Chem Biol 13(1):81–90, Epub 2006/01/24
78. Borah JC, Mujtaba S, Karakikes I, Zeng L, Muller M, Patel J et al (2011) A small molecule binding to the coactivator CREB-binding protein blocks apoptosis in cardiomyocytes. Chem Biol 18(4):531–541, Epub 2011/04/26
79. Demont E, Gosmini R (2011) Tetrahydroquinoline derivatives as bromodomain inhibitors. International patent WO/2011/054848. GLAXOSMITHKLINE LLC:1–199
80. Ito T, Umehara T, Sasaki K, Nakamura Y, Nishino N, Terada T et al (2011) Real-time imaging of histone H4K12-specific acetylation determines the modes of action of histone deacetylase and bromodomain inhibitors. Chem Biol 18(4):495–507, Epub 2011/04/26
81. Filippakopoulos P, Qi J, Picaud S, Shen Y, Smith WB, Fedorov O et al (2010) Selective inhibition of BET bromodomains. Nature 468(7327):1067–1073, Epub 2010/09/28
82. Nicodeme E, Jeffrey KL, Schaefer U, Beinke S, Dewell S, Chung CW et al (2010) Suppression of inflammation by a synthetic histone mimic. Nature 468(7327):1119–1123, Epub 2010/11/12
83. Chung CW, Coste H, White JH, Mirguet O, Wilde J, Gosmini RL et al (2011) Discovery and characterization of small molecule inhibitors of the BET family bromodomains. J Med Chem 54(11):3827–3838, Epub 2011/05/17

84. Filippakopoulos P, Picaud S, Fedorov O, Keller M, Wrobel M, Morgenstern O et al (2012) Benzodiazepines and benzotriazepines as protein interaction inhibitors targeting bromodomains of the BET family. Bioorg Med Chem 20(6):1878–1886, Epub 2011/12/06
85. Hewings DS, Wang M, Philpott M, Fedorov O, Uttarkar S, Filippakopoulos P et al (2011) 3,5-dimethylisoxazoles act as acetyl-lysine-mimetic bromodomain ligands. J Med Chem 54(19):6761–6770, Epub 2011/08/20
86. Bamborough P, Diallo H, Goodacre JD, Gordon L, Lewis A, Seal JT et al (2012) Fragment-based discovery of bromodomain inhibitors part 2: optimization of phenylisoxazole sulfonamides. J Med Chem 55(2):587–596, Epub 2011/12/06
87. French CA, Miyoshi I, Kubonishi I, Grier HE, Perez-Atayde AR, Fletcher JA (2003) BRD4-NUT fusion oncogene: a novel mechanism in aggressive carcinoma. Cancer Res 63(2):304–307, Epub 2003/01/25
88. French CA (2010) Demystified molecular pathology of NUT midline carcinomas. J Clin Pathol 63(6):492–496, Epub 2008/06/17
89. Dang TP, Gazdar AF, Virmani AK, Sepetavec T, Hande KR, Minna JD et al (2000) Chromosome 19 translocation, overexpression of Notch3, and human lung cancer. J Natl Cancer Inst 92(16):1355–1357, Epub 2000/08/17
90. Iyer NG, Ozdag H, Caldas C (2004) p300/CBP and cancer. Oncogene 23(24):4225–4231, Epub 2004/05/25
91. Borrow J, Stanton VP Jr, Andresen JM, Becher R, Behm FG, Chaganti RS et al (1996) The translocation t(8;16)(p11;p13) of acute myeloid leukaemia fuses a putative acetyltransferase to the CREB-binding protein. Nat Genet 14(1):33–41, Epub 1996/09/01
92. Panagopoulos I, Fioretos T, Isaksson M, Samuelsson U, Billstrom R, Strombeck B et al (2001) Fusion of the MORF and CBP genes in acute myeloid leukemia with the t(10;16)(q22;p13). Hum Mol Genet 10(4):395–404, Epub 2001/02/07
93. Lai JL, Jouet JP, Bauters F, Deminatti M (1985) Chronic myelogenous leukemia with translocation (8;22): report of a new case. Cancer Genet Cytogenet 17(4):365–366, Epub 1985/08/01
94. Kitabayashi I, Aikawa Y, Yokoyama A, Hosoda F, Nagai M, Kakazu N et al (2001) Fusion of MOZ and p300 histone acetyltransferases in acute monocytic leukemia with a t(8;22)(p11;q13) chromosome translocation. Leukemia 15(1):89–94, Epub 2001/03/13
95. Rowley JD, Reshmi S, Sobulo O, Musvee T, Anastasi J, Raimondi S et al (1997) All patients with the T(11;16)(q23;p13.3) that involves MLL and CBP have treatment-related hematologic disorders. Blood 90(2):535–541
96. Satake N, Ishida Y, Otoh Y, Hinohara S, Kobayashi H, Sakashita A et al (1997) Novel MLL-CBP fusion transcript in therapy-related chronic myelomonocytic leukemia with a t(11;16)(q23;p13) chromosome translocation. Genes Chromosomes Cancer 20(1):60–63, Epub 1997/09/18
97. Giles RH, Petrij F, Dauwerse HG, den Hollander AI, Lushnikova T, van Ommen GJ et al (1997) Construction of a 1.2-Mb contig surrounding, and molecular analysis of, the human CREB-binding protein (CBP/CREBBP) gene on chromosome 16p13.3. Genomics 42(1):96–114
98. Quintas-Cardama A, Qiu YH, Post SM, Zhang Y, Creighton CJ, Cortes J et al (2012) Reverse phase protein array profiling reveals distinct proteomic signatures associated with chronic myeloid leukemia progression and with chronic phase in the CD34-positive compartment. Cancer 118(21):5283–5292, Epub 2012/04/21
99. Ding WM, Yang L, Zheng Y, Sun HY, Liang H, Wang JJ (2012) Preparation of anti-BRDT-NY polyclonal antibody and expression of BRDT-NY protein in digestive tract tumors. Xi Bao Yu Fen Zi Mian Yi Xue Za Zhi 28(8):847–850, Epub 2012/08/07
100. Chen X, Cheung ST, So S, Fan ST, Barry C, Higgins J et al (2002) Gene expression patterns in human liver cancers. Mol Biol Cell 13(6):1929–1939, Epub 2002/06/12
101. Alizadeh AA, Eisen MB, Davis RE, Ma C, Lossos IS, Rosenwald A et al (2000) Distinct types of diffuse large B-cell lymphoma identified by gene expression profiling. Nature 403(6769):503–511, Epub 2000/02/17

102. van't Veer LJ, Dai H, van de Vijver MJ, He YD, Hart AA, Mao M et al (2002) Gene expression profiling predicts clinical outcome of breast cancer. Nature 415(6871):530–536, Epub 2002/02/02
103. Ma XJ, Salunga R, Tuggle JT, Gaudet J, Enright E, McQuary P et al (2003) Gene expression profiles of human breast cancer progression. Proc Natl Acad Sci USA 100(10):5974–5979, Epub 2003/04/26
104. Caron C, Lestrat C, Marsal S, Escoffier E, Curtet S, Virolle V et al (2010) Functional characterization of ATAD2 as a new cancer/testis factor and a predictor of poor prognosis in breast and lung cancers. Oncogene 29(37):5171–5181, Epub 2010/06/29
105. Ciro M, Prosperini E, Quarto M, Grazini U, Walfridsson J, McBlane F et al (2009) ATAD2 is a novel cofactor for MYC, overexpressed and amplified in aggressive tumors. Cancer Res 69(21):8491–8498, Epub 2009/10/22
106. Wong AK, Shanahan F, Chen Y, Lian L, Ha P, Hendricks K et al (2000) BRG1, a component of the SWI-SNF complex, is mutated in multiple human tumor cell lines. Cancer Res 60(21):6171–6177, Epub 2000/11/21
107. Muchardt C, Yaniv M (2001) When the SWI/SNF complex remodels…the cell cycle. Oncogene 20(24):3067–3075, Epub 2001/06/23
108. Medina PP, Romero OA, Kohno T, Montuenga LM, Pio R, Yokota J et al (2008) Frequent BRG1/SMARCA4-inactivating mutations in human lung cancer cell lines. Hum Mutat 29(5):617–622, Epub 2008/04/05
109. Cheng SW, Davies KP, Yung E, Beltran RJ, Yu J, Kalpana GV (1999) c-MYC interacts with INI1/hSNF5 and requires the SWI/SNF complex for transactivation function. Nat Genet 22(1):102–105, Epub 1999/05/13
110. Reisman DN, Sciarrotta J, Wang W, Funkhouser WK, Weissman BE (2003) Loss of BRG1/BRM in human lung cancer cell lines and primary lung cancers: correlation with poor prognosis. Cancer Res 63(3):560–566, Epub 2003/02/05
111. Decristofaro MF, Betz BL, Rorie CJ, Reisman DN, Wang W, Weissman BE (2001) Characterization of SWI/SNF protein expression in human breast cancer cell lines and other malignancies. J Cell Physiol 186(1):136–145, Epub 2001/01/09
112. Hill DA, Chiosea S, Jamaluddin S, Roy K, Fischer AH, Boyd DD et al (2004) Inducible changes in cell size and attachment area due to expression of a mutant SWI/SNF chromatin remodeling enzyme. J Cell Sci 117(Pt 24):5847–5854, Epub 2004/11/13
113. Becker TM, Haferkamp S, Dijkstra MK, Scurr LL, Frausto M, Diefenbach E et al (2009) The chromatin remodelling factor BRG1 is a novel binding partner of the tumor suppressor p16INK4a. Mol Cancer 8:4, Epub 2009/01/20

# Chapter 12
# Domain Specific Targeting of Cancer

**Pratik Chandrani and Amit Dutt**

**Abstract** Cancer is driven by sequential accumulation of genetic alterations leading to the gain- and loss-of-function of critical cellular proteins. Genes of several families (kinases, phosphatases, RAS, GPCRs, etc.) with similar functions, as predicted by similar coding sequences and protein domains, are known to be significantly altered in cancer. Targeted therapy directed against the activity of specific gene products that confer "oncogene addiction" to tumor cells minimizes general toxicity associated with the nonspecific chemotherapy treatments. The success though is restricted to a smaller subset of patients due to development of resistance. One solution to the enduring challenge of drug resistance is rational combinatorial targeted therapy targeting multiple domains across gene families.

**Keywords** Targeted therapy • Kinase • Phosphatase • Small GTPase • Resistance to targeted therapy

## 12.1 Introduction

Cancer, as we understand today, is a disease of the genome driven by specific alterations that may be inherited in the germ-line or acquired somatically. These genome alterations principally include copy number alterations, structural re-arrangements, nucleotide substitution mutations, and infection by microbial genomes leading to the gain- and loss-of-function of critical cellular proteins. Activated oncogenes conferring oncogenic addiction to cancer cells for maintenance of its malignant

P. Chandrani, M.Sc. • A. Dutt, Ph.D. (✉)
Advanced Centre for Treatment, Research and Education in Cancer,
Tata Memorial Center, Kharghar, Navi Mumbai, MA 410 210, India
e-mail: adutt@actrec.gov.in

R. Kumar (ed.), *Nuclear Signaling Pathways and Targeting Transcription in Cancer*, Cancer Drug Discovery and Development, DOI 10.1007/978-1-4614-8039-6_12,

phenotype, are the Achilles' heel for the cells. The cause of addiction could thus be an effective therapeutic target to kill the addicted cancer cells, a prototypical example being the use of Gleevec [1], a small-molecule inhibitor of the Bcr-Abl fusion protein in chronic myelogenous leukemia. Several additional examples of such alterations that potentially define therapeutic paradigm include: *ERBB2* amplification in breast cancer; *EML-ALK* fusion in lung cancer and *ALK* mutation in Neuroblastoma; point mutation, insertion or deletion of *EGFR* in lung cancer; *KIT* mutation in gastrointestinal tumors; *BRAF* mutation in melanoma; *JAK2* mutations in myeloproliferative disorders, *FGFR2* mutation in endometrial cancer, *IDH1* mutation in glioblastoma, and *PIK3CA* in several cancers. Effective development of targeted therapeutics that can interfere with the function of oncogenic molecular targets though remains unmet due to lack of our understanding of the molecular alterations that drive the tumorigenesis in different cancers. While each cancer subtype harbors its own unique pattern of genomic lesions, comparison of genomic lesions across cancer subtypes generate a global view of the disease associated genomic alterations. Besides informing therapeutic regimen in clinics, the elucidation of the function of cancer related genes has significantly enhanced our understanding of the developmental biology of normal cells as the fundamental biochemical pathways deregulated in cancer cells has uncovered various normal homeostatic roles such pathways play in non-transformed cells and tissues. A comprehensive catalog of all such genetic lesions, together with a description of the cancer types in which they occur and the combinations in which they co-occur, thus is necessary for the effective development of targeted cancer therapy. Interestingly, most frequently mutated genomic alterations which governs majority of cancer types lies into some of the most common pathways, for example oncogenic activations of diverse kinase proteins (*EGFR*, PDGFR, *ERBB2*, FGFR, *ABL1*, etc.) and downstream components all together leads to activation of RAS/RAF/MAPK/ERK pathway affecting deregulation of cell growth. Most of the inactivating alterations in tumor suppressor genes affect the CDKN2A/CDK/Rb/E2f and WNT/P53/CDK/Cyclin/ mediated cell cycle and apoptosis pathways. Thus, several pathways and pathway components have more profound effect on cellular deregulation. In this chapter, we will discuss these most frequently altered proteins, their functional domains and targeted therapy in cancer.

## 12.2 Major Protein Families Mutated in Human Cancer

The cellular necessity of a drug target to cancer versus normal cells dictates the width of the therapeutic window, as inhibiting a crucial cancer-specific target may as well jeopardize the viability of normal cells. The ideal drug target would be essential to tumors, yet non-essential to normal cells. Among the most widely recognized genetic alterations are those that lead to the activation of regulatory cascades that are governed by the biochemical activity of cellular kinases. These genetic alterations take the form of the direct mutation of kinase genes through translocation, missense mutation, nonsense mutation, gene amplification or gene deletion. Indeed, the very first oncogene, v-src, and the first genetic lesion identified in human cancer, t(9;22), targeted tyrosine kinases. Then, the *MYC* proto-oncogene was found to be activated by translocation as well as amplification, and amplification thus became recognized as an additional cardinal mechanism of cancer gene deregulation. One of the first pivotal discoveries of activating mutations was within *BRAF*, which encodes a serine/threonine kinase oncogene that transmits proliferative and survival signals downstream of *RAS* in the mitogen-activated protein (MAP) kinase cascade followed by receptor tyrosine kinases (*EGFR*, *ERBB2*, FGFR1, *FGFR2*, FGFR3, PDGFRA, PDGFRB, *ALK*, c-MET, IGF1R, c-KIT, FLT3, and RET), non-receptor tyrosine kinases (ABL, *JAK2*, and SRC), serine-threonine-lipid kinases (*BRAF*, Aura A and B kinases, mTOR, and PIK3), and DNA damage and repair genes (BRCA1 and BRCA2). Timeline for discovery of mutations in few of these therapeutically relevant genes is shown in Fig. 12.1.

Besides, a larger set in the human genome encode proteins termed the druggable genome. The human druggable genome consists of 5,520 different genes distributed through various gene families including serine/threonine and tyrosine kinases (22 %), G-protein-coupled receptors (15 %), cation channels (5 %) serine proteinases (4 %), protein phosphatases (4 %) and other (50 %). Other includes, but is not limited to, zinc peptidases, cytochrome P450s, and nuclear hormone receptors. While the ongoing discovery of the genetic basis of cancer has allowed us to leap forward in our understanding of cancer biology, these developments suggest strongly that a genetic footing for cancer coupled with a mechanistic and biological understanding of the downstream consequences can beget novel effective therapeutics. Following is an overview of few selected major protein families involved in human cancer:

### *12.2.1 RAS Family*

*RAS* family of oncogenes is the earliest oncogene family identified and validated for oncogenic addiction. Oncogenes *HRAS*, *KRAS* and *NRAS* are the founding member of *RAS* family. *RAS* is a membrane-associated guanine nucleotide-binding protein that is normally activated in response to the binding of extracellular signals such as growth factors to Receptor Tyrosine Kinases. *RAS* signaling controls many cellular functions including cell proliferation, apoptosis, migration and differentiation. *RAS* family includes 35 members of proteins, a subset of more than 150 members in GTPase family. All the *RAS* onco-proteins are reported to be having activating mutations (gain-of-function) in different types of cancers, mainly colorectal, lung, pancreatic and cervical cancers. The three *RAS* onco-proteins all together constitute oncogenic addiction in ~30 % of the cancers [2, 3]. Mutational screening of *RAS* genes in large number of tumor samples shows a repertoire of cancer associated mutations. Although there are several different mutations reported in all three oncogenic *RAS* proteins, there is a selective bias observed for some mutational hotspot sites (Fig. 12.2). As reported in cosmic database (as on date: December 2012),

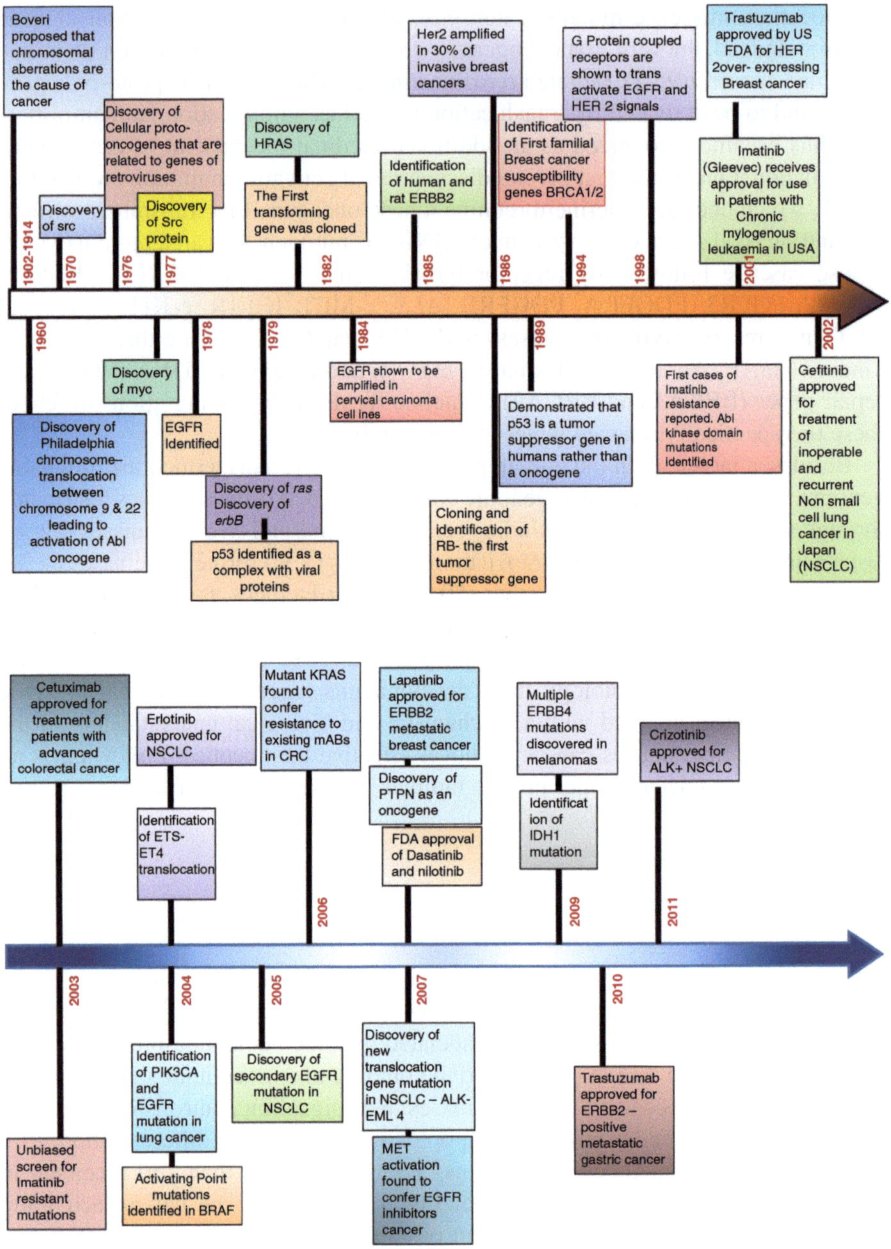


**Fig. 12.1** Schematic representation of timeline for discovery of domain specific mutations in human cancer

*RAS* proteins have three hot-spot positions, G12, G13 and Q61 which are hyper mutated. Selective pressure for these mutation sites is understandable since these sites are directly involved in activation of *RAS* protein. Complex interactions between G12 and Q61 with GTPase activating protein (GAP) induce

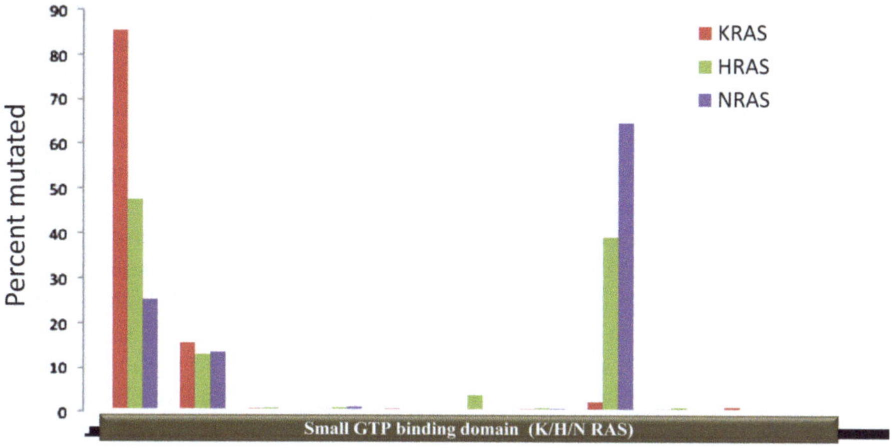


**Fig. 12.2** Domain specific mutation frequency among Ras family proteins in human cancer

conformational changes, making *RAS* into active conformation. Impaired ability of *RAS* mutants to hydrolyze GTP is responsible for the oncogenic nature of mutations at residues G12, G13, and Q61 in the active site. Having a conserved mutation patterns can be of great aid in therapeutic applications. Unlike tyrosine kinase targeted therapies, *RAS* targeted therapy has seen limited success. The most thoroughly investigated anti-RAS compounds are the farnesyl transferase inhibitors (FTIs) [4], which inhibit the post-translational modifications of *RAS* by a farnesyl isoprenoid group. Another distinct approach has been the compounds blocking *RAS* interaction with membrane, Trans-farnesyl thiosalicylic acid (FTS) [4], which blocks interaction of *RAS* from *RAS* anchorage domains in the cell membrane. The main reasons for limited success in *RAS* targeted therapy is given by nature of target, anti-RAS drugs need to restore GTPase functionality of mutated *RAS*. Other reasons for limited anti-RAS therapies are many other downstream pathways which might not be totally dependent on *RAS*, for example PI3-K pathway can be activated by both *RAS* dependent and independent mechanisms [2, 3]. Alternatively, scientists are taking alternative approach for targeting *RAS* by blocking downstream pathway and by taking novel approach of fragment based drug designing. This work is still at very early stage and confirmed reports are awaited.

### *12.2.2 Kinases*

Protein kinase (PK) is an enzyme, which modifies other proteins by adding phosphate group to them. Phosphorylation of target protein results in functional change in target protein by changing enzyme activity, cellular localization or protein-protein interactions. There are about 518 kinases and 106 kinase pseudo-genes in human genome (~2 % of human proteome) [5] which plays vital roles in regulation of proteins involved in cell growth and proliferation, cell cycle regulation, metabolism, apoptosis and various other signal transduction pathways. All human protein kinases constitute a complicated network of regulatory proteins with external and internal regulatory connections. Alteration in any node of this network is vulnerable to disruption of regulatory network due to the complexity and sophistication of the system. Alteration in kinases has been shown to be affecting several physiological processes resulting into malignancies including cancers.

Systematically kinases are divided into ten groups: (1) protein kinase A, G, and C family altogether called as AGC kinases, (2) cyclin-dependent kinases (CDKs), mitogen-activated protein kinases (MAP kinases), glycogen synthase kinases (GSK) and CDK-like kinases altogether make CMGC group of kinases, (3) Calmodulin/Calcium regulated kinases (CAMK), (4) Casein Kinase 1 (CK1) group, (5) STE group of kinases, (6) tyrosine kinases (TK), (7) Tyrosine kinase like (TKL) generally phosphorylating serine/threonine residue, (8) Receptor Guanylate Cyclases (RGC) group of kinases, (9) protein kinase like (PKL) proteins having fold similar to kinases and (10) atypical protein kinases (aPK) consist of proteins which do not share clear sequence similarity with other kinases but are functioning kinase like. We can also broadly classify all PKs as receptor protein kinases (RKs) which contains trans-membrane domain and cytoplasmic protein kinases (CKs) which are found in cytosol, nucleus and inner surface of plasma membrane.

Although PKs are much diversified and are divided into several groups, they all share a significant sequential and structural similarity that is understandable as they all transfer phosphate group of ATP to serine, threonine or tyrosine residue of target protein. All kinases share a common mechanism of catalysis where phospho-transfer is carried out by shared set of amino acids and ATP & active site divalent cation are bound in similar fashion. This similarity and divergence suggests that PKs share a common domain for general function of catalysis and harbor other domains for substrate specificity to the target protein. The domain architecture of representative receptor kinases and non-receptor kinases is much similar to each other.

Alterations in protein kinase' domains have been shown to be playing role in tumorigenesis by creating a genomic change which affects functionally important amino acids. The most commonly observed changes in PKs are activating mutations which lead to deregulation of cell growth and proliferation. Constitutive activation of these proteins gives advantage of uncontrolled growth to cancer cells, and so inhibition of these mutant proteins can be a therapeutic option. Small molecules or antibodies can be designed for mutant proteins and thus can be selective for mutant only. The first example of targeted therapeutic for a mutant protein is Imatinib (Gleevec) against *ABL1* gene in patients with refractory chronic myelogenous leukemia (CML) [1, 6]. Later, imatinib was used to treat advanced-stage gastrointestinal stromal tumors (GIST) carrying activating mutations in two tyrosine kinase receptors, *KIT* and PDGFR [7]. The relationship of imatinib response and *KIT* mutations developed a new paradigm which encouraged development of more targeted therapeutics. Another best model of oncogenic addiction and targeted cancer therapy is of tyrosine kinase inhibitors (TKIs) against epidermal growth factor receptor (*EGFR*) in non-small-cell lung cancer (NSCLC). *EGFR* is founding

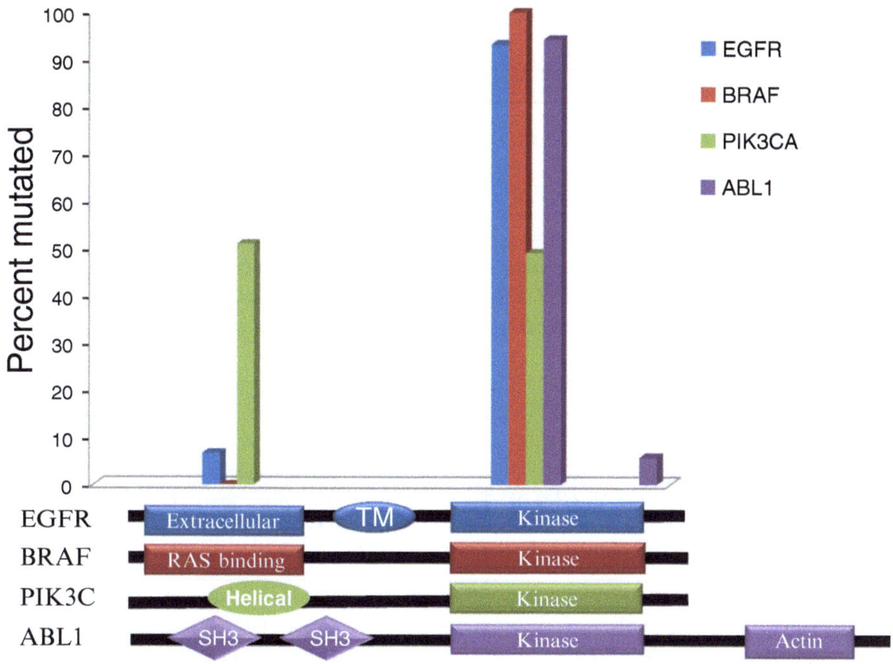


**Fig. 12.3** Domain specific mutation frequency among Kinases in human cancer

member of ErbB family, identified as an oncogene in 2004. Along with other three members of ErbB, *EGFR* has been of intense research for targeted therapeutics. Alterations in ErbB family of genes are reported in many cancers including lung, glioblastoma, thyroid, head and neck carcinoma etc. [8, 9]. Most of oncogenic *EGFR* mutations in NSCLC are located in kinase domains (exon 18-21) which constitutively signals the PI3K/Akt/mTor pathway to promotes cell growth, survival, and migration as well as resistance to apoptosis in response to *EGFR* mediated activation [10]. In the absence of therapy, patients with lung cancer who harbor an *EGFR* mutant tumor have reported better prognosis compared with patients with wild type *EGFR* [11]. Even though prevalence of *EGFR* mutations vary greatly between different ethnicity ranging from ~10 % in Europe and USA to ~35 % in Asia, *EGFR* specific inhibitors has shown successful response in clinical studies [12]. Erlotinib and gefitinib are the two targeted inhibitors designed for *EGFR* mutants which binds to ATP binding site of kinase domain and blocks downstream signaling pathway.

Genomic analysis of large number of patients across the globe shows highest density of mutation at selective sites of *EGFR*. As shown in Fig. 12.3, mutation frequency per amino acid position of *EGFR* (from COSMIC databases) shows clear preference for few selective sites. Maximum frequency of mutations are observed for L858, G719, TL861, T790 and small deletions of 4–6 base pairs in exon-19 (Fig. 12.3). This selective mutation pattern is also observed for other kinases, for

**Table 12.1** US FDA approved small molecule inhibitors and antibodies against various protein kinases for treatment of cancer

| Drug | Targets | US FDA approval for cancer subtype |
|---|---|---|
| Bosutinib | BCR-ABL | CML |
| Crizotinib | EML4-ALK | Lung cancer |
| Dasatinib | BCR–ABL | CML |
| Erlotinib | *EGFR* | Lung and pancreatic cancers |
| Everolimus | mTOR | Kidney cancer |
| Gefitinib | *EGFR* | Lung cancer |
| Imatinib | BCR-ABL, PDGFR and *KIT* | CML and GIST |
| Lapatinib | *EGFR* and *ERBB2* | Breast cancer |
| Nilotinib | BCR–ABL | CML |
| Pazopanib | VEGFR2, PDGFR and *KIT* | Kidney cancer |
| Sorafenib | VEGFR2 and PDGFR | Kidney and liver cancers |
| Sunitinib | VEGFR2, PDGFR and *KIT* | Kidney cancer and GIST |
| Temsirolimus | mTOR | Renal cancer |
| Vandetanib | *EGFR*, VEGFR, RET | Medullary thyroid cancer |
| Vemurafenib | *BRAF* | Melanoma |
| Antibodies | Targets | US FDA approval for cancer subtype |
| Bevacizumab | VEGF | Colorectal, lung and breast cancers |
| Cetuximab | *EGFR* | Colorectal, and head and neck cancers |
| Panitumumab | *EGFR* | Colorectal cancer |
| Pertuzumab | HER-2 | Breast cancer |
| Trastuzumab | *ERBB2* | Breast cancer |

*BCR* breakpoint cluster region, *ABL* abelson murine leukemia viral oncogene homolog, *CML* chronic myeloid leukaemia, *EGFR* epidermal growth factor receptor, *FDA* Food and Drug Administration, *GIST* gastrointestinal stromal tumour, *PDGFR* platelet-derived growth factor receptor, *VEGFR2* vascular endothelial growth factor receptor 2, *ERBB2* erythroblastic leukemia viral oncogene homolog, *RET* rearranged during transfection

instance *BRAF*, *PIK3CA* and *ABL1* all shows very high preference for mutations in kinase domain (Fig. 12.3). Selectivity of mutation sites in different members of same protein family shows a selective pressure on cancer cells which is satisfied only by mutations at some particular site. These high prevalent mutations of kinases make them feasible target for small molecule inhibitors. Several small molecule inhibitors and antibodies for kinase family are already approved by Food and Drug Administration (FDA) or are under clinical studies (Table 12.1).

### *12.2.3 Phosphatases*

A phosphatase is an enzyme which functions exactly opposite to kinases, by removing phosphate group from target protein. The state of phosphorylation of a protein at a time is the net result of the opposing activities of the relevant kinases and phosphatases. Phosphatases and kinases all together make a cycle of target protein activation-deactivation resulting into controlled cellular signaling pathways. Generally a balance of protein phosphorylation-dephosporylation cycle is skewed towards dephosphorylated state, making a constant check on active form (phosphorylated) of target proteins. Hence, phosphatases often play very specific, highly regulated, and very active roles in many cellular processes.

There are approximately 200 protein phosphatases predicted in human genome. Phosphatases can be broadly divided into six classes on the basis of sequence of catalytic domains: (1) protein tyrosine phosphatases (PTP), (2) protein serine/threonine phosphatases (PPP), (3) Protein phosphatase 2C-like phosphatases (PPM), (4) Haloacid dehalogenase-like phosphatases (HAD), (5) phosphatidic acid phosphatase, inositol monophosphatase and inositol polyphosphate-related phosphatase collectively called LP phosphatases and (6) NUDIX hydrolase (NUDT) phosphatases. All the phosphatases can also be grouped as receptor protein phosphatases and cellular protein phosphatases.

Role of phosphatases in cancer is not extensively studied like kinases but it's observed that most of the phosphatases act as tumor suppressors. With some exceptions to the rule, the main function of phosphatases is to control the generally active (phosphorylated) form of target proteins by dephosphorylating (generally inactive form) them. Thus, in normal conditions, phosphatases suppress the oncogenic behavior of certain proteins. Several alterations are observed in phosphatases, mostly in-activating alterations which lead to uncontrolled activation of target protein. Therapeutic applications of inactivating alterations are very limited with only options of replacing defective protein form by gene-therapy or other similar techniques which are difficult to effectively utilize in regular practice.

Interestingly, protein phosphatases are also reported to potentiate, rather than antagonize the oncogenic action of protein kinases. Overexpression of PTP has been observed to activate Src oncogene by dephosphorylating inhibitory T527 residues of Src [13, 14]. Direct role of PTP as oncogene was first reported for *PTPN11* which encodes Shp2 protein [15]. A number of activating (gain of function) mutations in SHP2 domain has been identified as the cause of the Noonan syndrome, several forms of leukemia and solid tumors [16, 17]. Another phosphatase known for oncogenic activities is cell cycle regulator protein Cdc25. *Cdc25* has been implicated in the oncogenesis of breast cancer, prostate cancer, and non- Hodgkin's lymphoma, in which elevated expression of *Cdc25* is thought to promote the loss of cell cycle check-point control resulting into uncontrolled cell proliferation, and genetic instability [18, 19].

Just like kinases, analysis of mutation distribution in different phosphatases shows selective hot-spot sites for mutations. As given in Fig. 12.4, mutation distribution in oncogene *PTPN11* (from COSMIC) shows mutational hot-spot in SH2 domains. Similar types of hot-spots are also observed for inactivating mutations in PTEN, a known tumor suppressor. Selective mutational hot-spot makes phosphatases an important target for cancer therapies.

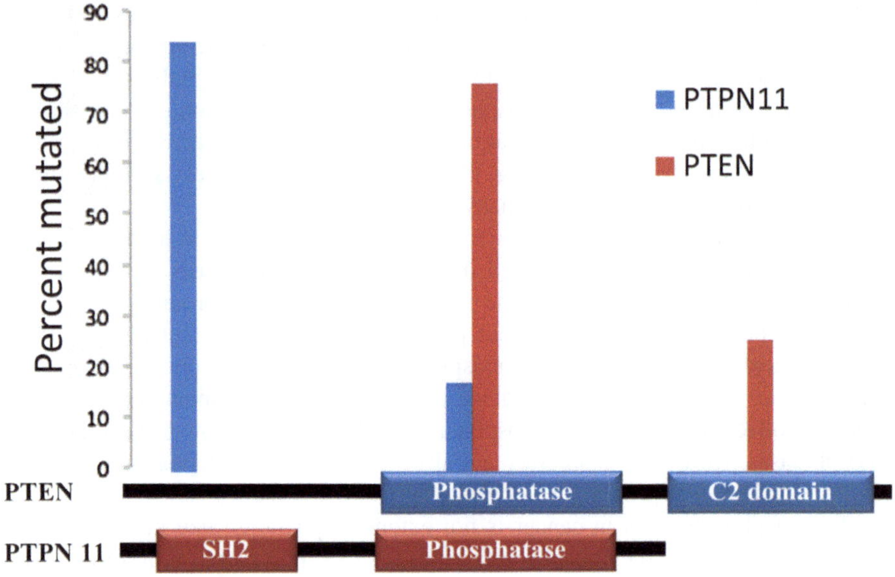


**Fig. 12.4** Domain specific mutation frequency among Phosphatases in human cancer

## 12.3 Acquired Resistance to Domain Specific Inhibitors

Many of the targeted small molecule inhibitors show successful response in tumors driven by single oncogenic mutation. However, many tumors are driven by interaction of multiple oncogenic mutations, and therefore benefit of mono-targeted therapies is limited. Another limitation of targeted therapy is acquired resistance in tumor. Higher genomic instability in cancer cells gives rise to acquired resistance for targeted therapeutic drug. Two known mechanisms of acquired resistance are (1) secondary mutations in target protein and (2) mutation in downstream pathway of target protein. Both types of acquired resistances are reported for *EGFR* targeted therapies. Lung adenocarcinoma patients having L858R mutations or small deletions in exon-19 (near codon 745-750) were reported to be having ~90 % response to targeted inhibitors erlotinib and gefitinib. Soon after first application of these targeted drugs, some patients showed reversion of tumor. Genomic analysis identified secondary mutation T790M in *EGFR* [8, 20]. This mutation changes threonine to methionine in ATP binding site, which leads to lesser binding affinity for drugs while ATP binding not affected. Later, another mechanism of resistance was observed for *EGFR* TKIs as amplification in *MET* gene. Amplification of *MET* leads to sustained phosphorylation of ERBB3, which in turn activates PI3K-Akt signals downstream. In another scenario, activating mutation in PI3K also activates PI3K-Akt pathway. Thus, even though TKIs inhibits *EGFR*, alteration in another

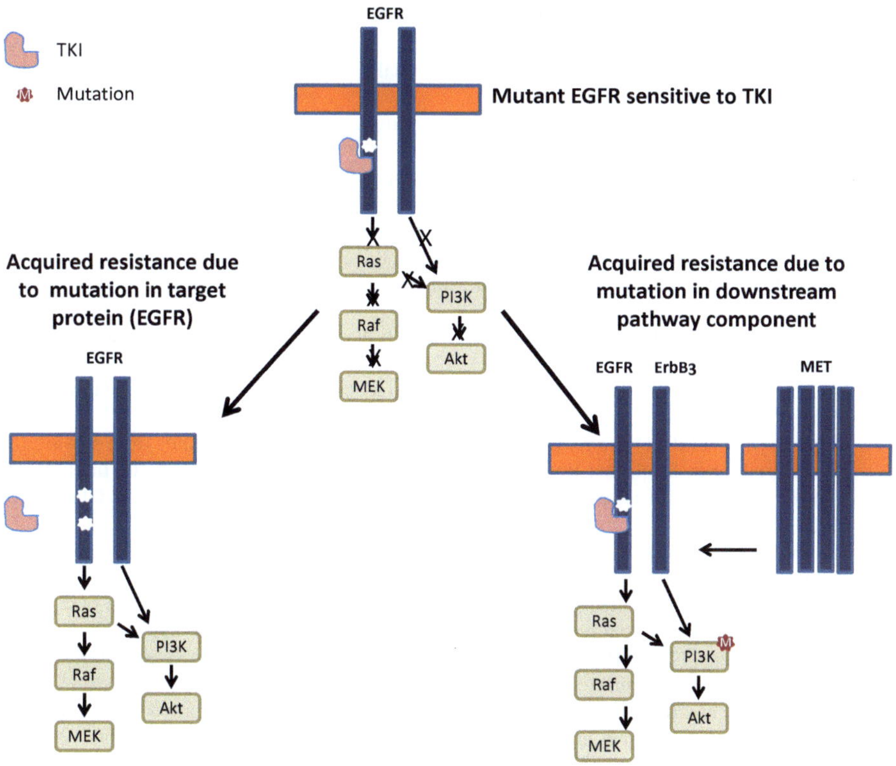


**Fig. 12.5** Mechanism of acquired resistance in human cancer

protein of the addicted pathway leads to resistance. Further studies reported acquired resistance in several other targeted therapies, calling it as unified limitation of targeted therapy (Fig. 12.5).

**Acknowledgements** The authors would like to thank Prajish Sundaram for his contribution to Fig. 12.1; Rohan Chaubal and Mukul Godbole for critical reading of the chapter. Amit Dutt is supported by an Intermediate fellowship from the Wellcome Trust/DBT India Alliance (IA/I/11/2500278), by a grant from DBT (BT/PR2372/AGR/36/696/2011), and intramural grants (Seed-In-Air 2897, TMH Plan Project 2712 and IRB 92).

## References

1. Druker BJ et al (1996) Effects of a selective inhibitor of the Abl tyrosine kinase on the growth of Bcr-Abl positive cells. Nat Med 2(5):561–566
2. Fernandez-Medarde A, Santos E (2011) Ras in cancer and developmental diseases. Genes Cancer 2(3):344–358

3. Prior IA, Lewis PD, Mattos C (2012) A comprehensive survey of Ras mutations in cancer. Cancer Res 72(10):2457–2467
4. Kloog Y, Cox AD (2000) RAS inhibitors: potential for cancer therapeutics. Mol Med Today 6(10):398–402
5. Manning G et al (2002) The protein kinase complement of the human genome. Science 298(5600):1912–1934
6. Druker BJ et al (2001) Efficacy and safety of a specific inhibitor of the BCR-ABL tyrosine kinase in chronic myeloid leukemia. N Engl J Med 344(14):1031–1037
7. Sierra JR, Cepero V, Giordano S (2010) Molecular mechanisms of acquired resistance to tyrosine kinase targeted therapy. Mol Cancer 9:75
8. Krause DS, Van Etten RA (2005) Tyrosine kinases as targets for cancer therapy. N Engl J Med 353(2):172–187
9. Yarden Y, Pines G (2012) The ERBB network: at last, cancer therapy meets systems biology. Nat Rev Cancer 12(8):553–563
10. Sharma SV et al (2007) Epidermal growth factor receptor mutations in lung cancer. Nat Rev Cancer 7(3):169–181
11. Fukuoka M et al (2011) Biomarker analyses and final overall survival results from a phase III, randomized, open-label, first-line study of gefitinib versus carboplatin/paclitaxel in clinically selected patients with advanced non-small-cell lung cancer in Asia (IPASS). J Clin Oncol 29(21):2866–2874
12. Ellison G et al (2013) EGFR mutation testing in lung cancer: a review of available methods and their use for analysis of tumour tissue and cytology samples. J Clin Pathol 66(2):79–89
13. Zheng XM, Wang Y, Pallen CJ (1992) Cell transformation and activation of pp 60c-src by overexpression of a protein tyrosine phosphatase. Nature 359(6393):336–339
14. den Hertog J et al (1993) Receptor protein tyrosine phosphatase alpha activates pp 60c-src and is involved in neuronal differentiation. EMBO J 12(10):3789–3798
15. Tartaglia M et al (2003) Somatic mutations in PTPN11 in juvenile myelomonocytic leukemia, myelodysplastic syndromes and acute myeloid leukemia. Nat Genet 34(2):148–150
16. Bentires-Alj M et al (2004) Activating mutations of the Noonan syndrome-associated SHP2/PTPN11 gene in human solid tumors and adult acute myelogenous leukemia. Cancer Res 64(24):8816–8820
17. Matozaki T et al (2009) Protein tyrosine phosphatase SHP-2: a proto-oncogene product that promotes Ras activation. Cancer Sci 100(10):1786–1793
18. Galaktionov K et al (1995) CDC25 phosphatases as potential human oncogenes. Science 269(5230):1575–1577
19. Lyon MA et al (2002) Dual-specificity phosphatases as targets for antineoplastic agents. Nat Rev Drug Discov 1(12):961–976
20. Kobayashi S et al (2005) EGFR mutation and resistance of non-small-cell lung cancer to gefitinib. N Engl J Med 352(8):786–792

# Part IV
# Targeted Cancer Therapy

# Chapter 13
# The Potential of Targeting Splicing for Cancer Therapy

**Ana Rita Grosso and Maria Carmo-Fonseca**

**Abstract** Many molecules currently used to treat cancer patients target proteins encoded by transcripts that are alternatively spliced. As a consequence, the treatment may simultaneously block isoforms with different and sometimes opposing biological activity, thus reducing its efficacy. Recent studies highlight the role of splicing regulation in cancer progression and the importance of the splicing machinery as a therapeutic target. In reviewing this emerging field of cancer biology, we describe very exciting novel findings that illustrate the range of scenarios in which alternative splicing can contribute to all cancer hallmarks, from avoidance of apoptosis to angiogenesis, invasion and acquired resistance to drug therapy. Finally, we address cancer-selective approaches that are being developed to interfere with the splicing machinery and modulate splicing decisions.

**Keywords** Alternative splicing • Splicing factors • Cancer biology • Cancer therapy • Drug targets

## 13.1 Introduction

Although cancer is a genetic disease, no single gene defect causes a tumor. Rather, it is only when several genes are altered that cancer arises. Moreover, cancer evolves through successive genetic changes that become advantageous to a cell. In essence, defective genes responsible for tumorigenesis belong to three groups: oncogenes, tumor-suppressor genes and genome stability genes [1]. Defects in oncogenes

---

A.R. Grosso, Ph.D. • M. Carmo-Fonseca, M.D., Ph.D. (✉)
Instituto de Medicina Molecular, Faculdade de Medicina, Universidade de Lisbon,
Av. Professor Egas Moniz, Lisbon 1649-028, Portugal
e-mail: carmo.fonseca@fm.ul.pt

R. Kumar (ed.), *Nuclear Signaling Pathways and Targeting Transcription in Cancer*, Cancer Drug Discovery and Development, DOI 10.1007/978-1-4614-8039-6_13,

render the gene constitutively active or active under conditions in which the normal gene is not, whereas tumor-suppressor and stability genes become inactivated. These defects in gene function can result from chromosomal translocations, deletions or insertions, amplifications or intragenic mutations. An additional, recently recognized mechanism of oncogene activation or tumor-suppressor and genome stability gene inactivation is alternative splicing. Most human genes produce multiple mRNA isoforms through alternative splicing, and altered splicing is a major contributor to cancer progression. This chapter focuses on the role of alternative splicing in cancer and highlights the therapeutic potential of targeting and modulating cancer-specific splicing isoforms.

## 13.2 Pre-mRNA Splicing and Its Regulation

RNA splicing was discovered in 1977 as a new mechanism for the biosynthesis of adenovirus mRNA in mammalian cells [2, 3]. Shortly after, cellular genes were also shown to be split into exons and introns. The first examples included the globin [4, 5], the ovalbumin [6] and the immunoglobulin [7] genes. Next, it was recognized that at each intron boundary there are consensus sequences common to vertebrate, plant and yeast cells, suggesting the splicing process was evolutionary conserved [8, 9].

Today we know that the vast majority of human protein-coding genes contain up to 90 % of non-coding sequence in the form of introns that must be spliced from the primary transcripts synthesized by RNA polymerase II (pre-mRNAs). There are over 200,000 different introns in the human genome, ranging in size from <100 to >700,000 nucleotides (nts), with median intron and exon lengths of 1,800 and 123 nts, respectively.

Excision of introns with single nucleotide precision relies on the spliceosome, one of the largest and most elaborate macromolecular machines in the cell [10]. The building blocks of the spliceosome are uridine-rich small nuclear RNAs (UsnRNAs) packaged as ribonucleoprotein particles (snRNPs) that function in conjunction with over 100 distinct non-snRNP auxiliary proteins [11]. The major spliceosomal small nuclear ribonucleoprotein particles comprise the U1, U2, U4, U5 and U6 snRNAs. In addition, human cells have a minor variant form of the spliceosome responsible for excision of about 800 so-called U12-dependent introns that are characterized by a distinct set of splice-site sequences [12–14]. Much of our current understanding of the role of snRNPs in splicing was triggered by studies using human autoantibodies from patients with systemic lupus erythematosus that selectively react with the spliceosomal RNA-protein complexes [15].

The spliceosomal snRNAs recognize, through base pairing, four short consensus sequences termed the exon-intron junctions (5′ splice site and 3′ splice site), the branch point sequence, and the polypyrimidine tract (Fig. 13.1a). Spliceosomes build anew on every intron that is synthesized and then disassemble for the next round of splicing (Fig. 13.1b–e). Assembly of the spliceosome starts with the ATP-independent binding of the U1 snRNP through base-pairing interactions of the 5′ end of the U1 snRNA to the 5′ splice site of the intron. This is followed by the binding of the SF1 protein and the heterodimeric U2 auxiliary factor (U2AF) to the

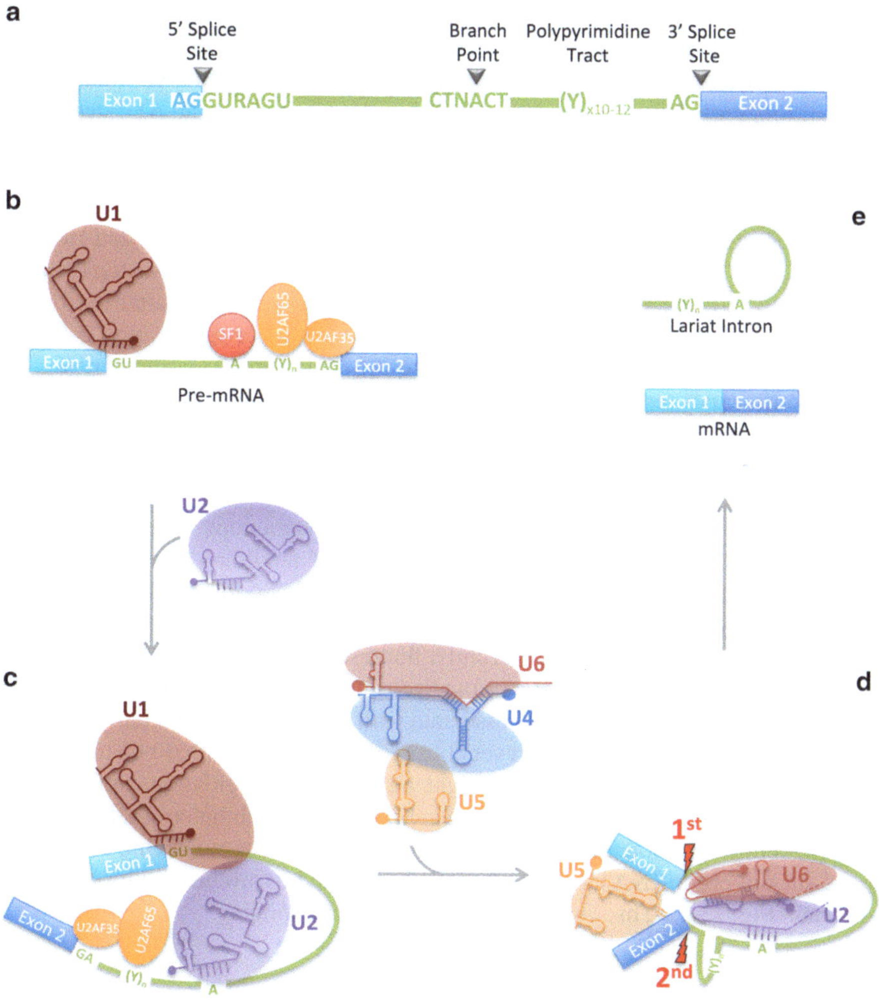


**Fig. 13.1 Intron removal by splicing.** (**a**) In humans, most introns are removed by the major spliceosome that recognizes conserved sequence elements located at the 5′ splice site, branch point (BP) and 3′ splice site. The polypyrimidine tract is a pyrimidine-rich stretch located between the BP and the 3′ splice site. The panel depicts two exons (*blue*) separated by an intron (*green*). N represents any nucleotide, R a purine, and Y a pyrimidine. (**b**) Spliceosome assembly initiates by binding of the U1 snRNP to the 5′ splice site and recruitment of the U2AF65/U2AF35 heterodimer to the 3′ splice site. The U2AF65 subunit binds to the polypyrimidine tract, the U2AF35 subunit to the AG dinucleotide at the 3′ splice site, and the SF1 protein binds to the branch point sequence. (**c**) In an ATP-dependent reaction, the U2 snRNP displaces SF1 and binds to the branch point. At this stage, the 5′ splice site, branch point sequence, and 3′ splice site, are in close spatial proximity. Bending of the polypyrimidine tract induced by interaction with U2AF brings the 3′ splice site into juxtaposition with the branch point sequence. Both the 5′ and 3′ splice sites are close to the 5′-end of the U2 snRNA, which later assembles with U6 snRNA forming the catalytic center of the spliceosome. (**d**) Catalytic activation occurs subsequent to addition of the U4/U6.U5 tri-snRNP and requires several rearrangements, including departure of the U1 and U4 snRNPs. The splicing reaction consists of two consecutive transesterification (replacement of one phosphodiester linkage for another) events. (**e**) After the two chemical steps of splicing are complete, the spliced exons are released, the spliceosome disassembles and the excised intron is degraded

branch point sequence and the downstream polypyrimidine tract, respectively. These proteins bind cooperatively, with SF1 interacting with the large subunit of U2AF (U2AF65), whereas the small subunit (U2AF35) binds the AG dinucleotide of the 3′ splice site (Fig. 13.1b). Next, the U2 snRNA engages in an ATP-dependent base-pairing interaction with the branch point sequence, displacing SF1 (Fig. 13.1c). Subsequently, the U4, U5 and U6snRNPs are recruited as a preassembled U4/U6.U5 tri-snRNP. With all snRNPs present, the spliceosome undergoes major conformational rearrangements that lead to release of U1 and U4. The spliceosome is now competent to catalyze the first transesterification step of splicing (Fig. 13.1d): the phosphodiester bond at the 5′ splice site is attacked by the 2′-hydroxyl of the adenosine at the branch point sequence, generating a free 5′ exon and an intron lariat-3′ exon intermediate. After additional rearrangements, the spliceosome catalyzes the second transesterification reaction: the 3′-hydroxyl of the 5′ exon attacks the phosphodiester bond at the 3′ splice site, leading to exon ligation and excision of the lariat intron. Then the spliceosome dissociates, releasing the mRNA (Fig. 13.1e).

Most of the functionally important RNA-RNA interactions formed within the spliceosome are weak and require the assistance of auxiliary proteins that bind weakly to specific sequences in exons and introns. This combination of multiple weak interactions is crucial for the flexibility of the spliceosome, in particular during regulated splicing decisions. Recently, fluorescence microscopy has been used to follow assembly of individual yeast spliceosomes in real time. The results indicate that spliceosomal components associate with pre-mRNA sequentially, but each step in the assembly pathway is reversible [16]. This implies that potentially any step during spliceosome formation might be subject to regulation. Spliceosome assembly is indeed highly regulated: depending on the combinatorial effect of proteins that either promote or repress the recognition of the core splicing sequences, splice sites in pre-mRNA can be differentially selected to produce multiple mRNA isoforms (Fig. 13.2). This process is called alternative splicing.

## 13.3 The Importance of Alternative Splicing

Diverse forms of mRNA are created by the differential use of splice sites (reviewed in [17]). Exons that are always included in the mRNA are called constitutive exons, and exons that are sometimes included and sometimes excluded from the mRNA are called cassette exons (Fig. 13.3). Some pre-mRNAs contain multiple cassette exons that are mutually exclusive, producing mRNAs that always include one of several possible exon choices. Exons can also be lengthened or shortened by altering the position of one of their splice sites alternative 5′ and alternative 3′ splice site selection; (Fig. 13.3). The 5′ and 3′-terminal exons can further be switched by combining alternative splicing with the use of alternative promoters or alternative polyadenylation sites, respectively (Fig. 13.3). Finally, certain intronic sequences may persist in the final mRNA, a splicing pattern called intron retention (Fig. 13.3).

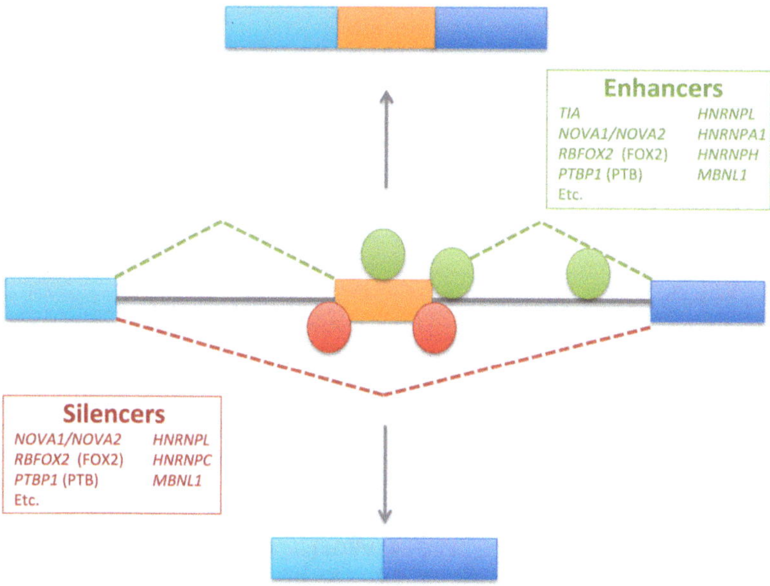


**Fig. 13.2 Splicing is regulated by positional binding of RNA-binding proteins.** The diagram depicts a model for mechanism of splicing activation or repression by RNA-binding proteins (RBPs). Two constitutively spliced exons (*blue*) are separated by an alternative or cassette exon (*orange*). Pending on the positions at which RBPs bind to the pre-mRNA, the alternative exon is either included (*top*) or excluded (*bottom*). Certain RBPs (depicted *red*) bind at intronic positions close to the 3′ and 5′ splice sites of the alternative exon to silence its inclusion. In contrast, binding of enhancer RBPs (*depicted green*) within the exon or in the downstream intron promotes inclusion of the alternative exon

Alternative splicing was first reported in 1980, when it was discovered that membrane-bound and secreted antibodies are encoded by the same gene [18, 19]. During the 1980s and 1990s many biologically important alternative splicing events were identified and characterized. Yet, the prevalence and general importance of this process was far from clear. More recently, the application of genome-wide profiling technologies coupled with bioinformatic approaches resulted in major advances in our understanding of alternative splicing. In particular, high-throughput massively parallel short-read sequencing provided for the first time unambiguous and unbiased detection of expressed RNA sequences. Compared to microarray-based systems for profiling alternative splicing, short-read sequencing offers a more accurate method for quantifying relative levels of different transcripts. Analysis of human tissue RNA sequencing (RNA-Seq) data revealed that approximately 95 % of human pre-mRNAs that contain more than one exon are spliced to yield multiple mRNAS, and that most isoforms display variable expression across tissues [20, 21]. Genes with few exons typically encode a small number of mRNA isoforms, while tremendously diverse mRNA repertoires can be produced from genes containing numerous exons. For example, the human gene *UTY* has 61 exons and can generate 129 mRNA isoforms (according to UCSC Known Gene annotations [22]).

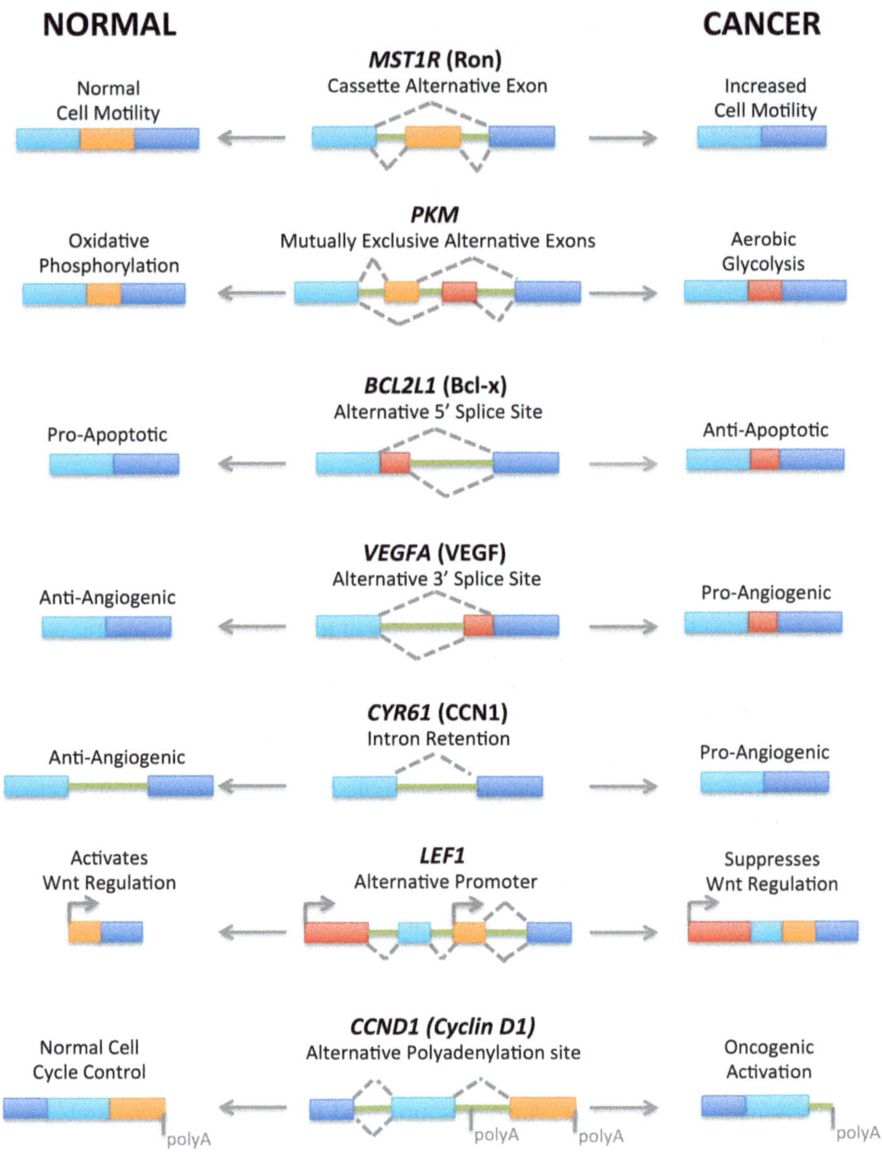


**Fig. 13.3 Alternative splicing events in cancer**. The basic types of alternative splicing include cassette-exon inclusion or exclusion, alternative 5′ or 3′ splice site selection, intron retention, alternative selection of transcription initiation (alternative promoter) and alternative selection of 3′ end processing sites (alternative polyadenylation). Alternative splicing events that have positive effects on cancer progression are shown. Constitutively spliced exons are depicted in *blue*

Overall, the high prevalence of alternative splicing combined with the finding that many mRNA isoforms represent low-abundant, non-conserved transcripts argue that they may be devoid of functional impact. Yet, recent studies are revealing a rapidly growing number of physiologically important splicing events.

Distinct splicing patterns of a given pre-mRNA can be observed pending on the cellular environment. For example, some mRNA isoforms are specifically expressed in certain tissues or developmental stages and other are triggered in response to external stimuli such as signaling pathways (reviewed in [23, 24]) or depolarization of neurons [25]. The diversity of mRNA isoforms appears to be higher in embryonic stem cells compared to differentiated cells [26] and some isoforms specifically detected in embryonic stem cells have recently been shown to play a key role in pluripotency [27–30]. Remarkably, a single splicing event can function to control an entire transcriptional network. This is well illustrated by alternative splicing of the transcription factor *FOXP1*. *FOXP1* mRNAs transcribed in embryonic stem cells contain a specific exon that is skipped in differentiated cells [30]. Inclusion of this exon determines the DNA binding properties of the encoded protein and is required for stimulating the expression of several pluripotency transcription factor genes [30].

Presumably, the intron-exon structure of genes played an important role in the generation of new genes during evolution. Moreover, alternative splicing seems to be rapidly evolving, particularly among physiologically equivalent organs from vertebrate species [31]. A remarkable example of how species-specific alternative splicing evolved was found in bats [32]. In order to detect warm-blooded prey, vampire bats express a splice isoform of the transient receptor potential cation channel V1 (*TRPV1*) gene. This isoform produces a channel with a truncated carboxy-terminal cytoplasmic domain capable of detecting infrared radiation. In contrast, the protein isoform expressed in fruit-feeding bats has a much higher thermal activation threshold [32].

Around 10–30 % of all alternatively spliced exons have inclusion levels that differ across tissues and are therefore called tissue-specific exons. Many of these exons play an important role in cell differentiation, for example during brain and heart development [33]. Recent large-scale computational analysis revealed that genes with tissue-specific exons tend to have more interaction partners compared to the other genes [34, 35]. Tissue-specific exons tend to encode flexible protein segments without a well-defined three-dimensional structure that likely form conserved interaction surfaces [34, 35]. Using a high-through put coimmunoprecipitation assay, inclusion of tissue-specific exons was shown to both promote and disrupt partner interactions [35]. Thus, tissue-specific splicing can potentially mediate new molecular interactions in a cell type-specific manner.

How human cells control more than 100,000 alternative splicing decisions remains incompletely understood. Clearly, there are multiple mechanisms involved, including RNA-binding proteins that interact with pre-mRNAs and modulate the efficiency of splice-site recognition by the spliceosome, formation of secondary structures in the RNA, the transcription rate and epigenetic modification of the template chromatin [36, 37]. A relatively small number of splicing regulators has been identified, and most are ubiquitously expressed although their relative abundances can fluctuate in different tissues [38]. A few, however, are tissue-specific

RNA-binding proteins, such as *NOVA1* and *NOVA2* [39], *PTBP2* (nPTB) [40, 41], *RBFOX1* (FOX-1) and *RBFOX2* (FOX-2) [42, 43], *ESRP1* and *ESRP2* [44] and *SRRM4* (nSR100) [45]. Currently, large-scale quantification of alternative splicing has been combined with genome-wide identification of in vivo binding sites of splicing regulators (Fig. 13.2) to create maps identifying all pre-mRNAs regulated by a single RNA-binding protein. RNA splicing maps are providing an unprecedented view of the global principles guiding splicing regulation [46].

## 13.4 Numerous Alterations in Splicing Occur in Cancer Cells

Recent high-throughput transcriptome sequencing studies revealed that different splicing variants are commonly found in cancer tissue compared to the normal surrounding tissue. This type of information has already proven useful in the classification of ovarian and breast cancer [47, 48]. In the case of prostate cancer, alternative splicing signatures are more reliable for diagnostic purposes than are gene expression signatures [49], and in osteosarcoma, changes in relative expression of splicing isoforms of the *TP53* (p53) inhibitor *MDM2* (HDMX) is a more effective prognostic biomarker than *TP53* mutation [50].

Direct causes of splicing alterations in cancer can be grouped into two main categories: *cis*-acting mutations in the pre-mRNA sequence and *trans*-acting changes in expression or activity of regulatory splicing factors. The first category encompasses mutations or polymorphisms in splice sites or regulatory sequence motifs. For example, in breast and ovarian cancer, mutations in the tumor suppressor breast cancer 1, early onset (*BRCA1*) often disrupt constitutive splice sites, leading to the production of inactive protein isoforms [51]. Splice site mutations in the *TP53* gene have also been described in various cancers [52]. In infant B-precursor leukemia, intronic mutations were found in the *CD22* gene that affect target motifs for splicing factors *HNRNPL* (hnRNP-L), *PTBP1* (PTB) and *PCBP1* leading to deletion of exon 12 and expression of a truncated and functionally defective *CD22* coreceptor protein unable to transmit apoptotic signals [53]. Recent systematic surveys identified 106 acquired somatic splice site mutations associated with aberrant splicing in lung cancer [54] and 158 essential splice site mutations in breast cancer [55].

In addition to inherited and acquired mutations, the human genome contains approximately 1,200 single nucleotide polymorphisms (SNPs) expected to modify splicing decisions [56]. A subset of these splicing-related SNPs may be functionally relevant in the context of cancer susceptibility and cancer progression, as shown by a recent study that identified a splicing polymorphism in the germline as predictor of response to targeted therapies [57]. A common intronic deletion polymorphism in the *BCL2L11* (BCL2-like 11, BIM) gene switches splicing from exon 4 to exon 3, leading to expression of *BCL2L11* isoforms that lack the pro-apoptotic BCL2-homology domain 3. The presence of this polymorphism explains why some individuals with chronic myeloid leukemia and epidermal growth factor receptor-mutated non-small-cell lung cancer have inferior responses to tyrosine kinase inhibitors [57].

Changes in expression or activity of *trans*-acting protein factors are caused by defects in components of the spliceosome or splicing regulatory factors. These can be induced by either genetic mutations and amplifications, or transcriptional and post-transcriptional mis-regulation. For example, *SRSF1* (SF2/ASF), a member of the SR protein family of splicing regulators, is frequently upregulated in many cancers. In some tumors, the gene locus is amplified accounting for the elevated levels of the protein [58]. Altered transcriptional regulation by *MYC* (c-Myc), which binds directly to E-boxes in the *SRSF1* gene promoter, is an alternative cause for *SRSF1* protein over expression in cancer [59]. *MYC* can also control expression of additional splicing regulators, namely hnRNP proteins [60]. In addition to *MYC*, other transcription factors control the expression of splicing proteins. Namely, mutations in the Wilms's tumor suppressor gene, *WT1*, abrogates binding of the *WT1* protein to the *SRPK1* promoter, causing over-expression of this SR-protein kinase and hyperphosphorylation of splicing regulator *SRSF1*; this in turn resulted in altered splicing of *VEGFA*, stimulating angiogenesis [61]. Direct binding of the transcription repressor Snail to the *ESRP1* promoter was also shown to cause reduced expression of this epithelial-specific splicing factor, thus promoting isoform switching of several genes involved in epithelial-to-mesenchymal transition [62]. Recurrent mutations in genes encoding essential components of the splicing machinery such as *U2AF1*, *ZRSR2*, *SRSF2*, *SF3A1* and *SF3B1* were recently found in patients with myelodysplastic syndromes [63–66]. More recently, mutations in *U2AF1*, *U2AF2*, and *SF3B1* genes were also detected in lung cancer patients [67]. Remarkably, most of these mutations affect proteins involved in 3′-splice site recognition during the early stages of spliceosome assembly (Fig. 13.1b, c). This, together with the finding that mutations were detected in a mutually exclusive manner, strongly suggests that the compromised function of early spliceosome complexes is a hallmark of cancer.

In general, factors involved in splicing regulation are RNA binding proteins that interact with particular sequence motifs, albeit with relatively low specificity. Therefore, most alternative splicing decisions are controlled by the cooperative binding of several protein factors to short redundant RNA motifs. As a consequence, each alternative splicing event is frequently regulated by multiple factors, and each factor may control several splicing events [46]. A striking example was recently reported for the tissue-specific splicing factors *RBFOX1* (FOX-1) and *RBFOX2* (FOX-2). Unlike most other known splicing regulators, the FOX proteins bind exclusively two defined sequence motifs: UGCAUG and AGCAUG [68]. The expression of *RBFOX2* was found downregulated in ovarian cancer and *RBFOX2* binding sites were detected downstream of one-third of the exons alternatively spliced in this type of cancer [69]. Importantly, reducing the expression of *RBFOX2* in cell lines recapitulated the cancer-associated splicing signature, suggesting that the reduced level of *RBFOX2* causes the changes in splicing [69]. This study further showed that ovarian and breast cancers share a common splicing signature. Although *RBFOX2* transcripts were not downregulated in breast cancer, they were alternatively spliced producing an inactive form of the protein [69]. This illustrates how changes in splicing of a splicing factor can change its regulatory activity, leading to further changes in splicing of its target pre-mRNAs.

Although splicing is primarily controlled by sequence elements in the pre-mRNA that recruit trans-acting splicing factors, recent work make it clear that alternative splicing is also sensitive to transcriptional rate and chromatin conformation (reviewed in [70]). Since many proteins that control DNA and histone modification show aberrant expression or altered activity in tumors [71], it is likely that epigenetics represents a third cause of splicing abnormalities in cancer. Of note in this regard, variations in the methylated status of the *MST1R* (RON) promoter correlate with transcription of a short isoform of the enzyme that is constitutively active and drives cell proliferation [72].

While the majority of cancer-specific alternative splicing events may have just coincidently occurred during tumor development, a few bestow a growth advantage on the tumor. To date, several splicing isoforms that are specifically expressed in cancer have been demonstrated to contribute to cellular malignant phenotypes such as avoidance of apoptosis, angiogenesis, limitless replication potential, and invasion [73], as detailed below (see also Table 13.1 and Fig. 13.3).

### 13.4.1 Apoptosis

Transcripts from numerous genes involved in apoptosis are alternatively spliced, often resulting in isoforms with opposing roles in promoting or preventing cell death. Well-characterized examples include the *BCL2L1* (Bcl-x), *CASP2* (Caspase-2), *CASP9* (Caspase-9), and *FAS* (Fas) genes. In general, isoforms that enhance survival tend to be up-regulated in cancer and correlated with clinical staging (reviewed in [73]). Another protein that promotes apoptosis upon DNA damage is *AIMP2* (Aminoacyl-tRNA synthetase-interacting multifunctional protein 2). A splicing variant of *AIMP2* was found highly expressed in lung cancer, leading to increased resistance to cell death, and the relative expression of this isoform correlated with cancer stage and survival of patients [75].

### 13.4.2 Angiogenesis

Primary transcripts encoding Vascular Endothelial Growth Factor (*VEGFA*), which plays a key role in promoting the formation of new blood vessels, undergo extensive alternative splicing. As a result, two families of splicing isoforms are produced with either pro-angiogenic or anti-angiogenic functions. Anti-angiogenic isoforms are expressed in normal tissues and are downregulated in many cancers (reviewed in [124, 125]). Another protein involved in angiogenesis that is regulated by alternative splicing is *CYR61* (cysteine rich 61, CCN1). While normal cells express a variant with retention of an intron that most likely targets the transcripts for degradation, in cancer cells the intron is spliced leading to an accumulation of active protein [85].

**Table 13.1** Alternative splicing (AS) events that favor cancer progression

| | Gene | AS event type | Altered regulatory proteins | Functional implications | References |
|---|---|---|---|---|---|
| DNA damage | *ABL1* | Cassette exon | *EWSR1* (EWS) | Response to DNA damage | [74] |
| | *CHEK2* | Cassette exon | *EWSR1* (EWS) | Response to DNA damage | [74] |
| | *MAP4K2* | Cassette exon | *EWSR1* (EWS) | Response to DNA damage | [74] |
| Apoptosis | *AIMP2* | Cassette exon | – | Inhibition of apoptosis | [75] |
| | *BCL2L1* (Bcl-x) | Alternative 5′ splice site | – | Inhibition of apoptosis | [76] |
| | *BCL2L11* (BIM) | Cassette exon | *SRSF1* (ASF/SF2) | Inhibition of apoptosis | [77] |
| | *CASP2* (Caspase-2) | Cassette exon | – | Inhibition of apoptosis | [78] |
| | *CASP9* (Caspase-9) | Cassette exon | *HNRNPL* (hnRNP L) | Inhibition of apoptosis | [79] |
| | *CD22* | Cassette exon | *HNRNPL*, *PTBP1*(*PTB*), *PCBP1* (PCB) | Inhibition of apoptosis | [53] |
| | *CFLAR* (FLIP) | Cassette exon | – | Inhibition of apoptosis | [80] |
| | *FAS* | Cassette exon | *TIA1*, *PTBP1* | Inhibition of apoptosis | [81] |
| | *STAT3* | Alternative 3′ splice site | – | Inhibition of apoptosis | [82] |
| | *TP53* | Alternative promoter | – | Inhibition of apoptosis | [83] |
| | *TP73* | Alternative promoter | – | Inhibition of apoptosis | [84] |
| Angiogenesis | *CYR61* | Intron retention | – | Inhibition of angiogenesis | [85] |
| | *FLT1* (VEGFR1) | Intron retention | – | Inhibition of angiogenesis | [86] |
| | *KDR* (VEGFR2) | Alternative polyadenylation | – | Inhibition of angiogenesis | [87] |
| | *VEGFA* | Alternative 3′ splice site | *SRSF1* | Activation of angiogenesis | [88] |
| Proliferative potential | *CCND1* (Cyclin D1) | Alternative polyadenylation | *EWSR1* | Oncogenic activation | [89] |
| | *CCND1* (Cyclin D1) | Alternative polyadenylation | *SRSF1* | Oncogenic activation | [90] |
| | *CD44* | Cassette exon | *ESRP1* | Evasion from oxidative stress | [91] |
| | *CYP19A1* | Alternative promoter | – | Activation of tumor growth | [92] |
| | *ERBB2* (HER2) | Cassette exon | – | Activation of tumor growth | [93] |
| | *ERBB4* | Cassette exon | – | Activation of tumor growth | [94] |
| | *FYN* | Mutually exclusive exons | – | Activation of tumor growth | [95] |
| | *H2AFY* (macroH2A.1) | Mutually exclusive exons | *QKI* | Activation of tumor growth | [96] |

(continued)

**Table 13.1** (continued)

| | Gene | AS event type | Altered regulatory proteins | Functional implications | References |
|---|---|---|---|---|---|
| | *KLF6* | Cassette exon | *SRSF1*, *SRSF3* | Activation of tumor growth | [97] |
| | *MDM2* | Cassette exon | *EWSR1* | Genotoxic stress response | [98] |
| | *MET* | Cassette exon | – | Activation of tumor growth | [99] |
| | *MTOR* | Cassette exon | – | Activation of tumor growth | [100] |
| | *LEF1* | Alternative promoter | – | Activation of tumor growth | [101] |
| | *PKM* | Mutually exclusive exons | *PTBP1*, *HNRNPA1*, *HNRNPA2B1* | Aerobic glycolysis | [60] |
| | *RASSF1* | Alternative promoter | – | Activation of tumor growth | [102] |
| | *RPS6KB1* (S6K1) | Cassette exon | *SRSF1* | Oncogenic activation | [58] |
| | *TP63* | Alternative promoter | – | Activation of tumor growth | [103] |
| | *TPM1* (tropomyosin) | Intron retention | *DEK* | Activation of tumor growth | [104] |
| Invasion and metastasis | *ADAM8* | Cassette exon | – | Invasive cell migration | [105] |
| | *CD44* | Cassette exon | *SRRM1* (SRm160) | Invasive cell migration | [106] |
| | *CD44* | Cassette exon | *ESRP1* | Induction of EMT | [107] |
| | *CPE* | Alternative promoter | – | Invasive cell migration | [108] |
| | *CTNND1* (p120) | Alternative promoter | – | Invasive cell migration | [109] |
| | *ENAH* (Mena) | Cassette exon | – | Invasive cell migration | [110] |
| | *FGFR2* | Mutually exclusive exons | *ESRP1*, *ESRP2* | Induction of EMT | [44] |
| | *GLI1* | Cassette exon | – | Invasive cell migration | [111] |
| | *MST1R* (Ron) | Cassette exon | *HNRNPA2B1* (hnRNPA2/B1) | Oncogenic activation | [112] |
| | *MST1R* (Ron) | Cassette exon | *SRSF1* | Oncogenic activation | [113] |
| | *MST1R* (Ron) and *MADD* (IG20) | Cassette exon | *HNRNPH1* (hnRNPH) | Oncogenic activation | [114] |
| | *NCOA3* (AIB1) | Cassette exon | – | Invasive cell migration | [115] |
| | *RAC1* | Cassette exon | *SRSF1*, *SRSF3* | Oncogenic activation | [116] |
| | *SENP7* | Cassette exon | – | Induction of EMT | [117] |
| | *TP53INP2* | Cassette exon | *HNRNPA2B1* | Invasive cell migration | [118] |
| Cancer therapy | *AR* | Cassette exon | – | Resistance to hormonal therapy | [119] |
| | *BCR* | Cassette exon | – | Resistance to drug therapy | [120] |
| | *BRAF* | Cassette exon | – | Resistance to drug therapy | [121] |
| | *MKNK2* (MNK2) | Mutually exclusive exons | *SRSF1* | Resistance to drug therapy | [122] |
| | *MS4A1* (CD20) | Cassette exon | – | Resistance to drug therapy | [123] |
| | *TP53* (p53) | Cassette exon | – | Resistance to chemotherapy | [52] |

### 13.4.3 Proliferative Potential

Proliferating cells reprogram their metabolism to engage in aerobic glycolysis (the Warburg effect), in part through alternative splicing of the pre-mRNA that encodes the enzyme pyruvate kinase M, *PKM*. Normal cells express the splicing isoform PKM1, whereas all tumors express PKM2. Importantly, replacing PKM2 with PKM1 in cancer cells reduced tumor growth (reviewed in [126]). Cancer-associated changes in alternative splicing can also result in activation of proto-oncogenes such as *CCND1* (Cyclin D1) [reviewed in 127].

### 13.4.4 Invasion and Metastasis

A significant reprogramming of alternative splicing occurs during the epithelial-to-mesenchymal transition (EMT), a process by which cancer cells acquire invasive capabilities and become metastatic. EMT-associated changes in splicing affect genes such as *MST1R* (Ron), *RAC1*, *CD44*, *FGFR2*, *CTNND1* (p120-catenin), and *ENAH* (Mena) (reviewed in [128, 129]). Among these, the *CD44* transmembrane protein was one of the first genes for which splicing variants were found associated with metastasis. The expression of specific *CD44* splicing variants correlates with aggressive behavior in several cancer cell types [130–132], and one particular isoform (CD44v8-10) potentiates the ability of cancer cells to defend themselves against reactive oxygen species [91, 133]. Alternative splicing can further contribute to regulate the onset of EMT in cancer cells. Indeed, it was recently found that normal breast epithelia express two splice variants of sentrin/small ubiquitin-like modifier (SUMO)-specific protease 7 (*SENP7*), and breast cancer cells express predominantly the isoform that promotes EMT initiation [117]. Additional examples of alternatively spliced isoforms that promote cancer cell migration and invasion via a gain-of-function mechanism include the truncated glioma-associated oncogene homolog 1, *GLI1* [111, 134], the steroid receptor coactivator 3 (*NCOA3*, AIB1) with a deletion of exon 4, SRC-3Δ4 [115, 135], truncated forms of *ADAM8* (a disintegrin and metalloprotease) [105] and *CPE* (carboxypeptidase E) [108], and alternative inclusion of an exon in the five untranslated region of tumor protein p53 inducible nuclear protein 2 *TP53INP2* [118].

In addition to generating protein isoforms with different biological activities, alternative splicing can also regulate gene expression level through inclusion of premature translation termination codons that target the mRNA for degradation by nonsense-mediated decay [136]. This mechanism causes downregulation of proteins involved in tumor development such as *NFAT5* transcription factor [137] and *CDH1* (E-cadherin) in chronic lymphocytic leukemia [138].

Finally, several lines of recent evidence reveal that splicing contributes for acquired resistance to chemotherapeutic drugs. For example, patients with metastatic melanoma are currently treated with vemurafenib, a newly approved drug that selectively binds monomers of the most prevalent oncogenic mutation of *BRAF* (B-RAF, V600E), inhibiting its kinase activity. However, most patients acquire resistance within a year of treatment. Different mechanisms have been identified that counteract vemurafenib effectiveness, and one of them consists in expression of truncated forms of the *BRAF* (V600E) protein generated by abnormal pre-mRNA splicing. These splicing isoforms lack the RAS-binding domain and dimerize in a RAS-independent manner therefore rendering the enzyme insensitive to RAF inhibitors [121]. Another example is gemcitabine, the drug used for pancreatic ductal adenocarcinoma. Gemcitabine induces overexpression of splicing factor *SRSF1*, leading to formation of a mitogen activated protein kinase interacting kinase *MKNK2* (MNK2) splicing variant that overrides upstream regulatory pathways and confers resistance to the drug [122]. In chronic myeloid leukemia, one of the mechanisms responsible for resistance to tyrosine kinase inhibitors is the expression of an alternatively spliced *BCR* (BCR-ABL) pre-mRNA that lacks the drug-targeted kinase domain [120, 139] and in B cell malignancies, a splicing isoform of *MS4A1* (CD20) produces a *truncated* protein that loses membrane anchorage and causes resistance to rituximab [123]. Splicing variants of the androgen receptor *AR* may also contribute to the development of castration-resistant prostate cancers [119], and in ovarian cancer expression of a particular *TP53* (p53) splicing isoform correlates with impaired response to primary platinum-based chemotherapy [52].

## 13.5 Targeting Splicing for Cancer Therapy

In cancer research, much effort is focused on the identification of molecular pathways that are specific to tumor cells and essential for their survival. Cancer-specific splice variants are therefore emerging as highly attractive therapeutic targets, since only cancer cells will be targeted with minimum toxicity towards normal cells. However, in contrast with diagnostic and prognostic purposes for which any discriminating isoform can be a valuable biomarker, the selection of splicing isoforms as drug targets requires detailed functional studies to evaluate their potential in ablating cancer cells. RNA interference (RNAi) screens specifically targeted to silence tumor-associated splicing variants currently represent a valuable tool for identification of isoforms essential for cancer cell survival. Recently, a systematic isoform-specific functional screen of 41 alternatively spliced variants associated with breast and ovarian cancer revealed that targeting the spleen tyrosine kinase *SYK* isoform induced apoptosis, whereas global knockdown of the same gene had no effect [140]. Clearly, the functional contribution of splicing isoforms to tumor behavior should be considered when designing anticancer strategies. This is well illustrated by the limited efficacy of currently used molecules like Bevacizumab that target angiogenesis but do not distinguish between the pro- and the anti-angiogenic splicing isoforms of *VEGFA* [125]. It remains to be studied whether targeting specifically the pro-angiogenic *VEGFA* isoform or treating patients with Bevacizumab only in cases where the anti-angiogenic isoforms are absent will be more beneficial [125].

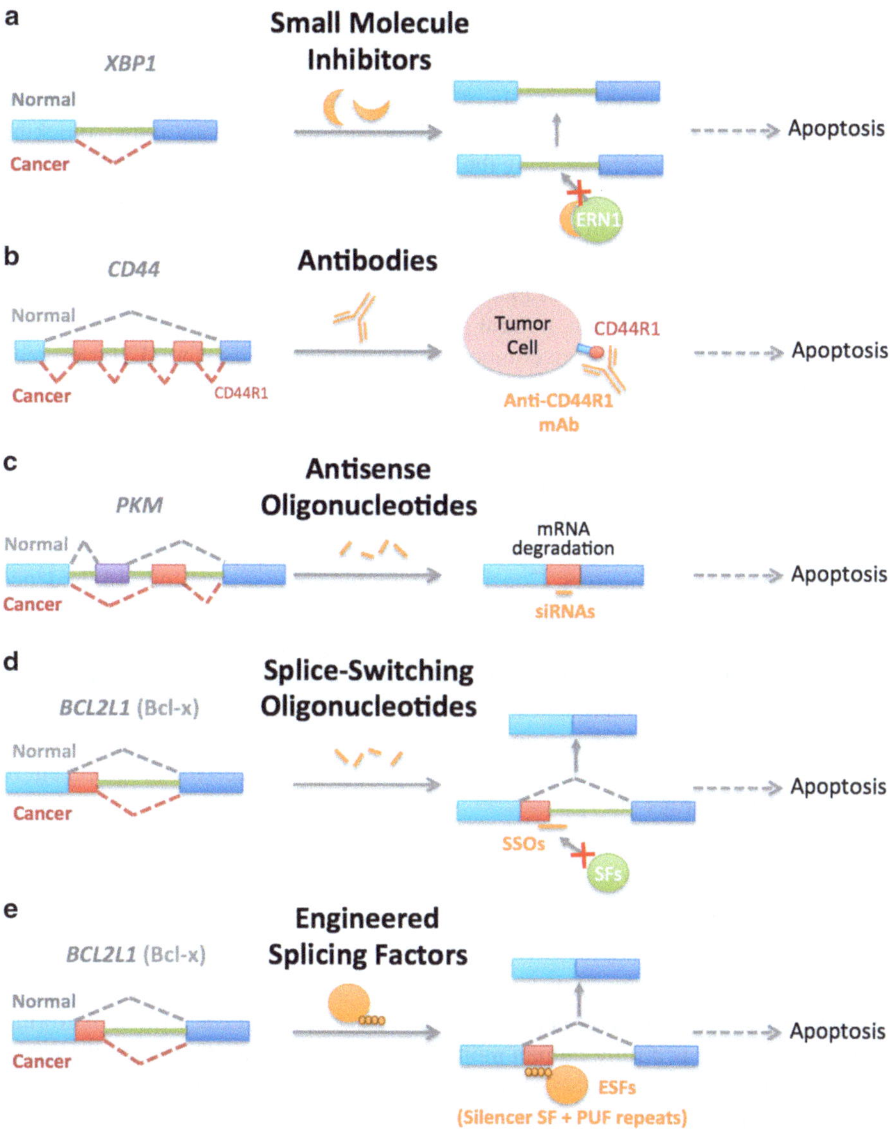


**Fig. 13.4 Strategies for splicing-targeted cancer therapies.** (**a**) Small molecule inhibitors capable of altering cancer-associated splicing. (**b**) Monoclonal antibodies that recognize unique protein epitopes encoded by cancer-associated mRNA isoforms. (**c**) Antisense oligonucleotides that induce RNAi-mediated knockdown of oncogenic mRNA isoforms. (**d**) Splice-switching oligonucleotides that redirect splicing decisions, thereby reducing oncogenic mRNA isoforms. (**e**) Instead of using oligonucleotides, splicing reprogramming can be achieved by engineered proteins that combine sequence-specific RNA-binding domains with functional domains that regulate splicing

Multiple strategies have been envisioned to therapeutically target cancer-associated splicing. These include small molecule inhibitors, antibodies, and antisense oligonucleotides, as described below in more detail (see Fig. 13.4).

### 13.5.1 Small Molecule Inhibitors

Different types of molecules capable of altering alternative splicing have emerged from several chemical screens. Many of these substances act by either blocking histone deacetylases or by inhibiting the kinases that phosphorylate SR splicing factors (reviewed in [141]). For example, amiloride can revert cancer-specific splicing events and this effect is likely mediated by changes in amount and phosphorylation status of SR proteins [142, 143]. Similarly, a small-molecule inhibitor of *XBP1* splicing may be a promising therapeutic option in multiple myeloma [144].

Using a different approach, natural products derived from distantly related bacteria were found to target a core component of the spliceosome, the SF3B1 protein, suggesting that interfering with splicing may be a mechanism by which bacteria compete with eukaryotes. Remarkably, mutations in the *SF3B1* gene were found in some cancers [66] and anti-SF3B1 compounds demonstrated dramatic, selective antitumor activity in human tumor xenograft models (reviewed in [145]). The mechanism responsible for such selective antitumor activity is unknown, but one intriguing possibility is that proliferating cancer cells are more vulnerable than normal cells to splicing inhibitors. Consistent with this view, several lines of evidence suggest that RNA splicing is functionally coupled to cell-cycle progression (see [146] and references therein). Moreover, interfering with the splicing machinery leads to activation of *TP53* [147] and induces an alternatively spliced isoform of *TP53* that promotes cellular senescence [148]. Altogether these observations suggest that activation of p53 may contribute to the observed selective anti-tumor activity. Thus, targeting the spliceosome might be a viable approach for development of novel anticancer drugs [145].

### 13.5.2 Antibodies

An alternative strategy that is being explored consists of raising antibodies against epitopes that are uniquely present in the cancer-associated protein isoforms. A recent example is a fully human monoclonal antibody that recognizes the extracellular domain of a *CD44* isoform expressed on the surface of various epithelial cancers [149].

### 13.5.3 Antisense Oligonucleotides

Oligonucleotides designed to bind defined sequence elements in the pre-mRNA can induce either an RNAi-mediated specific knockdown of a particular splicing isoform, or redirect splicing decisions. For example, alterations in glucose metabolism mediated by pyruvate kinase (*PKM*) activity are likely to confer a selective advantage for tumor cells to grow in hypoxic environments. Because PK activity is modulated by alternative splicing, inhibition of the PKM2 isoform that is commonly expressed at high levels in tumor cells appears as a promising target of broad therapeutic applicability. By screening a tiling siRNA library, Goldberg and Sharp recently identified sequences that discriminate between the M1 and M2 splicing isoforms of pyruvate kinase and produce a potent and specific knockdown of the M2 isoform. This resulted in decreased viability and increased apoptosis in multiple cancer cell lines but less so in normal fibroblasts or endothelial cells. Moreover, when the selected siRNAs were delivered as nanoparticles to established xenografts, a substantial reduction of tumor volume was observed [150]. Oligonucleotides can also be designed to redirect splicing decisions through blocking access to the transcript by the spliceosome. Splice-switching oligonucleotides (SSOs) are chemically modified to ensure stability and increase their binding affinity for the target sequence. SSOs have been applied to restore correct splicing of an aberrantly spliced transcript, induce expression of a novel splice variant with therapeutic value, or manipulate alternative splicing from one splice variant to another (reviewed in [151]). The latter mechanism can induce downregulation of a deleterious transcript while simultaneously upregulating expression of a preferred isoform, making it an attractive anti-cancer molecular therapy. Although the application of SSOs is still hindered by poor in vivo delivery to tumor cells, promising results were reported for antisense compounds directed at either inducing the pro-apoptotic splicing variant Bcl-x(S) at the expense of the more abundant survival Bcl-x(L) isoform of *BCL2L1* gene [152], or redirecting splicing of the signal transducer and activator of transcription 3 (*STAT3*) transcripts to produce a truncated isoform lacking the transactivation domain [82]. Enhanced delivery of SSOs to the cell nucleus can be achieved through aptamers that bind nucleolin, a protein that is found on the surface of rapidly proliferating tumor cells and traffics from the cell surface to the nucleus [153].

In principle, splicing can also be modulated using engineered proteins instead of antisense oligonucleotides. A recently proposed strategy relies on direct recognition of the pre-mRNA targets through protein-RNA interaction. Splicing reprogramming is achieved through engineering artificial splicing factors that combine sequence-specific RNA-binding domains with functional domains that regulate splicing [154]. Artificial factors targeted to the human endogenous gene *BCL2L1* (Bcl-X) increased the amount of the pro-apoptotic splicing isoform and promoted apoptosis of cancer cells [154].

## 13.6 Concluding Remarks

Molecular studies are increasingly used to guide therapeutic decisions for cancer patients, as clinical trials demonstrate superior efficacy of targeted treatments compared to "classical" chemotherapy. Recent advances in high-throughput DNA and RNA sequencing will ultimately lead to a comprehensive characterization of the genome and transcriptome of most cancers. New activating mutations and translocations in oncogenes will be discovered and pursued as drug targets. Additionally, a more insightful perspective on the contribution of post-transcriptional regulation of gene expression in cancer complexity and diversity will be gained. The complete landscape of splicing alterations will be described for each cancer type and their functional impact on cell growth, metabolism, viability, apoptosis, invasiveness, angiogenesis and drug resistance will be established with the help of RNAi screens. The molecular mechanisms responsible for the functionally relevant cancer-associated splicing events will also be identified. Doubtless, these studies will significantly increase the list of potential cancer therapeutic targets in the near future. The challenge ahead will be to further develop innovative approaches to selectively and efficiently interfere with the splicing machinery and modulate splicing decisions. There is growing optimism that this research area may enable new opportunities for cancer patients.

## References


1. Vogelstein B, Kinzler KW (2004) Cancer genes and the pathways they control. Nat Med 10(8):789–799
2. Berget SM, Moore C, Sharp PA (1977) Spliced segments at the 5′ terminus of adenovirus 2 late mRNA. Proc Natl Acad Sci USA 74(8):3171–3175
3. Chow LT et al (1977) An amazing sequence arrangement at the 5′ ends of adenovirus 2 messenger RNA. Cell 12(1):1–8
4. Jeffreys AJ, Flavell RA (1977) A physical map of the DNA regions flanking the rabbit beta-globin gene. Cell 12(2):429–439
5. Tilghman SM et al (1978) The intervening sequence of a mouse beta-globin gene is transcribed within the 15S beta-globin mRNA precursor. Proc Natl Acad Sci USA 75(3):1309–1313
6. Breathnach R, Mandel JL, Chambon P (1977) Ovalbumin gene is split in chicken DNA. Nature 270(5635):314–319
7. Tonegawa S et al (1978) Sequence of a mouse germ-line gene for a variable region of an immunoglobulin light chain. Proc Natl Acad Sci USA 75(3):1485–1489
8. Breathnach R, Chambon P (1981) Organization and expression of eucaryotic split genes coding for proteins. Annu Rev Biochem 50:349–383
9. Padgett RA et al (1986) Splicing of messenger RNA precursors. Annu Rev Biochem 55:1119–1150
10. Nilsen TW (2003) The spliceosome: the most complex macromolecular machine in the cell? Bioessays 25(12):1147–1149
11. Wahl MC, Will CL, Luhrmann R (2009) The spliceosome: design principles of a dynamic RNP machine. Cell 136(4):701–718
12. Hall SL, Padgett RA (1996) Requirement of U12 snRNA for in vivo splicing of a minor class of eukaryotic nuclear pre-mRNA introns. Science 271(5256):1716–1718
13. Tarn WY, Steitz JA (1996) A novel spliceosome containing U11, U12, and U5 snRNPs excises a minor class (AT-AC) intron in vitro. Cell 84(5):801–811
14. Levine A, Durbin R (2001) A computational scan for U12-dependent introns in the human genome sequence. Nucleic Acids Res 29(19):4006–4013
15. Lerner MR, Steitz JA (1979) Antibodies to small nuclear RNAs complexed with proteins are produced by patients with systemic lupus erythematosus. Proc Natl Acad Sci USA 76(11):5495–5499
16. Hoskins AA et al (2011) Ordered and dynamic assembly of single spliceosomes. Science 331(6022):1289–1295

17. Black DL (2003) Mechanisms of alternative pre-messenger RNA splicing. Annu Rev Biochem 72:291–336
18. Alt FW et al (1980) Synthesis of secreted and membrane-bound immunoglobulin mu heavy chains is directed by mRNAs that differ at their 3′ ends. Cell 20(2):293–301
19. Early P et al (1980) Two mRNAs can be produced from a single immunoglobulin mu gene by alternative RNA processing pathways. Cell 20(2):313–319
20. Pan Q et al (2008) Deep surveying of alternative splicing complexity in the human transcriptome by high-throughput sequencing. Nat Genet 40(12):1413–1415
21. Wang ET et al (2008) Alternative isoform regulation in human tissue transcriptomes. Nature 456(7221):470–476
22. Meyer LR et al (2013) The UCSC genome browser database: extensions and updates 2013. Nucleic Acids Res 41(D1):D64–D69
23. Lynch KW (2007) Regulation of alternative splicing by signal transduction pathways. Adv Exp Med Biol 623:161–174
24. Shin C, Manley JL (2004) Cell signalling and the control of pre-mRNA splicing. Nat Rev Mol Cell Biol 5(9):727–738
25. Xie J, Black DL (2001) A CaMK IV responsive RNA element mediates depolarization-induced alternative splicing of ion channels. Nature 410(6831):936–939
26. Wu JQ et al (2010) Dynamic transcriptomes during neural differentiation of human embryonic stem cells revealed by short, long, and paired-end sequencing. Proc Natl Acad Sci USA 107(11):5254–5259
27. Rao S et al (2010) Differential roles of Sall4 isoforms in embryonic stem cell pluripotency. Mol Cell Biol 30(22):5364–5380
28. Salomonis N et al (2010) Alternative splicing regulates mouse embryonic stem cell pluripotency and differentiation. Proc Natl Acad Sci USA 107(23):10514–10519
29. Mayshar Y et al (2008) Fibroblast growth factor 4 and its novel splice isoform have opposing effects on the maintenance of human embryonic stem cell self-renewal. Stem Cells 26(3):767–774
30. Gabut M et al (2011) An alternative splicing switch regulates embryonic stem cell pluripotency and reprogramming. Cell 147(1):132–146
31. Barbosa-Morais NL et al (2012) The evolutionary landscape of alternative splicing in vertebrate species. Science 338(6114):1587–1593
32. Gracheva EO et al (2011) Ganglion-specific splicing of TRPV1 underlies infrared sensation in vampire bats. Nature 476(7358):88–91
33. Kalsotra A, Cooper TA (2011) Functional consequences of developmentally regulated alternative splicing. Nat Rev Genet 12(10):715–729
34. Buljan M et al (2012) Tissue-specific splicing of disordered segments that embed binding motifs rewires protein interaction networks. Mol Cell 46(6):871–883
35. Ellis JD et al (2012) Tissue-specific alternative splicing remodels protein-protein interaction networks. Mol Cell 46(6):884–892
36. Nilsen TW, Graveley BR (2010) Expansion of the eukaryotic proteome by alternative splicing. Nature 463(7280):457–463
37. Luco RF et al (2010) Regulation of alternative splicing by histone modifications. Science 327(5968):996–1000
38. Grosso AR et al (2008) Tissue-specific splicing factor gene expression signatures. Nucleic Acids Res 36(15):4823–4832
39. Buckanovich RJ, Posner JB, Darnell RB (1993) Nova, the paraneoplastic Ri antigen, is homologous to an RNA-binding protein and is specifically expressed in the developing motor system. Neuron 11(4):657–672
40. Polydorides AD et al (2000) A brain-enriched polypyrimidine tract-binding protein antagonizes the ability of Nova to regulate neuron-specific alternative splicing. Proc Natl Acad Sci USA 97(12):6350–6355
41. Markovtsov V et al (2000) Cooperative assembly of an hnRNP complex induced by a tissue-specific homolog of polypyrimidine tract binding protein. Mol Cell Biol 20(20):7463–7479

42. Underwood JG et al (2005) Homologues of the *Caenorhabditis elegans* Fox-1 protein are neuronal splicing regulators in mammals. Mol Cell Biol 25(22):10005–10016
43. Jin Y et al (2003) A vertebrate RNA-binding protein Fox-1 regulates tissue-specific splicing via the pentanucleotide GCAUG. EMBO J 22(4):905–912
44. Warzecha CC et al (2009) ESRP1 and ESRP2 are epithelial cell-type-specific regulators of FGFR2 splicing. Mol Cell 33(5):591–601
45. Calarco JA et al (2009) Regulation of vertebrate nervous system alternative splicing and development by an SR-related protein. Cell 138(5):898–910
46. Witten JT, Ule J (2011) Understanding splicing regulation through RNA splicing maps. Trends Genet 27(3):89–97
47. Venables JP et al (2008) Identification of alternative splicing markers for breast cancer. Cancer Res 68(22):9525–9531
48. Klinck R et al (2008) Multiple alternative splicing markers for ovarian cancer. Cancer Res 68(3):657–663
49. Zhang C et al (2006) Profiling alternatively spliced mRNA isoforms for prostate cancer classification. BMC Bioinforma 7:202
50. Lenos K et al (2012) Alternate splicing of the p53 inhibitor HDMX offers a superior prognostic biomarker than p53 mutation in human cancer. Cancer Res 72(16):4074–4084
51. Tammaro C et al (2012) BRCA1 exon 11 alternative splicing, multiple functions and the association with cancer. Biochem Soc Trans 40(4):768–772
52. Hofstetter G et al (2010) Alternative splicing of p53 and p73: the novel p53 splice variant p53delta is an independent prognostic marker in ovarian cancer. Oncogene 29(13): 1997–2004
53. Uckun FM et al (2010) CD22 EXON 12 deletion as a pathogenic mechanism of human B-precursor leukemia. Proc Natl Acad Sci USA 107(39):16852–16857
54. Liu J et al (2012) Genome and transcriptome sequencing of lung cancers reveal diverse mutational and splicing events. Genome Res 22(12):2315–2327
55. Stephens PJ et al (2012) The landscape of cancer genes and mutational processes in breast cancer. Nature 486(7403):400–404
56. Faber K et al (2011) Genome-wide prediction of splice-modifying SNPs in human genes using a new analysis pipeline called AASsites. BMC Bioinformatics 12(Suppl 4):S2
57. Ng KP et al (2012) A common BIM deletion polymorphism mediates intrinsic resistance and inferior responses to tyrosine kinase inhibitors in cancer. Nat Med 18(4):521–528
58. Karni R et al (2007) The gene encoding the splicing factor SF2/ASF is a proto-oncogene. Nat Struct Mol Biol 14(3):185–193
59. Das S et al (2012) Oncogenic splicing factor SRSF1 is a critical transcriptional target of MYC. Cell Rep 1(2):110–117
60. David CJ et al (2010) HnRNP proteins controlled by c-Myc deregulate pyruvate kinase mRNA splicing in cancer. Nature 463(7279):364–368
61. Amin EM et al (2011) WT1 mutants reveal SRPK1 to be a downstream angiogenesis target by altering VEGF splicing. Cancer Cell 20(6):768–780
62. Reinke LM, Xu Y, Cheng C (2012) Snail represses the splicing regulator epithelial splicing regulatory protein 1 to promote epithelial-mesenchymal transition. J Biol Chem 287(43): 36435–36442
63. Graubert TA et al (2012) Recurrent mutations in the U2AF1 splicing factor in myelodysplastic syndromes. Nat Genet 44(1):53–57
64. Yoshida K et al (2011) Frequent pathway mutations of splicing machinery in myelodysplasia. Nature 478(7367):64–69
65. Zhang SJ et al (2012) Genetic analysis of patients with leukemic transformation of myeloproliferative neoplasms shows recurrent SRSF2 mutations that are associated with adverse outcome. Blood 119(19):4480–4485
66. Papaemmanuil E et al (2011) Somatic SF3B1 mutation in myelodysplasia with ring sideroblasts. N Engl J Med 365(15):1384–1395

67. Imielinski M et al (2012) Mapping the hallmarks of lung adenocarcinoma with massively parallel sequencing. Cell 150(6):1107–1120
68. Ponthier JL et al (2006) Fox-2 splicing factor binds to a conserved intron motif to promote inclusion of protein 4.1R alternative exon 16. J Biol Chem 281(18):12468–12474
69. Venables JP et al (2009) Cancer-associated regulation of alternative splicing. Nat Struct Mol Biol 16(6):670–676
70. Luco RF, Misteli T (2011) More than a splicing code: integrating the role of RNA, chromatin and non-coding RNA in alternative splicing regulation. Curr Opin Genet Dev 21(4):366–372
71. Elsasser SJ, Allis CD, Lewis PW (2011) Cancer. New epigenetic drivers of cancers. Science 331(6021):1145–1146
72. Angeloni D et al (2007) Hypermethylation of Ron proximal promoter associates with lack of full-length Ron and transcription of oncogenic short-Ron from an internal promoter. Oncogene 26(31):4499–4512
73. David CJ, Manley JL (2010) Alternative pre-mRNA splicing regulation in cancer: pathways and programs unhinged. Genes Dev 24(21):2343–2364
74. Paronetto MP, Minana B, Valcarcel J (2011) The Ewing sarcoma protein regulates DNA damage-induced alternative splicing. Mol Cell 43(3):353–368
75. Choi JW et al (2011) Cancer-associated splicing variant of tumor suppressor AIMP2/p38: pathological implication in tumorigenesis. PLoS Genet 7(3):e1001351
76. Taylor JK et al (1999) Induction of endogenous Bcl-xS through the control of Bcl-x pre-mRNA splicing by antisense oligonucleotides. Nat Biotechnol 17(11):1097–1100
77. Anczukow O et al (2012) The splicing factor SRSF1 regulates apoptosis and proliferation to promote mammary epithelial cell transformation. Nat Struct Mol Biol 19(2):220–228
78. Wang L et al (1994) Ich-1, an Ice/ced-3-related gene, encodes both positive and negative regulators of programmed cell death. Cell 78(5):739–750
79. Goehe RW et al (2010) hnRNP L regulates the tumorigenic capacity of lung cancer xenografts in mice via caspase-9 pre-mRNA processing. J Clin Invest 120(11):3923–3939
80. Irmler M et al (1997) Inhibition of death receptor signals by cellular FLIP. Nature 388(6638):190–195
81. Izquierdo JM et al (2005) Regulation of Fas alternative splicing by antagonistic effects of TIA-1 and PTB on exon definition. Mol Cell 19(4):475–484
82. Zammarchi F et al (2011) Antitumorigenic potential of STAT3 alternative splicing modulation. Proc Natl Acad Sci USA 108(43):17779–17784
83. Slatter TL et al (2011) Hyperproliferation, cancer, and inflammation in mice expressing a Delta133p53-like isoform. Blood 117(19):5166–5177
84. Muller M et al (2005) TAp73/Delta Np73 influences apoptotic response, chemosensitivity and prognosis in hepatocellular carcinoma. Cell Death Differ 12(12):1564–1577
85. Hirschfeld M et al (2009) Alternative splicing of Cyr61 is regulated by hypoxia and significantly changed in breast cancer. Cancer Res 69(5):2082–2090
86. Boeckel JN et al (2011) Jumonji domain-containing protein 6 (Jmjd6) is required for angiogenic sprouting and regulates splicing of VEGF-receptor 1. Proc Natl Acad Sci USA 108(8):3276–3281
87. Vorlova S et al (2011) Induction of antagonistic soluble decoy receptor tyrosine kinases by intronic poly A activation. Mol Cell 43(6):927–939
88. Nowak DG et al (2010) Regulation of vascular endothelial growth factor (VEGF) splicing from pro-angiogenic to anti-angiogenic isoforms: a novel therapeutic strategy for angiogenesis. J Biol Chem 285(8):5532–5540
89. Sanchez G et al (2008) Alteration of cyclin D1 transcript elongation by a mutated transcription factor up-regulates the oncogenic D1b splice isoform in cancer. Proc Natl Acad Sci USA 105(16):6004–6009
90. Olshavsky NA et al (2010) Identification of ASF/SF2 as a critical, allele-specific effector of the cyclin D1b oncogene. Cancer Res 70(10):3975–3984

91. Yae T et al (2012) Alternative splicing of CD44 mRNA by ESRP1 enhances lung colonization of metastatic cancer cell. Nat Commun 3:883
92. Harada N, Utsumi T, Takagi Y (1993) Tissue-specific expression of the human aromatase cytochrome P-450 gene by alternative use of multiple exons 1 and promoters, and switching of tissue-specific exons 1 in carcinogenesis. Proc Natl Acad Sci USA 90(23):11312–11316
93. Kwong KY, Hung MC (1998) A novel splice variant of HER2 with increased transformation activity. Mol Carcinog 23(2):62–68
94. Muraoka-Cook RS et al (2009) ErbB4 splice variants Cyt1 and Cyt2 differ by 16 amino acids and exert opposing effects on the mammary epithelium in vivo. Mol Cell Biol 29(18):4935–4948
95. Brignatz C et al (2009) Alternative splicing modulates autoinhibition and SH3 accessibility in the Src kinase Fyn. Mol Cell Biol 29(24):6438–6448
96. Novikov L et al (2011) QKI-mediated alternative splicing of the histone variant MacroH2A1 regulates cancer cell proliferation. Mol Cell Biol 31(20):4244–4255
97. Munoz U et al (2012) Hepatocyte growth factor enhances alternative splicing of the Kruppel-like factor 6 (KLF6) tumor suppressor to promote growth through SRSF1. Mol Cancer Res 10(9):1216–1227
98. Dutertre M et al (2010) Cotranscriptional exon skipping in the genotoxic stress response. Nat Struct Mol Biol 17(11):1358–1366
99. Lee JH et al (2006) An alternatively spliced form of Met receptor is tumorigenic. Exp Mol Med 38(5):565–573
100. Panasyuk G et al (2009) mTORbeta splicing isoform promotes cell proliferation and tumorigenesis. J Biol Chem 284(45):30807–30814
101. Li TW et al (2006) Wnt activation and alternative promoter repression of LEF1 in colon cancer. Mol Cell Biol 26(14):5284–5299
102. Amaar YG et al (2006) Ras association domain family 1C protein stimulates human lung cancer cell proliferation. Am J Physiol Lung Cell Mol Physiol 291(6):L1185–L1190
103. Hibi K et al (2000) AIS is an oncogene amplified in squamous cell carcinoma. Proc Natl Acad Sci USA 97(10):5462–5467
104. Babaei-Jadidi R et al (2011) FBXW7 influences murine intestinal homeostasis and cancer, targeting Notch, Jun, and DEK for degradation. J Exp Med 208(2):295–312
105. Hernandez I et al (2010) Novel alternatively spliced ADAM8 isoforms contribute to the aggressive bone metastatic phenotype of lung cancer. Oncogene 29(26):3758–3769
106. Cheng C, Sharp PA (2006) Regulation of CD44 alternative splicing by SRm160 and its potential role in tumor cell invasion. Mol Cell Biol 26(1):362–370
107. Brown RL et al (2011) CD44 splice isoform switching in human and mouse epithelium is essential for epithelial-mesenchymal transition and breast cancer progression. J Clin Invest 121(3):1064–1074
108. Lee TK et al (2011) An N-terminal truncated carboxypeptidase E splice isoform induces tumor growth and is a biomarker for predicting future metastasis in human cancers. J Clin Invest 121(3):880–892
109. Yanagisawa M et al (2008) A p120 catenin isoform switch affects Rho activity, induces tumor cell invasion, and predicts metastatic disease. J Biol Chem 283(26):18344–18354
110. Goswami S et al (2009) Identification of invasion specific splice variants of the cytoskeletal protein Mena present in mammary tumor cells during invasion in vivo. Clin Exp Metastasis 26(2):153–159
111. Lo HW et al (2009) A novel splice variant of GLI1 that promotes glioblastoma cell migration and invasion. Cancer Res 69(17):6790–6798
112. Golan-Gerstl R et al (2011) Splicing factor hnRNP A2/B1 regulates tumor suppressor gene splicing and is an oncogenic driver in glioblastoma. Cancer Res 71(13):4464–4472
113. Ghigna C et al (2005) Cell motility is controlled by SF2/ASF through alternative splicing of the Ron protooncogene. Mol Cell 20(6):881–890
114. Lefave CV et al (2011) Splicing factor hnRNPH drives an oncogenic splicing switch in gliomas. EMBO J 30(19):4084–4097

115. Reiter R, Wellstein A, Riegel AT (2001) An isoform of the coactivator AIB1 that increases hormone and growth factor sensitivity is overexpressed in breast cancer. J Biol Chem 276(43):39736–39741
116. Goncalves V, Matos P, Jordan P (2009) Antagonistic SR proteins regulate alternative splicing of tumor-related Rac1b downstream of the PI3-kinase and Wnt pathways. Hum Mol Genet 18(19):3696–3707
117. Bawa-Khalfe T et al (2012) Differential expression of SUMO-specific protease 7 variants regulates epithelial-mesenchymal transition. Proc Natl Acad Sci USA 109(43):17466–17471
118. Moran-Jones K et al (2009) hnRNP A2 regulates alternative mRNA splicing of TP53INP2 to control invasive cell migration. Cancer Res 69(24):9219–9227
119. Sun S et al (2010) Castration resistance in human prostate cancer is conferred by a frequently occurring androgen receptor splice variant. J Clin Invest 120(8):2715–2730
120. Laudadio J et al (2008) An intron-derived insertion/truncation mutation in the BCR-ABL kinase domain in chronic myeloid leukemia patients undergoing kinase inhibitor therapy. J Mol Diagn 10(2):177–180
121. Poulikakos PI et al (2011) RAF inhibitor resistance is mediated by dimerization of aberrantly spliced BRAF(V600E). Nature 480(7377):387–390
122. Adesso L et al (2013) Gemcitabine triggers a pro-survival response in pancreatic cancer cells through activation of the MNK2/eIF4E pathway. Oncogene 32:2848–2857
123. Henry C et al (2010) Identification of an alternative CD20 transcript variant in B-cell malignancies coding for a novel protein associated to rituximab resistance. Blood 115(12):2420–2429
124. Harper SJ, Bates DO (2008) VEGF-A splicing: the key to anti-angiogenic therapeutics? Nat Rev Cancer 8(11):880–887
125. Hilmi C, Guyot M, Pages G (2012) VEGF spliced variants: possible role of anti-angiogenesis therapy. J Nucleic Acids 2012:162692
126. Chen M, Zhang J, Manley JL (2010) Turning on a fuel switch of cancer: hnRNP proteins regulate alternative splicing of pyruvate kinase mRNA. Cancer Res 70(22):8977–8980
127. Witzel II, Koh LF, Perkins ND (2010) Regulation of cyclin D1 gene expression. Biochem Soc Trans 38(Pt 1):217–222
128. Warzecha CC, Carstens RP (2012) Complex changes in alternative pre-mRNA splicing play a central role in the epithelial-to-mesenchymal transition (EMT). Semin Cancer Biol 22(5–6):417–427
129. Biamonti G et al (2012) Making alternative splicing decisions during epithelial-to-mesenchymal transition (EMT). Cell Mol Life Sci 69(15):2515–2526
130. Gunthert U et al (1991) A new variant of glycoprotein CD44 confers metastatic potential to rat carcinoma cells. Cell 65(1):13–24
131. Klingbeil P et al (2009) CD44 variant isoforms promote metastasis formation by a tumor cell-matrix cross-talk that supports adhesion and apoptosis resistance. Mol Cancer Res 7(2):168–179
132. Wallach-Dayan SB et al (2001) CD44-dependent lymphoma cell dissemination: a cell surface CD44 variant, rather than standard CD44, supports in vitro lymphoma cell rolling on hyaluronic acid substrate and its in vivo accumulation in the peripheral lymph nodes. J Cell Sci 114(Pt 19):3463–3477
133. Ishimoto T et al (2011) CD44 variant regulates redox status in cancer cells by stabilizing the xCT subunit of system xc(-) and thereby promotes tumor growth. Cancer Cell 19(3):387–400
134. Cao X et al (2012) Upregulation of VEGF-A and CD24 gene expression by the tGLI1 transcription factor contributes to the aggressive behavior of breast cancer cells. Oncogene 31(1):104–115
135. Long W et al (2010) SRC-3Delta4 mediates the interaction of EGFR with FAK to promote cell migration. Mol Cell 37(3):321–332
136. Lewis BP, Green RE, Brenner SE (2003) Evidence for the widespread coupling of alternative splicing and nonsense-mediated mRNA decay in humans. Proc Natl Acad Sci USA 100(1):189–192

137. Germann S et al (2012) Dual role of the ddx5/ddx17 RNA helicases in the control of the pro-migratory NFAT5 transcription factor. Oncogene 31(42):4536–4549
138. Sharma S, Lichtenstein A (2009) Aberrant splicing of the E-cadherin transcript is a novel mechanism of gene silencing in chronic lymphocytic leukemia cells. Blood 114(19):4179–4185
139. Ma W et al (2011) Three novel alternative splicing mutations in BCR-ABL1 detected in CML patients with resistance to kinase inhibitors. Int J Lab Hematol 33(3):326–331
140. Prinos P et al (2011) Alternative splicing of SYK regulates mitosis and cell survival. Nat Struct Mol Biol 18(6):673–679
141. Sumanasekera C, Watt DS, Stamm S (2008) Substances that can change alternative splice-site selection. Biochem Soc Trans 36(Pt 3):483–490
142. Chang JG et al (2011) Small molecule amiloride modulates oncogenic RNA alternative splicing to devitalize human cancer cells. PLoS One 6(6):e18643
143. Chang WH et al (2011) Amiloride modulates alternative splicing in leukemic cells and resensitizes Bcr-AblT315I mutant cells to Imatinib. Cancer Res 71(2):383–392
144. Mimura N et al (2012) Blockade of XBP1 splicing by inhibition of IRE1alpha is a promising therapeutic option in multiple myeloma. Blood 119(24):5772–5781
145. Webb TR, Joyner AS, Potter PM (2013) The development and application of small molecule modulators of SF3b as therapeutic agents for cancer. Drug Discov Today 18(1–2):43–49
146. Ahn EY et al (2011) SON controls cell-cycle progression by coordinated regulation of RNA splicing. Mol Cell 42(2):185–198
147. Allende-Vega N et al (2013) p53 is activated in response to disruption of the pre-mRNA splicing machinery. Oncogene 32(1):1–14
148. Tang Y et al (2013) Downregulation of splicing factor SRSF3 induces p53beta, an alternatively spliced isoform of p53 that promotes cellular senescence. Oncogene 32:2792–2798
149. Masuko K et al (2012) Anti-tumor effect against human cancer xenografts by a fully human monoclonal antibody to a variant 8-epitope of CD44R1 expressed on cancer stem cells. PLoS One 7(1):e29728
150. Goldberg MS, Sharp PA (2012) Pyruvate kinase M2-specific siRNA induces apoptosis and tumor regression. J Exp Med 209(2):217–224
151. Spitali P, Aartsma-Rus A (2012) Splice modulating therapies for human disease. Cell 148(6):1085–1088
152. Bauman JA et al (2010) Anti-tumor activity of splice-switching oligonucleotides. Nucleic Acids Res 38(22):8348–8356
153. Kotula JW et al (2012) Aptamer-mediated delivery of splice-switching oligonucleotides to the nuclei of cancer cells. Nucleic Acid Ther 22(3):187–195
154. Wang Y et al (2009) Engineering splicing factors with designed specificities. Nat Methods 6(11):825–830

# Chapter 14
# Exploiting Cell Cycle Pathways in Cancer Therapy: New (and Old) Targets and Potential Strategies

**Angela Alexander and Khandan Keyomarsi**

**Abstract** There is a now a large body of evidence supporting the notion that cancer cells have vastly altered cell cycle networks that serve to maintain their high rate of proliferation. Consequently, targeting these pathways pharmacologically has been long studied, but only recently have some promising compounds progressed into the clinic. In this chapter, we review cell cycle function in both normal cells and describe how cancer cells deregulate this fundamental process. Next we describe in detail the development of different classes of CDK inhibitors and review the failures and successes so far, and provide insight into some future directions for research and clinical trials in order to exploit the ever-expanding molecular characterization of tumors with the drugs available and in the pipelines. In addition, we present a short overview of using differential cell cycle characteristics of normal and tumor cells as a way of protecting normal cells from cytotoxic chemotherapies. Finally we describe other potential targets such as regulating p27, inhibiting PIM and MELK kinases as well as some of the mitotic kinases.

**Keywords** Cell cycle • Cyclins • CDKs • Synthetic lethality • Mitosis • Combination therapy

## 14.1 Introduction

Hanahan and Weinberg recognized the importance of cell cycle and checkpoints in their original and updated "hallmarks of cancer" papers, which describe the key features that normal cells must acquire during transformation into a tumor [1, 2].

A. Alexander, Ph.D. • K. Keyomarsi, Ph.D. (✉)
Department of Experimental Radiation Oncology, Radiation Oncology, University of Texas, MD Anderson Cancer Center, 1515 Holcombe Blvd, Houston, TX 77030, USA
e-mail: kkeyomar@mdanderson.org

R. Kumar (ed.), *Nuclear Signaling Pathways and Targeting Transcription in Cancer*, Cancer Drug Discovery and Development, DOI 10.1007/978-1-4614-8039-6_14,

Cell cycle deregulation has long been appreciated as a fundamental early event during tumorigenesis, which contributes to several of these hallmarks, namely self-sufficiency in growth signals" and "insensitivity to anti-growth signals", and results in genomic instability, one of the newly added hallmarks. Since these alterations are almost universal among different tumor types, cancer biologists have expended considerable effort in interrogating these pathways as therapeutic targets for 20 years. In spite of the substantial body of literature focused on identifying the biological roles of many cell cycle pathway proteins in both normal development and in describing tumor-associated defects, the progress in the clinic has not been as rapid as desired. With this in mind, we felt that this chapter would be an ideal opportunity for us to review what is known about cell cycle deregulation in cancer, with a focus on personalized treatment strategies. Ultimately, we hope to suggest future directions for research and clinical trials to utilize the wealth of genomic knowledge we now have about cancer, and design more rational strategies likely to be effective in defined genetic contexts, as well as using the cell cycle as a means of protecting normal tissues from the chemotherapeutic insults. We will utilize a particularly promising strategy from our work as an example in a subsequent section.

## 14.2 Core Cell Cycle Proteins as Targets for Therapy

### *14.2.1 Cell Cycle Regulation*

Some of the best characterized cell cycle targets are key proteins that have been highly conserved throughout evolution in all eucaryotes. These include cyclins, cyclin-dependent kinases (CDKs) and cyclin-dependent kinase inhibitors (CDKi). Cyclins are master regulators of the cell cycle, via activating CDKs which in turn stimulate downstream signaling (Fig. 14.1).

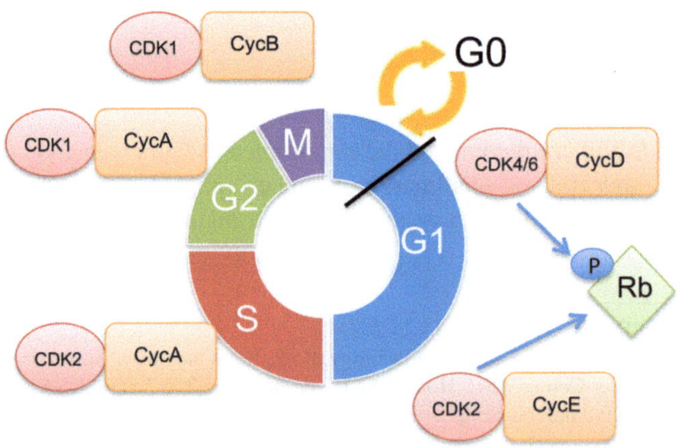


**Fig. 14.1** Cell cycle regulation by cyclins and CDKs

**Table 14.1** CDK knockout mouse phenotypes

| Gene | Phenotype | References |
|---|---|---|
| CDK1 | Not viable | [7] |
| CDK2 | Sterile but viable | [3, 4] |
| CDK3 | N/A – not active in most mouse strains | [210] |
| CDK4 | Impaired proliferation of pancreatic βcells, leading to diabetes | [5, 211] |
| CDK6 | Hematopoietic deficiency – anemia, thymic development delay | [6] |
| CDK4 and CDK6 | Embryonic lethal between E14.5 and E18.5, severe anemia | [6] |
| CDK2 and CDK6 | Viable but sterile and females are small | [6] |

Cell cycle progression is regulated by 4 major families of cyclins, cyclin D, E, A and B, and 4 respective CDKs (CDK4, 6, 2, or 1). Cyclin D is the first cyclin that is involved in the entry of cells from G0 into G1, in response to ample growth factors and other mitogens. Cyclin D exerts its activity via its catalytic partners CDK4 and CDK6, (and can also bind to CDK2 and CDK3) which phosphorylate many substrates. One of the most studied substrates of CDK4/6 is the retinoblastoma protein (Rb), which is a negative regulator of E2F transcriptional activity. Prior to phosphorylation by CDK4/6, Rb is in a hypophosphorylated state, and is bound to E2F and DP proteins, keeping E2F inactive. However CDK4/6 phosphorylation induces conformational change in Rb, releasing E2F to bind DNA and facilitate transcription. The next cyclins that are transcriptionally regulated are cyclin E and A. Cyclin E bound to CDK2 helps drive cells through G1/S transition by further phosphorylating Rb and other substrates involved in DNA replication such as cdc6. Later during S phase, CDK2 is also regulated by cyclin A levels. After DNA replication is complete, cells enter G2 phase where they prepare to enter mitosis by upregulating microtubule formation and other biosynthetic pathways necessary for chromosome segregation. Towards the end of G2 phase, CDK1 takes over as the predominant kinase, since cyclin B levels begin to rise and translocate to the nucleus to bind CDK1 to initiate the G2/M transition. This complex was first identified as the M-phase promoting factor since its main function is to break down the nuclear envelope and initiate prophase. Once mitosis is almost complete, CDK1 is deactivated via dephosphorylation, and a negative feedback loop is engaged via the anaphase-promoting complex which degrades cyclin B, allowing cells to exit mitosis.

The requirement for all of the cyclins and CDKs to control the cell cycle in normal cells as described in the previous paragraph has been recently challenged based on the findings from genetic studies in knockout mice. Tables 14.1 and 14.2 summarize the phenotypes observed in the knockout models. Each interphase CDK has been knocked out individually, and except for CDK1, all of the mice are viable. However, each mouse model has cell-type specific defects, which reveal tissue-specific roles for individual CDKs. For example, the CDK2-deficient model is viable (although born at slightly lower than predicted Mendelian ratio), but sterile due to an absolute requirement for CDK2 during meiosis in both male and female germ cells [3, 4]. Cell cycle analysis and proliferation rate of mouse embryonic fibroblasts from both CDK2 wild-type and knockout embryos showed no significant difference in cell cycle distribution, and a similar rate of proliferation for the first 4 days in culture, after which the knockout cells reached a plateau phase. Similarly

**Table 14.2** Cyclin knockout mouse phenotypes

| Gene | Phenotype | References |
|---|---|---|
| *D-type cyclins* | | |
| Cyclin D1 | Viable, but mice have neurological abnormalities, retinal hypoplasia, decreased body size, and during pregnancy only, impaired mammary epithelial cell proliferation | [212–215] |
| Cyclin D2 | Viable, but females are sterile, and males have decreased sperm counts and hypoplastic testes. Diabetes due to impaired pancreatic beta cell proliferation | [216, 217] |
| Cyclin D3 | Viable, but hypoplastic thymus | [218] |
| Cyclin D1 and Cyclin D2 | Viable until 3 weeks, but decreased body size and hypoplastic cerebellum | [219] |
| Cyclin D1 and Cyclin D3 | Some loss of viability by 4 weeks, most do not survive past 2 months due to respiratory failure and neurological defects | [219] |
| Cyclin D2 and Cyclin D3 | Embryonic lethal at E17.5-E18.5 due to megaloblastic anemia | [219] |
| Cyclin D1, Cyclin D2 and Cyclin D3 | Embryonic lethal at E16.5 due to megaloblastic anemia and defective fetal hematopoiesis | [220] |
| *E-type cyclins* | | |
| Cyclin E1 | Viable, no detected phenotype | [221, 222] |
| Cyclin E2 | Viable, males are infertile | [221, 222] |
| Cyclin E1 and Cyclin E2 | Embryonic lethal at E11.5. Placental failure due to lack of trophoblast-derived polyploid giant cells in the placenta | [221, 222] |
| *A-type cyclins* | | |
| Cyclin A1 | Viable, males are infertile | [223, 224] |
| Cyclin A2 | Embryonic lethal at implantation | [223] |
| *B-type cyclins* | | |
| Cyclin B1 | Embryonic lethal at E10.5, unknown reason | [225] |
| Cyclin B2 | Viable, no detected phenotype | [225] |

CDK4 and CDK6 are not necessary for cell cycle progression in most cells, although CDK4 is required for proliferation of pancreatic β-cells, leading to a diabetic phenotype in the knockout mouse [5]. Additionally, CDK6 is important in the hematopoietic system, both in the lymphocytes and erythrocytes. CDK6-deficient mice have small, less cellular thymi, since CDK6 is downstream of Notch and AKT signaling which is critical in early thymic T cell commitment to the T-cell lineage, and also have smaller spleens with less erythroid cells [6]. Double knockout of CDK4 and CDK6 induces late embryonic lethality, primarily due to the hematopoietic defects in erythroid cell production, however MEF cells from these embryos do proliferate and can become immortalized by continuous passage. CDK2 can partially compensate to phosphorylate Rb in these double-knockout cells by binding cyclin D, and therefore promote cell cycle progression, however this is not thought to fully explain the lack of cell cycle defects systemically. In stark contrast to the phenotypes seen in CDK2/4/6 knockout models, CDK1 deficiency causes cell cycle arrest and prevents embryos from developing beyond the 2-cell stage, demonstrating a lack of compensation between the mitotic CDK and the interphase CDKs [7].

Unlike normal cells that do not depend on any single cyclin or CDK for growth, there is beginning to be some evidence that in tumor cells, altered cellular wiring can lead to oncogenic addiction to CDK signaling. For example, in a mouse model of triple negative breast cancer driven by low-molecular weight cyclin E, tumors are highly dependent upon CDK2 signaling, even though CDK2 is dispensable in normal cells [8]. This type of variation on synthetic lethality involving CDK signaling is not limited to breast cancer but can be observed in a K-Ras mutant lung cancer model. In a K-Ras-driven mouse model of lung cancer, CDK2 and CDK6 knockout only partially inhibited tumor initiation, whereas CDK4 knockout significantly decreased tumorigenesis due to an immediate induction of senescence [9]. Even though the K-Ras-mutant transgene is expressed in several other epithelial tissues in this mouse model, none of these undergo hyperplasia or tumorigenesis, and senescence is not observed in these normal tissues. CDK4 was also shown to be essential for progression of established K-Ras driven NSCLC lesions, and pharmacological CDK4 inhibition significantly inhibited tumorigenesis. The reasons for lack of immediate compensation mechanisms involving other CDKs in tumors are not clear, but this phenomenon may allow us to turn this frequent observation into an Achilles heel in cancer cells if we carefully dissect true dependencies in well-planned genetic experiments.

## *14.2.2 CDK Inhibitors*

### 14.2.2.1 Pan-CDK Inhibitors

Since CDKs are the catalytically active drivers of cell cycle progression, targeting them pharmacologically has been a major effort. The early generation inhibitors, developed more than 15 years ago were pan-inhibitors targeting a large spectrum of CDKs. These drugs were somewhat promising based upon cell line and xenograft work, but when moved into early stage clinical studies failed to show considerable net benefit. The reasons for failure are likely multi-factorial, and include both biological issues as well clinical trial design flaws. In the forthcoming section we will describe the development of several generations of inhibitors and their related trials, and provide insights into future development of these classes of compounds. Figure 14.2 shows the structures of all the CDK inhibitors discussed in this section.

The most extensively tested pan-CDK inhibitor is flavopiridol, which inhibits all four of the interphase CDKs as well as CDK7 and is also the most potent known CDK9 inhibitor [10, 11]. CDK7 is both a cell cycle and a transcriptional CDK, since it is a part of the transcription factor IIH (TFIIH) complex with cyclin H and MAT [12, 13]. CDK7 promotes transcription elongation by phosphorylating the C-terminus of RNA polymerase II. CDK9 is also thought to be involved in transcriptional regulation, in complex with cyclin T, via phosphorylating different sites in the C-terminus of RNA polymerase II [14]. There is a profound response to flavopiridol in cells, that encompasses both cell cycle arrest in G1 and G2, but also transcriptional changes especially in mRNAs with short half lives such as early

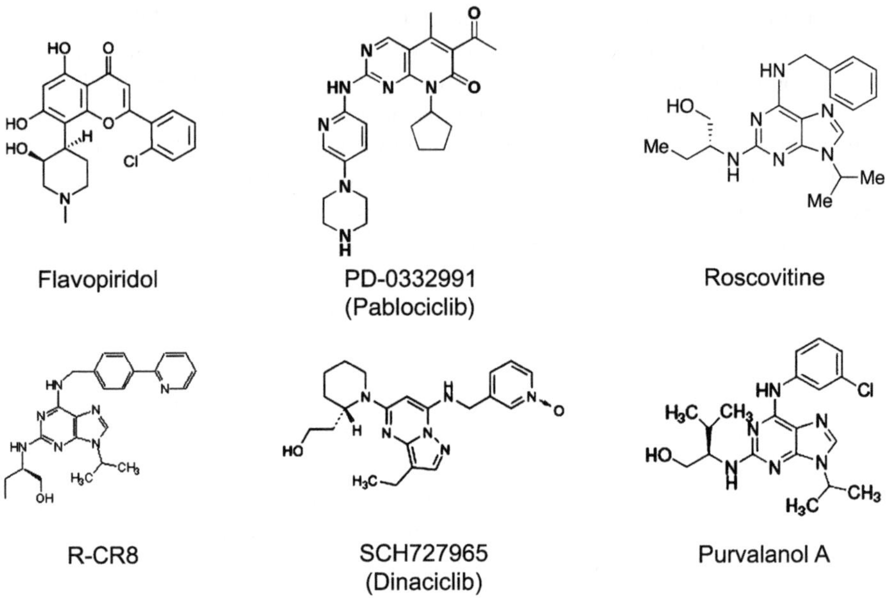


**Fig. 14.2** Structures of CDK inhibitors

response transcription factors, apoptosis regulating genes (like Mcl1) and NFκB responsive genes [11, 15, 16]. Whether these responses truly translated when this agent was tested in the clinic was not well studied.

Preclinical data had suggested that prolonged exposure to flavopiridol was necessary for maximal anti-tumor effect, so the two phase 1 trials that opened in 1994 used long infusions (72 h). The dose-limiting toxicities seen were primarily diarrhea, and at higher doses hypotension, anorexia and muscle weakness, and 21 % of patients had venous thromboses. Pharmaokinetic analysis of steady state plasma concentration revealed that 200–400nM was the range reached at the maximum tolerated dose [17]. The prior preclinical studies had found that for maximum activity, a higher concentration in the micromolar range would be desirable, so future studies attempted to reach these levels via bolus dosing on a 1 h per day for five consecutive days schedule. In these later trials, low micromolar peak concentrations were observed, and similar toxicities were observed [18]. However, when several phase 2 studies in solid tumors were analyzed, the enthusiasm for this agent waned, since no objective responses were seen in tumors ranging from melanoma to endometrial carcinoma [19–21]. In contrast, the results seen in hematopoietic malignancies appeared more promising [17]. For example, in chronic lymphocytic leukemia, 40 % of patients had partial responses, and the dose-limiting toxicity observed was tumor-lysis syndrome, indicative of strong anti-tumor activity of this agent [22].

In addition to studies using flavopiridol as a single agent, combination studies were pursued based on the hypothesis that flavopiridol may have benefit as a chemosensitizer. This hypothesis was generated based on the pre-clinical observation that

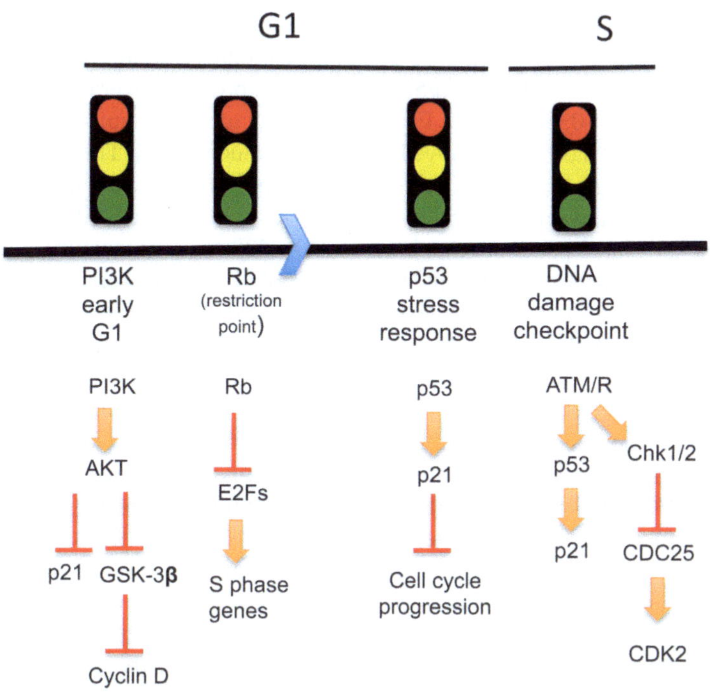


**Fig. 14.3** Key regulators of G1 to S phase transition

synchronizing the cells into S phase sensitized them to flavopiridol-induced cytotoxicity, resulting in E2F dependent cell death that is selective to transformed cells [23]. These studies used a variety of classes of cytotoxic drugs including platinum agents, anthracyclines, taxanes and 5-fluorouracil. These studies had more promising results, including a 30–40 % rate of partial responses in some studies [24–31].

In spite of some of these promising activities in both solid tumors and leukemias, recently, there has not been significant progress with this agent. The chemistry of the agent does have some challenges, since it binds to plasma proteins and also is poorly water soluble [32, 33]. There has been a novel liposomal formulation reported a few years ago, which aimed to improve the therapeutic index by slowly releasing the drug to effectively synchronize a large portion of the tumor cell population, while not being bound up in the circulation by plasma proteins [34]. As of writing, there have not been any clinical studies presented or registered using this formulation.

#### 14.2.2.2 Selective CDK4/6 Inhibitors

The G1-S checkpoint (see Fig. 14.3) is altered in close to 90 % of human tumors, by various mechanisms, indicating that this phenotype provides a selective advantage for proliferation and/or survival. With this in mind, selectively targeting CDK4 and

CDK6 has been considered as an alternative strategy in several diseases including breast cancer due to the prevalence of amplification/overexpression of cyclin D1 observed (15–20 % amplification/50–70 % overexpression overall) [35, 36]. One drug candidate, PD-0332991 (see Fig. 14.2 for structure), has quickly moved to the top of its class and has rapidly moved into clinical studies [37]. This compound was selected from a high-throughput screen of pyridopyrimidines, in which both potent anti-proliferative and selective inhibition of CDK4 were used as criteria [38]. When tested against a large panel of other kinases, PD-0332291 (Pablociclib) had a highly selectivity index towards CDK4 and CDK6 (IC50's 11/16nM, versus >8–10 μM for 36 other kinases tested). When tested in MDA-MB435 breast carcinoma cells, PD-0332991 induced a robust G1 arrest, and concomitant reduction of phosphorylation of Rb at the CDK4/6 phosphorylation sites (Ser 780–795). As expected, in Rb-negative cell lines, this compound has no activity, further demonstrating that its mechanism of action includes inhibition of CDK4/6 phosphorylation of Rb [39]. Another marker of resistance that has been identified is elevated p16 expression, since CDK4/6 is physically bound and unable to be inhibited [40].

Breast cancer is the model system which has been best studied so far in terms of understanding mechanism of action, potential synergistic combinations and predictors of resistance. At the molecular level, breast cancers can be divided into luminal or basal based on gene expression signatures. Most of the sensitive cell lines are luminal in nature, and all have intact Rb signaling, whereas the resistant cells tend to be basal-like and lack Rb activity [41, 42]. Intriguingly, basal cell lines, which retain Rb activity are still unresponsive to PD-0332991, and have hyperphosphorylated Rb. It is unclear at present what the precise mechanism is that drives hyperphosphorylation of Rb in these cell lines. It is possible that there is a greater dependence upon CDK2/CDK1, in which case these cells might respond to a combination of CDK2/1 and CDK4/6 inhibitors. Luminal tumors encompass both estrogen-receptor (ER) positive and many HER2-amplified tumors, so naturally combinations of ER antagonists or HER2 inhibiting drugs with CDK4/6 inhibitors were tested. In ER-positive cell lines, treatment with tamoxifen and PD-0332991 resulted in synergism and G1 arrest, and similarly trastuzumab and PD-0332991 are synergistic in HER2-amplified cell lines [41].

Apart from breast cancer cell lines, PD-0332991 has now been evaluated in a variety of other solid tumor types, including pancreatic neuroendocrine tumors, glioblastoma multiforme, rhabdomyosarcoma and mantle cell lymphoma with similar results [43–47]. In xenograft experiments, this drug is mostly cytostatic, with a few examples of cytotoxicity. In addition, PD-0332991 has been explored as a radiosensitizer in glioblastoma, due to its high penetrance of the blood-brain barrier, and preclinically appears to be useful in this scenario [47].

However not all tumor contexts are ideal candidates for such a strategy even if the underlying genetic changes would predict sensitivity. A recent paper described an unanticipated effect observed in pancreatic adenocarcinoma (PDAC). In PDAC cell lines examined, PD-0332991 had anti-proliferative activity, induced robust G1 arrest and hypophosphorylation of Rb [48]. However, gene expression analysis revealed that PD-0332991 upregulated genes involved in pro-angiogenic signaling, cell adhesion, cell migration/ECM remodeling, and inflammatory pathways. In addition, EMT was induced correlating with increased invasion via TGFβ-SMAD4 signaling, suggesting perhaps combinations of TGFβ inhibitors with CDK4/6 inhibitors might be a way forward in PDAC tumors expressing wild-type SMAD4. Genetic manipulation of CDK4/6 recapitulated this phenotype, ruling out a drug-mediated off-target kinase inhibition.

The first phase 1 study performed using PD-0332991 was recently published, and examined patients with Rb-positive advanced cancers [49]. This study showed that the drug was generally well tolerated, with the main toxicity being myelosuppression, consistent with other cell cycle targeted therapies. Pharmacokinetic analysis suggested favorable properties including slow absorption and elimination. The response rate was moderate (~27 %), however given all the usual caveats of generalizing based on phase 1 studies, the patients who derived some benefit (i.e. stable disease) could tolerate the drug well enough to remain on study for 10+ cycles. In breast cancer patients, a randomized phase I-II study utilizing PD-0332991 in combination with letrozole in ER-positive, HER2-normal post-menopausal patients has been completed (personal communication). In the phase 1 portion, there has been no biomarker selection, but in the phase II portion, the trial is specifically focused on patients with cyclin D1 amplification and/or loss of p16, since these are the patients predicted to respond best. So far, the clinical benefit rate in this combination trial was 70%, which resulted in statistically significant increase in progression-free survival and the adverse event profiles are very similar to what was reported in the single-agent phase 1 studies trials previously discussed i.e. this combination is generally well tolerated. The few patients so far in the study have been safely treated with some patients having partial responses. These well-designed trials with integrated biomarkers built in, are likely to provide more useful information not only about safety and pharmacokinetics but also pharmacogenomics information about responders and the biology behind responses. Caution must be taken though in considering combinations with chemotherapies that depend on actively cycling cells, based on a recent study that showed that PD-0332991 actually protected RB-positive MDA-MB-231 cells from paclitaxel-induced mitotic catastrophe [50]. Pre-treatment with PD-0322991 also resulted in a switch from homologous recombination (HR) DNA repair mechanism to non-homologous end-joining (NHEJ), which is an error prone pathway that could potentially induce further genomic instability in cancer cells. While this finding is intriguing and worthy of further mechanistic study in a broader panel of cell lines and normal mammary epithelial cells, it still remains to be determined how this result will be translated into ER+/HER2+ cell lines, which have been the focus of the majority of the breast cancer studies using this agent.

Going forward in hormone-receptor positive breast cancer, targeting CDK4/6 should eventually become a first line therapy in combination with endocrine therapy for a number of reasons. Certainly, as described above, the preclinical data regarding this combination is compelling, and this includes cell lines that acquired tamoxifen resistance being re-sensitized by PD-0332991. Secondly, genomic data have confirmed the relevance of this pathway in resistance to endocrine therapy alone, such as the fact that cyclin D1 is overexpressed/amplified in endocrine-therapy resistant tumors [51, 52]. In addition to deregulating the cell cycle, amplified cyclin

D can directly activate ER in a hormonally-independent manner that does not require CDK/Rb activity [53]. Thirdly, the fact that cyclin D1-CDK4 is downstream of multiple pathways that mediate resistance to anti-estrogens (e.g. EGFR/HER2, ERK, AKT, NFκB) may make this strategy useful regardless of which pathways are upregulated in any particular patient [54]. Despite being a good target however, resistance to CDK4/6 inhibition is likely to occur, since resistance arises to every targeted therapy tested so far. Indeed, there is evidence currently for activation of CDK2 due to p27 down regulation as a mechanism of resistance to these agents, which could potentially be targeted via CDK2 inhibitors as will be discussed in the next section [55].

Further clinical studies in other diseases are also underway, for example in mantle cell lymphoma. The single agent study showed some evidence of benefit, and now a subsequent study has been designed using PD-0332991 in combination with bortezomib in this patient subset.

#### 14.2.2.3 Selective CDK1/2 Inhibitors

The other subclass of CDK inhibitors that have been developed are more specific for CDK1 and CDK2 (versus CDK4/6), such as R-roscovitine (also known as CYC202 or seliciclib), SNS-032 and the newer agent SCH727965 (see Fig. 14.2 for the structures). Roscovitine, a 2,6,9-trisubstituted purine was generated in a screen of olomoucine-related analogues for CDK1/cyclin B inhibition, and found to potently inhibit CDK1 kinase activity (IC50 of 0.45 μM) [56]. Once olomoucine was shown to co-crystalize with CDK2, roscovitine was also confirmed to bind directly to CDK2 [57]. Several years later, roscovitine became the first orally bioavailable drug from this class to go into clinical trials based on the preclinical data showing that CDK1/2 inhibition causes S and G2 arrest followed by apoptosis in tumor cells [58–60]. Apart from the effects on these cell cycle CDKs, roscovitine also inhibits CDK7 and CDK9, thereby inhibiting transcription as well, via reducing key anti-apoptotic proteins such as Mcl1 [61–64].

Several studies have shown that apoptosis is further induced when CDK1/2 inhibitors are combined with most cytotoxic therapies including taxanes, anthracyclines as well as radiation. For example, a combination of purvalanol A and taxol caused profound apoptosis in Hela cells, when taxol was used first to stabilize microtubules then purvalanol A was added [60]. Intriguingly when the drugs were used in the reverse order, the response was decreased, demonstrating that synchronization of cells in mitosis (i.e. the end results of taxol) is important for the mechanism of CDK inhibitor-induced cell death. A similar synergistic combination strategy was demonstrated in MCF7 breast cancer xenografts using roscovitine and doxorubicin, however in this context, cell cycle synchronization in G2/M phase with roscovitine was used to prime the cells to respond to doxorubicin (versus the taxol→CDKi strategy in Hela cells) [65, 66]. Similar to the Hela cell study described above, the taxol-purvalanol A combination was found to be similarly effective in MCF7 xenografts [67]. These dichotomous results in two different systems just illustrate one of the challenges we have moving forward with sequential combination therapies that exploit mechanism of action of drugs. Likely a number of factors could contribute to which direction of treatment is likely to be best, including genomic factors (such as Rb status, p53 pathway status), timing of exposure to agents as well as which specific drugs under investigation. Clearly, further mechanistic work is still needed to dissect out these details in order to rationally match treatments to individual patients.

In spite of this incomplete understanding of mechanism of action of these agents in both solid tumors and hematopoietic malignancies, roscovitine was moved into clinical trials in the early 2000s [68]. The phase 1 studies demonstrated that this agent could be administered both in an intravenous formulation as well as orally [69, 70]. It has reasonable pharmacokinetic properties including high bioavailability, slow GI absorption, however it is rapidly metabolized to an inactive metabolite, making its half-life fairly short (~1 h). However, there were a number of dose-limiting side effects observed including liver and kidney toxicity, electrolyte disturbances, rashes and fatigue that accumulated over time, making repeated administration daily for more than 5 days too challenging for patients. Responses were unimpressive over a few phase 1 studies, with primarily stable disease and very few partial responses seen as monotherapy. Two phase 2 studies were undertaken in non-small cell lung cancer and nasopharngeal carcinoma which essentially replicated the results in the phase 1 studies [71]. Looking at the pharmacological and response data together, the researchers concluded that one of the major challenges was maintaining a plasma dose that is high enough for anti-tumor activity based on the preclinical studies, and even when white blood cells were used as a surrogate for tumor cells, Rb phosphorylation was not decreased, further supporting the claim of insufficient dose reaching tumor cells. One way of potentially overcoming this problem might be to use a more frequent dosing schedule such as 2–3 times a day, since the preclinical studies showed that 8–16 h of continued exposure is needed to effectively inhibit tumor growth. Whether this would actually work might not be known since the excitement regarding this agent has waned, in light of newer compounds that have been developed. One such potential compound is CR8, which is a N6-biaryl-substituted derivative of roscovitine that is 2–4 fold more potent at inhibiting CDK1, CDK2 and CDK7, which translated to 40–70 fold higher potency in cellular activity measures such as PARP cleavage and caspase activation [72–74]. Since discovery of this compound a few years ago, animal studies have not been published in cancer models, although a very recent paper utilizing CR8 in a mouse model of traumatic brain injury demonstrated that this drug could be delivered safely *in vivo*.

The other CDK inhibitor that targets CDK1 and CDK2 (as well as CDK5 and CDK9) that appears promising is SCH727965 (Dinaciclib) (see Fig. 14.2 for structure), a compound developed to address some of the issues with previous generation inhibitors with respect to therapeutic index [75]. Indeed, in direct comparison to flavopiridol this agent had > tenfold greater therapeutic index in the A2780 ovarian cancer xenograft model (defined as the ratio between MTD defined as 20 % body weight loss and minimal effective dose to cause 50 % inhibition of tumor growth when given i.p. once daily for 7 days). In addition, cell line studies showed that even a brief 2 h exposure to SCH727965 was sufficient to inhibit progression of cells into S phase. *In vivo*, this drug was at least as effective as paclitaxel in A2780 xenografts, and well tolerated with the main toxicity being reversible myelosuppression.

In addition to ovarian cancer, SCH727965 has been tested and found to be potentially effective in pancreatic cancer, melanoma and various forms of sarcoma including osteosarcoma [76–79]. The pancreatic study was particularly exciting, as it was performed using low-passage patient-derived xenografts (PDX) as opposed to cell line xenografts. These PDX models are thought to more faithfully recapitulate human tumorigenesis for multiple reasons including the fact that they maintain human stroma for multiple passages [80]. This intense desmoplastic stroma and hypovascular microenvironment which characterizes pancreatic cancer, is known to be a major barrier to chemotherapy drug access. Therefore the data showing efficacy in multiple mouse models with these characteristics bodes well for subsequent trials in humans. In addition to using these better disease models, the authors performed gene set enrichment analysis on the tumors to interrogate potential mechanisms of resistance, an area of research that is very undeveloped in the cell cycle field. They compared sensitive and resistant tumors, and found that in the most resistant tumors, the Notch and TGFβ pathways were upregulated, suggesting that perhaps combinations of these inhibitors may be future directions for research.

Preclinical studies in adult and pediatric leukemia are also underway, and a report of efficacy in CLL cells showed promise independent of high-risk genomic features (del 17p13.1 and gVHI unmutated). Short-term exposure of CLL cells directly taken from patients was sufficient to induce apoptosis [81]. Moreover, SCH727965 was shown to abrogate microenvironment-derived cytokine-induced survival signaling, in a PI3K-dependent mechanism, suggesting that a logical combination to explore might be SCH727965 in combination with PI3K inhibitors.

These pre-clinical studies provided strong rationale for moving this agent into clinical studies. A phase 1 study using SCH727965 dosed once every 3 weeks as a single agent in unselected adult patients has been performed which showed moderate responses (mainly stable disease) and similar to the preclinical studies, myelosuppression was the DLT [82]. Notably, unlike flavopiridol, there was no diarrhea, and much less fatigue, making this agent better tolerated. Since nausea and vomiting were common side effects as well, a subset of patients was given the anti-emetic drug aprepitant [83]. Aprepitant is known to weakly inhibit CYP3A4 which is one of the enzymes involved in metabolizing SCH727965, so the pharmacokinetic parameters were compared in patients treated with aprepitant with those not given it. These results showed no interaction between aprepitant and SCH727965, suggesting that use of this agent prophylactically in the clinic is a feasible and safe strategy moving forward to maximize use of SCH727965 in different patient populations.

Several phase 2 studies with SCH727965 as a single agent have also begun in both solid tumors and hematopoietic diseases, and some have been presented as abstracts at meetings. The first and most promising was performed in adult acute myeloid or lymphocytic leukemia patients and reported at ASH in 2010 [84]. The response rate was 60 %, and many patients had rapid decrease in their blast counts. Correlative studies including pharmacodynamics analyses showed that CDK activity (i.e. Rb phosphorylation and Mcl1 decreased expression) was effectively inhibited in the samples taken at 4 h post infusion, however these biomarkers returned almost to baseline by 24 h, suggesting a need for frequent dosing. A similar study is also underway in multiple myeloma, but no results are currently available [85].

With respect to solid tumors being examined in single-agent phase 2 trials, the progress has been slower. One single arm study has been reported in melanoma, which had ~72 patients enrolled [86]. The response profiles were very modest, with 22 % of patients with stable disease and no partial or complete responses, and toxicities were common. Another phase 1–2 study in unresectable melanoma is now approved and about to open. The only other trial that has been started is a multi-arm randomized phase 2 in breast cancer and NSCLC, in which SCH727965 is being compared to active treatments for each respective disease (oral capecitabine for breast cancer, and erlotinib for NSCLC). Importantly crossover from the control arm was allowed after disease progression, which is likely to make detection of a significant difference in overall survival extremely challenging. The study has been completed but no data has been presented as of writing (Nov 2012) [87].

Similar to the other CDK inhibitors, combinations with chemotherapy/other targeted agents are ultimately going to be necessary for optimal activity of SCH727965, and already a number of combination studies have been started. These trials include combinations with the PARP inhibitor Veliparib with or without carboplatin, rituximab in CLL, bortezomib and dexamethasone in myeloma, and our own trial with epirubicin in triple negative breast cancer [88–91]. Our trial differs from the others in that by limiting our patients to a specific subtype of breast cancer in which we have preclinical data showing dependence on CDK2 signaling because of LMW-E expression [92]. In addition, another group has shown elevated c-Myc expression is synthetically lethal with CDK inhibition in triple negative breast cancer [93]. By pre-selecting patients with a high likelihood of being oncogenically addicted to CDKs either via the LMW-E pathway or via amplification of c-Myc, we believe we will enrich for potential responders. Further biomarkers of response or markers predicting resistance will be needed in order to make this drug clinically useful. For example, basal like breast cancers or serous ovarian cancers with defects in DNA repair such as BRCA1/2 or ATM mutations may be more sensitive to CDK2 inhibition (in combination with chemotherapy), analogous to their propensity for sensitivity to PARP inhibition. BRCA1 or ATM knockdown sensitized various cell lines to CDK2 inhibitors, since CDK2 can regulate DNA repair independently of its effect on the cell cycle [94].

### *14.2.3 Targeting p27*

Moving away from directly targeting components of the cell cycle that promote growth, the other strategy that has been proposed is targeting negative regulators such as p27. p27 is a small nuclear protein involved in negatively regulating G1 to S phase transition via inhibiting cyclin E-CDK2, cyclin A-CDK2 and cyclin D-CDK4/6 activity [95]. However, in addition, it has novel functions when mislocalized in the cytoplasm that contribute to tumor cell survival and cancer progression [95]. For example, in response to stress, p27 is driven into the cytoplasm via AMPK-mediated phosphorylation at Thr198 which blocks apoptosis and induces a cytoprotective autophagy response [96, 97]. Other cytoplasmic functions also include increased invasion and metastatic potential, perhaps via binding to RhoA and stimulating changes in actin cytoskeleton formation [98]. A p27 knock-in mouse model where p27 has been mutated to be unable to bind cyclins and CDK2, was generated to determine whether there are cell-cycle independent roles for p27 [99]. One of the major findings from this model includes a role for cytoplasmic p27 in stem/progenitor cell enrichment, leading to lung tumorigenesis [100]. Taken together these results clearly show that p27 has pleiotropic functions and therefore merely upregulating it without understanding cellular context may not be very effective.

In many cancers p27 levels are decreased as a result of deregulation of transcriptional, translational or post-translational pathways. Much effort has been expended into understanding these mechanisms with the end goal being to determine ways of upregulating nuclear p27 to prevent cell cycle progression. To briefly summarize these studies, which have been reviewed extensively elsewhere, p27 transcriptional regulation is complex, and involves both repression via oncogenic transcription factors such as c-Myc and Id3, as well as activation via FOXO transcription factors and E2F1 [101]. In addition to transcriptional regulation, p27 has been shown to be regulated via miRNAs such as miR-221/222, which are overexpressed in some tumors [102]. p27 is also extensively regulated at a post-translational level, including multiple phosphorylation sites that dictate localization, as well as that phosphorylation sites that stabilize the protein via inhibiting ubiquitination. Some of these most important sites include Ser10, Tyr74/88/89, Thr157, Thr187, and Thr198.

Beyond phosphorylation, p27 localization can also be regulated via protein-protein interactions, such as Jab1, which binds to p27 and induces its nuclear export and degradation in the cytoplasm [103]. Jab1 is an interesting potential target for a number of reasons. Jab1 expression in tumors is inversely correlated with p27 levels and overexpression is correlated with poor prognosis [104–107]. This proto-oncogene has also been linked to radiation resistance and cisplatin-resistance due to inhibiting several apoptotic and DNA repair pathways [108]. Apart from p27, Jab1 can induce degradation of other tumor suppressors such as Smad4 and p53 as well as inhibit DNA repair via HR pathways that involve Rad51 [109–111].

p27 proteolysis via the proteasome is regulated by Skp2 [112]. Skp2 is part of a larger SCF complex, comprised of cullin1, Skp1, ROC1/Rbx1 and requires an adaptor protein called Cks1. Together, this complex functions as an E3-ligase that regulates a number of substrates including p27, p21, p57, FOXO1 and c-Myc [113–116]. Towards the end of G1 phase, Skp2 recognizes p27 phosphorylated at Thr187, which serves as a degradation signal, allowing cells to enter S phase. However, in cancer cells, Skp2 can be amplified and/or overexpressed, leading to decreased p27 levels constitutively [117–119]. As a result of this reciprocal relationship, the concept of targeting Skp2 has been proposed, however these studies have not progressed well. Genetic studies using a Skp2 deficient mouse model have demonstrated that acute inactivation of Skp2 in the context of Pten or Arf heterozygosity induced senescence but not in the Skp2-deficient mice without other oncogenic signals [120]. Senescence induction correlated with decreased tumorigenesis and p27 induction in the preneoplastic lesions that could be detected in some mice, providing additional rationale for the development of Skp2 inhibitors.

A high-throughput screen for inhibitors of Skp2 was performed using purified components of the complex and p27 as a substrate [121]. A compound designated CpdA was discovered to induce cell cycle arrest at low micromolar doses in multiple myeloma cells, and this led to caspase-independent cell death via autophagy. In addition, because low levels of p27 is associated with resistance to therapy in myeloma cells, CpdA was examined as a chemosensitizer, and was found to synergize with bortezomib, and overcome resistance to doxorubicin and melphalan. Despite this useful spectrum of activity however, further progress has not been made using this agent since it is difficult to make, and was not potent enough to use *in vivo* (required 5–10 μM dose in cell lines). More recently, another screen was performed using a chemical-genetics approach using automated microscopy to identify compounds that upregulate p27 in prostate cancer cell lines [122]. After several rounds of stringent validation, two compounds were identified – SMIP001 and SMIP004. These compounds are not broad proteasome inhibitors, but specifically target Skp2, resulting in elevated p27 and decreased CDK2 activity at low micromolar doses. At this point is it not known whether either of these compounds will be effective *in vivo*.

Targeting p27 presents a considerable challenge, despite the large body of knowledge regarding the mechanisms of its regulation and their redundancy. Some open questions include which of the upstream regulatory enzymes would be the best target to induce a sustained increase in p27 cells in tumor cells specifically. In some ways, the lack of complete specificity of substrates presents the largest conundrum. Assuming it would be possible to move one or more of the identified compounds (or a derivative) into the clinic, it is possible that some of the off-target effects may be undesirable e.g. in some contexts the upregulation of cyclin E or c-Myc may drive additional genomic instability. In fact there is data suggesting this might be the case using siRNA in A549 lung cancer cells targeting p27, Skp2 or the combination of both mRNAs [123]. In the dual-siRNA treated cells, there was an increase in centrosome number, abnormal mitoses/nuclear atypia, which could be attributed to an increase in both full length and low-molecular weight forms of cyclin E. In tumors, such as triple negative breast cancers, which already have LMW-E expression we propose that a Skp2-targeting strategy could be detrimental because of this off-target effect, unless combined with CDK inhibitors.

The other concern with Skp2 targeting relates to the specific genetic background of p27. As mentioned previously, p27 transcriptional silencing such as by miRNAs or promoter methylation are not uncommon events, and could co-exist with Skp2 overexpression. In this scenario even a potent Skp2 inhibitor would be ineffective at restoring p27 expression, with similar potential consequences as described above. In addition, because cyclin D-CDK complexes also bind p27, the level of p27 induction that may be necessary to slow down the cell cycle might be fairly high. Clearly, much work lies ahead in developing more potent inhibitors and understanding in greater depth the cellular contexts in which this strategy could be beneficial and which contexts to avoid.

## 14.3 Exploiting Normal and Cancer Cell Differences for Protection of Normal Tissue

One of the most challenging problems in cancer therapy involving cytotoxic chemotherapy is how to selectively kill tumor cells while sparing normal dividing cells. Previous work from our group has demonstrated two potential strategies that utilize cell cycle synchronization as a mechanism of selectively arresting normal cells. We proposed that using UCN-01, a staurosporine analog which was developed as a PKC inhibitor, but was later shown to inhibit other kinases including CDK1, CDK2 and CDK4 at low nanomolar concentrations could be used in this manner. UCN-01 has been found to induce a reversible G1 arrest in normal cells, while Rb-deficient tumor cells arrest in S phase instead [124]. Cytostatic doses of staurosporine (0.5–10nM) can also be used to arrest normal cells in G1 without any detectable effect on tumor cells [125]. Importantly, staurosporine priming does not compromise the ability for tumor cells to respond to cytotoxic therapies, while normal cells are arrested in G1 and therefore not responsive to chemotherapy that targets cycling cells.

More recently with the availability of specific CDK4/6 inhibitors that are more clinically relevant than staurosporine, this hypothesis has been revived and tested in mouse models using PD-0332991 [126]. Myelosuppression induced by platinum drugs and anthracyclines are one of the most life-threatening toxicities seen in cancer patients, due to both heightened risk of infection while immunosuppressed and also due to the subsequent chemotherapy delays or dose reduction, which can compromise treatment efficacy. A study was performed comparing two different mouse models of breast cancer, one with intact Rb signaling (the MMTV-neu model) and the other with inactive Rb (C3-Tag), to examine whether CDK4/6 inhibition could protect the hematopoietic progenitor cells from carboplatin-induced quiescence. The C3-Tag model which best resembles basal-like breast cancers, is Rb-deficient and, unsurprisingly, does not respond to CDK4/6 inhibition as a single agent. Carboplatin is highly active in basal-like breast cancer, though, and co-treatment with PD-0332991 did not protect tumor cells from death. However, these mice had reduced thrombocytopenia. In contrast, MMTV-neu mice, which have previously been shown to be dependent upon CDK4 and cyclin D, were sensitive to PD-0332991 as a single agent. In addition, when carboplatin was combined with PD-0332291, tumors grew back faster, indicating the CDK4/6 inhibition in this context protects tumor cells as well normal cells from toxicity, therefore not gaining any significant therapeutic index. These results suggest that CDK4/6 inhibitors may have a new utility – in tumors that are CDK4/6 sensitive, these drugs can be used for anti-tumor effect (and should be used separately from other cytotoxic therapies), and in other tumors that are insensitive (e.g. Rb deficient, p16 overexpressed), these drugs can be normal tissue protectors from other cytotoxic therapies. As a practical consideration as a result of these discoveries, we advocate for Rb mutation status to become one of the biomarkers tested routinely in the clinic.

## 14.4 Newer Cell Cycle-Related Targets for Therapy

In this last section, we briefly outline rationale for targeting other cell cycle-related proteins and review the state of drug development for each class of agent. Some of these proteins are intimately involved in mitosis regulation, and will be discussed separately.

### *14.4.1 PIM Family Kinases*

One of the most interesting emerging targets in the PIM family of kinases, which regulate multiple pathways including the cell cycle. The PIM family of serine/threonine kinases consists of three isoforms that have a significant degree of sequence homology, but differ in their tissue distribution [127]. Overall these proteins are expressed throughout in hematopoietic progenitors as well as liver, spleen and other epithelial and mesenchymal tissues, and have considerable functional redundancy. PIM1 is the best studied member of this family, and is thought to be the most widely relevant gene (of the 3 PIM isoforms) in cancer.

The PIM1 gene was identified in the 1980s as a frequent proviral integration site for Moloney murine-leukemia virus (MuLV) which induced T-cell lymphomas in transgenic mice [128]. Subsequently PIM1 was shown to cooperate with c-Myc in inducing lymphomas in utero or around birth, whereas Eμ-Myc transgenic mouse crossed onto a PIM1 and PIM2-deficient background had delayed lymphomagenesis [129–131]. More recently, PIM family kinases have been discovered to be overexpressed or mutated in other solid tumors such as prostate, pancreatic, ER-negative breast cancer and head and neck squamous carcinomas as well as many leukemias and lymphomas (AML and CLL), leading to the question of whether they could be targeted [132–137].

PIM kinases are unusual in that they are constitutively active, but are regulated largely at the transcriptional and translational level [127–138]. A wide range of cytokines and growth factors can activate PIM kinases, mainly via the JAK-STAT pathway and NFκB pathways [139, 140]. Since the mRNA has a short half-life, inhibitors of JAK-STAT could potentially be used to inhibit PIM signaling as well. An emerging paradigm places PIM1 at the center of a cellular stress response, since PIM1 can be induced by hypoxia and DNA damage via various mechanisms. For example hypoxia can induce PIM1 expression rapidly in a HIF1α independent mechanism, as well as induce nuclear translocation [141, 142]. PIM1 induction in response to hypoxia promotes cell survival via inhibition of apoptosis and is linked to chemoresistance under these conditions. PIM1 can also be induced in response to DNA damage by Kruppel-like factor 5 [143]. In a study of head and neck squamous carcinoma patients, upregulation of PIM1 in response to irradiation was shown to be associated with a poor response [144]. Since EGFR expression is also correlated

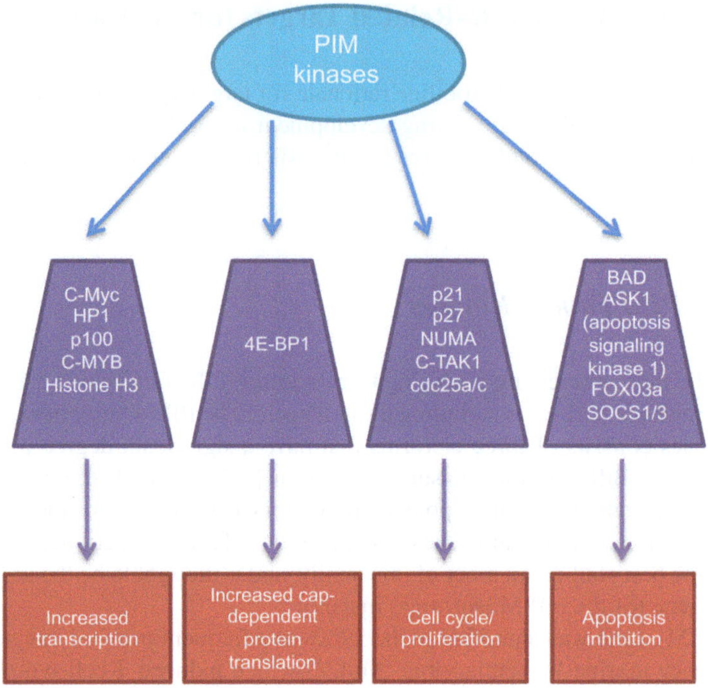


**Fig. 14.4** PIM1 kinase substrates and cellular functions

with radiation resistance and EGFR is autophosphorylated and nuclear localized after IR, the authors asked with EGFR can regulate PIM1 levels/activity. In cell lines, EGF-ligands induced PIM1 nuclear-translocation, and this effect could be blocked by the EGFR antibody cetuximab or tyrosine kinase inhibitor, gefitinib. Similarly, HNSCC cells that were irradiated had more nuclear PIM1, and PIM1 knockdown demonstrated the pro-survival role that PIM1 plays in this context. Taken together, these studies show that PIM1 may be a good target in a number of different cancer systems.

PIM1 plays a number of cellular functions that all contribute to tumorigenesis, as depicted in Fig. 14.4. One of these is regulation of the cell cycle via phosphorylating several substrates such as p21, p27, cdc25A, cdc25C and HP1 [145–151]. One of the most robust readouts of PIM1 activity is phosphorylation of the pro-apoptotic protein BAD at Ser 112, which inactivates it, therefore enhancing anti-apoptotic activity of Bcl2 [152]. As mentioned previously, PIM1/PIM2 cooperates with c-Myc to regulate lymphomagenesis, and one of the mechanisms it does so is via phosphorylation of c-Myc which stabilizes the protein [153]. Overexpression of PIM1 also induces genomic instability via deregulating the mitotic spindle checkpoint, which causes abnormal mitoses, centrosome amplification and aneuploidy [154]. In hematopoietic malignancies in which PIM2 is highly expressed, 4EBP1 is also a target that is involved in promoting cap-dependent translation initiation of

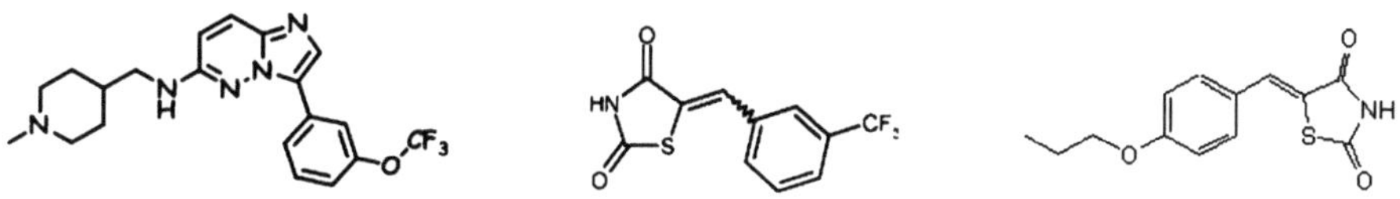


SGI-1776 SMI-4a SMI-16a

**Fig. 14.5** Structure of PIM1 kinase inhibitors

proteins that have growth promoting roles such as c-Myc and cyclin D1 [136]. In prostate cancer specifically, the androgen receptor is also a substrate of PIM1, and this phosphorylated form is transcriptionally inactivated and degraded [155, 156].

Structurally PIM kinases are distinct from other kinases in terms of how ATP binds to them, which has allowed chemists to design highly selective inhibitors. One of the most attractive features of PIM1 as a drug target is the lack of obvious phenotype in the knockout mouse, which is viable and fertile [157]. Compensation by other PIM family members is unlikely since compound knockout mice are also viable and fertile. The only phenotype that was observed in the $Pim1^{-/-}$ mouse is a subtle hematopoietic effect, such that the red blood cells are abnormally small but this did not lead to any physiological effects. When other potential hematopoietic functions were examined closely, it was found that bone-marrow-derived cells in culture had a significant impairment in IL-3 and IL-7 growth factor response [131].

The first compound that has been developed that has moved into cellular and *in vivo* studies is SGI-1776, which is an imidazo [1, 2-b] pyridazine compound that inhibits all three PIM kinases with IC50s of 7, 363 and 69nM (PIM1, PIM2 and PIM3 respectively), and has some activity against FLT3, another target in AML (see Fig. 14.5 for structure of this and other PIM inhibitors) [158]. In xenograft models of AML cells, this drug was highly active as an oral agent, inducing complete regression of blasts [159]. In addition, SGI-1776 can re-sensitize chemoresistant prostate cells to taxanes due to inhibiting multidrug resistance proteins including MDR1 [160]. Unfortunately when moved into phase 1 trials in humans (one trial was focused on prostate cancer and the other was non-Hodgkin's lymphoma) this drug was found to cause dose-limiting cardiac toxicity for reasons that are not clear, and the studies were stopped [161]. Another two structurally-related PIM1 inhibitors were identified in a chemical library screen called Smi-4a and Smi-16a which are benzylidene-thiazolidine-2, 4-diones [162]. When tested in vitro, these agents both had growth inhibitory activity in leukemia and prostate cancer cell lines, induced G1 arrest and induced p27 nuclear translocation. In addition, PIM1 inhibitors including Smi-4a synergize with both rapamycin in prostate cancer cells, and more recently with the Bcl2 inhibitor ABT-737 [163]. In our opinion despite these compounds having promising pre-clinical activity, the drug discovery market relating to PIM1 kinase is wide open right now. Special attention should be focused on the potential cardiac toxicity profile of future inhibitors (compared to SGI-1776), to try to understand what the off-target mechanisms are that underlie the Qt prolongation seen with SGI-1776 in patients.

### *14.4.2 Mitotic Kinases: Aurora Kinase Family*

In much of this chapter so far we have discussed targets that function early in the cell cycle in regulating G1 and S phases. However, G2 and M phases are also very kinase-rich and have tremendous potential as drug targets. Several classes of chemotherapies already target these processes, such as taxanes, which bind to tubulin and disrupts the assembly of the spindle. In order to design better therapies against proteins that act in G2 and M phase, we must understand their functions at a mechanistic level and how they contribute to the events that are necessary for progression through these stages. During G2 phase when cells are preparing for mitosis, the cell is very active in ensuring the DNA was replicated correctly, and dividing the other organelles. In addition, microtubule proteins are being synthesized in order to form the mitotic spindle along which the chromosomes will segregate during mitosis. G2-M phase targets include proteins that are involved in entering mitosis (such as Aurora kinase A), the spindle assembly checkpoint (such as BUB1), and mitotic exit (such as APC). Each of these proteins and processes could be the focus of entire chapters, so we will provide a high-level overview of each here, and point the readers to recent reviews on these proteins.

Aurora kinases, of which there are three highly related isoforms (A, B and C) in mammalian cells, are key regulators of mitosis that have been well conserved throughout eukaryotic organisms. All three isoforms have been the focus of drug development over the past several years, and most of the inhibitors target two or three of them due to the highly conserved catalytic domain [164]. Aurora kinase A (AURKA) and Aurora kinase B (AURKB) have the strongest evidence for a role in tumor cell growth, whereas Aurora kinase C has scant evidence. This may be because for many years Aurora kinase C was thought to be primarily expressed in the testes where it plays a role in spermatogenesis, by playing similar roles to AURKB [165]. However more recently it has also been found to be expressed at high levels in some cancer lines, and several point mutations have been found in lung tumors but relatively little is known about its function [166, 167].

AURKA is ubiquitously expressed and is the first of the family members to be activated starting in late S phase and working through to completion of mitosis. Many related processes are regulated by AURKA, including maturation and separation of the centrosomes, assembling the mitotic spindle, chromosome alignment and cytokinesis [168–170]. Regulation of AURKA levels is also important for mitotic exit, as either too much or too little activity leads to failure of cytokinesis and multinucleation [171]. AURKA regulation occurs at both transcriptional and posttranslational levels, including activation by autophosphorylation at Thr288 on the activation loop, and deactivation via protein phosphatase 1 [172]. In cancer, AURKA is frequently amplified/overexpressed due to various mechanisms, especially in higher grade tumors and has been demonstrated to be a poor prognostic factor [173–176]. It was established as a bona fide oncogene when it was shown to be capable of inducing rodent fibroblast cell transformation due to formation of multipolar mitotic spindles that induce genomic instability [177]. Interestingly, these chromosomal abnormalities that occur in AURKA overexpressing cells does not lead to cell death, because AURKA also promotes cell survival pathways including AKT-mTOR and nuclear accumulation of cyclin D1 [178]. NFκB is another anti-apoptotic pathway that is regulated by AURKA phosphorylation of its inhibitor, IκB [179]. AURKA has also been shown to interact with the p53 network, specifically via phosphorylating p53 inducing its degradation via MDM2, as well as phosphorylating p53 on Ser 215 which inhibits its DNA binding ability [180, 181]. These findings demonstrate the wide spectrum of roles that AURKA plays in cellular transformation and provide significant rationale for targeting this kinase.

Similar to Aurora kinase A, Aurora kinase B (AURKB) is also expressed in all proliferating cells, however it plays more limited roles as a chromosome passenger protein. AURKB is primarily expressed starting during prophase, where it is localized at the kinetochore to ensure correct chromosome alignment to the spindle and also helps ensure chromosomes segregate correctly [182]. In addition, AURKB phosphorylates Histone H3 at Ser10 and Ser28, which facilitates chromosome condensation [183, 184]. In cancer AURKB is not amplified, however it is still highly expressed in several tumor types [185–187]. Apart from regulating kinetochore-spindle interactions, in cancer cells, AURKB has been linked to degradation of p53 via phosphorylation at multiple sites [188], providing further rationale for targeting this kinase.

Inhibition of AURKA leads to G2 arrest, and has been shown to increase chemo- and radiosensitivity in cancer cells [189]. Quite a number of inhibitors have been generated by most of the major pharmaceutical companies and are currently being tested in early stage clinical trials. Many of these target both AURKA and AURKB, although Millennium has two AURKA specific compounds, MLN8054 and MLN8237. For a recent review with information about the clinical development of these agents, see [169]. Inhibition of AURKB are known as mitotic drivers, since they cause overriding of the mitotic checkpoints and results in aberrant mitosis and aneuploidy. This contrasts with AURKA inhibitors which block passage through mitosis. The question of whether inhibiting both AURKA and AURKB is better than either kinase alone has still not been answered. In preclinical genetic studies, the results have been equivocal. In one study in pancreatic cancer, antisense oligonucleotides to AURKA, AURKB or the combination were added to cells and responses compared [190]. The combination of both oligonucleotides was not better at inducing caspase activation, accumulating tetraploid cells or reducing formation in soft agar than either one alone. Targeting AURKA alone had a slightly better response overall versus AURKB alone, and this correlated with cells rounding up and detaching from the plate versus becoming large and multinucleated with the AURKB oligonucleotide. In contrast to this pancreatic study however, in colon cancer cells, AURKB inhibition was better than AURKA [191]. In order to move these targets forward, greater emphasis will have to be placed on understanding what contexts predict response to inhibition of each protein, and multiple readouts of each kinase inhibition should be analyzed since it is possible that each drug will have a slightly different profile. One clue that has already emerged is that p53-deficient cells more readily undergo apoptosis in response to the VX-680 inhibitor, however since this is a pan-aurora inhibitor, it is difficult to dissect out which target is most relevant in p53-deficient tumors [192]. Further molecular and pharmacodynamic characterization of sensitive and resistant patients in the many clinical trials in progress should elucidate more such factors, as well as more detailed preclinical work with patient derived xenograft models should be the *in vivo* assay of choice in these studies.

### *14.4.3 Other Mitotic Targets of Interest*

Moving forward as genomic studies are completed and more functional screens are performed it is likely more novel cell cycle targets will be found. Some examples of this nature that have been identified so far include MELK, Bub1, and Mps1. We will briefly summarize some of these proteins and how targeting them might be useful in cancer.

Maternal embryonic leucine zipper (MELK) is an atypical member of Snf1/AMPK family of kinases that has received only a little research attention so far. MELK is upregulated in several solid tumors including high-grade prostate cancer, astrocytoma, medulloblastoma and in breast cancer [193–195]. In addition, MELK is highly expressed in neural and breast cancer stem cells, making it a potentially attractive target to eradicate this population of cells thought to be the main drivers of drug resistance and eventual disease progression [196–198]. Expression of MELK is known to be increased in mitotically-arrested cells, and in prostate cancer cells is highly correlated with several other cell cycle/proliferation related genes including AURKB, cyclin B2 and DNA topoisomerase 2 alpha [193, 199, 200]. A few recent studies have suggested a role for MELK in radioresistance and chemoresistance, and have provided some in vitro evidence that knockdown can sensitize cancer cells to additional therapies [201, 202]. The only known pharmacological agent that targets MELK so far is the antibiotic siomycin A which reduces MELK expression and has been shown to decrease glioblastoma growth *in vivo* via targeting the neural stem cells [203].

The spindle assembly checkpoint (SAC) is a mechanism of delaying anaphase if the kinetochores are unattached to microtubules. There are at least 14 proteins involved in this process, four of which are kinases that are potentially targetable. These kinases are Bub1, BubR1, Mps1 and aurora B, although aurora B is dispensable for the checkpoint. If the SAC checkpoint is active, some of the components, such as Bub1, sequester Cdc20 which is the active part of the APC/C complex which degrades cyclin B. Bub1 may be a master regulator of the spindle assembly checkpoint by recruiting other important proteins involved such as BubR1, Mad1 and Mad2 [204]. The underlying concept behind targeting this checkpoint is that by preventing SAC activation, severe chromosome segregation occurs occur which causes cell death. Even partial inhibition of any of these essential mitotic checkpoint components can sensitize tumor cells to mitotic-targeting chemotherapies such as taxanes, whereas normal cells are not sensitized since normal cells can maintain a diploid population of cells [205]. Proof of this principle *in vitro* has been obtained for inhibitors of Mps1 [206, 207].

Cdc20 has also been proposed as a target in cancer due to the strong phenotypes seen in genetic studies from blocking mitotic exit. The cdc20 homozygous knockout mouse is embryonic lethal at the two-cell stage due to a metaphase arrest [208]. When an inducible knockout model was generated, a similar phenotype could be observed upon induction, and very high levels of cyclin B was observed in the cell, consistent with a defect in APC/C function [209]. When tumors of either epithelial or mesenchymal origin were induced in this model, and then cdc20 knockout was induced, the tumors rapidly regressed due to mitotic arrest and apoptosis.

In summary, there are various strategies that are being investigated to interfere with mitosis including delaying mitotic entry and spindle formation, preventing activation of the spindle assembly checkpoint or targeting mitotic exit via the APC/C-cdc20 complex. Such strategies may synergize with current chemotherapies that act in mitosis such as taxanes and vinca alkaloids, allowing lower doses of these agents to be administered. The question remains however whether a sufficient therapeutic index can be reached since normal cells also require these processes to be intact to undergo normal mitosis.

## References


1. Hanahan D, Weinberg RA (2000) The hallmarks of cancer. Cell 100(1):57–70, PubMed PMID: 10647931. eng
2. Hanahan D, Weinberg RA (2011) Hallmarks of cancer: the next generation. Cell 144(5):646–674, PubMed PMID: 21376230. eng
3. Berthet C, Aleem E, Coppola V, Tessarollo L, Kaldis P (2003) Cdk2 knockout mice are viable. Curr Biol 13(20):1775–1785, PubMed PMID: 14561402. eng
4. Ortega S, Prieto I, Odajima J, Martín A, Dubus P, Sotillo R et al (2003) Cyclin-dependent kinase 2 is essential for meiosis but not for mitotic cell division in mice. Nat Genet 35(1):25–31, PubMed PMID: 12923533. eng
5. Rane SG, Dubus P, Mettus RV, Galbreath EJ, Boden G, Reddy EP et al (1999) Loss of Cdk4 expression causes insulin-deficient diabetes and Cdk4 activation results in beta-islet cell hyperplasia. Nat Genet 22(1):44–52, PubMed PMID: 10319860. eng
6. Malumbres M, Sotillo R, Santamaría D, Galán J, Cerezo A, Ortega S et al (2004) Mammalian cells cycle without the D-type cyclin-dependent kinases Cdk4 and Cdk6. Cell 118(4):493–504, PubMed PMID: 15315761. eng
7. Malumbres M, Barbacid M (2005) Mammalian cyclin-dependent kinases. Trends Biochem Sci 30(11):630–641, PubMed PMID: 16236519. eng
8. Akli S, Van Pelt CS, Bui T, Meijer L, Keyomarsi K (2011) Cdk2 is required for breast cancer mediated by the low-molecular-weight isoform of cyclin E. Cancer Res 71(9):3377–3386, PubMed PMID: 21385896. Pubmed Central PMCID: PMC3085722. eng
9. Puyol M, Martín A, Dubus P, Mulero F, Pizcueta P, Khan G et al (2010) A synthetic lethal interaction between K-Ras oncogenes and Cdk4 unveils a therapeutic strategy for non-small cell lung carcinoma. Cancer Cell 18(1):63–73, PubMed PMID: 20609353. eng
10. Chao SH, Price DH (2001) Flavopiridol inactivates P-TEFb and blocks most RNA polymerase II transcription in vivo. J Biol Chem 276(34):31793–31799, PubMed PMID: 11431468. eng
11. Sedlacek H, Czech J, Naik R, Kaur G, Worland P, Losiewicz M et al (1996) Flavopiridol (L86 8275; NSC 649890), a new kinase inhibitor for tumor therapy. Int J Oncol 9(6):1143–1168, PubMed PMID: 21541623. eng

12. Fisher RP, Morgan DO (1994) A novel cyclin associates with MO15/CDK7 to form the CDK-activating kinase. Cell 78(4):713–724, PubMed PMID: 8069918. eng
13. Roy R, Adamczewski JP, Seroz T, Vermeulen W, Tassan JP, Schaeffer L et al (1994) The MO15 cell cycle kinase is associated with the TFIIH transcription-DNA repair factor. Cell 79(6):1093–1101, PubMed PMID: 8001135. eng
14. Kim YK, Bourgeois CF, Isel C, Churcher MJ, Karn J (2002) Phosphorylation of the RNA polymerase II carboxyl-terminal domain by CDK9 is directly responsible for human immunodeficiency virus type 1 Tat-activated transcriptional elongation. Mol Cell Biol 22(13):4622–4637, PubMed PMID: 12052871. Pubmed Central PMCID: PMC133925. eng
15. Lam LT, Pickeral OK, Peng AC, Rosenwald A, Hurt EM, Giltnane JM et al (2001) Genomic-scale measurement of mRNA turnover and the mechanisms of action of the anti-cancer drug flavopiridol. Genome Biol 2(10): RESEARCH0041 PubMed PMID: 11597333. Pubmed Central PMCID: PMC57796. eng http://genomebiology.com/content/pdf/gb-2001-2-10-research0041.pdf
16. Garriga J, Graña X (2004) Cellular control of gene expression by T-type cyclin/CDK9 complexes. Gene 337:15–23, PubMed PMID: 15276198. eng
17. Byrd JC, Lin TS, Dalton JT, Wu D, Phelps MA, Fischer B et al (2007) Flavopiridol administered using a pharmacologically derived schedule is associated with marked clinical efficacy in refractory, genetically high-risk chronic lymphocytic leukemia. Blood 109(2):399–404, PubMed PMID: 17003373. Pubmed Central PMCID: PMC1785084. eng
18. Tan AR, Headlee D, Messmann R, Sausville EA, Arbuck SG, Murgo AJ et al (2002) Phase I clinical and pharmacokinetic study of flavopiridol administered as a daily 1-hour infusion in patients with advanced neoplasms. J Clin Oncol 20(19):4074–4082, PubMed PMID: 12351605. eng
19. Burdette-Radoux S, Tozer RG, Lohmann RC, Quirt I, Ernst DS, Walsh W et al (2004) Phase II trial of flavopiridol, a cyclin dependent kinase inhibitor, in untreated metastatic malignant melanoma. Invest New Drugs 22(3):315–322, PubMed PMID: 15122079. eng
20. Grendys EC, Blessing JA, Burger R, Hoffman J (2005) A phase II evaluation of flavopiridol as second-line chemotherapy of endometrial carcinoma: a gynecologic oncology group study. Gynecol Oncol 98(2):249–253, PubMed PMID: 15978659. eng
21. Dispenzieri A, Gertz MA, Lacy MQ, Geyer SM, Fitch TR, Fenton RG et al (2006) Flavopiridol in patients with relapsed or refractory multiple myeloma: a phase 2 trial with clinical and pharmacodynamic end-points. Haematologica 91(3):390–393, PubMed PMID: 16503551. eng
22. Phelps MA, Lin TS, Johnson AJ, Hurh E, Rozewski DM, Farley KL et al (2009) Clinical response and pharmacokinetics from a phase 1 study of an active dosing schedule of flavopiridol in relapsed chronic lymphocytic leukemia. Blood 113(12):2637–2645, PubMed PMID: 18981292. Pubmed Central PMCID: PMC2661854. eng
23. Matranga CB, Shapiro GI (2002) Selective sensitization of transformed cells to flavopiridol-induced apoptosis following recruitment to S-phase. Cancer Res 62(6):1707–1717, PubMed PMID: 11912144. eng
24. Shah MA, Kortmansky J, Motwani M, Drobnjak M, Gonen M, Yi S et al (2005) A phase I clinical trial of the sequential combination of irinotecan followed by flavopiridol. Clin Cancer Res 11(10):3836–3845, PubMed PMID: 15897584. eng
25. Rathkopf D, Dickson MA, Feldman DR, Carvajal RD, Shah MA, Wu N et al (2009) Phase I study of flavopiridol with oxaliplatin and fluorouracil/leucovorin in advanced solid tumors. Clin Cancer Res 15(23):7405–7411, PubMed PMID: 19934304. Pubmed Central PMCID: PMC2787644. eng
26. George S, Kasimis BS, Cogswell J, Schwarzenberger P, Shapiro GI, Fidias P et al (2008) Phase I study of flavopiridol in combination with paclitaxel and carboplatin in patients with non-small-cell lung cancer. Clin Lung Cancer 9(3):160–165, PubMed PMID: 18621626. eng
27. El-Rayes BF, Gadgeel S, Parchment R, Lorusso P, Philip PA (2006) A phase I study of flavopiridol and docetaxel. Invest New Drugs 24(4):305–310, PubMed PMID: 16683073. eng
28. Fornier MN, Rathkopf D, Shah M, Patil S, O'Reilly E, Tse AN et al (2007) Phase I dose-finding study of weekly docetaxel followed by flavopiridol for patients with advanced solid tumors. Clin Cancer Res 13(19):5841–5846, PubMed PMID: 17908977. eng

29. Bible KC, Lensing JL, Nelson SA, Lee YK, Reid JM, Ames MM et al (2005) Phase 1 trial of flavopiridol combined with cisplatin or carboplatin in patients with advanced malignancies with the assessment of pharmacokinetic and pharmacodynamic end points. Clin Cancer Res 11(16):5935–5941, PubMed PMID: 16115936. eng
30. Karp JE, Passaniti A, Gojo I, Kaufmann S, Bible K, Garimella TS et al (2005) Phase I and pharmacokinetic study of flavopiridol followed by 1-beta-D-arabinofuranosylcytosine and mitoxantrone in relapsed and refractory adult acute leukemias. Clin Cancer Res 11(23): 8403–8412, PubMed PMID: 16322302. eng
31. Bible KC, Peethambaram PP, Oberg AL, Maples W, Groteluschen DL, Boente M et al (2012) A phase 2 trial of flavopiridol (alvocidib) and cisplatin in platin-resistant ovarian and primary peritoneal carcinoma: MC0261. Gynecol Oncol 127(1):55–62, PubMed PMID: 22664059. eng
32. Rudek MA, Bauer KS, Lush RM, Stinson SF, Senderowicz AM, Headlee DJ et al (2003) Clinical pharmacology of flavopiridol following a 72-hour continuous infusion. Ann Pharmacother 37(10):1369–1374, PubMed PMID: 14519054. eng
33. Li P, Tabibi SE, Yalkowsky SH (1999) Solubilization of flavopiridol by pH control combined with cosolvents, surfactants, or complexants. J Pharm Sci 88(9):945–947, PubMed PMID: 10479359. eng
34. Yang X, Zhao X, Phelps MA, Piao L, Rozewski DM, Liu Q et al (2009) A novel liposomal formulation of flavopiridol. Int J Pharm 365(1-2):170–174, PubMed PMID: 18778761. Pubmed Central PMCID: PMC3035394. eng
35. Santarius T, Shipley J, Brewer D, Stratton MR, Cooper CS (2010) A census of amplified and overexpressed human cancer genes. Nat Rev Cancer 10(1):59–64, PubMed PMID: 20029424. eng
36. Arnold A, Papanikolaou A (2005) Cyclin D1 in breast cancer pathogenesis. J Clin Oncol 23(18):4215–4224, PubMed PMID: 15961768. eng
37. Fry DW, Harvey PJ, Keller PR, Elliott WL, Meade M, Trachet E et al (2004) Specific inhibition of cyclin-dependent kinase 4/6 by PD 0332991 and associated antitumor activity in human tumor xenografts. Mol Cancer Ther 3(11):1427–1438, PubMed PMID: 15542782. eng
38. Toogood PL, Harvey PJ, Repine JT, Sheehan DJ, VanderWel SN, Zhou H et al (2005) Discovery of a potent and selective inhibitor of cyclin-dependent kinase 4/6. J Med Chem 48(7):2388–2406, PubMed PMID: 15801831. eng
39. Dean JL, Thangavel C, McClendon AK, Reed CA, Knudsen ES (2010) Therapeutic CDK4/6 inhibition in breast cancer: key mechanisms of response and failure. Oncogene 29(28):4018–4032, PubMed PMID: 20473330. eng
40. Ertel A, Dean JL, Rui H, Liu C, Witkiewicz AK, Knudsen KE et al (2010) RB-pathway disruption in breast cancer: differential association with disease subtypes, disease-specific prognosis and therapeutic response. Cell Cycle 9(20):4153–4163, PubMed PMID: 20948315. Pubmed Central PMCID: PMC3055199. eng
41. Finn RS, Dering J, Conklin D, Kalous O, Cohen DJ, Desai AJ et al (2009) PD 0332991, a selective cyclin D kinase 4/6 inhibitor, preferentially inhibits proliferation of luminal estrogen receptor-positive human breast cancer cell lines in vitro. Breast Cancer Res 11(5):R77 PubMed PMID: 19874578. Pubmed Central PMCID: PMC2790859. Eng
42. Dean JL, McClendon AK, Hickey TE, Butler LM, Tilley WD, Witkiewicz AK et al (2012) Therapeutic response to CDK4/6 inhibition in breast cancer defined by ex vivo analyses of human tumors. Cell Cycle 11(14):2756–2761, PubMed PMID: 22767154. Pubmed Central PMCID: PMC3409015. eng
43. Tang LH, Contractor T, Clausen R, Klimstra DS, Du YC, Allen PJ et al (2012) Attenuation of the retinoblastoma pathway in pancreatic neuroendocrine tumors due to increased cdk4/cdk6. Clin Cancer Res 18(17):4612–4620, PubMed PMID: 22761470. eng
44. Cen L, Carlson BL, Schroeder MA, Ostrem JL, Kitange GJ, Mladek AC et al (2012) p16-Cdk4-Rb axis controls sensitivity to a cyclin-dependent kinase inhibitor PD0332991 in glioblastoma xenograft cells. Neuro Oncol 14(7):870–881, PubMed PMID: 22711607. Pubmed Central PMCID: PMC3379801. eng
45. Saab R, Bills JL, Miceli AP, Anderson CM, Khoury JD, Fry DW et al (2006) Pharmacologic inhibition of cyclin-dependent kinase 4/6 activity arrests proliferation in myoblasts and

rhabdomyosarcoma-derived cells. Mol Cancer Ther 5(5):1299–1308, PubMed PMID: 16731763. eng
46. Leonard JP, LaCasce AS, Smith MR, Noy A, Chirieac LR, Rodig SJ et al (2012) Selective CDK4/6 inhibition with tumor responses by PD0332991 in patients with mantle cell lymphoma. Blood 119(20):4597–4607, PubMed PMID: 22383795. eng
47. Michaud K, Solomon DA, Oermann E, Kim JS, Zhong WZ, Prados MD et al (2010) Pharmacologic inhibition of cyclin-dependent kinases 4 and 6 arrests the growth of glioblastoma multiforme intracranial xenografts. Cancer Res 70(8):3228–3238, PubMed PMID: 20354191. Pubmed Central PMCID: PMC2855904. eng
48. Liu F, Korc M (2012) Cdk4/6 inhibition induces epithelial-mesenchymal transition and enhances invasiveness in pancreatic cancer cells. Mol Cancer Ther 11(10):2138–2148, PubMed PMID: 22869556. eng
49. Flaherty KT, Lorusso PM, Demichele A, Abramson VG, Courtney R, Randolph SS et al (2012) Phase I, dose-escalation trial of the oral cyclin-dependent kinase 4/6 inhibitor PD 0332991, administered using a 21-day schedule in patients with advanced cancer. Clin Cancer Res 18(2):568–576, PubMed PMID: 22090362. eng
50. Dean JL, McClendon AK, Knudsen ES (2012) Modification of the DNA damage response by therapeutic CDK4/6 inhibition. J Biol Chem 287(34):29075–29087, PubMed PMID: 22733811. Pubmed Central PMCID: PMC3436568. eng
51. Zwart W, Rondaij M, Jalink K, Sharp ZD, Mancini MA, Neefjes J et al (2009) Resistance to antiestrogen arzoxifene is mediated by overexpression of cyclin D1. Mol Endocrinol 23(9):1335–1345, PubMed PMID: 19477949. Pubmed Central PMCID: PMC2737554. eng
52. Kilker RL, Planas-Silva MD (2006) Cyclin D1 is necessary for tamoxifen-induced cell cycle progression in human breast cancer cells. Cancer Res 66(23):11478–11484, PubMed PMID: 17145896. eng
53. Zwijsen RM, Wientjens E, Klompmaker R, van der Sman J, Bernards R, Michalides RJ (1997) CDK-independent activation of estrogen receptor by cyclin D1. Cell 88(3):405–415, PubMed PMID: 9039267. eng
54. Musgrove EA, Sutherland RL (2009) Biological determinants of endocrine resistance in breast cancer. Nat Rev Cancer 9(9):631–643, PubMed PMID: 19701242. eng
55. Wang L, Wang J, Blaser BW, Duchemin AM, Kusewitt DF, Liu T et al (2007) Pharmacologic inhibition of CDK4/6: mechanistic evidence for selective activity or acquired resistance in acute myeloid leukemia. Blood 110(6):2075–2083, PubMed PMID: 17537993. eng
56. Meijer L, Raymond E (2003) Roscovitine and other purines as kinase inhibitors. From starfish oocytes to clinical trials. Acc Chem Res 36(6):417–425, PubMed PMID: 12809528. eng
57. De Azevedo WF, Leclerc S, Meijer L, Havlicek L, Strnad M, Kim SH (1997) Inhibition of cyclin-dependent kinases by purine analogues: crystal structure of human cdk2 complexed with roscovitine. Eur J Biochem 243(1-2):518–526, PubMed PMID: 9030780. eng
58. Raynaud FI, Whittaker SR, Fischer PM, McClue S, Walton MI, Barrie SE et al (2005) In vitro and in vivo pharmacokinetic-pharmacodynamic relationships for the trisubstituted aminopurine cyclin-dependent kinase inhibitors olomoucine, bohemine and CYC202. Clin Cancer Res 11(13):4875–4887, PubMed PMID: 16000586. eng
59. Wesierska-Gadek J, Gueorguieva M, Horky M (2003) Dual action of cyclin-dependent kinase inhibitors: induction of cell cycle arrest and apoptosis. A comparison of the effects exerted by roscovitine and cisplatin. Pol J Pharmacol 55(5):895–902, PubMed PMID: 14704484. eng
60. Whittaker SR, Walton MI, Garrett MD, Workman P (2004) The cyclin-dependent kinase inhibitor CYC202 (R-roscovitine) inhibits retinoblastoma protein phosphorylation, causes loss of cyclin D1, and activates the mitogen-activated protein kinase pathway. Cancer Res 64(1):262–272, PubMed PMID: 14729633. eng
61. McClue SJ, Blake D, Clarke R, Cowan A, Cummings L, Fischer PM et al (2002) In vitro and in vivo antitumor properties of the cyclin dependent kinase inhibitor CYC202 (R-roscovitine). Int J Cancer 102(5):463–468, PubMed PMID: 12432547. eng
62. Wang D, de la Fuente C, Deng L, Wang L, Zilberman I, Eadie C et al (2001) Inhibition of human immunodeficiency virus type 1 transcription by chemical cyclin-dependent kinase

inhibitors. J Virol 75(16):7266–7279, PubMed PMID: 11461999. Pubmed Central PMCID: PMC114962. eng

63. MacCallum DE, Melville J, Frame S, Watt K, Anderson S, Gianella-Borradori A et al (2005) Seliciclib (CYC202, R-roscovitine) induces cell death in multiple myeloma cells by inhibition of RNA polymerase II-dependent transcription and down-regulation of Mcl-1. Cancer Res 65(12):5399–5407, PubMed PMID: 15958589. eng
64. Fischer PM, Gianella-Borradori A (2003) CDK inhibitors in clinical development for the treatment of cancer. Expert Opin Investig Drugs 12(6):955–970, PubMed PMID: 12783600. eng
65. Appleyard MV, O'Neill MA, Murray KE, Paulin FE, Bray SE, Kernohan NM et al (2009) Seliciclib (CYC202, R-roscovitine) enhances the antitumor effect of doxorubicin in vivo in a breast cancer xenograft model. Int J Cancer 124(2):465–472, PubMed PMID: 19003963. eng
66. Nanos-Webb A, Jabbour NA, Multani AS, Wingate H, Oumata N, Galons H et al (2012) Targeting low molecular weight cyclin E (LMW-E) in breast cancer. Breast Cancer Res Treat 132(2):575–588, PubMed PMID: 21695458. eng
67. O'Connor DS, Wall NR, Porter AC, Altieri DC (2002) A p34(cdc2) survival checkpoint in cancer. Cancer Cell 2(1):43–54, PubMed PMID: 12150824. eng
68. Guzi T (2004) CYC-202 cyclacel. Curr Opin Investig Drugs 5(12):1311–1318, PubMed PMID: 15648953. eng
69. Benson C, White J, De Bono J, O'Donnell A, Raynaud F, Cruickshank C et al (2007) A phase I trial of the selective oral cyclin-dependent kinase inhibitor seliciclib (CYC202; R-roscovitine), administered twice daily for 7 days every 21 days. Br J Cancer 96(1):29–37, PubMed PMID: 17179992. Pubmed Central PMCID: PMC2360206. eng
70. Le Tourneau C, Faivre S, Laurence V, Delbaldo C, Vera K, Girre V et al (2010) Phase I evaluation of seliciclib (R-roscovitine), a novel oral cyclin-dependent kinase inhibitor, in patients with advanced malignancies. Eur J Cancer 46(18):3243–3250, PubMed PMID: 20822897. eng
71. Hsieh WS, Soo R, Peh BK, Loh T, Dong D, Soh D et al (2009) Pharmacodynamic effects of seliciclib, an orally administered cell cycle modulator, in undifferentiated nasopharyngeal cancer. Clin Cancer Res 15(4):1435–1442, PubMed PMID: 19228744. eng
72. Bettayeb K, Oumata N, Echalier A, Ferandin Y, Endicott JA, Galons H et al (2008) CR8, a potent and selective, roscovitine-derived inhibitor of cyclin-dependent kinases. Oncogene 27(44):5797–5807, PubMed PMID: 18574471. eng
73. Bettayeb K, Baunbæk D, Delehouze C, Loaëc N, Hole AJ, Baumli S et al (2010) CDK inhibitors roscovitine and CR8 trigger Mcl-1 down-regulation and apoptotic cell death in neuroblastoma cells. Genes Cancer 1(4):369–380, PubMed PMID: 21779453. Pubmed Central PMCID: PMC3092200. eng
74. Kabadi SV, Stoica BA, Hanscom M, Loane DJ, Kharebava G, Murray Ii MG et al (2012) CR8, a selective and potent CDK inhibitor, provides neuroprotection in experimental traumatic brain injury. Neurotherapeutics 9(2):405–421, PubMed PMID: 22167461. Pubmed Central PMCID: PMC3324621. eng
75. Parry D, Guzi T, Shanahan F, Davis N, Prabhavalkar D, Wiswell D et al (2010) Dinaciclib (SCH 727965), a novel and potent cyclin-dependent kinase inhibitor. Mol Cancer Ther 9(8):2344–2353, PubMed PMID: 20663931. eng
76. Feldmann G, Mishra A, Bisht S, Karikari C, Garrido-Laguna I, Rasheed Z et al (2011) Cyclin-dependent kinase inhibitor dinaciclib (SCH727965) inhibits pancreatic cancer growth and progression in murine xenograft models. Cancer Biol Ther 12(7):598–609, PubMed PMID: 21768779. Pubmed Central PMCID: PMC3218385. eng
77. Abdullah C, Wang X, Becker D (2011) Expression analysis and molecular targeting of cyclin-dependent kinases in advanced melanoma. Cell Cycle 10(6):977–988, PubMed PMID: 21358262. Pubmed Central PMCID: PMC3100877. eng
78. Fu W, Ma L, Chu B, Wang X, Bui MM, Gemmer J et al (2011) The cyclin-dependent kinase inhibitor SCH 727965 (dinacliclib) induces the apoptosis of osteosarcoma cells. Mol Cancer Ther 10(6):1018–1027, PubMed PMID: 21490307. eng

79. Gorlick R, Kolb EA, Houghton PJ, Morton CL, Neale G, Keir ST et al (2012) Initial testing (stage 1) of the cyclin dependent kinase inhibitor SCH 727965 (dinaciclib) by the pediatric preclinical testing program. Pediatr Blood Cancer 59(7):1266–1274, PubMed PMID: 22315240. Pubmed Central PMCID: PMC3349821. eng
80. Tentler JJ, Tan AC, Weekes CD, Jimeno A, Leong S, Pitts TM et al (2012) Patient-derived tumour xenografts as models for oncology drug development. Nat Rev Clin Oncol 9(6): 338–350, PubMed PMID: 22508028. eng
81. Johnson AJ, Yeh YY, Smith LL, Wagner AJ, Hessler J, Gupta S et al (2012) The novel cyclin-dependent kinase inhibitor dinaciclib (SCH727965) promotes apoptosis and abrogates microenvironmental cytokine protection in chronic lymphocytic leukemia cells. 12:2554–7, Leukemia.PubMed PMID: 22791353. ENG
82. Shapiro G, Bannerji R, Small K, Black S, Statkevich P, Abutarif M (eds) (2008) A phase I dose-escalation study of the safety, pharmacokinetics (PK) and pharmacodynamics (PD) of the novel cyclin-dependent kinase inhibitor SCH 727965 administered every 3 weeks in subjects with advanced malignancies. ASCO, Chicago, ASCO Annual Meeting
83. Zhang D, Mita M, Shapiro GI, Poon J, Small K, Tzontcheva A et al (2012) Effect of aprepitant on the pharmacokinetics of the cyclin-dependent kinase inhibitor dinaciclib in patients with advanced malignancies. Cancer Chemother Pharmacol. ASH stands for the American Society of Hematology. Headquarters are in Washington DC 6:891–8, PubMed PMID: 23053255. ENG
84. Gojo I, Walker AR, Cooper M, Feldman EJ, Padmanabhan S, Baer MR et al (eds) (2010) Phase II Study of the Cyclin-Dependent Kinase (CDK) Inhibitor Dinaciclib (SCH 727965) In Patients with Advanced Acute Leukemias ASH
85. Available from: http://clinicaltrials.gov/ct2/show/NCT01096342
86. Lao CD, Moon J, Fruehauf JP, Flaherty LE, Bury MJ, Ribas A et al (eds) (2012) SWOG S0826: A phase II trial of SCH 727965 (NSC 747135) in patients with stage IV melanoma. 2012 ASCO annual meeting, Chicago
87. SCH 727965 in Patients with advanced breast and lung cancers (Study P04716AM3) [Internet]. Available from: http://clinicaltrials.gov/ct2/show/NCT00732810
88. Veliparib and dinaciclib with or without carboplatin in treating patients with advanced solid tumors: http://clinicaltrials.gov/show/NCT01434316
89. A study of dinaciclib in combination with rituximab in participants with chronic lymphocytic leukemia and small lymphocytic lymphoma (P07974 AM1) http://clinicaltrials.gov/ct2/show/NCT01650727
90. Dinaciclib, bortezomib, and dexamethasone in treating patients with relapsed multiple myeloma [Internet]. Available from: http://www.clinicaltrials.gov/ct2/show/NCT01711528
91. Dinaciclib and epirubicin hydrochloride in treating patients with metastatic triple-negative breast cancer [Internet]. Available from: http://clinicaltrials.gov/show/NCT01624441
92. Akli S, Van Pelt CS, Bui T, Meijer L, Keyomarsi K (2011) Cdk2 Is required for breast cancer mediated by the Low-molecular-weight isoform of cyclin E. Cancer Res 71(9):3377–3386, PubMed PMID: 21385896. Pubmed Central PMCID: PMC3085722. eng
93. Horiuchi D, Kusdra L, Huskey NE, Chandriani S, Lenburg ME, Gonzalez-Angulo AM et al (2012) MYC pathway activation in triple-negative breast cancer is synthetic lethal with CDK inhibition. J Exp Med 209(4):679–696, PubMed PMID: 22430491. Pubmed Central PMCID: PMC3328367. eng
94. Deans AJ, Khanna KK, McNees CJ, Mercurio C, Heierhorst J, McArthur GA (2006) Cyclin-dependent kinase 2 functions in normal DNA repair and is a therapeutic target in BRCA1-deficient cancers. Cancer Res 66(16):8219–8226, PubMed PMID: 16912201. eng
95. Lee J, Kim SS (2009) The function of p27 KIP1 during tumor development. Exp Mol Med 41(11):765–771, PubMed PMID: 19887899. Pubmed Central PMCID: PMC2788730 eng
96. Liang J, Shao SH, Xu ZX, Hennessy B, Ding Z, Larrea M et al (2007) The energy sensing LKB1-AMPK pathway regulates p27(kip1) phosphorylation mediating the decision to enter autophagy or apoptosis. Nat Cell Biol 9(2):218–224, PubMed PMID: 17237771. eng
97. Short JD, Houston KD, Dere R, Cai SL, Kim J, Johnson CL et al (2008) AMP-activated protein kinase signaling results in cytoplasmic sequestration of p27. Cancer Res 68(16): 6496–6506, PubMed PMID: 18701472. Pubmed Central PMCID: PMC3011867. eng

98. Besson A, Gurian-West M, Schmidt A, Hall A, Roberts JM (2004) p27Kip1 modulates cell migration through the regulation of RhoA activation. Genes Dev 18(8):862–876, PubMed PMID: 15078817. Pubmed Central PMCID: PMC395846. eng
99. Besson A, Gurian-West M, Chen X, Kelly-Spratt KS, Kemp CJ, Roberts JM (2006) A pathway in quiescent cells that controls p27Kip1 stability, subcellular localization, and tumor suppression. Genes Dev 20(1):47–64, PubMed PMID: 16391232. Pubmed Central PMCID: PMC1356100. eng
100. Besson A, Hwang HC, Cicero S, Donovan SL, Gurian-West M, Johnson D et al (2007) Discovery of an oncogenic activity in p27Kip1 that causes stem cell expansion and a multiple tumor phenotype. Genes Dev 21(14):1731–1746, PubMed PMID: 17626791. Pubmed Central PMCID: PMC1920168. eng
101. Borriello A, Bencivenga D, Criscuolo M, Caldarelli I, Cucciolla V, Tramontano A et al (2011) Targeting p27Kip1 protein: its relevance in the therapy of human cancer. Expert Opin Ther Targets 15(6):677–693, PubMed PMID: 21355788. eng
102. Martínez-Sánchez A, Gebauer F (2010) Regulation of p27(kip1) mRNA expression by microRNAs. Prog Mol Subcell Biol 50:59–70, PubMed PMID: 19841881. eng
103. Tomoda K, Kubota Y, Kato J (1999) Degradation of the cyclin-dependent-kinase inhibitor p27Kip1 is instigated by Jab1. Nature 398(6723):160–165, PubMed PMID: 10086358. eng
104. Sui L, Dong Y, Ohno M, Watanabe Y, Sugimoto K, Tai Y et al (2001) Jab1 expression is associated with inverse expression of p27(kip1) and poor prognosis in epithelial ovarian tumors. Clin Cancer Res 7(12):4130–4135, PubMed PMID: 11751512. eng
105. Rassidakis GZ, Claret FX, Lai R, Zhang Q, Sarris AH, McDonnell TJ et al (2003) Expression of p27(Kip1) and c-Jun activation binding protein 1 are inversely correlated in systemic anaplastic large cell lymphoma. Clin Cancer Res 9(3):1121–1128, PubMed PMID: 12631617. eng
106. Kouvaraki MA, Rassidakis GZ, Tian L, Kumar R, Kittas C, Claret FX (2003) Jun activation domain-binding protein 1 expression in breast cancer inversely correlates with the cell cycle inhibitor p27(Kip1). Cancer Res 63(11):2977–2981, PubMed PMID: 12782606. eng
107. Fukumoto A, Ikeda N, Sho M, Tomoda K, Kanehiro H, Hisanaga M et al (2004) Prognostic significance of localized p27Kip1 and potential role of Jab1/CSN5 in pancreatic cancer. Oncol Rep 11(2):277–284, PubMed PMID: 14719054. eng
108. Pan Y, Zhang Q, Atsaves V, Yang H, Claret FX (2012) Suppression of Jab1/CSN5 induces radio- and chemo-sensitivity in nasopharyngeal carcinoma through changes to the DNA damage and repair pathways, Oncogene. PubMed PMID: 22797071. ENG
109. Wan M, Cao X, Wu Y, Bai S, Wu L, Shi X et al (2002) Jab1 Antagonizes TGF-beta signaling by inducing Smad4 degradation. EMBO Rep 3(2):171–176, PubMed PMID: 11818334. Pubmed Central PMCID: PMC1083965. eng
110. Lee EW, Oh W, Song J (2006) Jab1 As a mediator of nuclear export and cytoplasmic degradation of p53. Mol Cells 22(2):133–140, PubMed PMID: 17085963. eng
111. Tian L, Peng G, Parant JM, Leventaki V, Drakos E, Zhang Q et al (2010) Essential roles of Jab1 in cell survival, spontaneous DNA damage and DNA repair. Oncogene 29(46): 6125–6137, PubMed PMID: 20802511. Pubmed Central PMCID: PMC3495558. eng
112. Carrano AC, Eytan E, Hershko A, Pagano M (1999) SKP2 Is required for ubiquitin-mediated degradation of the CDK inhibitor p27. Nat Cell Biol 1(4):193–199, PubMed PMID: 10559916. eng
113. Kamura T, Hara T, Kotoshiba S, Yada M, Ishida N, Imaki H et al (2003) Degradation of p57Kip2 mediated by SCFSkp2-dependent ubiquitylation. Proc Natl Acad Sci U S A 100(18):10231–10236, PubMed PMID: 12925736. Pubmed Central PMCID: PMC193544. eng
114. Huang H, Regan KM, Wang F, Wang D, Smith DI, van Deursen JM et al (2005) Skp2 inhibits FOXO1 in tumor suppression through ubiquitin-mediated degradation. Proc Natl Acad Sci U S A 102(5):1649–1654, PubMed PMID: 15668399. Pubmed Central PMCID: PMC545492. eng
115. Kim SY, Herbst A, Tworkowski KA, Salghetti SE, Tansey WP (2003) Skp2 regulates Myc protein stability and activity. Mol Cell 11(88):1177–1188, PubMed PMID: 12769843. eng
116. von der Lehr N, Johansson S, Wu S, Bahram F, Castell A, Cetinkaya C et al (2003) The F-box protein Skp2 participates in c-Myc proteosomal degradation and acts as a cofactor for c-Myc-regulated transcription. Mol Cell 11(5):1189–1200, PubMed PMID: 12769844. eng

117. Latres E, Chiarle R, Schulman BA, Pavletich NP, Pellicer A, Inghirami G et al (2001) Role of the F-box protein Skp2 in lymphomagenesis. Proc Natl Acad Sci U S A 98(5):2515–2520, PubMed PMID: 11226270. Pubmed Central PMCID: PMC30169. eng
118. Yokoi S, Yasui K, Saito-Ohara F, Koshikawa K, Iizasa T, Fujisawa T et al (2002) A novel target gene, SKP2, within the 5p13 amplicon that is frequently detected in small cell lung cancers. Am J Pathol 161(1):207–216, PubMed PMID: 12107105. Pubmed Central PMCID: PMC1850681. eng
119. Gstaiger M, Jordan R, Lim M, Catzavelos C, Mestan J, Slingerland J et al (2001) Skp2 is oncogenic and overexpressed in human cancers. Proc Natl Acad Sci U S A 98(9):5043–5048, PubMed PMID: 11309491. Pubmed Central PMCID: PMC33160. eng
120. Lin HK, Chen Z, Wang G, Nardella C, Lee SW, Chan CH et al (2010) Skp2 Targeting suppresses tumorigenesis by Arf-p53-independent cellular senescence. Nature 464(9):374–379, PubMed PMID: 20237562. Pubmed Central PMCID: PMC2928066. eng
121. Chen Q, Xie W, Kuhn DJ, Voorhees PM, Lopez-Girona A, Mendy D et al (2008) Targeting the p27 E3 ligase SCF(Skp2) results in p27- and Skp2-mediated cell-cycle arrest and activation of autophagy. Blood 111(9):4690–4699, PubMed PMID: 18305219. Pubmed Central PMCID: PMC2343599. eng
122. Rico-Bautista E, Yang CC, Lu L, Roth GP, Wolf DA (2010) Chemical genetics approach to restoring p27Kip1 reveals novel compounds with antiproliferative activity in prostate cancer cells. BMC Biol 8:153, PubMed PMID: 21182779. Pubmed Central PMCID: PMC3025922. eng
123. Koutsami M, Velimezi G, Kotsinas A, Evangelou K, Papavassiliou AG, Kittas C et al (2008) Is exclusive Skp2 targeting always beneficial in cancer therapy? Blood 112(12):4777–4779, PubMed PMID: 19029458. eng
124. Chen X, Lowe M, Keyomarsi K (1999) UCN-01-mediated G1 arrest in normal but not tumor breast cells is pRb-dependent and p53-independent. Oncogene 18(41):5691, PubMed PMID: 10577141. eng
125. Chen X, Lowe M, Herliczek T, Hall MJ, Danes C, Lawrence DA et al (2000) Protection of normal proliferating cells against chemotherapy by staurosporine-mediated, selective, and reversible G(1) arrest. J Natl Cancer Inst 92(24):1999–2008, PubMed PMID: 11121462. eng
126. Roberts PJ, Bisi JE, Strum JC, Combest AJ, Darr DB, Usary JE et al (2012) Multiple roles of cyclin-dependent kinase 4/6 inhibitors in cancer therapy. J Natl Cancer Inst 104(6):476–487, PubMed PMID: 22302033. Pubmed Central PMCID: PMC3309128. eng
127. Nawijn MC, Alendar A, Berns A (2011) For better or for worse: the role of Pim oncogenes in tumorigenesis. Nat Rev Cancer 11(1):23–34, PubMed PMID: 21150935. eng
128. Cuypers HT, Selten G, Quint W, Zijlstra M, Maandag ER, Boelens W et al (1984) Murine leukemia virus-induced T-cell lymphomagenesis: integration of proviruses in a distinct chromosomal region. Cell 37(1):141–150, PubMed PMID: 6327049. eng
129. van Lohuizen M, Verbeek S, Krimpenfort P, Domen J, Saris C, Radaszkiewicz T et al (1989) Predisposition to lymphomagenesis in pim-1 transgenic mice: cooperation with c-myc and N-myc in murine leukemia virus-induced tumors. Cell 56(4):673–682, PubMed PMID: 2537153. eng
130. Verbeek S, van Lohuizen M, van der Valk M, Domen J, Kraal G, Berns A (1991) Mice bearing the E mu-myc and E mu-pim-1 transgenes develop pre-B-cell leukemia prenatally. Mol Cell Biol 11(2):1176–1179, PubMed PMID: 1990273. Pubmed Central PMCID: PMC359805. eng
131. Mikkers H, Nawijn M, Allen J, Brouwers C, Verhoeven E, Jonkers J et al (2004) Mice deficient for all PIM kinases display reduced body size and impaired responses to hematopoietic growth factors. Mol Cell Biol 24(13):6104–6115, PubMed PMID: 15199164. Pubmed Central PMCID: PMC480904. eng
132. Valdman A, Fang X, Pang ST, Ekman P, Egevad L (2004) Pim-1 expression in prostatic intraepithelial neoplasia and human prostate cancer. Prostate 60(4):367–371, PubMed PMID: 15264249. eng
133. Reiser-Erkan C, Erkan M, Pan Z, Bekasi S, Giese NA, Streit S et al (2008) Hypoxia-inducible proto-oncogene Pim-1 is a prognostic marker in pancreatic ductal adenocarcinoma. Cancer Biol Ther 7(9):1352–1359, PubMed PMID: 18708761. eng

134. Speers C, Tsimelzon A, Sexton K, Herrick AM, Gutierrez C, Culhane A et al (2009) Identification of novel kinase targets for the treatment of estrogen receptor-negative breast cancer. Clin Cancer Res 15(20):6327–6340, PubMed PMID: 19808870. Pubmed Central PMCID: PMC2763053. eng
135. Beier UH, Weise JB, Laudien M, Sauerwein H, Görögh T (2007) Overexpression of Pim-1 in head and neck squamous cell carcinomas. Int J Oncol 30(6):1381–1387, PubMed PMID: 17487358. eng
136. Tamburini J, Green AS, Bardet V, Chapuis N, Park S, Willems L et al (2009) Protein synthesis is resistant to rapamycin and constitutes a promising therapeutic target in acute myeloid leukemia. Blood 114(8):1618–1627, PubMed PMID: 19458359. eng
137. Cohen AM, Grinblat B, Bessler H, Kristt D, Kremer A, Schwartz A et al (2004) Increased expression of the hPim-2 gene in human chronic lymphocytic leukemia and non-Hodgkin lymphoma. Leuk Lymphoma 45(5):951–955, PubMed PMID: 15291354. eng
138. Qian KC, Wang L, Hickey ER, Studts J, Barringer K, Peng C et al (2005) Structural basis of constitutive activity and a unique nucleotide binding mode of human Pim-1 kinase. J Biol Chem 280(7):6130–6137, PubMed PMID: 15525646. eng
139. Amaravadi R, Thompson CB (2005) The survival kinases Akt and Pim as potential pharmacological targets. J Clin Invest 115(10):2618–2624, PubMed PMID: 16200194. Pubmed Central PMCID: PMC1236693. eng
140. Zhu N, Ramirez LM, Lee RL, Magnuson NS, Bishop GA, Gold MR (2002) CD40 Signaling in B cells regulates the expression of the Pim-1 kinase via the NF-kappa B pathway. J Immunol 168(2):744–754, PubMed PMID: 11777968. eng
141. Chen J, Kobayashi M, Darmanin S, Qiao Y, Gully C, Zhao R et al (2009) Hypoxia-mediated up-regulation of Pim-1 contributes to solid tumor formation. Am J Pathol 400(1):400–411, PubMed PMID: 19528349. Pubmed Central PMCID: PMC2708825. eng
142. Chen J, Kobayashi M, Darmanin S, Qiao Y, Gully C, Zhao R et al (2009) Pim-1 plays a pivotal role in hypoxia-induced chemoresistance. Oncogene 28(28):2581–2592, PubMed PMID: 19483729. Pubmed Central PMCID: PMC3358117. eng
143. Zhao Y, Hamza MS, Leong HS, Lim CB, Pan YF, Cheung E et al (2008) Kruppel-like factor 5 modulates p53-independent apoptosis through Pim1 survival kinase in cancer cells. Oncogene 27(1):1–8, PubMed PMID: 17603560. eng
144. Peltola K, Hollmen M, Maula SM, Rainio E, Ristamäki R, Luukkaa M et al (2009) Pim-1 kinase expression predicts radiation response in squamocellular carcinoma of head and neck and is under the control of epidermal growth factor receptor. Neoplasia 11(7):629–636, PubMed PMID: 19568408. Pubmed Central PMCID: PMC2697349. eng
145. Zhang Y, Wang Z, Magnuson NS (2007) Pim-1 kinase-dependent phosphorylation of p21Cip1/WAF1 regulates its stability and cellular localization in H1299 cells. Mol Cancer Res 5(9):909–922, PubMed PMID: 17855660. eng
146. Morishita D, Katayama R, Sekimizu K, Tsuruo T, Fujita N (2008) Pim kinases promote cell cycle progression by phosphorylating and down-regulating p27Kip1 at the transcriptional and posttranscriptional levels. Cancer Res 68(13):5076–5085, PubMed PMID: 18593906. eng
147. Mochizuki T, Kitanaka C, Noguchi K, Muramatsu T, Asai A, Kuchino Y (1999) Physical and functional interactions between Pim-1 kinase and Cdc25A phosphatase. Implications for the Pim-1-mediated activation of the c-Myc signaling pathway. J Biol Chem 274(26):18659–18666, PubMed PMID: 10373478. eng
148. Bachmann M, Hennemann H, Xing PX, Hoffmann I, Möröy T (2004) The oncogenic serine/threonine kinase Pim-1 phosphorylates and inhibits the activity of Cdc25C-associated kinase 1 (C-TAK1): a novel role for Pim-1 at the G2/M cell cycle checkpoint. J Biol Chem 279((46):48319–48328, PubMed PMID: 15319445. eng
149. Bachmann M, Kosan C, Xing PX, Montenarh M, Hoffmann I, Möröy T (2006) The oncogenic serine/threonine kinase Pim-1 directly phosphorylates and activates the G2/M specific phosphatase Cdc25C. Int J Biochem Cell Biol 38(3):430–443, PubMed PMID: 16356754. eng
150. Koike N, Maita H, Taira T, Ariga H, Iguchi-Ariga SM (2000) Identification of heterochromatin protein 1 (HP1) as a phosphorylation target by Pim-1 kinase and the effect of phosphorylation

on the transcriptional repression function of HP1(1). FEBS Lett 467(1):17–21, PubMed PMID: 10664448. eng
151. Zhao T, Heyduk T, Eissenberg JC (2001) Phosphorylation site mutations in heterochromatin protein 1 (HP1) reduce or eliminate silencing activity. J Biol Chem 276(12):9512–9518, PubMed PMID: 11121421. eng
152. Aho TL, Sandholm J, Peltola KJ, Mankonen HP, Lilly M, Koskinen PJ (2004) Pim-1 kinase promotes inactivation of the pro-apoptotic Bad protein by phosphorylating it on the Ser112 gatekeeper site. FEBS Lett 571(1-3):43–49, PubMed PMID: 15280015. eng
153. Zhang Y, Wang Z, Li X, Magnuson NS (2008) Pim kinase-dependent inhibition of c-Myc degradation. Oncogene 27(35):4809–4819, PubMed PMID: 18438430. eng
154. Roh M, Gary B, Song C, Said-Al-Naief N, Tousson A, Kraft A et al (2003) Overexpression of the oncogenic kinase Pim-1 leads to genomic instability. Cancer Res 63(23):8079–8084, PubMed PMID: 14678956. Eng
155. Ha S, Iqbal NJ, Mita P, Ruoff R, Gerald WL, Lepor H et al (2012) Phosphorylation of the androgen receptor by PIM1 in hormone refractory prostate cancer. Oncogene. PubMed PMID: 22986532. ENG
156. Linn DE, Yang X, Xie Y, Alfano A, Deshmukh D, Wang X et al (2012) Differential regulation of androgen receptor by PIM-1 kinases via phosphorylation-dependent recruitment of distinct ubiquitin E3 ligases. J Biol Chem 287(27):22959–22968, PubMed PMID: 22584579. Pubmed Central PMCID: PMC3391098. Eng
157. Laird PW, van der Lugt NM, Clarke A, Domen J, Linders K, McWhir J et al (1993) Nucleic Acids Res 21(20):4750–4755, PubMed PMID: 8233823. Pubmed Central PMCID: PMC331501. eng
158. Chen LS, Redkar S, Bearss D, Wierda WG, Gandhi V (2009) Pim kinase inhibitor, SGI-1776, induces apoptosis in chronic lymphocytic leukemia cells. Blood 114(19):4150–4157, PubMed PMID: 19734450. Pubmed Central PMCID: PMC2774551. eng
159. Chen LS, Redkar S, Taverna P, Cortes JE, Gandhi V (2011) Mechanisms of cytotoxicity to Pim kinase inhibitor, SGI-1776, in acute myeloid leukemia. Blood 118(3):693–702, PubMed PMID: 21628411. Pubmed Central PMCID: PMC3142906. eng
160. Mumenthaler SM, Ng PY, Hodge A, Bearss D, Berk G, Kanekal S et al (2009) Pharmacologic inhibition of Pim kinases alters prostate cancer cell growth and resensitizes chemoresistant cells to taxanes. Mol Cancer Ther 8(10):2882–2893, PubMed PMID: 19825806. Pubmed Central PMCID: PMC2808126. eng
161. SuperGen discontinues clinical development of SGI-1776. Press release 18 Nov 2012. Available from: http://www.businesswire.com/news/home/20101110007218/en/SuperGen-Discontinues-Clinical-Development-SGI-1776
162. Beharry Z, Zemskova M, Mahajan S, Zhang F, Ma J, Xia Z et al (2009) Novel benzylidene-thiazolidine-2,4-diones inhibit Pim protein kinase activity and induce cell cycle arrest in leukemia and prostate cancer cells. Mol Cancer Ther 8(6):1473–1483, PubMed PMID: 19509254. Pubmed Central PMCID: PMC3415237. eng
163. Song JH, Kraft AS (2012) Pim kinase inhibitors sensitize prostate cancer cells to apoptosis triggered by Bcl-2 family inhibitor ABT-737. Cancer Res 72(1):294–303, PubMed PMID: 22080570. Pubmed Central PMCID: PMC3251634. eng
164. Dar AA, Goff LW, Majid S, Berlin J, El-Rifai W (2010) Aurora kinase inhibitors – rising stars in cancer therapeutics? Mol Cancer Ther 9(2):268–278, PubMed PMID: 20124450. Pubmed Central PMCID: PMC2820587. eng
165. Kimmins S, Crosio C, Kotaja N, Hirayama J, Monaco L, Höög C et al (2007) Differential functions of the Aurora-B and Aurora-C kinases in mammalian spermatogenesis. Mol Endocrinol 21(3):726–739, PubMed PMID: 17192404. eng
166. Khan J, Ezan F, Crémet JY, Fautrel A, Gilot D, Lambert M et al (2011) Overexpression of active Aurora-C kinase results in cell transformation and tumour formation. PLoS One 6(10):e26512, PubMed PMID: 22046298. Pubmed Central PMCID: PMC3203144. eng
167. Kimura M, Matsuda Y, Yoshioka T, Okano Y (1999) Cell cycle-dependent expression and centrosome localization of a third human aurora/Ipl1-related protein kinase, AIK3. J Biol Chem 274(11):7334–7340, PubMed PMID: 10066797. eng

168. Glover DM, Leibowitz MH, McLean DA, Parry H (1995) Mutations in aurora prevent centrosome separation leading to the formation of monopolar spindles. Cell 81(1):95–105, PubMed PMID: 7720077. eng
169. Kelly KR, Ecsedy J, Mahalingam D, Nawrocki ST, Padmanabhan S, Giles FJ et al (2011) Targeting aurora kinases in cancer treatment. Curr Drug Targets 12(14):2067–2078, PubMed PMID: 21777198. eng
170. Sasai K, Parant JM, Brandt ME, Carter J, Adams HP, Stass SA et al (2008) Targeted disruption of Aurora A causes abnormal mitotic spindle assembly, chromosome misalignment and embryonic lethality. Oncogene 27(29):4122–4177, PubMed PMID: 18345035. eng
171. Marumoto T, Honda S, Hara T, Nitta M, Hirota T, Kohmura E et al (2003) Aurora-A kinase maintains the fidelity of early and late mitotic events in HeLa cells. J Biol Chem 278(51):51786–51795, PubMed PMID: 14523000. eng
172. Littlepage LE, Wu H, Andresson T, Deanehan JK, Amundadottir LT, Ruderman JV (2002) Identification of phosphorylated residues that affect the activity of the mitotic kinase Aurora-A. Proc Natl Acad Sci U S A 99(24):15440–15445, PubMed PMID: 12422018. Pubmed Central PMCID: PMC137735. eng
173. Tanaka T, Kimura M, Matsunaga K, Fukada D, Mori H, Okano Y (1999) Centrosomal kinase AIK1 is overexpressed in invasive ductal carcinoma of the breast. Cancer Res 59(9): 2041–2044, PubMed PMID: 10232583. eng
174. Lehman NL, O'Donnell JP, Whiteley LJ, Stapp RT, Lehman TD, Roszka KM et al (2012) Aurora A is differentially expressed in gliomas, is associated with patient survival in glioblastoma and is a potential chemotherapeutic target in gliomas. Cell Cycle 11(3):489–502, PubMed PMID: 22274399. Pubmed Central PMCID: PMC3315093. eng
175. Lo Iacono M, Monica V, Saviozzi S, Ceppi P, Bracco E, Papotti M et al (2011) Aurora kinase a expression is associated with lung cancer histological-subtypes and with tumor de-differentiation. J Transl Med 9:100, PubMed PMID: 21718475. Pubmed Central PMCID: PMC3148570. eng
176. Lassus H, Staff S, Leminen A, Isola J, Butzow R (2011) Aurora-A overexpression and aneuploidy predict poor outcome in serous ovarian carcinoma. Gynecol Oncol 120(1):11–17, PubMed PMID: 20937525. eng
177. Zhou H, Kuang J, Zhong L, Kuo WL, Gray JW, Sahin A et al (1998) Tumour amplified kinase STK15/BTAK induces centrosome amplification, aneuploidy and transformation. Nat Genet 20(2):189–193, PubMed PMID: 9771714. eng
178. Dar AA, Zaika A, Piazuelo MB, Correa P, Koyama T, Belkhiri A et al (2008) Frequent overexpression of Aurora kinase A in upper gastrointestinal adenocarcinomas correlates with potent antiapoptotic functions. Cancer 112(8):1688–1698, PubMed PMID: 18311783. eng
179. Chefetz I, Holmberg JC, Alvero AB, Visintin I, Mor G (2011) Inhibition of Aurora-A kinase induces cell cycle arrest in epithelial ovarian cancer stem cells by affecting NFκB pathway. Cell Cycle 10(13):2206–2214, PubMed PMID: 21623171. Pubmed Central PMCID: PMC3154367. eng
180. Katayama H, Sasai K, Kawai H, Yuan ZM, Bondaruk J, Suzuki F et al (2004) Phosphorylation by aurora kinase A induces Mdm2-mediated destabilization and inhibition of p53. Nat Genet 36(1):55–62, PubMed PMID: 14702041. eng
181. Liu Q, Kaneko S, Yang L, Feldman RI, Nicosia SV, Chen J et al (2004) Aurora-A abrogation of p53 DNA binding and transactivation activity by phosphorylation of serine 215. J Biol Chem 279(50):52175–52182, PubMed PMID: 15469940. eng
182. Hauf S, Cole RW, LaTerra S, Zimmer C, Schnapp G, Walter R et al (2003) The small molecule hesperadin reveals a role for Aurora B in correcting kinetochore-microtubule attachment and in maintaining the spindle assembly checkpoin. J Cell Biol 161(2):281–294, PubMed PMID: 12707311. Pubmed Central PMCID: PMC2172906. eng
183. Crosio C, Fimia GM, Loury R, Kimura M, Okano Y, Zhou H et al (2002) Mitotic phosphorylation of histone H3: spatio-temporal regulation by mammalian Aurora kinases. Mol Cell Biol 22(3):874–885, PubMed PMID: 11784863. Pubmed Central PMCID: PMC133550. eng

184. Goto H, Yasui Y, Nigg EA, Inagaki M (2002) Aurora-B phosphorylates histone H3 at serine28 with regard to the mitotic chromosome condensation. Genes Cells 7(1):11–17, PubMed PMID: 11856369. eng
185. Hetland TE, Nymoen DA, Holth A, Brusegard K, Flørenes VA, Kærn J et al (2012) Aurora B expression in metastatic effusions from advanced-stage ovarian serous carcinoma is predictive of intrinsic chemotherapy resistance. Hum Pathol. 5:777–85, PubMed PMID: 23114921. ENG
186. Hartsink-Segers SA, Zwaan CM, Exalto C, Luijendijk MW, Calvert VS, Petricoin EF et al (2012) Aurora kinases in childhood acute leukemia: the promise of aurora B as therapeutic target. Leukemia. 3:560–8, PubMed PMID: 22940834. ENG
187. Buczkowicz P, Zarghooni M, Bartels U, Morrison A, Misuraca KL, Chan T, et al. (2012) Aurora kinase B is a potential therapeutic target in pediatric diffuse intrinsic pontine glioma. Brain Pathol. 3:244–53, PubMed PMID: 22971244. ENG
188. Gully CP, Velazquez-Torres G, Shin JH, Fuentes-Mattei E, Wang E, Carlock C et al (2012) Aurora B kinase phosphorylates and instigates degradation of p53. Proc Natl Acad Sci U S A 109(24):E1513–E1522, PubMed PMID: 22611192. Pubmed Central PMCID: PMC3386093. eng
189. Hata T, Furukawa T, Sunamura M, Egawa S, Motoi F, Ohmura N et al (2005) RNA interference targeting aurora kinase a suppresses tumor growth and enhances the taxane chemosensitivity in human pancreatic cancer cells. Cancer Res 65(7):2899–2905, PubMed PMID: 15805292. eng
190. Warner SL, Munoz RM, Stafford P, Koller E, Hurley LH, Von Hoff DD et al (2006) Comparing Aurora A and Aurora B as molecular targets for growth inhibition of pancreatic cancer cells. Mol Cancer Ther 5(10):2450–2458, PubMed PMID: 17041088. eng
191. Girdler F, Gascoigne KE, Eyers PA, Hartmuth S, Crafter C, Foote KM et al (2006) Validating aurora B as an anti-cancer drug target. J Cell Sci 119(Pt 17):3664–3675, PubMed PMID: 16912073. eng
192. Gizatullin F, Yao Y, Kung V, Harding MW, Loda M, Shapiro GI (2006) The Aurora kinase inhibitor VX-680 induces endoreduplication and apoptosis preferentially in cells with compromised p53-dependent postmitotic checkpoint function. Cancer Res 66(15):7668–7677, PubMed PMID: 16885368. eng
193. Kuner R, Fälth M, Pressinotti NC, Brase JC, Puig SB, Metzger J et al (2012) The maternal embryonic leucine zipper kinase (MELK) is upregulated in high-grade prostate cancer. J Mol Med (Berl). 2:237–48, PubMed PMID: 22945237. ENG
194. Marie SK, Okamoto OK, Uno M, Hasegawa AP, Oba-Shinjo SM, Cohen T et al (2008) Maternal embryonic leucine zipper kinase transcript abundance correlates with malignancy grade in human astrocytomas. Int J Cancer 122(4):807–815, PubMed PMID: 17960622. eng
195. Pickard MR, Green AR, Ellis IO, Caldas C, Hedge VL, Mourtada-Maarabouni M et al (2009) Dysregulated expression of Fau and MELK is associated with poor prognosis in breast cancer. Breast Cancer Res 11(4):60, PubMed PMID: 19671159. Pubmed Central PMCID: PMC2750122. eng
196. Nakano I, Paucar AA, Bajpai R, Dougherty JD, Zewail A, Kelly TK et al (2005) Maternal embryonic leucine zipper kinase (MELK) regulates multipotent neural progenitor proliferation. J Cell Biol 170(3):413–427, PubMed PMID: 16061694. Pubmed Central PMCID: PMC2171475. eng
197. Nakano I, Masterman-Smith M, Saigusa K, Paucar AA, Horvath S, Shoemaker L et al (2008) J Neurosci Res 86(1):48–60, PubMed PMID: 17722061. eng
198. Hebbard LW, Maurer J, Miller A, Lesperance J, Hassell J, Oshima RG et al (2010) Maternal embryonic leucine zipper kinase is upregulated and required in mammary tumor-initiating cells in vivo. Cancer Res 70(21):8863–8873, PubMed PMID: 20861186. eng
199. Badouel C, Körner R, Frank-Vaillant M, Couturier A, Nigg EA, Tassan JP (2006) M-phase MELK activity is regulated by MPF and MAPK. Cell Cycle 5(8):883–889, PubMed PMID: 16628004. eng
200. Badouel C, Chartrain I, Blot J, Tassan JP (2010) Maternal embryonic leucine zipper kinase is stabilized in mitosis by phosphorylation and is partially degraded upon mitotic exit. Exp Cell Res 316(13):2166–2173, PubMed PMID: 20420823. eng

201. Choi S, Ku JL (2011) Resistance of colorectal cancer cells to radiation and 5-FU is associated with MELK expression. Biochem Biophys Res Commun 412(2):207–213, PubMed PMID: 21806965. eng
202. Liu G, Yuan X, Zeng Z, Tunici P, Ng H, Abdulkadir IR et al (2006) Analysis of gene expression and chemoresistance of CD133+ cancer stem cells in glioblastoma. Mol Cancer 5:67, PubMed PMID: 17140455. Pubmed Central PMCID: PMC1697823. eng
203. Nakano I, Joshi K, Visnyei K, Hu B, Watanabe M, Lam D et al (2011) Siomycin A targets brain tumor stem cells partially through a MELK-mediated pathway. Neuro Oncol 13(6): 622–634, PubMed PMID: 21558073. Pubmed Central PMCID: PMC3107094. eng
204. Klebig C, Korinth D, Meraldi P (2009) Bub1 regulates chromosome segregation in a kinetochore-independent manner. J Cell Biol 185(5):841–858, PubMed PMID: 19487456. Pubmed Central PMCID: PMC2711590. eng
205. Janssen A, Kops GJ, Medema RH (2009) Elevating the frequency of chromosome missegregation as a strategy to kill tumor cells. Proc Natl Acad Sci U S A 106(45):19108–19113, PubMed PMID: 19855003. Pubmed Central PMCID: PMC2776415. eng
206. Kwiatkowski N, Jelluma N, Filippakopoulos P, Soundararajan M, Manak MS, Kwon M et al (2010) Small-molecule kinase inhibitors provide insight into Mps1 cell cycle function. Nat Chem Biol 6(5):359–368, PubMed PMID: 20383151. Pubmed Central PMCID: PMC2857554. eng
207. Hewitt L, Tighe A, Santaguida S, White AM, Jones CD, Musacchio A et al (2010) Sustained Mps1 activity is required in mitosis to recruit O-Mad2 to the Mad1-C-Mad2 core complex. J Cell Biol 190(1):25–34, PubMed PMID: 20624899. Pubmed Central PMCID: PMC2911659. eng
208. Li M, York JP, Zhang P (2007) Loss of Cdc20 causes a securin-dependent metaphase arrest in two-cell mouse embryos. Mol Cell Biol 27(9):3481–3488, PubMed PMID: 17325031. Pubmed Central PMCID: PMC1899968. eng
209. Manchado E, Guillamot M, de Cárcer G, Eguren M, Trickey M, García-Higuera I et al (2010) Argeting mitotic exit leads to tumor regression in vivo: modulation by Cdk1, Mastl, and the PP2A/B55α,δ phosphatase. Cancer Cell 18(6):641–654, PubMed PMID: 21156286. eng
210. Ye X, Zhu C, Harper JW (2001) A premature-termination mutation in the Mus musculus cyclin-dependent kinase 3 gene. Proc Natl Acad Sci U S A 98(4):1682–1686, PubMed PMID: 11172011. Pubmed Central PMCID: PMC29317. eng
211. Tsutsui T, Hesabi B, Moons DS, Pandolfi PP, Hansel KS, Koff A et al (1999) Targeted disruption of CDK4 delays cell cycle entry with enhanced p27(Kip1) activity. Mol Cell Biol 19(10):7011–7019, PubMed PMID: 10490638. Pubmed Central PMCID: PMC84696. eng
212. Atanasoski S, Shumas S, Dickson C, Scherer SS, Suter U (2001) Differential cyclin D1 requirements of proliferating Schwann cells during development and after injury. Mol Cell Neurosci 18(6):581–592, PubMed PMID: 11749035. eng
213. Ciemerych MA, Sicinski P (2005) Cell cycle in mouse development. Oncogene 24(17): 2877–2898, PubMed PMID: 15838522. eng
214. Fantl V, Stamp G, Andrews A, Rosewell I, Dickson C (1995) Mice lacking cyclin D1 are small and show defects in eye and mammary gland development. Genes Dev 9(19): 2364–2372, PubMed PMID: 7557388. eng
215. Sicinski P, Donaher JL, Parker SB, Li T, Fazeli A, Gardner H et al (1995) Cyclin D1 provides a link between development and oncogenesis in the retina and breast. Cell 82(4):621–630, PubMed PMID: 7664341. eng
216. Georgia S, Bhushan A (2004) Beta cell replication is the primary mechanism for maintaining postnatal beta cell mass. J Clin Invest 114(7):963–968, PubMed PMID: 15467835. Pubmed Central PMCID: PMC518666. eng
217. Sicinski P, Donaher JL, Geng Y, Parker SB, Gardner H, Park MY et al (1996) Cyclin D2 is an FSH-responsive gene involved in gonadal cell proliferation and oncogenesis. Nature 384(6608):470–474, PubMed PMID: 8945475. Eng
218. Sicinska E, Aifantis I, Le Cam L, Swat W, Borowski C, Yu Q et al (2003) Requirement for cyclin D3 in lymphocyte development and T cell leukemias. Cancer Cell 4(6):451–461, PubMed PMID: 14706337. eng

219. Ciemerych MA, Kenney AM, Sicinska E, Kalaszczynska I, Bronson RT, Rowitch DH et al (2002) Development of mice expressing a single D-type cyclin. Genes Dev 16(24): 3277–3289, PubMed PMID: 12502747. Pubmed Central PMCID: PMC187507. Eng
220. Kozar K, Ciemerych MA, Rebel VI, Shigematsu H, Zagozdzon A, Sicinska E et al (2004) Mouse development and cell proliferation in the absence of D-cyclins. Cell 118(4):477–491, PubMed PMID: 15315760. eng
221. Geng Y, Yu Q, Sicinska E, Das M, Schneider JE, Bhattacharya S et al (2003) Cyclin E ablation in the mouse. Cell 114(4):431–443, PubMed PMID: 12941272. eng
222. Parisi T, Beck AR, Rougier N, McNeil T, Lucian L, Werb Z et al (2003) Cyclins E1 and E2 are required for endoreplication in placental trophoblast giant cells. EMBO J 22(18): 4794–4803, PubMed PMID: 12970191. Pubmed Central PMCID: PMC212738. eng
223. Liu D, Matzuk MM, Sung WK, Guo Q, Wang P, Wolgemuth DJ (1998) Cyclin A1 is required for meiosis in the male mouse. Nat Genet 20(4):377–380, PubMed PMID: 9843212. eng
224. Liu D, Liao C, Wolgemuth DJ (2000) A role for cyclin A1 in the activation of MPF and G2-M transition during meiosis of male germ cells in mice. Dev Biol 224(2):388–400, PubMed PMID: 10926775. eng
225. Brandeis M, Rosewell I, Carrington M, Crompton T, Jacobs MA, Kirk J et al (1998) Cyclin B2-null mice develop normally and are fertile whereas cyclin B1-null mice die in utero. Proc Natl Acad Sci U S A 95(8):4344–4349, PubMed PMID: 9539739. Pubmed Central PMCID: PMC22491. eng

# Chapter 15
# Histone Demethylases in Prostate Cancer

**Ling-Yu Wang, Wenchang Guo, Kevin Kim, Mamata Pochampalli, Chiu-Lien Hung, Yoshihiro Izumiya, and Hsing-Jien Kung**

**Abstract** Accumulating evidence has suggested that epigenetic alternations are as important as genetic mutations in cancer development. It is proposed that tumors are arisen by "malignant reprogramming" driven by a combination of both genetic and epigenetic changes. It therefore comes as no surprise that histone demethylases, the newest members of the histone modifying enzymes, are found to be targets of mutations and dysregulation in cancer cells. In this review article, we provide an overview of the types of histone demethylases whose genetic structure or expression is altered in cancers, the action of histone demethylases in cancer development and their potential inhibitors. Special emphasis is placed on the roles of histone demethylases in prostate cancer progression.

**Keywords** Histone demethylase • Inhibitor • Prostate cancer • Mutation • Epigenetics

---

L.-Y. Wang, Ph.D. • W. Guo, Ph.D. • M. Pochampalli, Ph.D. • C.-L. Hung, Ph.D.
Cancer Center, University of California, Davis, Sacramento, CA 95817, USA
e-mail: lywang@ucdavis.edu; wenchangguo@yahoo.com; mpochampalli@ucdavis.edu; joyhung0825@gmail.com

K. Kim, BS
Biochemistry, Molecular, Cellular and Developmental Biology, University of California, Davis, Davis, CA 95616, USA
e-mail: kvkim@ucdavis.edu

Y. Izumiya, DVM, Ph.D.
Dermatology, University of California, Davis, School of Medicine, Sacramento, CA 95616, USA
e-mail: yizumiya@ucdavis.edu

H.-J. Kung, Ph.D. (✉)
Cancer Center, University of California, Davis, Sacramento, CA 95817, USA

Biochemistry & Molecular Medicine, University of California, Davis, Sacramento, CA 95817, USA
e-mail: hkung@ucdavis.edu

R. Kumar (ed.), *Nuclear Signaling Pathways and Targeting Transcription in Cancer*, Cancer Drug Discovery and Development, DOI 10.1007/978-1-4614-8039-6_15,

## Abbreviations

| | |
|---|---|
| AR | Androgen receptor |
| ARE | Androgen responsive elements |
| CDK | Cyclin-dependent kinase |
| CGH | Comparative genomic hybridization |
| CRPC | Castration-resistant prostate cancer |
| ER | Estrogen receptor |
| FAD | Flavin adenine dinucleotide |
| HIF | Hypoxia-induced transcription factor |
| HP1 | Heterochromatin protein 1 |
| KDM | Histone lysine demethylase |
| MEF | Mouse embryonic fibroblast |
| MAO | Monoamine oxidase |
| NOG | N-oxalylglycine |
| PRC | Polycomb repressive complex |
| SAHF | Senescence-associated heterochromatic foci |

## 15.1 Histone Lysine Demethylase and Cancer

Chromatin is a highly ordered structure of eukaryotic DNA, which is packed into nucleosomes by core histone protein octamers: H2A, H2B, H3 and H4. Post-translational modifications of histone N-terminal basic tails cause conformational change of the nucleosome, allowing access of regulatory machineries to the DNA for transcription, replication, and repair [1, 2]. Several lysine residues on the histone tails can be mono-, di- or trimethylated. Depending on the position and degree of lysine methylation, the biological outcome is different. For example, histone marks such as H3K9me2, -me3 and H3K27me3, are involved in heterochromatin formation and gene silencing; while H3K4me3 is associated with actively transcribed genes. Because of the importance in gene expression, histone methylation and demethylation play critical roles in several biological processes, and altered histone methylation patterns are linked to human diseases such as neurological disorders and cancer.

Previously, histone methylation was believed to be irreversible. It was not until recently that the enzymes capable of removing methyl groups from histone were identified. As summarized in Table 15.1, histone lysine-specific demethylases (KDMs) have exquisite substrate specificities toward particular lysines on histones. KDMs can be broadly classified into two families: LSD1 (KDM1) and Jumonji C domain-containing (JmjC) family (KDM2 to KDM8). LSD1 family is amine oxidase which catalyzes demethylation through a flavin adenine dinucleotide (FAD)-dependent reaction. The JmjC family demethylase contains a conserved JmjC catalytic domain, which coordinates an electron shuffle between the methylated

**Table 15.1** Histone demethylases with specific substrates identified

| Name | Synonyms | Substrate |
|---|---|---|
| KDM1A | AOF2/BHC110/LSD1 | H3K4me2/me1 H3K9me2/me1 |
| KDM1B | AOF1/LSD2 | H3K4me2/me1 |
| KDM2A | JHDM1A/FBXL11 | H3K36me2/me1 |
| KDM2B | JHDM1B/FBXL10 | H3K36me2/me1 H3K4me3 |
| KDM3A | JHDM2A/JMJD1A/TSGA | H3K9me2/me1 |
| KDM3B | JMJD1B | H3K9me2 |
| KDM4A | JHDM3A/JMJD2A | H3K9me3/me2 H3K36me3/me2 |
| KDM4B | JHDM3B/JMJD2B | H3K9me3/me2 H3K36me3/me2 |
| KDM4C | JHDM3C/JMJD2C/GASC1 | H3K9me3/me2 H3K36me3/me2 |
| KDM4D | JHDM3D/JMJD2D | H3K9me3/me2 |
| KDM4E | KDM4DL | H3K9me3/me2 |
| KDM5A | JARID1A/RBP2 | H3K4me3/me2 |
| KDM5B | JARID1B/PLU1 | H3K4me3/me2 |
| KDM5C | JARID1C/SMCX | H3K4me3/me2 |
| KDM5D | JARID1D/SMCY | H3K4me3/me2 |
| KDM6A | UTX | H3K27me3/me2 |
| KDM6B | JMJD3 | H3K27me3/me2 |
| KDM7 | JHDM1D/KIAA1718 | H3K9me2/me1 H3K27me2/me1 |
| KDM8 | JMJD5 | H3K36me2 |
| PHF2 | JHDM1E | H3K9me2 |
| PHF8 | JHDM1F | H3K9me2/me1 H3K27me2 H4K20me1 |
| JMJD6 | PSR/PTDSR | H3R2 H4R3 |

lysine with the co-factors Fe(II), α-keto-glutarate, and molecular oxygen. This reaction ultimately results in the removal of the methyl group. Dysregulation of these enzymes alters the chromatin conformation and reprograms gene expression, which sometimes leads to malignant transformation of the cells. Table 15.2 presents a summary of recent literature on the topics of KDMs dysregulation in cancer. There are several other excellent reviews on this subject [3–5] which the readers may wish to refer to.

### 15.1.1 Dysregulation of Histone Demethylases in Cancer

Global alteration of histone methylation such as the loss of H3K4me2 and H4K20me3 are hallmarks of many cancers and are associated with poor prognosis [6–9]. Aside from the global changes, alterations of histone methylation patterns within specific cancer-causing loci also have consequences in cancer cell proliferation and survival. It is thus not surprising that mutations and aberrant expression of histone demethylases (Table 15.2) have been associated with carcinogenesis.

**Table 15.2** Alterations of histone demethylases in cancer cell

| Name | Alteration | Cancer associated | Target/Mechanism | References |
|---|---|---|---|---|
| KDM1A | Overexpression | Breast (ER-) | Regulates p21, ERBB2 and Cyclin A2 expression | [43] |
| | Overexpression | Non-small cell Lung | Regulates expression of EMT markers | [122] |
| | Overexpression | Prostate | AR coactivation | [98–101] |
| | Overexpression | Bladder | | [123] |
| | Overexpression | Neuroblastoma | Maintains un-differentiated, malignant phenotype | [124] |
| | Overexpression | Nasopharyngeal carcinoma | | [125] |
| | Underexpression | Hepatocellular carcinoma | | [16] |
| | Underexpression | Breast | Inhibits invasion and metastasis through TGFβ signaling | [17] |
| KDM2A | Underexpression | Prostate | Represses transcription of centromeric satellite repeats. Sustain genomic stability | [13] |
| KDM2B | Overexpression | Leukemia | Silences p15(Ink4b) | [18] |
| | Underexpression | Brain, glioblastoma | Represses transcription of rRNA. | [14] |
| KDM3A | Overexpression | Prostate | AR coactivation | [103, 104] |
| | Overexpression | Bladder, Lung | Positively regulates G1/S transition via HoxA1 gene expression | [40] |
| | Overexpression | Hepatocellular carcinoma | | [126] |
| | Overexpression | Kidney | | [127] |
| | Overexpression | Renal | Regulates hypoxic gene expression | [68] |
| KDM3B | Overexpression | Leukemia | Induces leukemogenic oncogene expression | [128] |
| | Overexpression | Prostate | | [104] |
| KDM4A | Overexpression | Breast | ER coactivation. Positively regulates Cyclin D1 | [36] |
| | Overexpression | Prostate | AR coactivation | [12, 104, 106] |
| | Underexpression | Bladder | | [129] |
| KDM4B | Overexpression | Colorectal cancer | Regulates hypoxia-inducible gene expression | [130] |
| | Overexpression | Gastric | Regulates p53 and p21 | [52] |
| | Overexpression | Prostate | | [12] |
| | Overexpression | Bladder, Lung | Promotes G1/S transition by positively regulating Cdk6 | [37] |
| | Overexpression | Breast | ER coactivation. Positively regulates Cyclin D1 and MYC | [38] |
| | Overexpression | Renal | Regulates hypoxic gene expression | [68] |

| | | | | |
|---|---|---|---|---|
| KDM4C | 9p23-24 amplification | Breast | Regulates NOTCH1 expression | [23, 131] |
| | Overexpression | Prostate | AR coactivation | [12, 101] |
| | 9p24 amplification | Lymphoma | Regulates MYC expression | [24] |
| | Overexpression | Oesophageal, Medulloblastoma | | [22, 132] |
| KDM5A | Gene amplification | Breast (drug resistance) | Negatively regulates CDK inhibitors and apoptotic genes | [25] |
| | Overexpression | Gastric | Negatively regulates CDK inhibitors | [44] |
| | Translocation | Acute myeloid leukaemia | NUP98-PHD fusion. Forced transcription of developmental loci | [26, 27] |
| KDM5B | Overexpression | Prostate | AR coactivation | [104, 107] |
| | Overexpression | Breast | | [133, 134] |
| | Overexpression | Bladder | E2F1 and E2F2 expression. | [49] |
| KDM5C | Inactivating mutation | Clear-cell renal cell carcinoma | | [29] |
| | Overexpression | Prostate | Suppresses TGF-β/Smad3 signaling pathway | [113] |
| KDM6A | Inactivating mutation | Clear cell renal cell carcinoma | | [29, 30] |
| | Inactivating mutation | Leukemia, Breast, Colorectal, Endometrium, Gliobastoma, Lymphoma, Myeloma, Lung, Esophageal, Pancreas | | [30] |
| | Inactivating mutation | Bladder | | [31] |
| KDM6B | Overexpression | Hodgkin's lymphoma | | [135] |
| | Overexpression | Prostate | | [136] |
| KDM8 | Overexpression | Breast | Regulates cyclin A1 expression. | [42] |
| | Overexpression | Leukemia | Negatively regulates p53 expression. | [46] |
| PHF8 | Overexpression | Prostate | | [104] |

#### 15.1.1.1 Aberrant Expression

While some histone demethylases behave like oncogenes, others play tumor suppressive roles. KDM4 family members that are widely overexpressed in several tumor types are believed to be putative oncogenes because of their demethylation activity toward H3K9me3/me2, a histone mark crucial to maintaining the heterochromatin structure. The maintenance of heterochromatic structure not only is essential for gene expression regulation but also plays an important role for protection of chromosome integrity. Narita et al. reported that during senescence, the levels of H3K9me3/me2 are increased at senescence-associated heterochromatic foci (SAHFs), concomitant with the increased binding of Heterochromatin protein 1 (HP1) [10]. The authors showed that some of the E2F target promoters in fact, acquire heterochromatic features during senescence, resulting in a permanent shutdown of these genes. Therefore, dissociation of heterochromatin could result in re-expression of E2F target genes and an escape from cellular senescence. Peters' study on the other hand, revealed that decrease of H3K9me3 at pericentric chromatin results in loss of the heterochromatin structure, leading to severe chromosome mis-segregation and genomic instability [11]. Together, as is found in prostate cancer [12], dysregulation of KDM4 family therefore may function as oncogenes and contributes to tumorigenesis. Other oncogenic roles such as alteration of cellular signaling pathways, promotion of cell cycle, expression of oncogenes/repression of tumor suppressors, are often linked with overexpression of various demethylases in tumor (Table 15.2). In addition to KDM4, KDM1A, KDM3 and KDM5 families are found to be overexpressed in several types of tumor.

While fewer cases, down-regulation of demethylases with tumor suppressive roles in cancer has also been reported. KDM2A plays a crucial role in sustaining heterochromatin structure and genome stability by repressing transcription of the centromeric satellite repeats [13]. KDM2B on the other hand, transcriptionally represses ribosomal RNA genes whose expressions are in high demand for proliferating cancer cells [14, 15]. Underexpression of KDM2B results in increased cell sizes and proliferation in tumor, suggesting it being a tumor suppressor [14].

Making things more complicated were the observations that the roles of histone demethylases as oncogenes or tumor suppressors are cell context dependent. As described above, KDM1A is overexpressed in many cancer types (Table 15.2), and appears to play oncogenic roles. However, KDM1A is also found to be down-regulated in liver and breast cancer, where it inhibits tumor invasion and metastasis [16, 17]. Similarly, while KDM2B is down-regulated in brain and glioblastoma, and was proposed as a putative tumor suppressor [14], it is found to be highly expressed in various leukemias with an oncogenic function [18]. These findings suggest that the functions of histone demethylases in cancer are dictated by the loci they act upon and the cell-type specific cofactors they associated with. As such, the results have strong therapeutic implications and underscore the importance in understanding the target gene profiles and the associated mechanisms of histone demethylases in particular cancer types.

The mechanisms of dysregulation of histone demethylase expression are multitude including alterations at the level of transcription and post-transcriptional modifications or genomic alterations. Below we will discuss the genetic alterations of KDMs in cancers.

#### 15.1.1.2 Gene Amplification

Comparative genomic hybridization (CGH) analysis revealed that the 9p23-24 region is frequently amplified in several tumors including esophageal cancer, breast cancer and lymphoma; while KDM4C (GASC1/JMJD2C) gene located at the 9p23-24 amplicon is overexpressed in these tumors [19–24]. Overexpression of KDM4C results in tumorigenic phenotypes, and is found to be associated with aggressive breast tumors [23]. Similarly, KDM5A gene located at 12p11 is found to be amplified in several tumors including breast cancer, and overexpression of which, contributes to cancer proliferation and drug resistance [25].

#### 15.1.1.3 Gene Translocation

van Zutven et al. first reported chromosome rearrangements involving KDM5A (JARID1A) and NUP98 in acute leukemias [26]. This translocation results in a fusion product consisting of the N-terminus of NUP98 and the C-terminal PHD finger domain of KDM5A. In doing so, the PHD domain targets NUP98 to active chromatin region with H3K4me3 histone mark and prevents the recruitment of the repressive polycomb complex (PRC2) to the promoter, thus, enabling active transcription of the crucial developmental loci and eventually leading to the development of leukemia [27]. In this case, it is not the catalytic of demethylase, but rather the chromatin binding domain which participates in the oncogenic transformation.

#### 15.1.1.4 Gene Mutation

Inactivating mutations in histone demethylases have been identified. H3K4me3 demethylase KDM5C (JARID1C), suggested to be a tumor suppressor [28], has several nonsense and missense mutations in clear cell renal cell carcinoma patients [29]. Similarly, systematic mutational screen of KDM6A (UTX) reveals that nonsense, frameshift, and deletion mutations are often identified in cancers including leukemia, lymphoma, myeloma, glioblastoma, breast, colorectal, endometrial, lung, esophageal, pancreatic, bladder and clear cell renal cell carcinoma [29–31]. These findings suggest that KDM6A is a tumor suppressor, consistent with its ability to demethylate H3K27me3, a histone mark whose elevation is often associated with malignancy, and to positively regulate Rb tumor suppressor and antagonize Notch signaling pathway [32–35].

## 15.1.2 *The Functional Roles of Histone Lysine Demethylases in Cancer*

Histone lysine demethylases affect a wide spectrum of cellular pathways. Based on the literature cited in Table 15.2, the following oncogenic pathways appear to be the most frequently dysregulated by histone demethylases.

### 15.1.2.1 Cell Cycle Regulation

One of the major oncogenic properties of histone lysine demethylases is their role in cell cycle regulation. Overexpression of several demethylases is linked to promoting G1-S progression and inducing tumor cell proliferation through positive regulation of S-phase cyclins and cyclin-dependent kinases (CDKs), and/or negatively regulating CDK inhibitors. Demethylases that are reported to induce expression of the S-phase cyclins, Cyclin D1, include KDM3A, KDM4A, KDM4B and KDM5B [36–40]; and those for Cyclin A1 expression includes KDM4B and KDM8 [41, 42]. Those which negatively regulate CDK inhibitors such as $p21^{Cip1/Waf1}$, are KDM1A, KDM5A, KDM5B and KDM8 [25, 43–46]; and $p27^{Kip1}$ by KDM5A and KDM5B [25, 47]. E2F transcription factors that play crucial roles in G1-S transition are also common targets of histone demethylases [43, 48, 49]. KDM1A positively regulates E2F1 gene expression, and also regulates its transcriptional activity by destabilizing the Rb regulator MYPT1 [43, 50]. PHF8 functions as a co-activator of E2F1 during G1-S transition by forming a complex with it, and upon recruitment to E2F1 target promoters, PHF8 removes the repressive H4K20me1 mark and consequently activates expression of the target genes [51].

Histone demethylase-mediated cell cycle control and proliferation is also found to channel through the p53-Rb axis [46, 48, 52–55] and senescence regulation (see below). In addition to transcriptional regulation, histone demethylase such as KDM4A directly regulates DNA replication by removing heterochromatin marks and increasing chromatin accessibility for replication machinery [56].

### 15.1.2.2 Senescence

Senescence is a process of irreversible cell growth arrest important for preventing excessive proliferation and functions as a suppression mechanism of tumorigenesis [57, 58]. *INK4b-ARF-INK4a* locus encodes $p15^{INK4B}$, $p16^{INK4A}$ and $p14^{ARF}$ proteins that sense stress signals and function as key regulators of senescence [59]. The *INK4b-ARF-INK4a* locus is normally silenced with H3K27 methylation by the polycomb complexes PRC1 and PRC2. When cells undergo senescence, the PRC complexes and H3K27me3 are lost from the locus, leading to expression of *INK4A*, *INK4B* and *ARF* [60]. Studies on KDM2B revealed a double regulatory mechanism for senescence and cell proliferation. KDM2B regulates senescence in part, by directly binding to the *INK4b-ARF-INK4a* locus and demethylating the locus-associated H3K36me2 and H3K4me3 which results in the suppression of *INK4a* and *INK4b* [61, 62]. KDM2B also modulates the expression of H3K27 tri-methyltransferase EZH2 by negatively regulating tumor suppressor miRNAs let-7b and miR-101. KDM2B-mediated up-regulation of EZH2 increases the suppressive histone mark of H3K27me3 on *INK4b-ARF-INK4a* locus and further contributes to silencing. When primary mouse embryonic fibroblasts (MEFs) undergo senescence, the KDM2B-let7-EZH2 pathway presents a feed-forward mechanism to ensure senescence: KDM2B is down-regulated, leading to expression of let-7b and miR-101 which in turn, represses EZH2 expression [63]. The ability of KDM2B overexpression in promoting immortalization and sustained proliferation of both *wild type* and *INK4a-ARF* null MEFs suggests its important roles in tumor initiation and development.

Rb and p53 tumor suppressors also play essential roles in senescence. They induce senescence by regulating the expression of CDK inhibitor p21, and formation of heterochromatin on E2F-responsive promoters [10, 64]. During senescence, global changes in histone modifications include increase of transcriptional silencing marks H3K9me3, H3K27me3 and H4K20me3; and loss of active histone mark H3K4me3/me2 [64]. As suggested above, overexpression of demethylases that are involved in the Rb-p53-p21 pathway, or in removing the transcriptional silencing marks globally or loci-specific, may contribute to the loss of the tumor-suppressing senescence mechanism.

#### 15.1.2.3 Hypoxia

Hypoxia is a stage when a cellular demand of oxygen for metabolism exceeds the local blood supply. It occurs in tumor tissues where cell proliferation outgrows angiogenesis, resulting in local low concentration of oxygen in the high-cell dense regions of tumors. Tumor hypoxia has been reported to associate with poor prognosis and with resistance to radiotherapy and chemotherapy. Therefore understanding the survival mechanisms of tumor cells under hypoxic conditions is of significant importance in the design of therapeutic strategy. Hypoxia-induced transcription factor (HIF) complexes, consisting of $\alpha$- (HIF-1$\alpha$, HIF-2$\alpha$ or HIF-3$\alpha$) and $\beta$- subunits, are the predominant hypoxia-responsive regulators that modulate adaptive gene expression aiming at restoring cellular oxygen homeostasis. Recent studies suggest cooperative actions between epigenetic regulators and HIF in response to hypoxia. These regulators include enzymes involved in DNA methylation, chromatin remodeling and histone modifications at the HIF promoter regions. Under hypoxic conditions, global levels of H3K4me3 and H3K9me2 have been demonstrated to be increased, whereas H3K27me3 level decreases [65, 66]. It is speculated that these chromatin modification signatures indicate flexible access of other chromatin modifiers and transcriptional regulators to promoters, in turn facilitating reprogramming of gene expression under transient hypoxia-reoxygenation conditions that often occur in tumor [65]. Several JmjC histone demethylases such as KDM3A, KDM4B,

KDM4C, KDM5A and KDM5B are found to be activated in the HIF signaling pathway during hypoxic conditions [67–70]. The importance of hypoxia-induced histone demethylases is best illustrated by KDM3A (JMJD1A/JHDM2A). Upon induction under hypoxia, KDM3A is recruited to HIF target genes, and by removing the repressive H3K9me2 marks, KDM3A facilitates hypoxic gene expression such as adrenomedullin (ADM), differentiation factor 15 (GDF15), and GLUT3 (SLC2A3) [67, 71]. Mimura et al. further demonstrated that the recruitment of KDM3A to the GLUT3 promoter depends on physical interaction between KDM3A and HIF1 [71]. In addition to HIF, other transcription factors such as NFκB, CREB and EGR-1 are involved in hypoxia responses [72–74]. Given that many non-histone proteins including NFκB are found to be methylated, the roles of lysine demethylases in hypoxia may be broader than previously thought. Indeed, KDM2A is found to demethylate NFκB [75] and affects its transcriptional potential.

## 15.2 Inhibitors of Histone Demethylases

Given the strong implications of KDMs in cancer, small molecule inhibitors targeting their enzyme activities are being actively investigated in recent years (Table 15.3). Among them, inhibitors for KDM1A are the most extensively pursued; 12 such inhibitors (Compound 1 to 12) are listed in Table 15.3. KDM1A belongs to a superfamily of FAD-containing amine oxidases and the developed inhibitors include substrate analogues, MAO (monoamine oxidase) inhibitor analogues and polyamine analogues. Culhane et al. first reported compound 1 as a suicide inhibitor of KDM1A through substitution of lysine4 in H3 [76]. Through coupling of compound 1 with FAD, the mechanism of KDM1A demethylation was revealed ([77, 78], compound 2). The prototype of amino oxidases is the antidepressant MAO, based on which several inhibitors have been developed. Analogues of MAO inhibitors such as trans-2-phenylcyclopropylamine (compound 3-8) were shown to be effective in inhibiting KDM1A [79–85]. Another class of KDM1A inhibitors is polyamine analogues (compound 9-11) [86–88], inhibition of KDM1A in colon cancer cell by this class of compounds resulted in reexpression of aberrantly silenced genes [86, 87]. Compound 12, a new KDM1A inhibitor, was found to selectively target cancer cells with pluripotent stem cell properties [89]. Although these inhibitors are very useful in exploring the structure and biological function of KDM1A, their IC50s for cell killing are in the range of micromolar to millimolar, too high to be appropriate for clinical trials and hence further optimization is needed.

Other KDMs are JmjC containing demethylases, which are α-ketoglutarate dependent oxygenases. The design of most inhibitors is based on the scaffold of this cofactor. Compound 13, a NOG (N-oxalylglycine) analogue, was identified as a KDM4A inhibitor with IC50 of 3 mM. This compound is not cell permeable and methylation of the two hydroxyl groups is required for cell penetration [90]. Other KDM inhibitors include hydroxamate analogs, pyridine dicarboxylates and bipyridil compounds. Hamada et al. identified a series of hydroxamate analogues as KDM4A

**Table 15.3** Histone demethylase inhibitors

| Compound | Inhibitor structure | Targeted KDMs | $IC_{50}$ (μM) | References |
|---|---|---|---|---|
| 1 | 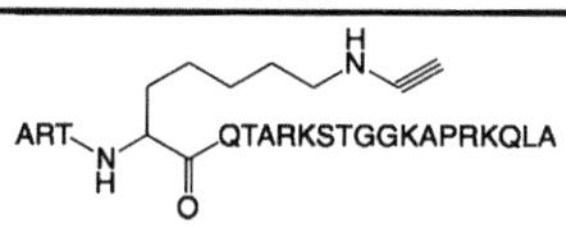 | KDM1A | Suicide | [76] |
| 2 | 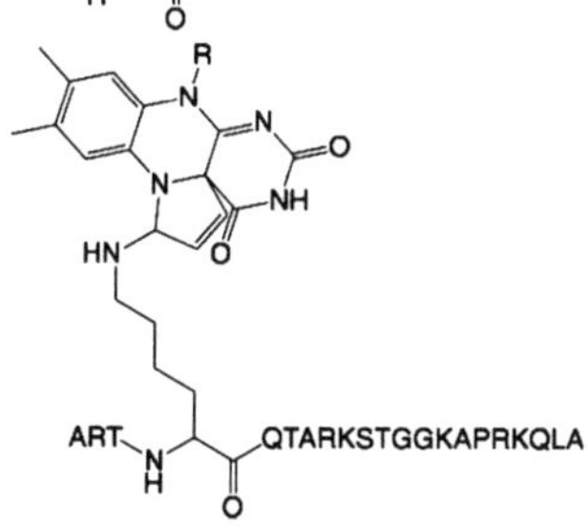 | KDM1A | Suicide | [77, 78] |
| 3 | 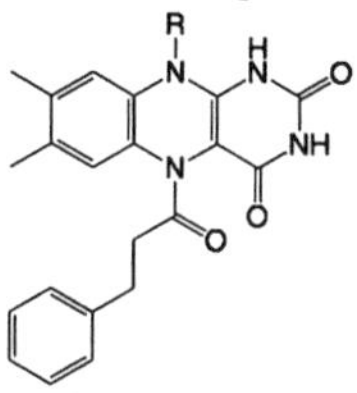 | KDM1A | 2 | [79, 80] |
| 4 | 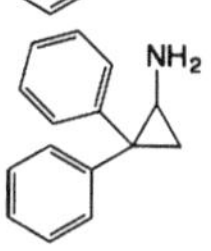 | KDM1A | ~250 | [81] |
| 5 | 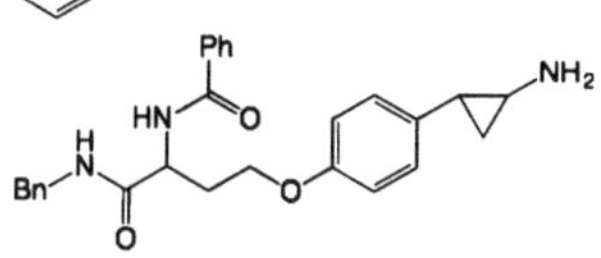 | KDM1A | 1.9 | [82] |
| 6 | 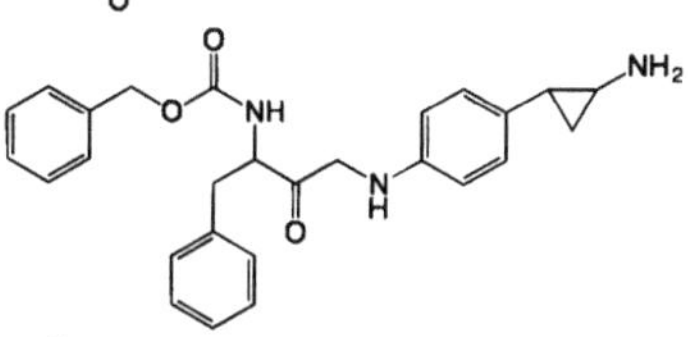 | KDM1A | 1.3 | [83] |
| 7 | 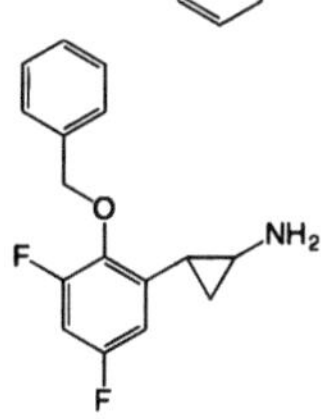 | KDM1A | 1.0 | [84] |
| 8 | 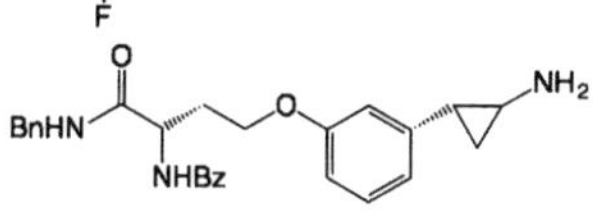 | KDM1A | 1.6 | [85] |

(continued)

**Table 15.3** (continued)

| Compound | Inhibitor structure | Targeted KDMs | $IC_{50}$ (μM) | References |
|---|---|---|---|---|
| 9 | 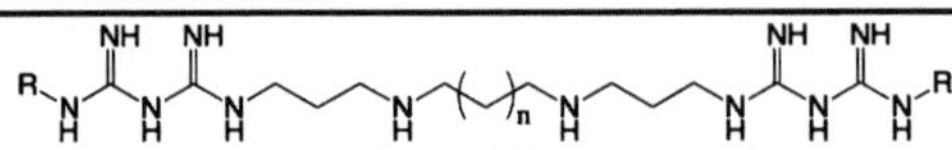 | KDM1A | 2.5 | [86] |
| 10 | 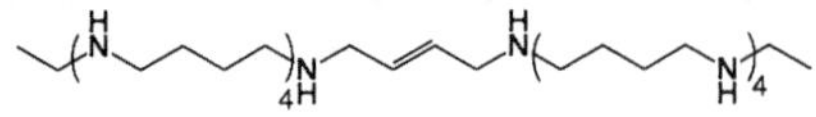 | KDM1A | ~10 | [87] |
| 11 | 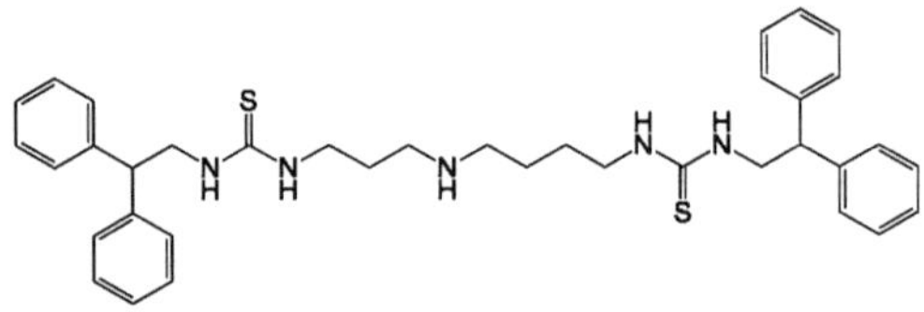 | KDM1A | 17.1 | [88] |
| 12 | 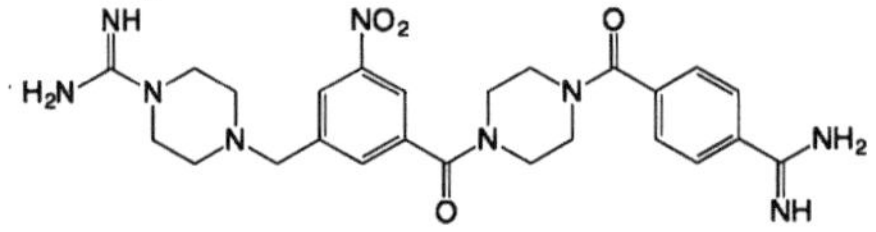 | KDM1A | 10.5 | [89] |
| 13 | 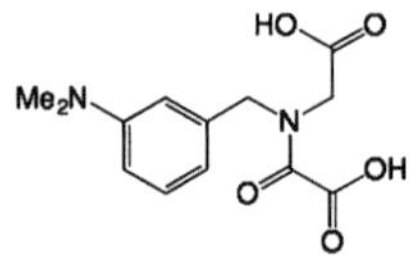 | KDM4A | 3,000 | [90] |
| 14 | 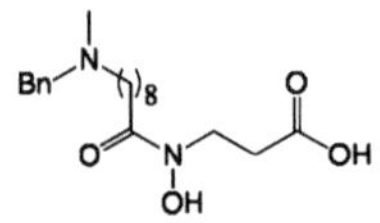 | KDM4A<br>KDM4C | 2.0<br>1.3 | [91] |
| 15 | 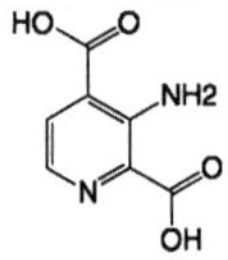 | KDM4E | 0.11 | [92] |
| 16 | 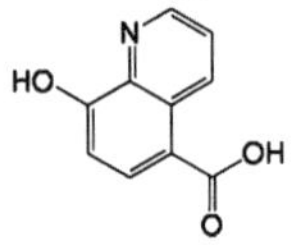 | KDM4A<br>KDM4E | 1.7<br>2.4 | [93] |
| 17 | 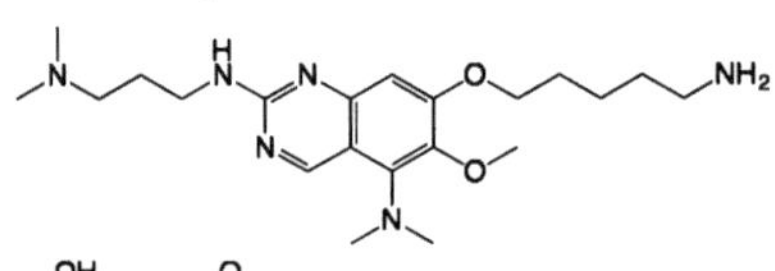 | KDM7A | 3.4 | [94] |
| 18 | 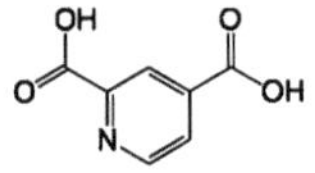 | KDM5B<br>KDM4A<br>KDM4E | 3<br>0.6<br>1.4 | [93, 95] |
| 19 | 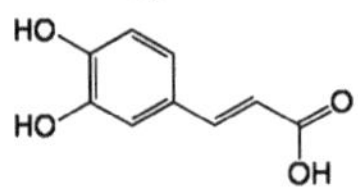 | KDM4C<br>KDM6A | 13.7<br>5.5 | [96] |
| 20 | 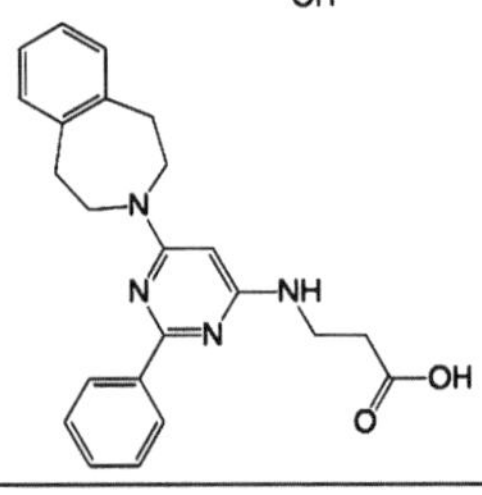 | KDM6B | 0.06 | [97] |

and KDM4C inhibitors in low micromole range ([91], compound 14), whereas 3-substituted pyridine 2,4-dicarboxylate was found to be a potent inhibitor of KDM4E ([92], compound 15). Through high-throughput screening, 8-hydroxyquinolines were identified as cell-active KDM4A and KDM4E inhibitors ([93], compound 16). Compound 17, a diazepin-quinazoline-amine derivative, selectively inhibits KDM7A ([94]). Interestingly, 2,4-pyridine-dicarboxylic acid (compound 18) was found to be an active inhibitor of KDM4A, KDM4E and KDM5A, indicating substrate similarities of these KDMs [93, 95]. Recently, catechols were reported to be active KDM4C and KDM6A inhibitors ([96], compound 19). The IC50s of most of these inhibitors are in millimolar or micromolar range, Kruidenier et al. identified compound 20 as a selective KDM6B inhibitor in nanomolar range. Interestingly, this inhibitor modulates the proinflammatory macrophage response ([97], compound 20). Although these inhibitors showed variable activities against purified KDMs, none of these inhibitors inhibit cancer cell growth below micromole range and none has reached clinical trials. Given the important function of KDMs in cancer, development of more potent KDM inhibitors is highly desirable.

## 15.3 Histone Lysine Demethylases in Prostate Cancer

Using the Oncomine database, we performed a comparison of several prostate cancer studies to identify histone demethylases that are differentially expressed in normal and tumor samples (Fig. 15.1). Our analysis agrees well with those reported in the literature (Table 15.2), and can serve as a future guide for developing demethylase-targeted therapies. The mechanisms regarding how the demethylases serve as progression factors in prostate cancer are being actively investigated, and below is a summary of the most relevant signaling pathways targeted by histone lysine demethylases.

### *15.3.1 Targeting Androgen Signaling*

Androgen/androgen receptor (AR) signaling is essential for early stage of prostate cancer cell growth and survival, which can be managed by anti-androgen hormone therapy. At a later stage, the majority of the tumors are transitioned into hormone-independent or castration-resistant state (CRPC), for which there is thus far no effective cure. Although most of the CRPCs no longer depend on the external androgen for growth and survival, the AR activity which is often aberrantly activated is still required. Several histone demethylases overexpressed in prostate cancer (Table 15.2) contribute to the aberrant activation of AR and AR associated signaling. This is understandable, as nuclear hormone receptors such as AR are known to form complex with "co-activators" to exert their transcriptional function; these coactivators are often histone-modifying enzymes and chromatin remodeling

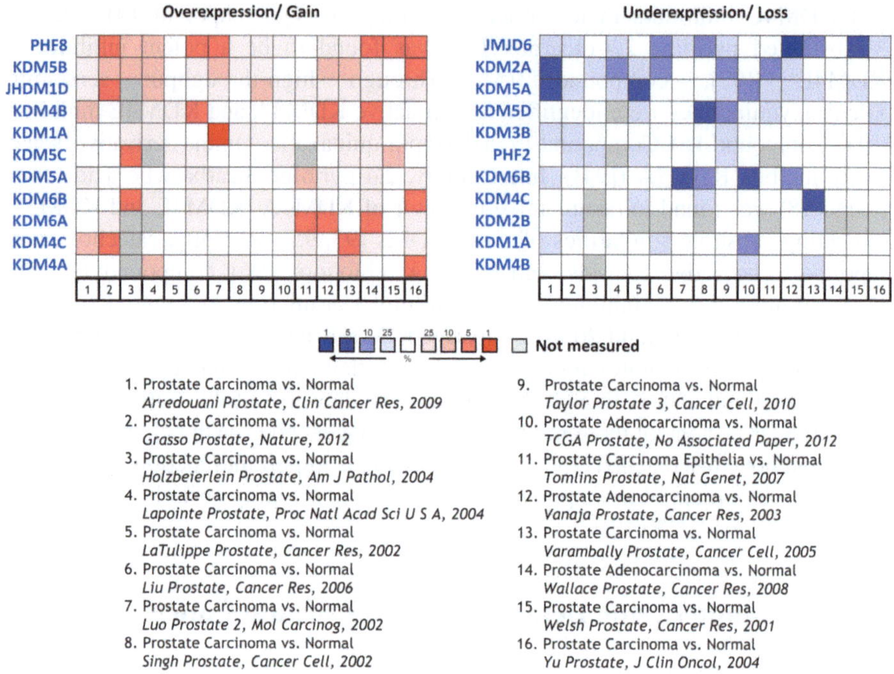


**Fig. 15.1** Expression of histone demethylases that are differentially expressed in prostate cancer versus normal tissues are compared among 16 independent studies from Oncomine databases. In each dataset, the expression data was either based upon mRNA or DNA copy number analysis. *Left panel* shows the top 11 histone demethylase (half of total demethylases with substrates identified) that are gained/overexpressed in prostate tumors compared to that in normal tissue; *right panel* illustrates those that are loss/underexpressed in tumor. As described in Oncomine Differential Expression Analyses, each gene in each individual dataset is ranked to indicate its statistical significance of over- or under-expression. The ranking percentile of each demethylase relative to the ranks of all other genes in individual dataset is indicated with different color gradient: top 1 % of the ranks are shown with the darkest color (*red*, overexpression; *blue*, underexpression), the lower the ranking percentile, the lighter the color to be indicated. Briefly, the demethylases that are shown to be present with higher ranks (*darker color*) in more datasets suggests a higher probability of them being overexpressed or underexpressed in prostate tumors

proteins, to which histone demethylases belong. Their general functions are to generate an open chromatin conformation allowing RNA polymerase and transcriptional complex to engage and to transcribe the target gene. Aberrantly expressed histone methylases thus can cause aberrant activation of AR.

At least seven histone lysine demethylases are found to promote AR transcriptional activity: KDM1A, KDM3A, KDM4A, KDM4C, KDM4D, KDM5B and KDM8. KDM1A was originally described as a specific "eraser" of the active histone mark, H3K4me2/me1, and thus, functions as a transcriptional repressor. Interestingly, when it complexes with AR, and after H3T6 is phosphorylated by PKCβ1, KDM1A switches its demethylation specificity from H3K4me2/me1 to the repressive histone mark H3K9me2/me1 [98, 99], thereby enhancing AR activity on the target genes. Inhibition or silencing of KDM1A results in reduced androgen-dependent proliferation and PSA (an AR target gene) transcription [98, 100]. These findings suggest that KDM1A functions as a coactivator for AR, and plays an important role in prostate cancer. Indeed, KDM1A is up-regulated in prostate tumor tissues, and overexpression of which is associated with higher relapse risk [100]. Wissmann et al. later identified a single complex consisting of KDM1A, KDM4C and AR, and reported KDM4C also as an AR coactivator. KDM4C and KDM1A bind to androgen responsive elements (ARE) located at promoter and enhancer of AR target genes, and upon hormone stimulation, they cooperatively remove the repressive tri-, di- and mono-methylated H3K9 marks [101]. This cooperation action synergistically enhances AR transcriptional activity on PSA enhancer. It is worth noting that different to the scenario of KDM1A-mediated estrogen receptor (ER) coactivation, where recruitment of KDM1A to ER target genes is ligand-dependent [102], chromatin binding of KDM1A and KDM4C to AR targets occurs in the absence of androgen treatment, while their demethylation activity on H3K9 depends on androgen signaling. KDM3A by contrast, displays hormone-dependent interaction with AR as well as chromatin recruitment to AREs [103]. Binding of KDM3A in turn, catalyzes loci specific demethylation of mono- and di-methylated H3K9. Similar to KDM1A, KDM3A and KDM4C are essential for hormone-induction of AR targets and hormone-dependent proliferation [101, 103]. Given their overexpression in prostate tumors and contributing to AR activation [12, 104], KDM3A and KDM4C are potential therapeutic targets for prostate cancer.

As exemplified by the analysis of AR and ER, recent studies have suggested that removal of the repressive H3K9 methylation marks at the promoter of target genes is crucial for nuclear receptor-mediated gene expression [105]. Aside from the KDM1A-KDM4C complex, H3K9me3/me2 demethylases KDM4A and KDM4D are also shown to interact with AR, and their overexpression enhance AR activity on the PSA enhancer [106]. Surprisingly, KDM5B (JARID1B) that demethylates the active histone mark H3K4me3/me2, is found to be overexpressed in prostate cancer and also serves as a coactivator for AR [104, 107]. While the detailed mechanism associated with KDM5B being an AR coactivator is not clear, at least two possibilities can be considered. First, as has been reported in different contexts [108, 109], H3K4me3/me2 may function as a repressive mark in a loci-specific way, and demethylation by KDM5B would activate AR target gene transcription. Second, similar to the case of KDM1A discussed above, association with AR could alter KDM5B's substrate specificity. The fact that KDM5B is significantly overexpressed in metastatic prostate cancer cells [107] indicates that the KDM5B-mediated AR activation is likely to bypass the androgen requirement, and KDM5B could serve as a potential target for late stage therapeutics. Our unpublished data showed that several other histone demethylases also interact with AR. We found, for instance, KDM8 (JMJD5), a H3K36me2 demethylase [42, 46], is overexpressed in high-grade prostate cancer and forms a complex with KDM4A and AR on chromatin. Ectopic expression of KDM8 and KDM4A synergistically enhanced AR activity with concomitant decrease of H3K36me2 at the target promoter. Because KDM4A is also capable of catalyzing demethylation on H3K36me3/me2 [110, 111], the combination of KDM8 and KDM4A is expected to potently demethylate both H3K36me3/me2 and H3K9me3, allowing effective H3 acetylation. In addition to KDM8, we also identified interactions between KDM1B, KDM2A, KDM4B, KDM5A, KDM5D and AR. The data taken together suggest that histone demethylases either singly or in combination can serve as coactivator of AR to change the chromatin landscape of the AR target genes, thereby augmenting the transcription.

Finally, in addition to changing the local chromatin structure for AR target genes, histone demethylases are found to directly regulate AR. For instance, KDM1A is recruited by AR to an intronic enhancer of AR locus and represses AR expression via the removal of H3K4me3 mark [112]. Unlike the situation in PSA promoter, KDM1A in this case is not complexed with KDM4C and serves as a transcriptional repressor. This autoregulation takes place only when there is abundant androgen and acts as a feedback mechanism to shut off androgen signal. Under androgen-depleted conditions as in the case of CRPC, AR expression is usually increased. The discussion above indicates that histone demethylases, like histone acetylases and deacetylases, are partners of AR, and may play significant role in the dysregulation of AR activity during transition to hormone refractory prostate cancer. What discussed above is almost certainly only the tip of iceberg. More histone demethylases which directly or indirectly affect androgen receptor signaling are likely to be uncovered in the coming years.

### *15.3.2 Targeting Other Oncogenic Signals*

KDM5C (SMCX/JARID1C) is a H3K4me3/me2 demethylase also found to be overexpressed in prostate cancer. KDM5C physically interacts with TGFβ-downstream transcription factor Smad3, and overexpression of which inhibits Smad3 activity independently of its demethylase activity [113]. Since TGFβ signaling acts as a tumor suppressive pathway in early prostate cancer [114], antagonizing the TGFβ-Smad3 pathway by KDM5C may therefore promote prostate tumor initiation. Another strongly overexpressed demethylase observed in clinical prostate cancer samples is PHF8, whose expression is correlated with high Gleason grade and poor prognosis [104]. PHF8 can demethylate multiple substrates including H3K9me2/me1, H3K27me2 and H4K20me1. Although the mechanism remains unclear, knockdown of PHF8 inhibits proliferation, migration and invasion ability of prostate cancer cells, indicating PHF8 as a potent oncogene for prostate cancer [104].

While a number of demethylases seem to exhibit oncogenic potential, KDM2A was found to be underexpressed in prostate cancer and functions as a tumor suppressor. Frescas et al. showed that KDM2A is required to maintain the centromeric heterochromatin state and also sustain genomic integrity. Underexpression of KDM2A in prostate cancer may thus cause genomic instability, contributing to cellular transformation [13].

### *15.3.3 Histone Methylation as Biomarkers for CRPC*

Seligson et al. first reported that global levels of histone modification can be used to predict clinical outcome of prostate cancer patients with low Gleason grade [115]. Elevated H3K4me3 and H3K27me3, and reduced level of H3K9me2 in prostate tumor tissue are found to associate with poor prognosis [7, 116, 117]. While H3K4me1, H3K4me2 and H3K4me3 levels are significantly increased in CRPC, higher level of H3K4me1 is more likely to develop recurrence [118]. One of the mechanisms underlying the altered histone methylation-associated malignancy and prostate tumor recurrence is AR -mediated activation of proto-oncogenes and repression of tumor suppressors. Genome-wide analysis revealed that in CRPC cells, H3K4me1 and H3K4me2 are selectively enriched at enhancers of oncogenes such as *UBE2C* and *CDK1*, facilitating recruitment of AR for their transcription. Up-regulation of these cell cycle genes in turn, promotes growth of CRPC cells [119]. Similarly, increased H3K4me3 in prostate cancer cell correlates with the expression of oncogenes including *FGFR1*, *BCL2* and *HOXC5*[120]. By contrast, H3K27me3 mark is enriched at the promoters of tumor suppressor genes, leading to their silencing in metastatic prostate cancer cell [121]. Together, emerging studies have suggested that histone modifications can serve as prognostic markers to predict outcome of prostate cancer. The intervention potential of the possible demethylases and methyltransferases that are responsible for the altered histone methylation is worthy of further consideration.

## 15.4 Concluding Remarks

In the past 8 years since the discovery of the first histone demethylase, KDM1A, extraordinary progress has been made in understanding their modes of action on histones and their connections to epigenetic regulation of carcinogenesis. Epigenetic regulation of cancer is important not only during transformation and metastasis processes, but also during therapeutic resistance. As master programmers of epigenetic regulation, histone demethylases are potential targets for intervention. Attentions to this group of genes, especially on understanding of their up- and down-stream signal pathways will only increase in the coming years. A few comments on the future direction of these research activities are provided here. First, the early literatures on histone demethylases have mostly focused on their actions on histone. Yet, we now know that KDMs may have other cellular substrates whose demethylation fuel the carcinogenesis processes. Identification of non-histone substrates of KDMs will be important to fully appreciate KDMs' modes of action. Second, as exemplified by KDM5C, KDMs may exert their function in a demethylation-independent manner. Hence, small molecules targeting the enzymatic activity may not work in this case. Third, paradoxically, in some cancers, KDMs and their counteracting histone methylases can both be overexpressed and serve as progression factors (e.g., KDM8 and

NSD2). This suggests that it is not the global level of the particular histone marks, but rather the loci-specific epigenetic landscape which determines the final outcome. This makes the measurement of therapeutic responses more challenging. The development of histone demethylase inhibitors is still at very early stage; however several promising leads have already surfaced (e.g., KDM1A for prostate cancer). Given the wide range of activities and biological outcomes of histone demethylases, one can envision a tremendous surge of research activities in the related areas with an intensity which may rival those for tyrosine kinases.

**Acknowledgement** This work is supported by NIH and DOD grants to H.J.K. H.J.K. acknowledges the generous support from Auburn Community Cancer Endowment Fund.

## References

1. Berger SL (2007) The complex language of chromatin regulation during transcription. Nature 447(7143):407–412
2. Kouzarides T (2007) Chromatin modifications and their function. Cell 128(4):693–705
3. Varier RA, Timmers HT (2011) Histone lysine methylation and demethylation pathways in cancer. Biochim Biophys Acta 1815(1):75–89
4. Pedersen MT, Helin K (2010) Histone demethylases in development and disease. Trends Cell Biol 20(11):662–671
5. Geutjes EJ, Bajpe PK, Bernards R (2012) Targeting the epigenome for treatment of cancer. Oncogene 31(34):3827–3844
6. Fraga MF, Ballestar E, Villar-Garea A, Boix-Chornet M, Espada J, Schotta G et al (2005) Loss of acetylation at Lys16 and trimethylation at Lys20 of histone H4 is a common hallmark of human cancer. Nat Genet 37(4):391–400
7. Seligson DB, Horvath S, McBrian MA, Mah V, Yu H, Tze S et al (2009) Global levels of histone modifications predict prognosis in different cancers. Am J Pathol 174(5):1619–1628
8. Elsheikh SE, Green AR, Rakha EA, Powe DG, Ahmed RA, Collins HM et al (2009) Global histone modifications in breast cancer correlate with tumor phenotypes, prognostic factors, and patient outcome. Cancer Res 69(9):3802–3809
9. Barlesi F, Giaccone G, Gallegos-Ruiz MI, Loundou A, Span SW, Lefesvre P et al (2007) Global histone modifications predict prognosis of resected non small-cell lung cancer. J Clin Oncol 25(28):4358–4364
10. Narita M, Nunez S, Heard E, Lin AW, Hearn SA, Spector DL et al (2003) Rb-mediated heterochromatin formation and silencing of E2F target genes during cellular senescence. Cell 113(6):703–716
11. Peters AH, O'Carroll D, Scherthan H, Mechtler K, Sauer S, Schofer C et al (2001) Loss of the Suv39h histone methyltransferases impairs mammalian heterochromatin and genome stability. Cell 107(3):323–337
12. Cloos PA, Christensen J, Agger K, Maiolica A, Rappsilber J, Antal T et al (2006) The putative oncogene GASC1 demethylates tri- and dimethylated lysine 9 on histone H3. Nature 442(7100):307–311
13. Frescas D, Guardavaccaro D, Kuchay SM, Kato H, Poleshko A, Basrur V et al (2008) KDM2A represses transcription of centromeric satellite repeats and maintains the heterochromatic state. Cell Cycle 7(22):3539–3547
14. Frescas D, Guardavaccaro D, Bassermann F, Koyama-Nasu R, Pagano M (2007) JHDM1B/FBXL10 is a nucleolar protein that represses transcription of ribosomal RNA genes. Nature 450(7167):309–313

15. Ruggero D, Pandolfi PP (2003) Does the ribosome translate cancer? Nat Rev Cancer 3(3):179–192
16. Magerl C, Ellinger J, Braunschweig T, Kremmer E, Koch LK, Holler T et al (2010) H3K4 dimethylation in hepatocellular carcinoma is rare compared with other hepatobiliary and gastrointestinal carcinomas and correlates with expression of the methylase Ash2 and the demethylase LSD1. Hum Pathol 41(2):181–189
17. Wang Y, Zhang H, Chen Y, Sun Y, Yang F, Yu W et al (2009) LSD1 is a subunit of the NuRD complex and targets the metastasis programs in breast cancer. Cell 138(4):660–672
18. He J, Nguyen AT, Zhang Y (2011) KDM2b/JHDM1b, an H3K36me2-specific demethylase, is required for initiation and maintenance of acute myeloid leukemia. Blood 117(14):3869–3880
19. Han W, Jung EM, Cho J, Lee JW, Hwang KT, Yang SJ et al (2008) DNA copy number alterations and expression of relevant genes in triple-negative breast cancer. Genes Chromosom Cancer 47(6):490–499
20. Italiano A, Attias R, Aurias A, Perot G, Burel-Vandenbos F, Otto J et al (2006) Molecular cytogenetic characterization of a metastatic lung sarcomatoid carcinoma: 9p23 neocentromere and 9p23-p24 amplification including JAK2 and JMJD2C. Cancer Genet Cytogenet 167(2):122–130
21. Savelyeva L, Claas A, An H, Weber RG, Lichter P, Schwab M (1999) Retention of polysomy at 9p23-24 during karyotypic evolution in human breast cancer cell line COLO 824. Genes Chromosomes Cancer 24(1):87–93
22. Yang ZQ, Imoto I, Fukuda Y, Pimkhaokham A, Shimada Y, Imamura M et al (2000) Identification of a novel gene, GASC1, within an amplicon at 9p23-24 frequently detected in esophageal cancer cell lines. Cancer Res 60(17):4735–4739
23. Liu G, Bollig-Fischer A, Kreike B, van de Vijver MJ, Abrams J, Ethier SP et al (2009) Genomic amplification and oncogenic properties of the GASC1 histone demethylase gene in breast cancer. Oncogene 28(50):4491–4500
24. Rui L, Emre NC, Kruhlak MJ, Chung HJ, Steidl C, Slack G et al (2010) Cooperative epigenetic modulation by cancer amplicon genes. Cancer Cell 18(6):590–605
25. Hou J, Wu J, Dombkowski A, Zhang K, Holowatyj A, Boerner JL et al (2012) Genomic amplification and a role in drug-resistance for the KDM5A histone demethylase in breast cancer. Am J Transl Res 4(3):247–256
26. van Zutven LJ, Onen E, Velthuizen SC, van Drunen E, von Bergh AR, van den Heuvel-Eibrink MM et al (2006) Identification of NUP98 abnormalities in acute leukemia: JARID1A (12p13) as a new partner gene. Genes Chromosomes Cancer 45(5):437–446
27. Wang GG, Song J, Wang Z, Dormann HL, Casadio F, Li H et al (2009) Haematopoietic malignancies caused by dysregulation of a chromatin-binding PHD finger. Nature 459(7248):847–851
28. Niu X, Zhang T, Liao L, Zhou L, Lindner DJ, Zhou M et al (2012) The von Hippel-Lindau tumor suppressor protein regulates gene expression and tumor growth through histone demethylase JARID1C. Oncogene 31(6):776–786
29. Dalgliesh GL, Furge K, Greenman C, Chen L, Bignell G, Butler A et al (2010) Systematic sequencing of renal carcinoma reveals inactivation of histone modifying genes. Nature 463(7279):360–363
30. van Haaften G, Dalgliesh GL, Davies H, Chen L, Bignell G, Greenman C et al (2009) Somatic mutations of the histone H3K27 demethylase gene UTX in human cancer. Nat Genet 41(5):521–523
31. Gui Y, Guo G, Huang Y, Hu X, Tang A, Gao S et al (2011) Frequent mutations of chromatin remodeling genes in transitional cell carcinoma of the bladder. Nat Genet 43(9):875–878
32. Wang JK, Tsai MC, Poulin G, Adler AS, Chen S, Liu H et al (2010) The histone demethylase UTX enables RB-dependent cell fate control. Genes Dev 24(4):327–332
33. Herz HM, Madden LD, Chen Z, Bolduc C, Buff E, Gupta R et al (2010) The H3K27me3 demethylase dUTX is a suppressor of Notch- and Rb-dependent tumors in Drosophila. Mol Cell Biol 30(10):2485–2497

34. Tsai MC, Wang JK, Chang HY (2010) Tumor suppression by the histone demethylase UTX. Cell Cycle 9(11):2043–2044
35. Terashima M, Ishimura A, Yoshida M, Suzuki Y, Sugano S, Suzuki T (2010) The tumor suppressor Rb and its related Rbl2 genes are regulated by Utx histone demethylase. Biochem Biophys Res Commun 399(2):238–244
36. Berry WL, Shin S, Lightfoot SA, Janknecht R (2012) Oncogenic features of the JMJD2A histone demethylase in breast cancer. Int J Oncol 41(5):1701–1706
37. Toyokawa G, Cho HS, Iwai Y, Yoshimatsu M, Takawa M, Hayami S et al (2011) The histone demethylase JMJD2B plays an essential role in human carcinogenesis through positive regulation of cyclin-dependent kinase 6. Cancer Prev Res (Phila) 4(12):2051–2061
38. Kawazu M, Saso K, Tong KI, McQuire T, Goto K, Son DO et al (2011) Histone demethylase JMJD2B functions as a co-factor of estrogen receptor in breast cancer proliferation and mammary gland development. PLoS One 6(3):e17830
39. Mitra D, Das PM, Huynh FC, Jones FE (2011) Jumonji/ARID1 B (JARID1B) protein promotes breast tumor cell cycle progression through epigenetic repression of microRNA let-7e. J Biol Chem 286(47):40531–40535
40. Cho HS, Toyokawa G, Daigo Y, Hayami S, Masuda K, Ikawa N et al (2012) The JmjC domain-containing histone demethylase KDM3A is a positive regulator of the G1/S transition in cancer cells via transcriptional regulation of the HOXA1 gene. Int J Cancer 131(3): E179–E189
41. Yang J, Jubb AM, Pike L, Buffa FM, Turley H, Baban D et al (2010) The histone demethylase JMJD2B is regulated by estrogen receptor alpha and hypoxia, and is a key mediator of estrogen induced growth. Cancer Res 70(16):6456–6466
42. Hsia DA, Tepper CG, Pochampalli MR, Hsia EY, Izumiya C, Huerta SB et al (2010) KDM8, a H3K36me2 histone demethylase that acts in the cyclin A1 coding region to regulate cancer cell proliferation. Proc Natl Acad Sci U S A 107(21):9671–9676
43. Lim S, Janzer A, Becker A, Zimmer A, Schule R, Buettner R et al (2010) Lysine-specific demethylase 1 (LSD1) is highly expressed in ER-negative breast cancers and a biomarker predicting aggressive biology. Carcinogenesis 31(3):512–520
44. Zeng J, Ge Z, Wang L, Li Q, Wang N, Bjorkholm M et al (2010) The histone demethylase RBP2 Is overexpressed in gastric cancer and its inhibition triggers senescence of cancer cells. Gastroenterology 138(3):981–992
45. Ishimura A, Minehata K, Terashima M, Kondoh G, Hara T, Suzuki T (2012) Jmjd5, an H3K36me2 histone demethylase, modulates embryonic cell proliferation through the regulation of Cdkn1a expression. Development 139(4):749–759
46. Oh S, Janknecht R (2012) Histone demethylase JMJD5 is essential for embryonic development. Biochem Biophys Res Commun 420(1):61–65
47. Dey BK, Stalker L, Schnerch A, Bhatia M, Taylor-Papidimitriou J, Wynder C (2008) The histone demethylase KDM5b/JARID1b plays a role in cell fate decisions by blocking terminal differentiation. Mol Cell Biol 28(17):5312–5327
48. Nijwening JH, Geutjes EJ, Bernards R, Beijersbergen RL (2011) The histone demethylase Jarid1b (Kdm5b) is a novel component of the Rb pathway and associates with E2f-target genes in MEFs during senescence. PLoS One 6(9):e25235
49. Hayami S, Yoshimatsu M, Veerakumarasivam A, Unoki M, Iwai Y, Tsunoda T et al (2010) Overexpression of the JmjC histone demethylase KDM5B in human carcinogenesis: involvement in the proliferation of cancer cells through the E2F/RB pathway. Mol Cancer 9:59
50. Cho HS, Suzuki T, Dohmae N, Hayami S, Unoki M, Yoshimatsu M et al (2011) Demethylation of RB regulator MYPT1 by histone demethylase LSD1 promotes cell cycle progression in cancer cells. Cancer Res 71(3):655–660
51. Liu W, Tanasa B, Tyurina OV, Zhou TY, Gassmann R, Liu WT et al (2010) PHF8 mediates histone H4 lysine 20 demethylation events involved in cell cycle progression. Nature 466(7305):508–512
52. Li W, Zhao L, Zang W, Liu Z, Chen L, Liu T et al (2011) Histone demethylase JMJD2B is required for tumor cell proliferation and survival and is overexpressed in gastric cancer. Biochem Biophys Res Commun 416(3–4):372–378

53. Tsai WW, Nguyen TT, Shi Y, Barton MC (2008) p53-targeted LSD1 functions in repression of chromatin structure and transcription in vivo. Mol Cell Biol 28(17):5139–5146
54. Kim TD, Oh S, Shin S, Janknecht R (2012) Regulation of tumor suppressor p53 and HCT116 cell physiology by histone demethylase JMJD2D/KDM4D. PLoS One 7(4):e34618
55. Ishimura A, Terashima M, Kimura H, Akagi K, Suzuki Y, Sugano S et al (2009) Jmjd2c histone demethylase enhances the expression of Mdm2 oncogene. Biochem Biophys Res Commun 389(2):366–371
56. Black JC, Allen A, Van Rechem C, Forbes E, Longworth M, Tschop K et al (2010) Conserved antagonism between JMJD2A/KDM4A and HP1gamma during cell cycle progression. Mol Cell 40(5):736–748
57. Braig M, Lee S, Loddenkemper C, Rudolph C, Peters AH, Schlegelberger B et al (2005) Oncogene-induced senescence as an initial barrier in lymphoma development. Nature 436(7051):660–665
58. Collado M, Gil J, Efeyan A, Guerra C, Schuhmacher AJ, Barradas M et al (2005) Tumour biology: senescence in premalignant tumours. Nature 436(7051):642
59. Lowe SW, Sherr CJ (2003) Tumor suppression by Ink4a-Arf: progress and puzzles. Curr Opin Genet Dev 13(1):77–83
60. Bracken AP, Kleine-Kohlbrecher D, Dietrich N, Pasini D, Gargiulo G, Beekman C et al (2007) The Polycomb group proteins bind throughout the INK4A-ARF locus and are disassociated in senescent cells. Genes Dev 21(5):525–530
61. He J, Kallin EM, Tsukada Y, Zhang Y (2008) The H3K36 demethylase Jhdm1b/Kdm2b regulates cell proliferation and senescence through p15(Ink4b). Nat Struct Mol Biol 15(11):1169–1175
62. Pfau R, Tzatsos A, Kampranis SC, Serebrennikova OB, Bear SE, Tsichlis PN (2008) Members of a family of JmjC domain-containing oncoproteins immortalize embryonic fibroblasts via a JmjC domain-dependent process. Proc Natl Acad Sci U S A 105(6): 1907–1912
63. Tzatsos A, Paskaleva P, Lymperi S, Contino G, Stoykova S, Chen Z et al (2011) Lysine-specific demethylase 2B (KDM2B)-let-7-enhancer of zester homolog 2 (EZH2) pathway regulates cell cycle progression and senescence in primary cells. J Biol Chem 286(38): 33061–33069
64. Chicas A, Kapoor A, Wang X, Aksoy O, Evertts AG, Zhang MQ et al (2012) H3K4 demethylation by Jarid1a and Jarid1b contributes to retinoblastoma-mediated gene silencing during cellular senescence. Proc Natl Acad Sci U S A 109(23):8971–8976
65. Johnson AB, Denko N, Barton MC (2008) Hypoxia induces a novel signature of chromatin modifications and global repression of transcription. Mutat Res 640(1–2):174–179
66. Chen H, Yan Y, Davidson TL, Shinkai Y, Costa M (2006) Hypoxic stress induces dimethylated histone H3 lysine 9 through histone methyltransferase G9a in mammalian cells. Cancer Res 66(18):9009–9016
67. Krieg AJ, Rankin EB, Chan D, Razorenova O, Fernandez S, Giaccia AJ (2010) Regulation of the histone demethylase JMJD1A by hypoxia-inducible factor 1 alpha enhances hypoxic gene expression and tumor growth. Mol Cell Biol 30(1):344–353
68. Beyer S, Kristensen MM, Jensen KS, Johansen JV, Staller P (2008) The histone demethylases JMJD1A and JMJD2B are transcriptional targets of hypoxia-inducible factor HIF. J Biol Chem 283(52):36542–36552
69. Sar A, Ponjevic D, Nguyen M, Box AH, Demetrick DJ (2009) Identification and characterization of demethylase JMJD1A as a gene upregulated in the human cellular response to hypoxia. Cell Tissue Res 337(2):223–234
70. Xia X, Lemieux ME, Li W, Carroll JS, Brown M, Liu XS et al (2009) Integrative analysis of HIF binding and transactivation reveals its role in maintaining histone methylation homeostasis. Proc Natl Acad Sci U S A 106(11):4260–4265
71. Mimura I, Nangaku M, Kanki Y, Tsutsumi S, Inoue T, Kohro T et al (2012) Dynamic change of chromatin conformation in response to hypoxia enhances the expression of GLUT3 (SLC2A3) by cooperative interaction of hypoxia-inducible factor 1 and KDM3A. Mol Cell Biol 32(15):3018–3032

72. Koong AC, Chen EY, Giaccia AJ (1994) Hypoxia causes the activation of nuclear factor kappa B through the phosphorylation of I kappa B alpha on tyrosine residues. Cancer Res 54(6):1425–1430
73. Leonard MO, Howell K, Madden SF, Costello CM, Higgins DG, Taylor CT et al (2008) Hypoxia selectively activates the CREB family of transcription factors in the in vivo lung. Am J Respir Crit Care Med 178(9):977–983
74. Rong Y, Hu F, Huang R, Mackman N, Horowitz JM, Jensen RL et al (2006) Early growth response gene-1 regulates hypoxia-induced expression of tissue factor in glioblastoma multiforme through hypoxia-inducible factor-1-independent mechanisms. Cancer Res 66(14): 7067–7074
75. Lu T, Jackson MW, Wang B, Yang M, Chance MR, Miyagi M et al (2010) Regulation of NF-kappaB by NSD1/FBXL11-dependent reversible lysine methylation of p65. Proc Natl Acad Sci U S A 107(1):46–51
76. Culhane JC, Szewczuk LM, Liu X, Da G, Marmorstein R, Cole PA (2006) A mechanism-based inactivator for histone demethylase LSD1. J Am Chem Soc 128(14):4536–4537
77. Szewczuk LM, Culhane JC, Yang M, Majumdar A, Yu H, Cole PA (2007) Mechanistic analysis of a suicide inactivator of histone demethylase LSD1. Biochemistry 46(23): 6892–6902
78. Yang M, Culhane JC, Szewczuk LM, Gocke CB, Brautigam CA, Tomchick DR et al (2007) Structural basis of histone demethylation by LSD1 revealed by suicide inactivation. Nat Struct Mol Biol 14(6):535–539
79. Mimasu S, Sengoku T, Fukuzawa S, Umehara T, Yokoyama S (2008) Crystal structure of histone demethylase LSD1 and tranylcypromine at 2.25 A. Biochem Biophys Res Commun 366(1):15–22
80. Yang M, Culhane JC, Szewczuk LM, Jalili P, Ball HL, Machius M et al (2007) Structural basis for the inhibition of the LSD1 histone demethylase by the antidepressant trans-2-phenylcyclopropylamine. Biochemistry 46(27):8058–8065
81. Gooden DM, Schmidt DM, Pollock JA, Kabadi AM, McCafferty DG (2008) Facile synthesis of substituted trans-2-arylcyclopropylamine inhibitors of the human histone demethylase LSD1 and monoamine oxidases A and B. Bioorg Med Chem Lett 18(10):3047–3051
82. Ueda R, Suzuki T, Mino K, Tsumoto H, Nakagawa H, Hasegawa M et al (2009) Identification of cell-active lysine specific demethylase 1-selective inhibitors. J Am Chem Soc 131(48):17536–17537
83. Binda C, Valente S, Romanenghi M, Pilotto S, Cirilli R, Karytinos A et al (2010) Biochemical, structural, and biological evaluation of tranylcypromine derivatives as inhibitors of histone demethylases LSD1 and LSD2. J Am Chem Soc 132(19):6827–6833
84. Mimasu S, Umezawa N, Sato S, Higuchi T, Umehara T, Yokoyama S (2010) Structurally designed trans-2-phenylcyclopropylamine derivatives potently inhibit histone demethylase LSD1/KDM1. Biochemistry 49(30):6494–6503
85. Ogasawara D, Suzuki T, Mino K, Ueda R, Khan MN, Matsubara T et al (2011) Synthesis and biological activity of optically active NCL-1, a lysine-specific demethylase 1 selective inhibitor. Bioorg Med Chem 19(12):3702–3708
86. Huang Y, Greene E, Murray Stewart T, Goodwin AC, Baylin SB, Woster PM et al (2007) Inhibition of lysine-specific demethylase 1 by polyamine analogues results in reexpression of aberrantly silenced genes. Proc Natl Acad Sci U S A 104(19):8023–8028
87. Huang Y, Stewart TM, Wu Y, Baylin SB, Marton LJ, Perkins B et al (2009) Novel oligoamine analogues inhibit lysine-specific demethylase 1 and induce reexpression of epigenetically silenced genes. Clin Cancer Res 15(23):7217–7228
88. Sharma SK, Wu Y, Steinbergs N, Crowley ML, Hanson AS, Casero RA et al (2010) (Bis)urea and (bis)thiourea inhibitors of lysine-specific demethylase 1 as epigenetic modulators. J Med Chem 53(14):5197–5212
89. Wang J, Lu F, Ren Q, Sun H, Xu Z, Lan R et al (2011) Novel histone demethylase LSD1 inhibitors selectively target cancer cells with pluripotent stem cell properties. Cancer Res 71(23):7238–7249

90. Hamada S, Kim TD, Suzuki T, Itoh Y, Tsumoto H, Nakagawa H et al (2009) Synthesis and activity of N-oxalylglycine and its derivatives as Jumonji C-domain-containing histone lysine demethylase inhibitors. Bioorg Med Chem Lett 19(10):2852–2855
91. Hamada S, Suzuki T, Mino K, Koseki K, Oehme F, Flamme I et al (2010) Design, synthesis, enzyme-inhibitory activity, and effect on human cancer cells of a novel series of jumonji domain-containing protein 2 histone demethylase inhibitors. J Med Chem 53(15):5629–5638
92. Thalhammer A, Mecinovic J, Loenarz C, Tumber A, Rose NR, Heightman TD et al (2011) Inhibition of the histone demethylase JMJD2E by 3-substituted pyridine 2,4-dicarboxylates. Org Biomol Chem 9(1):127–135
93. King ON, Li XS, Sakurai M, Kawamura A, Rose NR, Ng SS et al (2010) Quantitative high-throughput screening identifies 8-hydroxyquinolines as cell-active histone demethylase inhibitors. PLoS One 5(11):e15535
94. Upadhyay AK, Rotili D, Han JW, Hu R, Chang Y, Labella D et al (2012) An analog of BIX-01294 selectively inhibits a family of histone H3 lysine 9 Jumonji demethylases. J Mol Biol 416(3):319–327
95. Kristensen LH, Nielsen AL, Helgstrand C, Lees M, Cloos P, Kastrup JS et al (2012) Studies of H3K4me3 demethylation by KDM5B/Jarid1B/PLU1 reveals strong substrate recognition in vitro and identifies 2,4-pyridine-dicarboxylic acid as an in vitro and in cell inhibitor. FEBS J 279(11):1905–1914
96. Nielsen AL, Kristensen LH, Stephansen KB, Kristensen JB, Helgstrand C, Lees M et al (2012) Identification of catechols as histone-lysine demethylase inhibitors. FEBS Lett 586(8):1190–1194
97. Kruidenier L, Chung CW, Cheng Z, Liddle J, Che K, Joberty G et al (2012) A selective jumonji H3K27 demethylase inhibitor modulates the proinflammatory macrophage response. Nature 488(7411):404–408
98. Metzger E, Wissmann M, Yin N, Muller JM, Schneider R, Peters AH et al (2005) LSD1 demethylates repressive histone marks to promote androgen-receptor-dependent transcription. Nature 437(7057):436–439
99. Metzger E, Imhof A, Patel D, Kahl P, Hoffmeyer K, Friedrichs N et al (2010) Phosphorylation of histone H3T6 by PKCbeta(I) controls demethylation at histone H3K4. Nature 464(7289):792–796
100. Kahl P, Gullotti L, Heukamp LC, Wolf S, Friedrichs N, Vorreuther R et al (2006) Androgen receptor coactivators lysine-specific histone demethylase 1 and four and a half LIM domain protein 2 predict risk of prostate cancer recurrence. Cancer Res 66(23):11341–11347
101. Wissmann M, Yin N, Muller JM, Greschik H, Fodor BD, Jenuwein T et al (2007) Cooperative demethylation by JMJD2C and LSD1 promotes androgen receptor-dependent gene expression. Nat Cell Biol 9(3):347–353
102. Garcia-Bassets I, Kwon YS, Telese F, Prefontaine GG, Hutt KR, Cheng CS et al (2007) Histone methylation-dependent mechanisms impose ligand dependency for gene activation by nuclear receptors. Cell 128(3):505–518
103. Yamane K, Toumazou C, Tsukada Y, Erdjument-Bromage H, Tempst P, Wong J et al (2006) JHDM2A, a JmjC-containing H3K9 demethylase, facilitates transcription activation by androgen receptor. Cell 125(3):483–495
104. Bjorkman M, Ostling P, Harma V, Virtanen J, Mpindi JP, Rantala J et al (2012) Systematic knockdown of epigenetic enzymes identifies a novel histone demethylase PHF8 overexpressed in prostate cancer with an impact on cell proliferation, migration and invasion. Oncogene 31(29):3444–3456
105. Stratmann A, Haendler B (2012) Histone demethylation and steroid receptor function in cancer. Mol Cell Endocrinol 348(1):12–20
106. Shin S, Janknecht R (2007) Activation of androgen receptor by histone demethylases JMJD2A and JMJD2D. Biochem Biophys Res Commun 359(3):742–746
107. Xiang Y, Zhu Z, Han G, Ye X, Xu B, Peng Z et al (2007) JARID1B is a histone H3 lysine 4 demethylase up-regulated in prostate cancer. Proc Natl Acad Sci U S A 104(49):19226–19231

108. Pinskaya M, Gourvennec S, Morillon A (2009) H3 lysine 4 di- and tri-methylation deposited by cryptic transcription attenuates promoter activation. EMBO J 28(12):1697–1707
109. Shi X, Hong T, Walter KL, Ewalt M, Michishita E, Hung T et al (2006) ING2 PHD domain links histone H3 lysine 4 methylation to active gene repression. Nature 442(7098):96–99
110. Chen Z, Zang J, Whetstine J, Hong X, Davrazou F, Kutateladze TG et al (2006) Structural insights into histone demethylation by JMJD2 family members. Cell 125(4):691–702
111. Hillringhaus L, Yue WW, Rose NR, Ng SS, Gileadi C, Loenarz C et al (2011) Structural and evolutionary basis for the dual substrate selectivity of human KDM4 histone demethylase family. J Biol Chem 286(48):41616–41625
112. Cai C, He HH, Chen S, Coleman I, Wang H, Fang Z et al (2011) Androgen receptor gene expression in prostate cancer is directly suppressed by the androgen receptor through recruitment of lysine-specific demethylase 1. Cancer Cell 20(4):457–471
113. Kim TD, Shin S, Janknecht R (2008) Repression of Smad3 activity by histone demethylase SMCX/JARID1C. Biochem Biophys Res Commun 366(2):563–567
114. Buijs JT, Henriquez NV, van Overveld PG, van der Horst G, ten Dijke P, van der Pluijm G (2007) TGF-beta and BMP7 interactions in tumour progression and bone metastasis. Clin Exp Metastasis 24(8):609–617
115. Seligson DB, Horvath S, Shi T, Yu H, Tze S, Grunstein M et al (2005) Global histone modification patterns predict risk of prostate cancer recurrence. Nature 435(7046):1262–1266
116. Yu J, Rhodes DR, Tomlins SA, Cao X, Chen G, Mehra R et al (2007) A polycomb repression signature in metastatic prostate cancer predicts cancer outcome. Cancer Res 67(22): 10657–10663
117. Zhou LX, Li T, Huang YR, Sha JJ, Sun P, Li D (2010) Application of histone modification in the risk prediction of the biochemical recurrence after radical prostatectomy. Asian J Androl 12(2):171–179
118. Ellinger J, Kahl P, von der Gathen J, Rogenhofer S, Heukamp LC, Gutgemann I et al (2010) Global levels of histone modifications predict prostate cancer recurrence. Prostate 70(1): 61–69
119. Wang Q, Li W, Zhang Y, Yuan X, Xu K, Yu J et al (2009) Androgen receptor regulates a distinct transcription program in androgen-independent prostate cancer. Cell 138(2): 245–256
120. Ke XS, Qu Y, Rostad K, Li WC, Lin B, Halvorsen OJ et al (2009) Genome-wide profiling of histone h3 lysine 4 and lysine 27 trimethylation reveals an epigenetic signature in prostate carcinogenesis. PLoS One 4(3):e4687
121. Kondo Y, Shen L, Cheng AS, Ahmed S, Boumber Y, Charo C et al (2008) Gene silencing in cancer by histone H3 lysine 27 trimethylation independent of promoter DNA methylation. Nat Genet 40(6):741–750
122. Harris WJ, Huang X, Lynch JT, Spencer GJ, Hitchin JR, Li Y et al (2012) The histone demethylase KDM1A sustains the oncogenic potential of MLL-AF9 leukemia stem cells. Cancer Cell 21(4):473–487
123. Hayami S, Kelly JD, Cho HS, Yoshimatsu M, Unoki M, Tsunoda T et al (2011) Overexpression of LSD1 contributes to human carcinogenesis through chromatin regulation in various cancers. Int J Cancer 128(3):574–586
124. Schulte JH, Lim S, Schramm A, Friedrichs N, Koster J, Versteeg R et al (2009) Lysine-specific demethylase 1 is strongly expressed in poorly differentiated neuroblastoma: implications for therapy. Cancer Res 69(5):2065–2071
125. Liao QL, Chen XD, Zhao L, Ding YQ (2008) [Comparative proteomics of the serum in patients with nasopharyngeal carcinoma: a study with two-dimensional electrophoresis and MALDI-TOF-MS]. Nan Fang Yi Ke Da Xue Xue Bao 28(2):154–8
126. Yamada D, Kobayashi S, Yamamoto H, Tomimaru Y, Noda T, Uemura M et al (2012) Role of the hypoxia-related gene, JMJD1A, in hepatocellular carcinoma: clinical impact on recurrence after hepatic resection. Ann Surg Oncol 19(Suppl 3):S355–S364
127. Guo X, Shi M, Sun L, Wang Y, Gui Y, Cai Z et al (2011) The expression of histone demethylase JMJD1A in renal cell carcinoma. Neoplasma 58(2):153–157

128. Kim JY, Kim KB, Eom GH, Choe N, Kee HJ, Son HJ et al (2012) KDM3B is the H3K9 demethylase involved in transcriptional activation of lmo2 in leukemia. Mol Cell Biol 32(14):2917–2933
129. Kauffman EC, Robinson BD, Downes MJ, Powell LG, Lee MM, Scherr DS et al (2011) Role of androgen receptor and associated lysine-demethylase coregulators, LSD1 and JMJD2A, in localized and advanced human bladder cancer. Mol Carcinog 50(12):931–944
130. Fu L, Chen L, Yang J, Ye T, Chen Y, Fang J (2012) HIF-1alpha-induced histone demethylase JMJD2B contributes to the malignant phenotype of colorectal cancer cells via an epigenetic mechanism. Carcinogenesis 33(9):1664–1673
131. Wu J, Liu S, Liu G, Dombkowski A, Abrams J, Martin-Trevino R et al (2012) Identification and functional analysis of 9p24 amplified genes in human breast cancer. Oncogene 31(3):333–341
132. Ehrbrecht A, Muller U, Wolter M, Hoischen A, Koch A, Radlwimmer B et al (2006) Comprehensive genomic analysis of desmoplastic medulloblastomas: identification of novel amplified genes and separate evaluation of the different histological components. J Pathol 208(4):554–563
133. Barrett A, Madsen B, Copier J, Lu PJ, Cooper L, Scibetta AG et al (2002) PLU-1 nuclear protein, which is upregulated in breast cancer, shows restricted expression in normal human adult tissues: a new cancer/testis antigen? Int J Cancer 101(6):581–588
134. Lu PJ, Sundquist K, Baeckstrom D, Poulsom R, Hanby A, Meier-Ewert S et al (1999) A novel gene (PLU-1) containing highly conserved putative DNA/chromatin binding motifs is specifically up-regulated in breast cancer. J Biol Chem 274(22):15633–15645
135. Anderton JA, Bose S, Vockerodt M, Vrzalikova K, Wei W, Kuo M et al (2011) The H3K27me3 demethylase, KDM6B, is induced by Epstein-Barr virus and over-expressed in Hodgkin's lymphoma. Oncogene 30(17):2037–2043
136. Xiang Y, Zhu Z, Han G, Lin H, Xu L, Chen CD (2007) JMJD3 is a histone H3K27 demethylase. Cell Res 17(10):850–857

# Chapter 16
# Therapeutic Significance of Chromatin Remodeling Complexes in Cancer

**Da-Qiang Li and Rakesh Kumar**

**Abstract** The compaction of genomic DNA into chromatin has far-ranging consequences for almost all aspects of DNA metabolism activity. ATP-dependent chromatin remodeling complexes (CRCs) enable DNA-binding proteins access to nucelosomal DNA by altering chromatin structure through distinct mechanisms including nucleosome sliding, nucleosome assembly, and histone exchanges, in an energy-dependent manner. Consequently, CRCs play critical roles in diverse cellular processes that are dependent on chromatin template, including transcription, replication, and DNA repair. Thus, an aberration in these chromatin remodeling proteins leads to human diseases including cancer. In this chapter, we discuss the functional roles of CRCs in the regulation of gene transcription, DNA damage response, and its potential connection with cancer development as well as tumor therapeutics.

**Keywords** Chromatin remodeling • Transcription • DNA damage response • Cancer • Cancer development and progression • Cancer therapeutics

## 16.1 Introduction

It is increasingly accepted that cancer is a genetic disease. A precise understanding of how genetic alternations contribute to tumor development and progression is the key to develop effective strategies for winning the fight against cancer. The genetic

D.-Q. Li, M.D.
Department of Biochemistry and Molecular Medicine, The George Washington University, Washington, DC 20037, USA

R. Kumar, Ph.D. (✉)
Department of Biochemistry and Molecular Medicine, School of Medicine and Health Sciences, The George Washington University, Washington, DC 20037, USA
e-mail: bcmrxk@gwu.edu

R. Kumar (ed.), *Nuclear Signaling Pathways and Targeting Transcription in Cancer*, Cancer Drug Discovery and Development, DOI 10.1007/978-1-4614-8039-6_16,

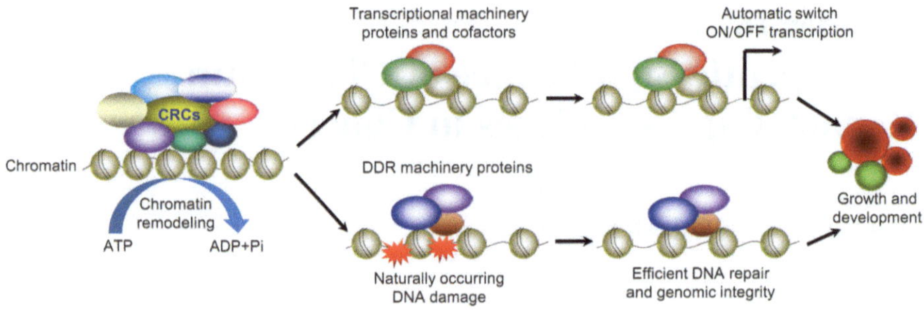


**Fig. 16.1** Functional role of the CRCs in transcription and DDR. The CRCs alter chromatin structure in an energy-dependent manner. Consequently, the transcriptional and DDR machinery proteins get access to nucleosomal DNA and facilitate gene transcription and efficient DNA repair

material is stored into the nucleus in the form of chromatin. The fundamental building block of chromatin is the nucleosome core particle, which is made of approximately 147 base pairs of DNA wrapped around a histone octamer consisting of two copies of each of the four core histones H2A, H2B, H3, and H4 [1, 2]. The repeating nucleosome cores are connected by 20–80 base pairs of linker DNA and further assembled into hierarchically folded higher-order structures with the linker histone H1, nonhistone proteins and divalent metal ions [1–3]. By its very nature, the highly condensed structure of chromatin generally limits the accessibility of DNA binding proteins to the DNA, thus exerting an inhibitory effect on many critical DNA metabolizing activities, such as transcription, DNA replication, recombination and repair. To counteract this repressive barrier imposed by nucleosome architecture, eukaryotes have developed multiple intricate mechanisms to remodel nucleosomes, thus allowing DNA binding proteins such as transcription factors and DNA repair proteins access to the DNA. One of such mechanisms is involved in the ATP-dependent chromatin remodeling complexes (CRCs) that hydrolyze ATP to alter histone-DNA contacts through several mechanisms including nucleosome sliding, histone exchange, and nucleosome/histone eviction [4]. To date, four families of CRCs have been characterized in eukaryotes based on their compositions and functional domains, including the SWI/SNF (switching/sucrose non-fermenting), ISWI (imitation switch), Mi-2/NuRD (nucleosome remodeling and histone deacetylase), and INO80 (inositol requiring) [5, 6]. All CRCs are multisubunit complexes that contain an ATPase subunit [6] and most of them are conserved from yeast to humans. These CRCs play essential roles in many basic biological processes, including gene expression, DNA damage repair, and cell division (Fig. 16.1) [6–11]. Consequently, aberrations in these chromatin remodeling proteins are associated with a variety of human diseases including cancer. Thus, targeting the components of chromatin remodeling signaling pathways is currently being evaluated as a major therapeutic strategy in the prevention and treatment of human cancers. In the following sections, we focus on discussing the emerging role of CRCs in gene transcription, DNA damage response (DDR), and tumor development as well as its potential implication in cancer therapeutics.

## 16.2 Characterization of the CRC Family

The first discovered CRC family is the SWI/SNF, which was initially identified in independent screens for mutants affecting mating-type switching and growth on sucrose in 1994 [12–14]. The SWI/SNF family is composed of 8–14 subunits and is characterized by a bromodomain in its ATPase catalytic subunits, BRG1 (also known as SMARCA4 or BAF190A) and BRM (also known as SMARCA2 or BAF190B) http://www.genecards.org/ [15]. The bromodomain preferentially interacts with acetylated histones, which play both positive and negative roles in regulating the activity of the SWI/SNF remodeling factors [16]. In *S. cerevisiae*, this family contains the founding member SWI/SNF complex as well as the highly related RSC (remodel the structure of chromatin) complex [17]. Both SWI/SNF and RSC complexes exhibit a DNA-dependent ATPase activity to perturb nucleosome structure [17, 18], and contain nuclear actin-related proteins Arp7 and Arp9 [19, 20]. Arp7 and Arp9 form a stable heterodimer relying on their actin-related regions for heterodimerization, and function with DNA binding proteins to facilitate proper chromatin architecture and complex-complex interactions [19, 20]. Human complexes of this family have also been identified, including the BRG1-associated factor (BAF) complex and the polybromo BRG1-assocaited factor (PBAF) complex [21, 22]. With regard to homology, BAF1 is similar to the yeast SWI/SNF and PBAF is more like yeast RSC complex [21, 22].

The second family of CRCs is ISWI. One distinguishing feature of this family is that its ATPase subunit contains a carboxyl-terminal SANT and a SLIDE (SANT-like ISWI domain) domain, which together form a nucleosome recognition module that binds histone tails and linker DNA [23, 24]. The founding member of this family is the *Drosophila* NURF (nucleosome remodeling factor) complex, which was identified in 1995 by assaying the ability of drosophila embryo extracts to generate a nuclease-hypersensitive site within an array of nucleosomes [25]. NURF is composed of four distinct subunits, including the 140-kD ISWI ATPase subunit (NURF140) [26], a 55-kD WD repeat protein (NURF55) [27], the smallest NURF38 component [28], and the large NURF301 subunit [29]. In contrast to the SWI/SNF complex, the ATPase activity of NURF requires nucleosomes rather than free DNA or histones [25]. Thus, NURF acts directly on a nucleosome to alter chromatin structure by catalyzing nucleosome sliding, thereby exposing DNA sequences associated with nucleosomes [25, 30, 31]. Interestingly, the N-terminal histone tails are functionally important for modulating nucleosome mobility and regulating ATP-dependent nucleosome sliding by NURF [30]. In addition to NURF, the *drosophila* ISWI complex also contains the CHRAC (chromatin remodeling and assembly complex) [32] and ACF (ATP-utilizing chromatin remodeling and assembly factor) complexes [33]. Both exhibit chromatin assembly and nucleosome sliding activity in an ATP-dependent mechanism [32, 33]. In mammals, two highly related ATPase subunits of the ISWI CRC have been identified, including SNF2L (SNF2 like) and SNF2H (SNF2 homologue) [34]. Biochemical analysis revealed that the SNF2H ATPase catalytic subunit is contained in multiple complexes including ACF, CHRAC, RSF (remodeling and spacing factor), NoRC (nucleosome-remodeling complex), WICH (WSTF-ISWI chromatin remodeling), and WCRF (WSTF-related chromatin-remodeling factor) [34–41]. In contrast, only a small number of complexes, such as the human NURF and CERF (CECR2-containing remodeling factor) complexes, contain the SNF2L ATPase subunit [35, 42].

The third family of CRCs is the Mi-2/NuRD complex, which was identified in1998 from several independent groups and processes both ATP-dependent chromatin remodeling and histone deacetylase activities [43–46]. The complex contains the histone deacetylases HDAC1/2, the histone-binding proteins RbAp46/48, the dermatomyositis-specific autoantigen Mi-2, the metastasis-associated proteins MTA1/2/3, and the methyl-CpG-binding domain proteins MBD2/3 [43, 44, 47]. Notably, Mi-2 subunit contains a chromodomain and plant homeo domain (PHD)-type zinc finger, and functions as a DNA-dependent, nucleosome-stimulated ATPase that remodels nucleosomes in an ATP-dependent manner [48]. In particular, Mi-2 lacking its chromodomains fails to bind or remodel nucleosomes [23].

The fourth family of CRCs is the evolutionarily conserved INO80 subfamily, which includes the INO80 complex and SWR1 complex [49]. The subfamily is characterized by a split ATPase domain and the presence of two RuvB-like proteins Rvb1 and Rvb2 [49]. The INO80 complex was initially purified from *S. cerevisiae* in 2000, consisting of about 12 subunits including Arp4, Arp5, Arp8, actin, and the Rvb1 and Rvb2 helicase proteins, and displays nucleosome-stimulated ATPase activity and ATP-dependent chromatin remodeling activities [50]. Deletion of Arp5 in yeast strains impairs INO80 ATPase activity, DNA binding, and nucleosome mobilization [51]. Similarly, Arp8 forms a complex with nucleosomes via the H3 and H4 histones [51, 52] and is essential for activity of INO80. In this context, deletion of Arp8 results in loss of INO80 function with multiple effects on cellular processes such as double-strand break (DSB) repair and chromosome alignment [52–55]. In contrast, Rvbp1p/Rvb2p is required for the complete assembly of a functional INO80 complex and for recruiting Arp5p to the INO80 complex in an ATP dependent manner [56]. The highly-related SWR1 complex was identified from *S. cerevisiae* in 2004 [57] and its human counterparts, termed Snf2-related CREBBP activator protein (SRCAP) and p400, were also identified afterward [58, 59]. In *S. cerevisiae*, this SWR1 complex contains Swr1p, a putative Swi2/Snf2-related ATPase, and 12 additional subunits. Among them, several subunits including Act1, Arp4, Rvb1 and Rvb2 are common to the INO80 complex [50]. Despite highly related to INO80, the SWR1 complex is unique in its ability to catalyze the incorporation of the histone variant H2AZ (Htz1 in *S. cerevisiae*) into nucleosomes [57, 60, 61], and this occurs in vitro in a stepwise and unidirectional fashion and requires dual activation with histone H2AZ and canonical nucleosome [62].

## 16.3 CRCs in Gene Transcription

The compaction of DNA into chromatin in the eukarytotic nucleus poses many obstacles to transcription [63]. The CRCs bind directly to nucleosomes and disrupt histone-DNA interactions using the energy of ATP hydrolysis, thus facilitating the access of the core transcription machinery proteins and general cofactors to nucleosomal DNA.

As a result, CRCs play a fundamental role in modulating transcription in yeast and higher eukaryotes. Notably, these CRCs have a range of specific and context-dependent roles in control of gene expression depending on the circumstance.

### *16.3.1 The SWI/SNF Complex*

The SWI/SNF complex is involved in a variety of functionally distinct complexes and exerts diverse roles in gene regulation and genome function [64]. One outstanding example is that the SWI/SNF CRC participates in promoting transcriptional activation by nuclear receptors. The androgen receptor (AR) is a ligand-dependent transcription factor whose activity is tightly regulated by interacting cofactors and cofactor complexes and is a key player in prostate cancer development and progression [65, 66]. Considerable evidence has pointed out that the SWI/SNF CRC directs AR-mediated transcriptional activation, and different AR targets show disparity in the requirement for SWI/SNF [65–67]. A case in point is the BAF57 (also known as SMARCE1) subunit, which directly binds to the AR and is recruited to endogenous AR targets upon ligand activation, thus regulating AR activity, coactivator function, and AR-dependent proliferation [68]. Similarly, the BAF57 subunit specifically regulates estrogen receptor alpha (ERα)-dependent gene expression and proliferation in human breast cancer cells [69, 70]. Consequently, mutations in BAF57 deregulate several oncogenic signaling pathways, thus contributing to the development of breast cancer [71, 72].

In addition to BAF57, the BRG1subunit is a critical modulator of transcriptional regulation in various tissues and pathophysiological conditions [73]. For instance, BRG1predominantly interacts with Smad2 and Smad3 and is specifically required for transforming growth factor β-induced expression of endogenous Smad2/3 target genes through recruitment to Smad-dependent promoters [74, 75]. BRG1, as well as BRM, associates with the *CD44* and *E-cadherin* promoters and promotes their transcriptional activation in cancer cells through deceasing DNA methylation at their promoters [76]. In addition, SNF5 (also known as Ini1, BAF47, SNR1, or SMARCB1) mediates BRG1 recruitment to the $p15^{INK4b}$ and $p16^{INK4a}$ promoters and activates their expression through eviction of polycomb group silencing complex and extensive chromatin reprogramming [77].

Although the SWI/SNF CRC is generally associated with transcriptional activation, emerging evidence points out its additional role in transcriptional silencing pathway [78]. For instance, SWI3B, an essential subunit of the SWI/SNF complex, physically interacts with a long noncoding RNA (lncRNA)-binding protein, IDN2, and contributes to lncRNA-mediated transcriptional silencing [78]. Human BRM is functionally linked with the methyl-CpG binding protein MeCP2-depenendent transcriptional silencing [79]. Both BRG-1 and SNF5 subunits repress transcription of *cyclin D1* gene through the direct recruitment of histone deacetylase (HDAC) activity to its promoter, thereby exerting their tumor suppressor functions [80, 81]. More interestingly, BRG1 and BRM can switch their mode of function at same promoter between activation and repression through the context-dependent reprogramming of the SWI/SNF complex [82].

### 16.3.2 The ISWI Complex

The ISWI complex can space nucleosomes, thus affecting a variety of nuclear processes including transcription. Genome-wide analysis demonstrates that ISWI binds both genic and intergenic regions, and remarkably, binds genes near their promoters causing specific alterations in nucleosome positioning at the level of the transcription start sites [83]. Accumulating evidence suggests that the ISWI containing NURF complex is able to facilitate transcriptional activation via remodeling of chromatin in vitro and in vivo [29, 84, 85]. However, NURF also functions as a co-repressor of a large set of JAK/STAT target genes in *drosophila* to regulate innate immunity network [86]. Similarly, ISWI and ACF1 directly repress Wingless transcriptional targets in *drosophila* [87]. In *S. cerevisiae*, Isw1 also functions in stress-induced gene repression under normal growth conditions [88]. In contrast, the Isw2 complex represses transcription of early meiotic genes during mitotic growth and this repressor function of lsw2 complex is largely dependent upon Ume6p, which recruits the complex to target genes [89, 90]. Subsequent studies further demonstrate that Isw2 acts as a transcriptional repressor by altering nucleosome positions, and loss of Isw2 activity results in the generation of both coding and noncoding transcripts due to inappropriate transcription [91].

### 16.3.3 The Mi-2/NuRD Complex

Accumulating evidence has uncovered a number of interesting connections between the Mi-2/NuRD complex and gene regulation [9, 92, 93]. A case in point is the metastasis-associated protein 1 (MTA1), the founding member of the MTA family, which was isolated by differential cDNA library screening using a rat mammary adenocarcinoma metastatic system [94]. MTA1 functions not only as a transcriptional repressor of estrogen receptor α [95], p21WAF1 [96], breast cancer type 1 susceptibility protein [97], RING finger protein 144A [98], phosphatase and tensin homolog [99], transforming growth factor β signaling component SMAD7 [100], guanine nucleotide-binding protein G(i) subunit alpha-2 [101], and homeobox protein SIX3 [102], but also as a transcriptional activator for certain genes, such as the breast cancer-amplified sequence 3 [103] paired box 5 [104], tumor suppressor alternative reading frame [105], cell surface oncogenic protein hyaluronan-mediated motility receptor [106], proto-oncogene protein Wnt-1 [107], and tyrosine hydroxylase [108]. One unanswered question in this field is what is the underlying mechanism for the physiologic switch between coactivator and corepressor functions of MTA1. It is becoming increasingly clear that post-translational modifications might play a role in the regulation of MTA1 function in transcription. In this context, SUMOylation and SUMO-interacting motif of MTA1 synergistically regulate its co-repressor activity on *PS2* transcription [109]. Similarly, acetylation status of MTA1 might also be crucial for its corepressor function on a negative modifier of

Ras activation and its oncogenic activity [101]. More interestingly, methylation of lysine 532 in MTA1 protein seemly represents a molecular switch between coactivator and corepressor [110]. In this context, methylated MTA1 is required for the NuRD repressor complex, while demethylated MTA1 recognizes the active histone mark and recruits coactivator complex onto its target gene promoters in a signaling-dependent manner [110].

### *16.3.4 The INO80 Complex*

Involvement of the INO80 complex in transcription was first discovered in *S. cerevisiae*, in which INO80 facilitates transcription in vitro and in vivo [50, 111]. Subsequent studies further demonstrate that its mammalian orthologue also promotes transcription with transcription factor Yin-Yang-1 (YY1) [112]. In contrast, TBP-interacting protein 49b (TIP49b), a component of the INO80 complex, inhibits transcription factor 2 (ATF2) transcriptional activities in response to stress and DNA damage [113].

## 16.4 CCRs in the DDR

In response to DNA damage, chromatin undergoes a marked reorganization in an energy dependent manner, thus facilitating the DDR machinery proteins to recognize and repair the damaged DNA [114]. In addition to their putative roles in transcription, CCRs are intimately linked with the DDR.

### *16.4.1 The SWI/SNF Complex*

The SWI/SNF complex is required for DNA replication [115, 116], somatic recombination [117], nucleotide excision repair (NER) [118, 119], and DSB repair [120]. SWI/SNF also regulates checkpoint activation after ultraviolet (UV) damage via regulation of the proliferating cell nuclear antigen-binding proteins Gadd45a and p21 [121]. The highly related RSC complex is also linked with efficient DSB repair [122, 123]. Interestingly, two isoforms of this complex, defined by the presence of either Rsc1 or Rsc2, play distinct roles in DDR and that at least part of the functional specificity is dictated by the bromo-adjacent homology (BAH) domains [124]. Moreover, the RSC and SWI/SNF chromatin remodelers play distinct roles in DSB repair; SWI/SNF is required during the early steps of homologous recombination (HR), while RSC is important upon the completion of the repair process [125].

### 16.4.2 The ISWI Complex

SNF2H, the catalytic subunit of ISWI complex, is rapidly recruited to DSBs in a poly(ADP-ribose) polymerase 1 (PARP1)-dependent manner and facilitates the RNF168-dependent signaling and repair of DSBs [126]. Similarly, the ACF1 chromatin remodeling factor accumulates at UV-induced DNA damage sites immediately following UV radiation [127] and promotes NER of UV-induced DNA lesions [128]. Similarly, the ACF1 complex accumulates rapidly at DSBs and is also required for non-homologous end joining (NHEJ) repair of DSBs in human cells [129]. Rsf-1 (also known as HBXAP) protein interacts with SNF2H to form an ISWI complex, RSF, and has been reported as an amplified gene in human cancers, including the highly aggressive ovarian serous carcinoma [130]. Emerging evidence shows that Rsf-1induces DNA damage and promotes genomic instability [131], and consequently, high-grade ovarian serous carcinomas, especially those with Rsf-1 overexpression, exhibit high levels of the DDR [132]. These findings highlight that increased Rsf-1 expression in tumors can induce chromosomal instability probably through DDR [131].

### 16.4.3 The Mi2/NuRD Complex

The initial link between the NuRD complex and DDR was found in 1999 [133], when Schmidt and colleagues discovered that ataxia telangiectasia and Rad3-related protein (ATR), a master regulator of the DDR, associates with multiple components of the NuRD complex, including MTA1, MTA2, HDAC1, HDAC2, and CHD4 [133]. Afterward, van Haaften G et al. in 2006 defined a role for *C. elegans* early growth response protein 1 (Egr-1), the homologue of human *MTA2* gene, in cellular sensitivity to ionizing radiation (IR) using a genome-wide RNA interference screening [134]. In 2009, Li et al. further discovered a previously unknown role for MTA1in IR-induced DSB repair and cell survival using MTA1-knockout fibroblasts [135]. In 2010, several studies from four different groups simultaneously reported a conserved role of the NuRD complex, including MTA1, MTA2, CHD4, HDAC1, and HDAC2 in DDR and DNA repair in multiple model systems [136–140].

The PARP family of proteins has been implicated in recruitment of proteins to sites of damage and is known to localize rapidly to sites of damage [136, 141]. In support of our early findings, emerging evidence shows that MTA1 is recruited to sites of DNA damage in a PARP-dependent manner, and depletion of MTA1 by siRNAs renders cells sensitive to IR, further highlighting its importance in promoting DNA repair [136]. The human homologue of *egr-1*, MTA2, also protects human cells against IR, suggesting its conserved role in the DDR [139]. CHD4 is rapidly recruited to DSBs in a PARP-dependent manner [136, 138], where it promotes RNF8/RNF168-mediated histone ubiquitylation and the ubiquitin-dependent accumulation of RNF168 and BRCA1 at sites of DNA lesions [137, 139]. CHD4 also acts as an important regulator of the G1/S cell-cycle transition by controlling p53 deacetylation [138]. Consequently, loss of CHD4 causes defects in DNA repair and checkpoint activation, resulting in accumulation of spontaneous DNA damage and increased IR sensitivity [138, 139]. Furthermore, human HDAC1 and HDAC2 also function in the DDR to promote NHEJ repair [140]. Consistently, HDAC inhibitors block the activity of HDAC1 and HDAC2, resulting in defects in the DDR and hypersensitivity to the DSB-inducing agents [140]. Taken together, the NuRD chromatin–remodeling complex is a novel DDR factor that helps to preserve genome stability by regulating signaling and repair of DNA damage [11, 142]. Interestingly, recent studies pointed out that multiple NuRD components are lost during premature and normal ageing, resulting in accumulation of DNA damage during ageing [143], which could contribute to aging-related genomic instability and cancer [144].

### *16.4.4 The INO80 Complex*

In addition to their well-established role in regulating transcriptional processes, accumulating evidence shows that INO80 and SWR1 chromatin remodeling components are essential for maintaining genomic integrity [10]. The INO80 complex is recruited to sites of DSBs through a specific interaction with the DNA damage-induced phosphorylated histone H2A (termed γH2AX) [145, 146], and mediates DSB repair through its role in DNA end strand resection [147]. INO80 is also recruited to sites of UV lesion repair through interactions with the NER apparatus and promotes the removal of UV lesions by the NER pathway [148, 149]. Moreover, INO80 is required for the restoration of chromatin structure after repair in response to UV-induced damage [149]. Interestingly, INO80 also shapes the DNA replication landscape. In this context, INO80 complexes are enriched at sites of replication and are required for efficient replication of late-replicating regions during replication stress through regulating S-phase checkpoint activity [4, 150]. INO80 also regulates the threshold of DNA damage during replication phase via modifying PCNA ubiquitination and Rad51-mediated processing of recombination intermediates at impeded replication forks [151–153].

## 16.5 CRCs in Cancer Development and Progression

### *16.5.1 The SWI/SNF Complex*

Given its central function in epigenetic chromatin remodeling mechanisms, it is not surprising that alternation of the SWI/SNF CRC plays an important role in tumor development and progression. A substantial body of evidence indicates that several components of the SWI/SNF complexes function as tumor suppressors or negative regulator of cellular proliferation [21, 154, 155]. One such example is the SNF5 core subunit, which has been documented to be mutated or inactivated in a number of human cancers including rhabdoid, rhabdomyosarcoma, epithelioid sarcoma, chronic myeloid leukemia, medulloblastomas, choroid plexus carcinomas, and melanoma [156–161]. In support of this notion, haploinsufficiency of SNF5 predisposes to malignant rhabdoid tumors in mice, and loss of SNF5 results in highly penetrant cancer predisposition with 100 % of mice developing T cell lymphoma or rhabdoid tumors with a median onset of only 11 weeks [162]. Collective evidence establishes that the tumor suppressor activity of SNF5 depends on its regulation of cell cycle progression, cell survival and senescence [163–168]. Inactivation of the SNF5 tumor suppressor stimulates cell cycle progression and cooperates with p53 loss to accelerate oncogenic transformation and tumor growth in mice [169, 170]. The inhibition of RhoA-dependent migration is another crucial tumor suppressor function of hSNF5, and its loss-of-function may lead to increased invasiveness and metastatic potential of cancer cells [171].

Another example is the ARID1A (also known as BAF250A, SMARCF1, p270, or hOSA1), which encodes a human homolog of yeast SWI1. The significance of ARID1A loss or mutation in cancer is now subject to intensive investigation. In this context, mutation of the *ARID1A* gene has been widely described in a broad array of tumor types, including gynecologic ovarian and endotrial carcinomas, pediatric Burkitt lymphoma, gastric carcinoma, breast cancer, and hepatitis B virus-associated hepatocellular carcinoma [172–178]. Consistently, restoring wild-type ARID1A expression in cancer cells that harbor ARID1A mutations is sufficient to suppress cell proliferation and tumor growth in mice [172, 175]. In contrast, ARID1A knockdown significantly promotes the proliferation, migration and invasion of cancer cells [173]. Functional evidence further points out that ARID1A collaborates with p53 to regulate p21WAF1 and SMAD family member 3 [179]. Together, accumulating genomic and functional evidence strongly supports classification of ARID1A as a tumor suppressor [177]. Similar to ARID1A, ARID1B (also known as BAF250B, or hOsa2) also inhibits cell growth and regulates cell cycle arrest through differentially regulating *c-myc* and *p21WAF1* gene expression [180].

In addition, loss or inactivation of BRG1, BRM, BAF155/SMARCC1, BAF180, and BAF200/ARID2 expression represents another mechanism for SWI/SNF complex in the development in human cancers, including hepatitis C infection-related liver cancer, melanoma, lung, pancreatic, skin, and breast cancers [21, 181–188]. Notably, BRG1 and BRM are silenced by different mechanisms. BRG1 is commonly silenced by loss-of-function mutations, whereas epigenetic silencing is a major mechanism for the loss of BRM in human cancer cells [188].

### *16.5.2 The ISWI Complex*

A well-studied example is the Rsf-1, which plays an important role in cellular growth, survival, and oncogenic transformation, and its up-regulation is closely associated with disease aggressiveness and poor prognosis in patients with various types of human cancers including bladder, colon, nasopharyngeal, gallbladder, oral, and ovarian carcinomas [130, 189–197]. A mechanistic study demonstrates that Rsf-1 interacts and collaborates with cyclin E1 in neoplastic transformation and p53 mutations are a prerequisite for tumour-promoting functions of the RSF/cyclin E1 complex [194]. In contrast, overexpression of Rsf-1 is rare in breast cancer, indicating that Rsf-1 is not a critical gene in breast cancer development [130, 198]. In contrast, SNF2L, a mammalian ISWI ortholog, suppresses cell proliferation and migration in human HeLa cells by attenuating Wnt signaling [199].

### *16.5.3 The Mi2/NuRD Complex*

Of all the NuRD complex subunits, the MTA family members are best studied in the context of cancer development [92, 93]. MTA1, the founding member of the MTA1 family, has been documented to be overexpressed in a variety of human cancers and is significantly associated with tumor progression and poor clinical outcome [92, 93]. In contrast, the information concerning the expression of MTA2 and MTA3 in human cancers is limited. Like MTA1, increased expression of HDAC1 and HDAC2 has been documented in a variety of human cancers and linked with therapeutic resistance [200–202]. In contrast, lysine-specific demethylase 1, a newly identified component of the Mi-2/NuRD complex, inhibits the invasion of breast cancer cells in vitro and suppresses breast cancer metastatic potential in vivo [203].

### *16.5.4 The INO80 Complex*

Although the function of the INO80 complex in transcription and DDR, its connection with human cancers is rarely reported. The SRCAP, a homolog of Swr1 in human cells, modulates expression of prostate specific antigen and cellular proliferation in prostate cancer cells [204]. Similarly, p400, another Swr1 homolog, inhibits p53-mediated *p21WAF1* transcription and the development of premature senescence [205]. P400 is an essential E1A transformation target that plays a major role in the E1A transforming process [206].

## 16.6 CRCs in Cancer Therapeutics

Glucocorticoids are used in the curative treatment of acute lymphoblastic leukemia (ALL) and resistance to glucocorticoids is an important adverse prognostic factor in newly diagnosed ALL patients [207]. Emerging evidence suggests that decreased expression of the BRG1, ARID1A, and SNF5 subunits appears to be associated

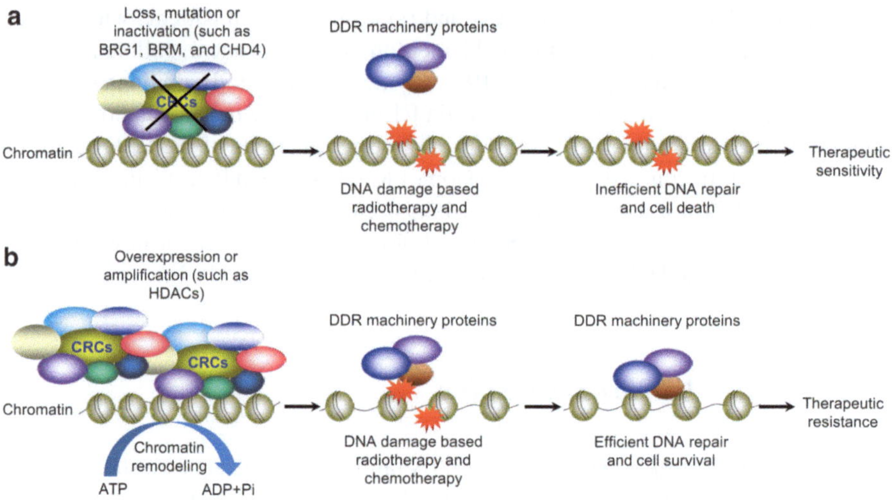


**Fig. 16.2** Implication of CRCs in cancer therapeutics. Cancer cells with loss, mutation, or inactivation of the CRC components such as of BRG1, BRM and CHD4 are sensitive to DNA damage based radiotherapy and chemotherapy due to impaired DNA repair (**a**). In contrast, the CRC subunits such as HDACs are over expressed or amplified in cancer cells and promote efficient DNA repair, thus contributing to therapeutic resistance to DNA damage based radiotherapy and chemotherapy (**b**)

with glucocorticoid resistance in primary ALL cells [207]. Similarly, knockdown of BRG1 and BRM enhances cellular sensitivity to chemotherapy drug cisplatin by regulating efficient repair of the cisplatin DNA lesions [208]. Thus, cisplatin chemotherapy could be more effective in BRG1- and BRM-negative or inactivated tumors (Fig. 16.2a). Consistent with these findings, depletion of CHD4 renders cell significantly hypersensitive to DSB-inducing agents and PARP inhibitors as a consequence of impaired HR repair (Fig. 16.2a) [209]. As loss or mutation of BRG1, BRM and CHD4 has been observed in a variety of human cancers [186, 188, 210, 211], it is highly interesting to examine whether these tumors are sensitive to PAPR inhibitors or other DNA-damaging agents.

In contrast, overexpression of some components of CRCs is linked with therapeutic resistance (Fig. 16.2b). For instance, Rsf-1 overexpression confers paclitaxel resistance in ovarian cancer cells [212] and is associated with poor therapeutic response in rectal cancer patients treated with neoadjuvant chemoradiation therapy [191] and associated with incomplete response to radiotherapy in patients with nasopharyngeal carcinoma [192]. Notably, the Rsf-1-hSNF2H interaction is essential for developing resistance phenotype in tumors overexpressing Rsf-1 [212]. Thus, inhibition of Rsf-1 activity or disruption of the Rsf-1-hSNF2H interaction has the potential to sensitize cells to paclitaxel in human cancers with Rsf-1 amplication or overexpression. Similarly, HDAC2 is highly expressed in pancreatic ductal adenocarcinoma (PDAC) and confers resistance towards the topoisomerase II inhibitor etoposide in PDAC cells [201]. Consistently, selective inhibition of HDACs synergises with etoposide to induce apoptosis in PDAC cells [201]. In a broader perspective, targeting the CRC-mediated DNA repair pathways might provide unique potential therapeutic avenues for human cancers when used in combination with DNA-damaging chemotherapeutic drugs [213, 214].

## 16.7 Conclusions and Perspectives

During the past decades, it has been made great progress in our understanding of the functional roles for ATP-dependent CRCs in transcription and DDR, and it has been increasingly recognized that these CRCs show remarkable diversity and specifity in their contributions to these biological processes. However, it remains unknown why the transcription and DDR pathways need multiple CCRs and whether or how these CRCs exert their functions in these processes in an integrated manner at molecular levels. In addition, the detailed mechanisms by which these CRCs regulate transcription and DDR and drive tumorigenesis and progression are largely unclear.

From a translational perspective, the importance of the CRCs in cancer causation and progression provides new avenues to improve cancer management by targeting the chromatin remodeling machinery. One example is that the CRCs predominantly function in the DNA repair pathways, which may contribute to therapeutic resistance in patients with cancers by enabling cancer cells to survival DNA damage induced by chemotherapeutic agents and radiotherapy [214]. Thus, targeting the CRC components and related DNA repair signaling pathways in human cancers could be efficacious as monotherapy or in combination with DNA-damaging agents [213, 214]. A case in point is the HDACs, whose inhibitors are emerging as promising drugs for cancer therapy that selectively kill cancer cells and sensitize cancer cells to DSB-inducing agents [200]. On the other hand, as components of the CRCs are frequently mutated in human cancers, this unique property of cancer cells gives a great opportunity to screen appropriate patients in clinic for optimum personal therapy using DNA-damaging radiotherapy and chemotherapy. Together, further work that directs to understand the in vivo function and mechanism of action of these CRCs will definitely provide opportunities to discover new therapeutic targets and therapeutic strategies for the treatment of cancer as well as other CRC-related diseases.

**Acknowledgement** We sincerely apologize to all those colleagues whose important work has not been cited owing to space limitations. We would like to thank all members of the Kumar laboratory for helpful discussion and comments.

## References

1. Luger K, Mader AW, Richmond RK, Sargent DF, Richmond TJ (1997) Crystal structure of the nucleosome core particle at 2.8 A resolution. Nature 389:251–260
2. McGhee JD, Felsenfeld G (1980) Nucleosome structure. Annu Rev Biochem 49:1115–1156

3. Luger K (2003) Structure and dynamic behavior of nucleosomes. Curr Opin Genet Dev 13:127–135
4. Au TJ, Rodriguez J, Vincent JA, Tsukiyama T (2011) ATP-dependent chromatin remodeling factors tune S phase checkpoint activity. Mol Cell Biol 31:4454–4463
5. Bao Y, Shen X (2007) SnapShot: chromatin remodeling complexes. Cell 129:632
6. Tsukiyama T (2002) The in vivo functions of ATP-dependent chromatin-remodelling factors. Nat Rev Mol Cell Biol 3:422–429
7. Euskirchen G, Auerbach RK, Snyder M (2012) SWI/SNF chromatin-remodeling factors: multiscale analyses and diverse functions. J Biol Chem 287:30897–30905
8. Tsukiyama T, Palmer J, Landel CC, Shiloach J, Wu C (1999) Characterization of the imitation switch subfamily of ATP-dependent chromatin-remodeling factors in *Saccharomyces cerevisiae*. Genes Dev 13:686–697
9. Denslow SA, Wade PA (2007) The human Mi-2/NuRD complex and gene regulation. Oncogene 26:5433–5438
10. Morrison AJ, Shen X (2009) Chromatin remodelling beyond transcription: the INO80 and SWR1 complexes. Nat Rev Mol Cell Biol 10:373–384
11. Li DQ, Kumar R (2010) Mi-2/NuRD complex making inroads into DNA-damage response pathway. Cell Cycle 9:2071–2079
12. Cairns BR, Kim YJ, Sayre MH, Laurent BC, Kornberg RD (1994) A multisubunit complex containing the SWI1/ADR6, SWI2/SNF2, SWI3, SNF5, and SNF6 gene products isolated from yeast. Proc Natl Acad Sci U S A 91:1950–1954
13. Cote J, Quinn J, Workman JL, Peterson CL (1994) Stimulation of GAL4 derivative binding to nucleosomal DNA by the yeast SWI/SNF complex. Science 265:53–60
14. Imbalzano AN, Kwon H, Green MR, Kingston RE (1994) Facilitated binding of TATA-binding protein to nucleosomal DNA. Nature 370:481–485
15. Tamkun JW, Deuring R, Scott MP, Kissinger M, Pattatucci AM, Kaufman TC et al (1992) Brahma: a regulator of Drosophila homeotic genes structurally related to the yeast transcriptional activator SNF2/SWI2. Cell 68:561–572
16. Kanno T, Kanno Y, Siegel RM, Jang MK, Lenardo MJ, Ozato K (2004) Selective recognition of acetylated histones by bromodomain proteins visualized in living cells. Mol Cell 13:33–43
17. Cairns BR, Lorch Y, Li Y, Zhang M, Lacomis L, Erdjument-Bromage H et al (1996) RSC, an essential, abundant chromatin-remodeling complex. Cell 87:1249–1260
18. Saha A, Wittmeyer J, Cairns BR (2002) Chromatin remodeling by RSC involves ATP-dependent DNA translocation. Genes Dev 16:2120–2134
19. Szerlong H, Saha A, Cairns BR (2003) The nuclear actin-related proteins Arp7 and Arp9: a dimeric module that cooperates with architectural proteins for chromatin remodeling. EMBO J 22:3175–3187
20. Cairns BR, Erdjument-Bromage H, Tempst P, Winston F, Kornberg RD (1998) Two actin-related proteins are shared functional components of the chromatin-remodeling complexes RSC and SWI/SNF. Mol Cell 2:639–651
21. Wilson BG, Roberts CW (2011) SWI/SNF nucleosome remodellers and cancer. Nat Rev Cancer 11:481–492
22. Xue Y, Canman JC, Lee CS, Nie Z, Yang D, Moreno GT et al (2000) The human SWI/SNF-B chromatin-remodeling complex is related to yeast RSC and localizes at kinetochores of mitotic chromosomes. Proc Natl Acad Sci U S A 97:13015–13020
23. Clapier CR, Cairns BR (2009) The biology of chromatin remodeling complexes. Annu Rev Biochem 78:273–304
24. Boyer LA, Latek RR, Peterson CL (2004) The SANT domain: a unique histone-tail-binding module? Nat Rev Mol Cell Biol 5:158–163
25. Tsukiyama T, Wu C (1995) Purification and properties of an ATP-dependent nucleosome remodeling factor. Cell 83:1011–1020
26. Tsukiyama T, Daniel C, Tamkun J, Wu C (1995) ISWI, a member of the SWI2/SNF2 ATPase family, encodes the 140 kDa subunit of the nucleosome remodeling factor. Cell 83:1021–1026

27. Martinez Balbas MA, Tsukiyama T, Gdula D, Wu C (1998) Drosophila NURF 55, a WD repeat protein involved in histone metabolism. Proc Natl Acad Sci U S A 95:132–137
28. Gdula DA, Sandaltzopoulos R, Tsukiyama T, Ossipow V, Wu C (1998) Inorganic pyrophosphatase is a component of the Drosophila nucleosome remodeling factor complex. Genes Dev 12:3206–3216
29. Badenhorst P, Voas M, Rebay I, Wu C (2002) Biological functions of the ISWI chromatin remodeling complex NURF. Genes Dev 16:3186–3198
30. Hamiche A, Kang JG, Dennis C, Xiao H, Wu C (2001) Histone tails modulate nucleosome mobility and regulate ATP dependent nucleosome sliding by NURF. Proc Natl Acad Sci U S A 98:14316–14321
31. Hamiche A, Sandaltzopoulos R, Gdula DA, Wu C (1999) ATP dependent histone octamer sliding mediated by the chromatin remodeling complex NURF. Cell 97:833–842
32. Varga Weisz PD, Wilm M, Bonte E, Dumas K, Mann M, Becker PB (1997) Chromatin remodelling factor CHRAC contains the ATPases ISWI and topoisomerase II. Nature 388:598–602
33. Ito T, Bulger M, Pazin MJ, Kobayashi R, Kadonaga JT (1997) ACF, an ISWI containing and ATP utilizing chromatin assembly and remodeling factor. Cell 90:145–155
34. Bozhenok L, Poot R, Collins N, Varga Weisz P (2004) Functional analysis of ISWI complexes in mammalian cells. Methods Enzymol 377:376–389
35. Barak O, Lazzaro MA, Lane WS, Speicher DW, Picketts DJ, Shiekhattar R (2003) Isolation of human NURF: a regulator of Engrailed gene expression. EMBO J 22:6089–6100
36. Poot RA, Dellaire G, Hulsmann BB, Grimaldi MA, Corona DF, Becker PB et al (2000) HuCHRAC, a human ISWI chromatin remodelling complex contains hACF1 and two novel histone fold proteins. EMBO J 19:3377–3387
37. Strohner R, Nemeth A, Jansa P, Hofmann Rohrer U, Santoro R, Langst G et al (2001) NoRC–a novel member of mammalian ISWI-containing chromatin remodeling machines. EMBO J 20:4892–4900
38. Bozhenok L, Wade PA, Varga-Weisz P (2002) WSTF-ISWI chromatin remodeling complex targets heterochromatic replication foci. EMBO J 21:2231–2241
39. LeRoy G, Orphanides G, Lane WS, Reinberg D (1998) Requirement of RSF and FACT for transcription of chromatin templates in vitro. Science 282:1900–1904
40. Bochar DA, Savard J, Wang W, Lafleur DW, Moore P, Cote J et al (2000) A family of chromatin remodeling factors related to Williams syndrome transcription factor. Proc Natl Acad Sci U S A 97:1038–1043
41. LeRoy G, Loyola A, Lane WS, Reinberg D (2000) Purification and characterization of a human factor that assembles and remodels chromatin. J Biol Chem 275:14787–14790
42. Banting GS, Barak O, Ames TM, Burnham AC, Kardel MD, Cooch NS et al (2005) CECR2, a protein involved in neurulation, forms a novel chromatin remodeling complex with SNF2L. Hum Mol Genet 14:513–524
43. Zhang Y, LeRoy G, Seelig HP, Lane WS, Reinberg D (1998) The dermatomyositis-specific autoantigen Mi2 is a component of a complex containing histone deacetylase and nucleosome remodeling activities. Cell 95:279–289
44. Xue Y, Wong J, Moreno GT, Young MK, Cote J, Wang W (1998) NURD, a novel complex with both ATP-dependent chromatin-remodeling and histone deacetylase activities. Mol Cell 2:851–861
45. Tong JK, Hassig CA, Schnitzler GR, Kingston RE, Schreiber SL (1998) Chromatin deacetylation by an ATP-dependent nucleosome remodelling complex. Nature 395:917–921
46. Wade PA, Jones PL, Vermaak D, Wolffe AP (1998) A multiple subunit Mi-2 histone deacetylase from Xenopus laevis cofractionates with an associated Snf2 superfamily ATPase. Curr Biol 8:843–846
47. Zhang Y, Ng HH, Erdjument-Bromage H, Tempst P, Bird A, Reinberg D (1999) Analysis of the NuRD subunits reveals a histone deacetylase core complex and a connection with DNA methylation. Genes Dev 13:1924–1935

48. Wang HB, Zhang Y (2001) Mi2, an auto-antigen for dermatomyositis, is an ATP-dependent nucleosome remodeling factor. Nucleic Acids Res 29:2517–2521
49. Bao Y, Shen X (2007) INO80 subfamily of chromatin remodeling complexes. Mutat Res 618:18–29
50. Shen X, Mizuguchi G, Hamiche A, Wu C (2000) A chromatin remodelling complex involved in transcription and DNA processing. Nature 406:541–544
51. Shen X, Ranallo R, Choi E, Wu C (2003) Involvement of actin-related proteins in ATP-dependent chromatin remodeling. Mol Cell 12:147–155
52. Saravanan M, Wuerges J, Bose D, McCormack EA, Cook NJ, Zhang X et al (2012) Interactions between the nucleosome histone core and Arp8 in the INO80 chromatin remodeling complex. Proc Natl Acad Sci U S A 109:20883–20888
53. van Attikum H, Fritsch O, Gasser SM (2007) Distinct roles for SWR1 and INO80 chromatin remodeling complexes at chromosomal double-strand breaks. EMBO J 26:4113–4125
54. Tsukuda T, Lo YC, Krishna S, Sterk R, Osley MA, Nickoloff JA (2009) INO80-dependent chromatin remodeling regulates early and late stages of mitotic homologous recombination. DNA Repair 8:360–369
55. Aoyama N, Oka A, Kitayama K, Kurumizaka H, Harata M (2008) The actin-related protein hArp8 accumulates on the mitotic chromosomes and functions in chromosome alignment. Exp Cell Res 314:859–868
56. Jonsson ZO, Jha S, Wohlschlegel JA, Dutta A (2004) Rvb1p/Rvb2p recruit Arp5p and assemble a functional Ino80 chromatin remodeling complex. Mol Cell 16:465–477
57. Mizuguchi G, Shen X, Landry J, Wu WH, Sen S, Wu C (2004) ATP-driven exchange of histone H2AZ variant catalyzed by SWR1 chromatin remodeling complex. Science 303:343–348
58. Ruhl DD, Jin J, Cai Y, Swanson S, Florens L, Washburn MP et al (2006) Purification of a human SRCAP complex that remodels chromatin by incorporating the histone variant H2A.Z into nucleosomes. Biochemistry 45:5671–5677
59. Gevry N, Chan HM, Laflamme L, Livingston DM, Gaudreau L (2007) p21 transcription is regulated by differential localization of histone H2A.Z. Genes Dev 21:1869–1881
60. Krogan NJ, Keogh MC, Datta N, Sawa C, Ryan OW, Ding H et al (2003) A Snf2 family ATPase complex required for recruitment of the histone H2A variant Htz1. Mol Cell 12:1565–1576
61. Kobor MS, Venkatasubrahmanyam S, Meneghini MD, Gin JW, Jennings JL, Link AJ et al (2004) A protein complex containing the conserved Swi2/Snf2-related ATPase Swr1p deposits histone variant H2A.Z into euchromatin. PLoS Biol 2:E131
62. Luk E, Ranjan A, Fitzgerald PC, Mizuguchi G, Huang Y, Wei D et al (2010) Stepwise histone replacement by SWR1 requires dual activation with histone H2A.Z and canonical nucleosome. Cell 143:725–736
63. Edmondson DG, Roth SY (1996) Chromatin and transcription. FASEB J 10:1173–1182
64. Euskirchen GM, Auerbach RK, Davidov E, Gianoulis TA, Zhong G, Rozowsky J et al (2011) Diverse roles and interactions of the SWI/SNF chromatin remodeling complex revealed using global approaches. PLoS Genet 7:e1002008
65. Marshall TW, Link KA, Petre-Draviam CE, Knudsen KE (2003) Differential requirement of SWI/SNF for androgen receptor activity. J Biol Chem 278:30605–30613
66. van de Wijngaart DJ, Dubbink HJ, Molier M, de Vos C, Trapman J, Jenster G (2009) Functional screening of FxxLF-like peptide motifs identifies SMARCD1/BAF60a as an androgen receptor cofactor that modulates TMPRSS2 expression. Mol Endocrinol 23:1776–1786
67. Inoue H, Furukawa T, Giannakopoulos S, Zhou S, King DS, Tanese N (2002) Largest subunits of the human SWI/SNF chromatin-remodeling complex promote transcriptional activation by steroid hormone receptors. J Biol Chem 277:41674–41685
68. Link KA, Burd CJ, Williams E, Marshall T, Rosson G, Henry E et al (2005) BAF57 governs androgen receptor action and androgen-dependent proliferation through SWI/SNF. Mol Cell Biol 25:2200–2215

69. Garcia-Pedrero JM, Kiskinis E, Parker MG, Belandia B (2006) The SWI/SNF chromatin remodeling subunit BAF57 is a critical regulator of estrogen receptor function in breast cancer cells. J Biol Chem 281:22656–22664
70. Belandia B, Orford RL, Hurst HC, Parker MG (2002) Targeting of SWI/SNF chromatin remodelling complexes to estrogen-responsive genes. EMBO J 21:4094–4103
71. Villaronga MA, Lopez-Mateo I, Markert L, Espinosa E, Fresno Vara JA, Belandia B (2011) Identification and characterization of novel potentially oncogenic mutations in the human BAF57 gene in a breast cancer patient. Breast Cancer Res Treat 128:891–898
72. Kiskinis E, Garcia-Pedrero JM, Villaronga MA, Parker MG, Belandia B (2006) Identification of BAF57 mutations in human breast cancer cell lines. Breast Cancer Res Treat 98:191–198
73. Trotter KW, Archer TK (2008) The BRG1 transcriptional coregulator. Nucl Recept Signal 6:e004
74. Xi Q, He W, Zhang XH, Le HV, Massague J (2008) Genome-wide impact of the BRG1 SWI/SNF chromatin remodeler on the transforming growth factor beta transcriptional program. J Biol Chem 283:1146–1155
75. Ross S, Cheung E, Petrakis TG, Howell M, Kraus WL, Hill CS (2006) Smads orchestrate specific histone modifications and chromatin remodeling to activate transcription. EMBO J 25:4490–4502
76. Banine F, Bartlett C, Gunawardena R, Muchardt C, Yaniv M, Knudsen ES et al (2005) SWI/SNF chromatin-remodeling factors induce changes in DNA methylation to promote transcriptional activation. Cancer Res 65:3542–3547
77. Kia SK, Gorski MM, Giannakopoulos S, Verrijzer CP (2008) SWI/SNF mediates polycomb eviction and epigenetic reprogramming of the INK4b-ARF-INK4a locus. Mol Cell Biol 28:3457–3464
78. Zhu Y, Rowley MJ, Bohmdorfer G, Wierzbicki AT (2013) A SWI/SNF chromatin-remodeling complex acts in noncoding RNA-mediated transcriptional silencing. Mol Cell 49:298–309
79. Harikrishnan KN, Chow MZ, Baker EK, Pal S, Bassal S, Brasacchio D et al (2005) Brahma links the SWI/SNF chromatin-remodeling complex with MeCP2-dependent transcriptional silencing. Nat Genet 37:254–264
80. Rao M, Casimiro MC, Lisanti MP, D'Amico M, Wang C, Shirley LA et al (2008) Inhibition of cyclin D1 gene transcription by Brg-1. Cell Cycle 7:647–655
81. Zhang ZK, Davies KP, Allen J, Zhu L, Pestell RG, Zagzag D et al (2002) Cell cycle arrest and repression of cyclin D1 transcription by INI1/hSNF5. Mol Cell Biol 22:5975–5988
82. Zhang B, Chambers KJ, Faller DV, Wang S (2007) Reprogramming of the SWI/SNF complex for co-activation or co-repression in prohibitin-mediated estrogen receptor regulation. Oncogene 26:7153–7157
83. Sala A, Toto M, Pinello L, Gabriele A, Di Benedetto V, Ingrassia AM et al (2011) Genome-wide characterization of chromatin binding and nucleosome spacing activity of the nucleosome remodelling ATPase ISWI. EMBO J 30:1766–1777
84. Mizuguchi G, Tsukiyama T, Wisniewski J, Wu C (1997) Role of nucleosome remodeling factor NURF in transcriptional activation of chromatin. Mol Cell 1:141–150
85. Song H, Spichiger-Haeusermann C, Basler K (2009) The ISWI-containing NURF complex regulates the output of the canonical Wingless pathway. EMBO Rep 10:1140–1146
86. Kwon SY, Xiao H, Glover BP, Tjian R, Wu C, Badenhorst P (2008) The nucleosome remodeling factor (NURF) regulates genes involved in Drosophila innate immunity. Dev Biol 316:538–547
87. Liu YI, Chang MV, Li HE, Barolo S, Chang JL, Blauwkamp TA et al (2008) The chromatin remodelers ISWI and ACF1 directly repress Wingless transcriptional targets. Dev Biol 323:41–52
88. Lindstrom KC, Vary JC Jr, Parthun MR, Delrow J, Tsukiyama T (2006) Isw1 functions in parallel with the NuA4 and Swr1 complexes in stress-induced gene repression. Mol Cell Biol 26:6117–6129
89. Fazzio TG, Kooperberg C, Goldmark JP, Neal C, Basom R, Delrow J et al (2001) Widespread collaboration of Isw2 and Sin3-Rpd3 chromatin remodeling complexes in transcriptional repression. Mol Cell Biol 21:6450–6460

90. Goldmark JP, Fazzio TG, Estep PW, Church GM, Tsukiyama T (2000) The Isw2 chromatin remodeling complex represses early meiotic genes upon recruitment by Ume6p. Cell 103:423–433
91. Whitehouse I, Rando OJ, Delrow J, Tsukiyama T (2007) Chromatin remodelling at promoters suppresses antisense transcription. Nature 450:1031–1035
92. Lai AY, Wade PA (2011) Cancer biology and NuRD: a multifaceted chromatin remodelling complex. Nat Rev Cancer 11:588–596
93. Li DQ, Pakala SB, Nair SS, Eswaran J, Kumar R (2012) Metastasis-associated protein 1/nucleosome remodeling and histone deacetylase complex in cancer. Cancer Res 72:387–394
94. Toh Y, Pencil SD, Nicolson GL (1994) A novel candidate metastasis-associated gene, mta1, differentially expressed in highly metastatic mammary adenocarcinoma cell lines. cDNA cloning, expression, and protein analyses. J Biol Chem 269:22958–22963
95. Mazumdar A, Wang RA, Mishra SK, Adam L, Bagheri-Yarmand R, Mandal M et al (2001) Transcriptional repression of oestrogen receptor by metastasis-associated protein 1 corepressor. Nat Cell Biol 3:30–37
96. Li DQ, Pakala SB, Reddy SD, Ohshiro K, Peng SH, Lian Y et al (2010) Revelation of p53-independent function of MTA1 in DNA damage response via modulation of the p21 WAF1-proliferating cell nuclear antigen pathway. J Biol Chem 285:10044–10052
97. Molli PR, Singh RR, Lee SW, Kumar R (2008) MTA1-mediated transcriptional repression of BRCA1 tumor suppressor gene. Oncogene 27:1971–1980
98. Marzook H, Li DQ, Nair VS, Mudvari P, Reddy SD, Pakala SB et al (2012) Metastasis-associated protein 1 drives tumor cell migration and invasion through transcriptional repression of RING finger protein 144A. J Biol Chem 287:5615–5626
99. Reddy SD, Pakala SB, Molli PR, Sahni N, Karanam NK, Mudvari P et al (2012) Metastasis-associated protein 1/histone deacetylase 4-nucleosome remodeling and deacetylase complex regulates phosphatase and tensin homolog gene expression and function. J Biol Chem 287:27843–27850
100. Salot S, Gude R (2013) MTA1-mediated transcriptional repression of SMAD7 in breast cancer cell lines. Eur J Cancer 49:492–499
101. Ohshiro K, Rayala SK, Wigerup C, Pakala SB, Natha RS, Gururaj AE et al (2010) Acetylation-dependent oncogenic activity of metastasis-associated protein 1 co-regulator. EMBO Rep 11:691–697
102. Manavathi B, Peng S, Rayala SK, Talukder AH, Wang MH, Wang RA et al (2007) Proc Natl Acad Sci U S A 104:13128–13133
103. Gururaj AE, Singh RR, Rayala SK, Holm C, den Hollander P, Zhang H et al (2006) MTA1, a transcriptional activator of breast cancer amplified sequence 3. Proc Natl Acad Sci U S A 103:6670–6675
104. Balasenthil S, Gururaj AE, Talukder AH, Bagheri-Yarmand R, Arrington T, Haas BJ et al (2007) Identification of Pax5 as a target of MTA1 in B-cell lymphomas. Cancer Res 67:7132–7138
105. Li DQ, Pakala SB, Reddy SD, Ohshiro K, Zhang JX, Wang L et al (2011) Bidirectional autoregulatory mechanism of metastasis-associated protein 1-alternative reading frame pathway in oncogenesis. Proc Natl Acad Sci U S A 108:8791–8796
106. Sankaran D, Pakala SB, Nair VS, Sirigiri DN, Cyanam D, Ha NH et al (2012) Mechanism of MTA1 protein overexpression-linked invasion: MTA1 regulation of hyaluronan-mediated motility receptor (HMMR) expression and function. J Biol Chem 287:5483–5491
107. Kumar R, Balasenthil S, Manavathi B, Rayala SK, Pakala SB (2010) Metastasis-associated protein 1 and its short form variant stimulates Wnt1 transcription through promoting its derepression from Six3 corepressor. Cancer Res 70:6649–6658
108. Reddy SD, Rayala SK, Ohshiro K, Pakala SB, Kobori N, Dash P et al (2011) Multiple coregulatory control of tyrosine hydroxylase gene transcription. Proc Natl Acad Sci U S A 108:4200–4205
109. Cong L, Pakala SB, Ohshiro K, Li DQ, Kumar R (2011) SUMOylation and SUMO-interacting motif (SIM) of metastasis tumor antigen 1 (MTA1) synergistically regulate its transcriptional repressor function. J Biol Chem 286:43793–43808

110. Nair SS, Li DQ, Kumar R (2013) A core chromatin remodeling factor instructs global chromatin signaling through multivalent reading of nucleosome codes. Mol Cell 49:704–718
111. Ebbert R, Birkmann A, Schuller HJ (1999) The product of the SNF2/SWI2 paralogue INO80 of *Saccharomyces cerevisiae* required for efficient expression of various yeast structural genes is part of a high-molecular-weight protein complex. Mol Microbiol 32:741–751
112. Cai Y, Jin J, Yao T, Gottschalk AJ, Swanson SK, Wu S et al (2007) YY1 functions with INO80 to activate transcription. Nat Struct Mol Biol 14:872–874
113. Cho SG, Bhoumik A, Broday L, Ivanov V, Rosenstein B, Ronai Z (2001) TIP49b, a regulator of activating transcription factor 2 response to stress and DNA damage. Mol Cell Biol 21:8398–8413
114. Kruhlak MJ, Celeste A, Nussenzweig A (2006) Spatio-temporal dynamics of chromatin containing DNA breaks. Cell Cycle 5:1910–1912
115. Flanagan JF, Peterson CL (1999) A role for the yeast SWI/SNF complex in DNA replication. Nucleic Acids Res 27:2022–2028
116. Driscoll R, Cimprich KA (2009) HARPing on about the DNA damage response during replication. Genes Dev 23:2359–2365
117. Kwon J, Morshead KB, Guyon JR, Kingston RE, Oettinger MA (2000) Histone acetylation and hSWI/SNF remodeling act in concert to stimulate V(D)J cleavage of nucleosomal DNA. Mol Cell 6:1037–1048
118. Hara R, Sancar A (2002) The SWI/SNF chromatin-remodeling factor stimulates repair by human excision nuclease in the mononucleosome core particle. Mol Cell Biol 22:6779–6787
119. Ray A, Mir SN, Wani G, Zhao Q, Battu A, Zhu Q et al (2009) Human SNF5/INI1, a component of the human SWI/SNF chromatin remodeling complex, promotes nucleotide excision repair by influencing ATM recruitment and downstream H2AX phosphorylation. Mol Cell Biol 29:6206–6219
120. Park JH, Park EJ, Lee HS, Kim SJ, Hur SK, Imbalzano AN et al (2006) Mammalian SWI/SNF complexes facilitate DNA double-strand break repair by promoting gamma-H2AX induction. EMBO J 25:3986–3997
121. Gong F, Fahy D, Liu H, Wang W, Smerdon MJ (2008) Role of the mammalian SWI/SNF chromatin remodeling complex in the cellular response to UV damage. Cell Cycle 7:1067–1074
122. Liang B, Qiu J, Ratnakumar K, Laurent BC (2007) RSC functions as an early double-strand-break sensor in the cell's response to DNA damage. Curr Biol 17:1432–1437
123. Oum JH, Seong C, Kwon Y, Ji JH, Sid A, Ramakrishnan S et al (2011) RSC facilitates Rad59-dependent homologous recombination between sister chromatids by promoting cohesin loading at DNA double-strand breaks. Mol Cell Biol 31:3924–3937
124. Chambers AL, Brownlee PM, Durley SC, Beacham T, Kent NA, Downs JA (2012) The two different isoforms of the RSC chromatin remodeling complex play distinct roles in DNA damage responses. PLoS One 7:e32016
125. Chai B, Huang J, Cairns BR, Laurent BC (2005) Distinct roles for the RSC and Swi/Snf ATP-dependent chromatin remodelers in DNA double-strand break repair. Genes Dev 19:1656–1661
126. Smeenk G, Wiegant WW, Marteijn JA, Luijsterburg MS, Sroczynski N, Costelloe T et al (2013) Poly(ADP-ribosyl)ation links the chromatin remodeler SMARCA5/SNF2H to RNF168-dependent DNA damage signaling. J Cell Sci 126:889–903
127. Erdel F, Rippe K (2011) Binding kinetics of human ISWI chromatin-remodelers to DNA repair sites elucidate their target location mechanism. Nucleus 2:105–112
128. Ura K, Araki M, Saeki H, Masutani C, Ito T, Iwai S et al (2001) ATP-dependent chromatin remodeling facilitates nucleotide excision repair of UV-induced DNA lesions in synthetic dinucleosomes. EMBO J 20:2004–2014
129. Lan L, Ui A, Nakajima S, Hatakeyama K, Hoshi M, Watanabe R et al (2010) The ACF1 complex is required for DNA double-strand break repair in human cells. Mol Cell 40:976–987
130. Mao TL, Hsu CY, Yen MJ, Gilks B, Sheu JJ, Gabrielson E et al (2006) Expression of Rsf-1, a chromatin-remodeling gene, in ovarian and breast carcinoma. Hum Pathol 37:1169–1175

131. Sheu JJ, Guan B, Choi JH, Lin A, Lee CH, Hsiao YT et al (2010) Rsf-1, a chromatin remodeling protein, induces DNA damage and promotes genomic instability. J Biol Chem 285:38260–38269
132. Kshirsagar M, Jiang W, Shih IM (2012) DNA damage response is prominent in ovarian high-grade serous carcinomas, especially those with Rsf-1 (HBXAP) overexpression. J Oncol 2012:621685
133. Schmidt DR, Schreiber SL (1999) Molecular association between ATR and two components of the nucleosome remodeling and deacetylating complex, HDAC2 and CHD4. Biochemistry 38:14711–14717
134. van Haaften G, Romeijn R, Pothof J, Koole W, Mullenders LH, Pastink A et al (2006) Identification of conserved pathways of DNA-damage response and radiation protection by genome-wide RNAi. Curr Biol 16:1344–1350
135. Li DQ, Ohshiro K, Reddy SD, Pakala SB, Lee MH, Zhang Y et al (2009) E3 ubiquitin ligase COP1 regulates the stability and functions of MTA1. Proc Natl Acad Sci U S A 106:17493–17498
136. Chou DM, Adamson B, Dephoure NE, Tan X, Nottke AC, Hurov KE et al (2010) A chromatin localization screen reveals poly (ADP ribose)-regulated recruitment of the repressive polycomb and NuRD complexes to sites of DNA damage. Proc Natl Acad Sci U S A 107:18475–18480
137. Larsen DH, Poinsignon C, Gudjonsson T, Dinant C, Payne MR, Hari FJ et al (2010) The chromatin-remodeling factor CHD4 coordinates signaling and repair after DNA damage. J Cell Biol 190:731–740
138. Polo SE, Kaidi A, Baskcomb L, Galanty Y, Jackson SP (2010) Regulation of DNA-damage responses and cell-cycle progression by the chromatin remodelling factor CHD4. EMBO J 29:3130–3139
139. Smeenk G, Wiegant WW, Vrolijk H, Solari AP, Pastink A, van Attikum H (2010) The NuRD chromatin-remodeling complex regulates signaling and repair of DNA damage. J Cell Biol 190:741–749
140. Miller KM, Tjeertes JV, Coates J, Legube G, Polo SE, Britton S et al (2010) Human HDAC1 and HDAC2 function in the DNA-damage response to promote DNA nonhomologous end-joining. Nat Struct Mol Biol 17:1144–1151
141. Huber A, Bai P, de Murcia JM, de Murcia G (2004) PARP-1, PARP-2 and ATM in the DNA damage response: functional synergy in mouse development. DNA Repair 3:1103–1108
142. Smeenk G, van Attikum H (2011) NuRD alert! NuRD regulates the DNA damage response. Epigenomics 3:133–135
143. Pegoraro G, Kubben N, Wickert U, Gohler H, Hoffmann K, Misteli T (2009) Ageing-related chromatin defects through loss of the NURD complex. Nat Cell Biol 11:1261–1267
144. Zhang Y (2011) Biology of the Mi-2/NuRD Complex in SLAC (Stemness, Longevity/Ageing, and Cancer). Gene Regul Syst Bio 5:1–26
145. Morrison AJ, Highland J, Krogan NJ, Arbel-Eden A, Greenblatt JF, Haber JE et al (2004) INO80 and gamma-H2AX interaction links ATP-dependent chromatin remodeling to DNA damage repair. Cell 119:767–775
146. Downs JA, Allard S, Jobin-Robitaille O, Javaheri A, Auger A, Bouchard N et al (2004) Binding of chromatin-modifying activities to phosphorylated histone H2A at DNA damage sites. Mol Cell 16:979–990
147. Gospodinov A, Vaissiere T, Krastev DB, Legube G, Anachkova B, Herceg Z (2011) Mammalian Ino80 mediates double-strand break repair through its role in DNA end strand resection. Mol Cell Biol 31:4735–4745
148. Jiang Y, Wang X, Bao S, Guo R, Johnson DG, Shen X et al (2010) INO80 chromatin remodeling complex promotes the removal of UV lesions by the nucleotide excision repair pathway. Proc Natl Acad Sci U S A 107:17274–17279
149. Sarkar S, Kiely R, McHugh PJ (2010) The Ino80 chromatin-remodeling complex restores chromatin structure during UV DNA damage repair. J Cell Biol 191:1061–1068
150. Vincent JA, Kwong TJ, Tsukiyama T (2008) ATP-dependent chromatin remodeling shapes the DNA replication landscape. Nat Struct Mol Biol 15:477–484

151. Niimi A, Chambers AL, Downs JA, Lehmann AR (2012) A role for chromatin remodellers in replication of damaged DNA. Nucleic Acids Res 40:7393–7403
152. Kato D, Waki M, Umezawa M, Aoki Y, Utsugi T, Ohtsu M et al (2012) Phosphorylation of human INO80 is involved in DNA damage tolerance. Biochem Biophys Res Commun 417:433–438
153. Falbo KB, Alabert C, Katou Y, Wu S, Han J, Wehr T et al (2009) Involvement of a chromatin remodeling complex in damage tolerance during DNA replication. Nat Struct Mol Biol 16:1167–1172
154. Gunduz E, Gunduz M, Nagatsuka H, Beder L, Demircan K, Tamamura R et al (2006) Epigenetic alterations of BRG1 leads to cancer development through its nuclear-cytoplasmic shuttling abnormalities. Med Hypotheses 67:1313–1316
155. Reisman D, Glaros S, Thompson EA (2009) The SWI/SNF complex and cancer. Oncogene 28:1653–1668
156. Hornick JL, Dal Cin P, Fletcher CD (2009) Loss of INI1 expression is characteristic of both conventional and proximal-type epithelioid sarcoma. Am J Surg Pathol 33:542–550
157. Grand F, Kulkarni S, Chase A, Goldman JM, Gordon M, Cross NC (1999) Frequent deletion of hSNF5/INI1, a component of the SWI/SNF complex, in chronic myeloid leukemia. Cancer Res 59:3870–3874
158. Decristofaro MF, Betz BL, Rorie CJ, Reisman DN, Wang W, Weissman BE (2001) Characterization of SWI/SNF protein expression in human breast cancer cell lines and other malignancies. J Cell Physiol 186:136–145
159. Lin H, Wong RP, Martinka M, Li G (2009) Loss of SNF5 expression correlates with poor patient survival in melanoma. Clin Cancer Res 15:6404–6411
160. Modena P, Lualdi E, Facchinetti F, Galli L, Teixeira MR, Pilotti S et al (2005) SMARCB1/INI1 tumor suppressor gene is frequently inactivated in epithelioid sarcomas. Cancer Res 65:4012–4019
161. Sevenet N, Lellouch-Tubiana A, Schofield D, Hoang-Xuan K, Gessler M, Birnbaum D et al (1999) Spectrum of hSNF5/INI1 somatic mutations in human cancer and genotype-phenotype correlations. Hum Mol Genet 8:2359–2368
162. Roberts CW, Leroux MM, Fleming MD, Orkin SH (2002) Highly penetrant, rapid tumorigenesis through conditional inversion of the tumor suppressor gene Snf5. Cancer Cell 2:415–425
163. Sansam CG, Roberts CW (2006) Epigenetics and cancer: altered chromatin remodeling via Snf5 loss leads to aberrant cell cycle regulation. Cell Cycle 5:621–624
164. Xu Y, Yan W, Chen X (2010) SNF5, a core component of the SWI/SNF complex, is necessary for p53 expression and cell survival, in part through eIF4E. Oncogene 29:4090–4100
165. Medjkane S, Novikov E, Versteege I, Delattre O (2004) The tumor suppressor hSNF5/INI1 modulates cell growth and actin cytoskeleton organization. Cancer Res 64:3406–3413
166. Oruetxebarria I, Venturini F, Kekarainen T, Houweling A, Zuijderduijn LM, Mohd-Sarip A et al (2004) P16INK4a is required for hSNF5 chromatin remodeler-induced cellular senescence in malignant rhabdoid tumor cells. J Biol Chem 279:3807–3816
167. Lee S, Cimica V, Ramachandra N, Zagzag D, Kalpana GV (2011) Aurora A is a repressed effector target of the chromatin remodeling protein INI1/hSNF5 required for rhabdoid tumor cell survival. Cancer Res 71:3225–3235
168. Reincke BS, Rosson GB, Oswald BW, Wright CF (2003) INI1 expression induces cell cycle arrest and markers of senescence in malignant rhabdoid tumor cells. J Cell Physiol 194:303–313
169. DelBove J, Kuwahara Y, Mora-Blanco EL, Godfrey V, Funkhouser WK, Fletcher CD et al (2009) Inactivation of SNF5 cooperates with p53 loss to accelerate tumor formation in Snf5(+/−);p53(+/−) mice. Mol Carcinog 48:1139–1148
170. Isakoff MS, Sansam CG, Tamayo P, Subramanian A, Evans JA, Fillmore CM et al (2005) Inactivation of the Snf5 tumor suppressor stimulates cell cycle progression and cooperates with p53 loss in oncogenic transformation. Proc Natl Acad Sci U S A 102:17745–17750
171. Caramel J, Quignon F, Delattre O (2008) RhoA-dependent regulation of cell migration by the tumor suppressor hSNF5/INI1. Cancer Res 68:6154–6161

172. Mamo A, Cavallone L, Tuzmen S, Chabot C, Ferrario C, Hassan S et al (2012) An integrated genomic approach identifies ARID1A as a candidate tumor-suppressor gene in breast cancer. Oncogene 31:2090–2100
173. Huang J, Deng Q, Wang Q, Li KY, Dai JH, Li N et al (2012) Exome sequencing of hepatitis B virus-associated hepatocellular carcinoma. Nat Genet 44:1117–1121
174. Abe H, Maeda D, Hino R, Otake Y, Isogai M, Ushiku AS et al (2012) ARID1A expression loss in gastric cancer: pathway-dependent roles with and without Epstein-Barr virus infection and microsatellite instability. Virchows Arch 461:367–377
175. Wiegand KC, Shah SP, Al-Agha OM, Zhao Y, Tse K, Zeng T et al (2010) ARID1A mutations in endometriosis-associated ovarian carcinomas. N Engl J Med 363:1532–1543
176. Wiegand KC, Lee AF, Al-Agha OM, Chow C, Kalloger SE, Scott DW et al (2011) Loss of BAF250a (ARID1A) is frequent in high-grade endometrial carcinomas. J Pathol 224:328–333
177. Wu JN, Roberts CW (2013) ARID1A mutations in cancer: another epigenetic tumor suppressor? Cancer Dis 3:35–43
178. Giulino-Roth L, Wang K, Macdonald TY, Mathew S, Tam Y, Cronin MT et al (2012) Targeted genomic sequencing of pediatric Burkitt lymphoma identifies recurrent alterations in anti-apoptotic and chromatin-remodeling genes. Blood 120:5181–5184
179. Guan B, Wang TL, Shih IM (2011) ARID1A, a factor that promotes formation of SWI/SNF-mediated chromatin remodeling, is a tumor suppressor in gynecologic cancers. Cancer Res 71:6718–6727
180. Inoue H, Giannakopoulos S, Parkhurst CN, Matsumura T, Kono EA, Furukawa T et al (2011) Target genes of the largest human SWI/SNF complex subunit control cell growth. Biochem J 434:83–92
181. DelBove J, Rosson G, Strobeck M, Chen J, Archer TK, Wang W et al (2011) Identification of a core member of the SWI/SNF complex, BAF155/SMARCC1, as a human tumor suppressor gene. Epigenetics 6:1444–1453
182. Hodis E, Watson IR, Kryukov GV, Arold ST, Imielinski M, Theurillat JP et al (2012) A landscape of driver mutations in melanoma. Cell 150:251–263
183. Manceau G, Letouze E, Guichard C, Didelot A, Cazes A, Corte H et al (2013) Recurrent inactivating mutations of ARID2 in non-small cell lung carcinoma. Int J Cancer 132: 2217–2221
184. Shain AH, Giacomini CP, Matsukuma K, Karikari CA, Bashyam MD, Hidalgo M et al (2012) Convergent structural alterations define SWItch/Sucrose NonFermentable (SWI/SNF) chromatin remodeler as a central tumor suppressive complex in pancreatic cancer. Proc Natl Acad Sci U S A 109:E252–E259
185. Bock VL, Lyons JG, Huang XX, Jones AM, McDonald LA, Scolyer RA et al (2011) BRM and BRG1 subunits of the SWI/SNF chromatin remodelling complex are downregulated upon progression of benign skin lesions into invasive tumours. Br J Dermatol 164:1221–1227
186. Rodriguez-Nieto S, Canada A, Pros E, Pinto AI, Torres-Lanzas J, Lopez-Rios F et al (2011) Massive parallel DNA pyrosequencing analysis of the tumor suppressor BRG1/SMARCA4 in lung primary tumors. Hum Mutat 32:E1999–E2017
187. Xia W, Nagase S, Montia AG, Kalachikov SM, Keniry M, Su T et al (2008) BAF180 is a critical regulator of p21 induction and a tumor suppressor mutated in breast cancer. Cancer Res 68:1667–1674
188. Glaros S, Cirrincione GM, Muchardt C, Kleer CG, Michael CW, Reisman D (2007) The reversible epigenetic silencing of BRM: implications for clinical targeted therapy. Oncogene 26:7058–7066
189. Liang PI, Wu LC, Sheu JJ, Wu TF, Shen KH, Wang YH et al (2012) Rsf-1/HBXAP overexpression is independent of gene amplification and is associated with poor outcome in patients with urinary bladder urothelial carcinoma. J Clin Pathol 65:802–807
190. Fang FM, Li CF, Huang HY, Lai MT, Chen CM, Chiu IW et al (2011) Overexpression of a chromatin remodeling factor, RSF-1/HBXAP, correlates with aggressive oral squamous cell carcinoma. Am J Pathol 178:2407–2415

191. Lin CY, Tian YF, Wu LC, Chen LT, Lin LC, Hsing CH et al (2012) Rsf-1 expression in rectal cancer: with special emphasis on the independent prognostic value after neoadjuvant chemoradiation. J Clin Pathol 65:687–692
192. Tai HC, Huang HY, Lee SW, Lin CY, Sheu MJ, Chang SL et al (2012) Associations of Rsf-1 overexpression with poor therapeutic response and worse survival in patients with nasopharyngeal carcinoma. J Clin Pathol 65:248–253
193. Chen TJ, Huang SC, Huang HY, Wei YC, Li CF (2011) Rsf-1/HBXAP overexpression is associated with disease-specific survival of patients with gallbladder carcinoma. APMIS 119:808–814
194. Sheu JJ, Choi JH, Guan B, Tsai FJ, Hua CH, Lai MT et al (2013) Rsf-1, a chromatin remodelling protein, interacts with cyclin E1 and promotes tumour development. J Pathol 229: 559–568
195. Maeda D, Chen X, Guan B, Nakagawa S, Yano T, Taketani Y et al (2011) Rsf-1 (HBXAP) expression is associated with advanced stage and lymph node metastasis in ovarian clear cell carcinoma. Int J Gynecol Pathol 30:30–35
196. Davidson B, Trope CG, Wang TL, Shih IM (2006) Expression of the chromatin remodeling factor Rsf-1 is upregulated in ovarian carcinoma effusions and predicts poor survival. Gynecol Oncol 103:814–819
197. Shih Ie M, Sheu JJ, Santillan A, Nakayama K, Yen MJ, Bristow RE et al (2005) Amplification of a chromatin remodeling gene, Rsf-1/HBXAP, in ovarian carcinoma. Proc Natl Acad Sci U S A 102:14004–14009
198. Davidson B, Wang TL, Shih Ie M, Berner A (2008) Expression of the chromatin remodeling factor Rsf-1 is down-regulated in breast carcinoma effusions. Hum Pathol 39:616–622
199. Eckey M, Kuphal S, Straub T, Rummele P, Kremmer E, Bosserhoff AK et al (2012) Nucleosome remodeler SNF2L suppresses cell proliferation and migration and attenuates Wnt signaling. Mol Cell Biol 32:2359–2371
200. Minucci S, Pelicci PG (2006) Histone deacetylase inhibitors and the promise of epigenetic (and more) treatments for cancer. Nat Rev Cancer 6:38–51
201. Fritsche P, Seidler B, Schuler S, Schnieke A, Gottlicher M, Schmid RM et al (2009) HDAC2 mediates therapeutic resistance of pancreatic cancer cells via the BH3-only protein NOXA. Gut 58:1399–1409
202. Weichert W, Roske A, Niesporek S, Noske A, Buckendahl AC, Dietel M et al (2008) Class I histone deacetylase expression has independent prognostic impact in human colorectal cancer: specific role of class I histone deacetylases in vitro and in vivo. Clin Cancer Res 14:1669–1677
203. Wang Y, Zhang H, Chen Y, Sun Y, Yang F, Yu W et al (2009) LSD1 is a subunit of the NuRD complex and targets the metastasis programs in breast cancer. Cell 138:660–672
204. Slupianek A, Yerrum S, Safadi FF, Monroy MA (2010) The chromatin remodeling factor SRCAP modulates expression of prostate specific antigen and cellular proliferation in prostate cancer cells. J Cell Physiol 224:369–375
205. Chan HM, Narita M, Lowe SW, Livingston DM (2005) The p400 E1A-associated protein is a novel component of the p53 –>p21 senescence pathway. Genes Dev 19:196–201
206. Fuchs M, Gerber J, Drapkin R, Sif S, Ikura T, Ogryzko V et al (2001) The p400 complex is an essential E1A transformation target. Cell 106:297–307
207. Pottier N, Yang W, Assem M, Panetta JC, Pei D, Paugh SW et al (2008) The SWI/SNF chromatin-remodeling complex and glucocorticoid resistance in acute lymphoblastic leukemia. J Natl Cancer Inst 100:1792–1803
208. Kothandapani A, Gopalakrishnan K, Kahali B, Reisman D, Patrick SM (2012) Downregulation of SWI/SNF chromatin remodeling factor subunits modulates cisplatin cytotoxicity. Exp Cell Res 318:1973–1986
209. Pan MR, Hsieh HJ, Dai H, Hung WC, Li K, Peng G et al (2012) Chromodomain helicase DNA-binding protein 4 (CHD4) regulates homologous recombination DNA repair, and its deficiency sensitizes cells to poly(ADP-ribose) polymerase (PARP) inhibitor treatment. J Biol Chem 287:6764–6772

210. Kim MS, Chung NG, Kang MR, Yoo NJ, Lee SH (2011) Genetic and expressional alterations of CHD genes in gastric and colorectal cancers. Histopathology 58:660–668
211. Le Gallo M, O'Hara AJ, Rudd ML, Urick ME, Hansen NF, O'Neil NJ et al (2012) Exome sequencing of serous endometrial tumors identifies recurrent somatic mutations in chromatin-remodeling and ubiquitin ligase complex genes. Nat Genet 44:1310–1315
212. Choi JH, Sheu JJ, Guan B, Jinawath N, Markowski P, Wang TL et al (2009) Functional analysis of 11q13.5 amplicon identifies Rsf-1 (HBXAP) as a gene involved in paclitaxel resistance in ovarian cancer. Cancer Res 69:1407–1415
213. Curtin NJ (2012) DNA repair dysregulation from cancer driver to therapeutic target. Nat Rev Cancer 12:801–817
214. Helleday T, Petermann E, Lundin C, Hodgson B, Sharma RA (2008) DNA repair pathways as targets for cancer therapy. Nat Rev Cancer 8:193–204

# Index



R. Kumar (ed.), *Nuclear Signaling Pathways and Targeting Transcription in Cancer*, Cancer Drug Discovery and Development, DOI 10.1007/978-1-4614-8039-6,

**D**

## O



## P

MIX
Papier aus verantwortungsvollen Quellen
Paper from responsible sources
FSC® C105338

If you have any concerns about our products,
you can contact us on
**ProductSafety@springernature.com**

In case Publisher is established outside the EU,
the EU authorized representative is:
**Springer Nature Customer Service Center GmbH**
**Europaplatz 3, 69115 Heidelberg, Germany**

Printed by Libri Plureos GmbH
in Hamburg, Germany